KV-064-687

Food Structure—Its Creation and Evaluation

ANDERSONIAN LIBRARY
WITHDRAWN FROM LIBRARY STOCK
UNIVERSITY OF STRATHCLYDE

*Proceedings of Previous Easter Schools in Agricultural Science, published by Butterworths, London*

*SOIL ZOOLOGY Edited by D.K.McE. Kevan (1955)
*THE GROWTH OF LEAVES Edited by F.L. Milthorpe (1956)
*CONTROL OF THE PLANT ENVIRONMENT Edited by J.P. Hudson (1957)
*NUTRITION OF THE LEGUMES Edited by E.G. Hallsworth (1958)
*THE MEASUREMENT OF GRASSLAND PRODUCTIVITY Edited by J.D. Ivins (1959)
*DIGESTIVE PHYSIOLOGY AND NUTRITION OF THE RUMINANT Edited by D. Lewis (1960)
*NUTRITION OF PIGS AND POULTRY Edited by J.T. Morgan and D. Lewis (1961)
*ANTIBIOTICS IN AGRICULTURE Edited by M. Woodbine (1962)
*THE GROWTH OF THE POTATO Edited by J.D. Ivins and F.L. Milthorpe (1963)
*EXPERIMENTAL PEDOLOGY Edited by E.G. Hallsworth and D.V. Crawford (1964)
*THE GROWTH OF CEREALS AND GRASSES Edited by F.L. Milthorpe and S.D. Ivins (1965)
*REPRODUCTION IN THE FEMALE MAMMAL Edited by G.E. Lamming and E.C. Amoroso (1967)
*GROWTH AND DEVELOPMENT OF MAMMALS Edited by G.A. Lodge and G.E. Lamming (1968)
*ROOT GROWTH Edited by W.J. Whittington (1968)
*PROTEINS AS HUMAN FOOD Edited by R.A. Lawrie (1970)
*LACTATION Edited by I.R. Falconer (1971)
*PIG PRODUCTION Edited by D.J.A. Cole (1972)
*SEED ECOLOGY Edited by W. Heydecker (1973)
HEAT LOSS FROM ANIMALS AND MAN: ASSESSMENT AND CONTROL Edited by J.L. Monteith and L.E. Mount (1974)
*MEAT Edited by D.J.A. Lawrie (1975)
*PRINCIPLES OF CATTLE PRODUCTION Edited by Henry Swan and W.H. Broster (1976)
*LIGHT AND PLANT DEVELOPMENT Edited by H. Smith (1976)
PLANT PROTEINS Edited by G. Norton (1977)
ANTIBIOTICS AND ANTIBIOSIS IN AGRICULTURE Edited by M. Woodbine (1977)
CONTROL OF OVULATION Edited by D.B. Crighton, N.B. Haynes, G.R. Foxcroft and G.E. Lamming (1978)
POLYSACCHARIDES IN FOOD Edited by J.M.V. Blanshard and J.R. Mitchell (1979)
SEED PRODUCTION Edited by P.D. Hebblethwaite (1980)
PROTEIN DEPOSITION IN ANIMALS Edited by P.J. Buttery and D.B. Lindsay (1981)
PHYSIOLOGICAL PROCESSES LIMITING PLANT PRODUCTIVITY Edited by C. Johnson (1981)
ENVIRONMENTAL ASPECTS OF HOUSING FOR ANIMAL PRODUCTION Edited by J.A. Clark (1981)
EFFECTS OF GASEOUS AIR POLLUTION IN AGRICULTURE AND HORTICULTURE Edited by M.H. Unsworth and D.P. Ormrod (1982)
CHEMICAL MANIPULATION OF CROP GROWTH AND DEVELOPMENT Edited by J.S. McLaren (1982)
CONTROL OF PIG REPRODUCTION Edited by D.J.A. Cole and G.R. Foxcroft (1982)
SHEEP PRODUCTION Edited by W. Haresign (1983)
UPGRADING WASTE FOR FEEDS AND FOOD Edited by D.A. Ledward, A.J. Taylor and R.A. Lawrie (1983)
FATS IN ANIMAL NUTRITION Edited by J. Wiseman (1984)
IMMUNOLOGICAL ASPECTS OF REPRODUCTION IN MAMMALS Edited by D.B. Crighton (1984)
ETHYLENE AND PLANT DEVELOPMENT Edited by J.A. Roberts and G.A. Tucker (1985)
THE PEA CROP Edited by P.D. Hebblethwaite, M.C. Heath and T.C.K. Dawkins (1985)
PLANT TISSUE CULTURE AND ITS AGRICULTURAL APPLICATIONS Edited by Lyndsey A. Withers and P.G. Alderson (1986)
CONTROL AND MANIPULATION OF ANIMAL GROWTH Edited by P.J. Buttery, D.B. Lindsay and N.N. Haynes (1986)
COMPUTER APPLICATIONS IN AGRICULTURAL ENVIRONMENTS Edited by J.A. Clark, K. Gregson and R.A. Saffell (1986)
MANIPULATION OF FLOWERING Edited by J.G. Atherton (1987)

* *These titles are now out of print but are available in microfiche editions*

ML

This book is to be returned on or before
the last date stamped below.

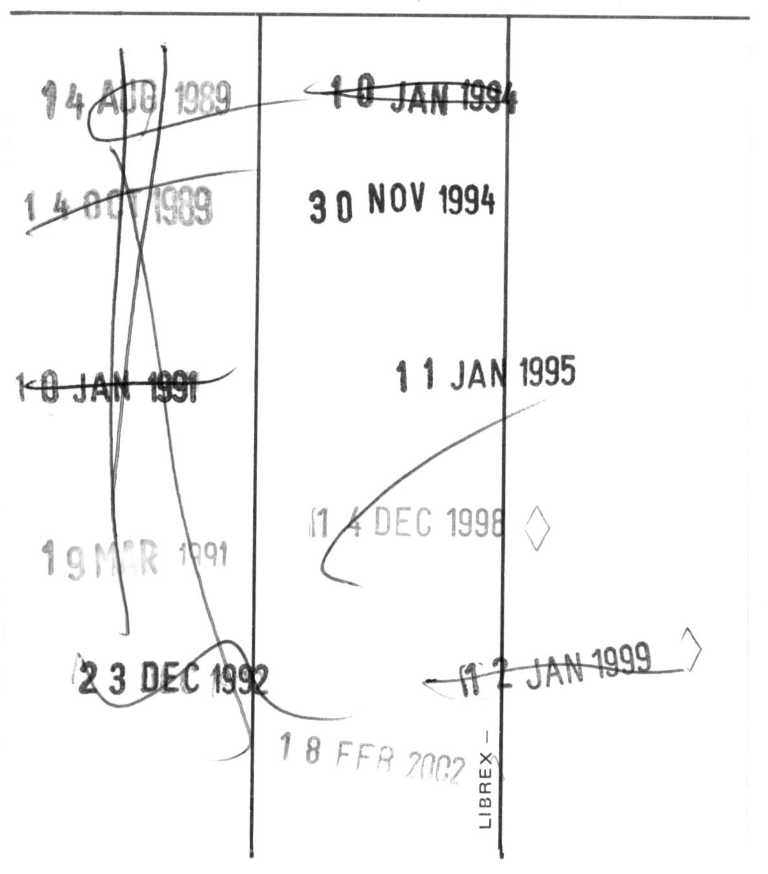

#17108335

# Food Structure–Its Creation and Evaluation

J. M. V. BLANSHARD
J. R. MITCHELL
*Faculty of Agricultural Science, University of Nottingham, UK*

Butterworths
London   Boston   Singapore   Sydney   Toronto   Wellington

All rights reserved. No part of this publication may be reproduced or transmitted in any form or by any means, including photocopying and recording, without the written permission of the copyright holder, application for which should be addressed to the Publishers. Such written permission must also be obtained before any part of this publication is stored in a retrieval system of any nature.

This book is sold subject to the Standard Conditions of Sale of Net Books and may not be re-sold in the UK below the net price given by the Publisher in their current price list.

First published, 1988

© The several contributors named in the list of contents, 1988

**British Library Cataloguing in Publication Data**

Food Structure: its creation and
evaluation.—(Easter schools).
1. Food 2. Rheology 3. Food texture
I. Blanshard, J. M. V. II. Mitchell, J. R.
III. Series
664 TP368
ISBN 0-408-02950-1

**Library of Congress Cataloging in Publication Data**

Food structure.

Bibliography: p.
Includes index.
1. Food texture. 2. Food processing. 3. Food-Sensory evaluation. I. Blanshard, J. M. V.
II. Mitchell, J. R.
TX531.F674 1987 664'.07 87-32012
ISBN 0-408-02950-1

D
664·07
FOO

Photoset by Latimer Trend & Company Ltd, Plymouth
Printed and bound in Great Britain by Butler and Tanner, Frome, Somerset

# PREFACE

One of the greatest challenges faced by the food scientist is man himself with his highly sensitive oral equipment for evaluating food texture. The recognition of this challenge leads immediately to two questions: 'How can we characterize the texture of foodstuffs in a manner that will reflect human perceptions?' and, then, 'How may we preserve, formulate and process products in a way that will be acceptable to the consumer?'

With unlimited time and resources the ideal scientific approach is to build from a molecular understanding through the microscopic to, ultimately, the macroscopic structure and behaviour. Such a programme has only been successfully pursued in a few simple systems and the financial backing and commitment are substantial. For the majority of foodstuffs it is impossible to contemplate the completion of such a comprehensive analysis in the next decade. In practice, therefore, two approaches have been widely used.

In the first, the major thrust has been to understand and exploit the role of ingredients in building food structures. In the second, the starting point has been the known finished product and the objective has been to examine the contribution and function of the ingredients within that structure.

The symposium, of which this book is a record, outlines research and many exciting developments in these areas. The reader will find not only state of the art reviews of molecular interactions in homogeneous and heterogeneous multicomponent systems, but also reviews of the macroscopic physics of their mechanical properties, of the more phenomenological approach necessary as systems increase in complexity and, finally, of the variety of techniques and strategies necessary to evaluate their properties if they are to be acceptable to the consumer.

We are extremely grateful to the distinguished international group of speakers who so willingly contributed to the success of the symposium, not only by lecturing but by providing the manuscripts which are the substance of this volume. A notable feature of the meeting was the enthusiasm of the participants. To all these the organizers extend their warm thanks.

J. R. Mitchell
J. M. V. Blanshard

## ACKNOWLEDGEMENTS

We should like to express our sincere gratitude to the following organizations who very generously provided financial support for the meeting.

Dalgety
Kelco International
Kellogg Company of Great Britain
Pedigree Petfoods
Rowntree Mackintosh
Unilever
United Biscuits

# CONTENTS

Preface v

1 THE RELEVANCE OF FOOD STRUCTURE — A DENTAL CLINICAL PERSPECTIVE 1
M.R. Heath, *Department of Prosthetic Dentistry, London Hospital Medical College Dental School, London, UK*

2 MIXED AND FILLED GELS—MODELS FOR FOODS 7
G.J. Brownsey and V.J. Morris, *AFRC Institute of Food Research, Norwich, UK*

3 GEL STRUCTURE OF FOOD BIOPOLYMERS 25
Anne-Marie Hermansson, *SIK–The Swedish Food Institute, Göteborg, Sweden*

4 THE STRUCTURE AND STABILITY OF EMULSIONS 41
E. Dickinson, *Procter Department of Food Science, University of Leeds, UK*

5 STRUCTURE AND PROPERTIES OF LIQUID AND SOLID FOAMS 59
G. Jeronimidis, *Department of Engineering, University of Reading, UK*

6 THE POLYMER/WATER RELATIONSHIP—ITS IMPORTANCE FOR FOOD STRUCTURE 75
P.J. Lillford, *Unilever Research, Colworth House, Sharnbrook, Bedfordshire, UK*

7 POLYMER FRACTURE 93
D.P. Isherwood, *Department of Mechanical Engineering, Imperial College of Science and Technology, London, UK*

8 STRUCTURAL STABILITY OF INTERMEDIATE MOISTURE FOODS—A NEW UNDERSTANDING? 115
Louise Slade and H. Levine, *General Foods Corporation, New York, USA*

9 'COLLAPSE' PHENOMENA—A UNIFYING CONCEPT FOR INTERPRETING THE BEHAVIOUR OF LOW MOISTURE FOODS 149
H. Levine and Louise Slade, *General Foods Corporation, New York, USA*

10 CREATION OF FIBROUS STRUCTURES BY SPINNERETLESS SPINNING 181
V.B. Tolstoguzov, *Institute of Organoelement Compounds, USSR Academy of Science, Moscow, USSR*

11 DRY SPINNING OF MILK PROTEINS 197
J. Visser, *Unilever Research Laboratory, Vlaardingen, The Netherlands*

12 PROTEIN EXTRUSION—MORE QUESTIONS THAN ANSWERS? 219
D.A. Ledward and J.R. Mitchell, *Faculty of Agricultural Science, University of Nottingham, UK*

13 REFORMED MEAT PRODUCTS—FUNDAMENTAL CONCEPTS AND NEW DEVELOPMENTS 231
P.D. Jolley and P.P. Purslow, *AFRC Institute of Food Research, Bristol, UK*

14 SURIMI-BASED FOODS—THE GENERAL STORY AND THE NORWEGIAN APPROACH 265
K. Fretheim, *The Innovation Centre, Oslo*, B. Egelandsdal, *Norwegian Food Research Institute, Ås*, E. Langmyhr, *Norwegian Herring Oil and Meal Industry Research Institute, Bergen*, O. Eide and R. Ofstad, *Institute of Fishery Technology Research, Tromsø, Norway*

15 STRUCTURED FAT SYSTEMS 279
G.G. Jewell and J.F. Heathcock, *Cadbury Schweppes, Reading, Berkshire, UK*

16 STRUCTURED SUGAR SYSTEMS 297
M.G. Lindley, *Tate and Lyle, Reading, Berkshire, UK*

17 ELEMENTS OF CEREAL PRODUCT STRUCTURE 313
J.M.V. Blanshard, *University of Nottingham School of Agriculture, Sutton Bonington, UK*

18 EXTRUSION AND CO-EXTRUSION OF CEREALS 331
R.C.E. Guy and A. W. Horne, *Flour Milling and Baking Research Association, Chorleywood, Hertfordshire, UK*

19 THE EVALUATION OF FOOD STRUCTURE BY LIGHT MICROSCOPY 351
F.O. Flint, *Department of Food Science, University of Leeds, UK*

20 AN ELECTRON MICROSCOPIST'S VIEW OF FOODS 367
D.F. Lewis, *Leatherhead Food Research Association, Leatherhead, Surrey, UK*

21 SMALL DEFORMATION MEASUREMENTS 387
S.B. Ross-Murphy, *Unilever Research, Colworth House, Sharnbrook, Bedfordshire, UK*

22 BEHAVIOUR OF FOODS IN LARGE DEFORMATION 401
E. B. Bagley and D. D. Christianson, *US Department of Agriculture, Peoria, Illinois, USA*

23 THE SENSORY–RHEOLOGICAL INTERFACE 417
P. Sherman, *Department of Food and Nutritional Sciences, King's College, University of London, UK*

24 EVALUATION OF CRISPNESS 433
Z. M. Vickers, *Department of Food Science and Nutrition, University of Minnesota, USA*

25 BEYOND THE TEXTURE PROFILE 449
E. Larmond, *Agriculture Canada, Ottawa, Canada*

26 ORAL PERCEPTION OF TEXTURE 465
M.R. Heath, *Department of Prosthetic Dentistry, London Hospital Medical College Dental School, London, UK* and P.W. Lucas, *Department of Anatomy, National University of Singapore, Singapore*

List of Participants 483

Index 489

# 1

# THE RELEVANCE OF FOOD STRUCTURE—A DENTAL CLINICAL PERSPECTIVE

M.R. HEATH
*Department of Prosthetic Dentistry, London Hospital Medical College Dental School, London UK*

Food scientists do not need to be told that most people take great pleasure in eating. Satisfaction drives appetite which has survival value for a species in the wild and persists despite modern food technology. To some extent, pleasure from texture is by association with familiar tastes and smells. Obviously the smoothness of a sauce is a merit but the dental perspective for this symposium stems from man's satisfaction in chewing resistant foods. Despite the skill of the dental profession many patients still complain of difficulty in chewing some of the foods which they used to enjoy. This is not surprising because in 1978 nearly one-third of all adults in England and Wales had no natural teeth; the incidence was 74% for those over the age of 65 and 87% for those over 75 years (Todd and Walker, 1980). This incidence is already falling as shown by comparison with 1968 data (Gray *et al.*, 1970), but food structure is going to remain relevant to the chewing ability of many UK residents (*Figure 1.1*). In a small survey of 75 housebound pensioners the majority never ate several of the more difficult foods. Nuts, tough meat, hard biscuits, celery and toffee were most frequently excluded (*Figure 1.2*; Heath, 1972).

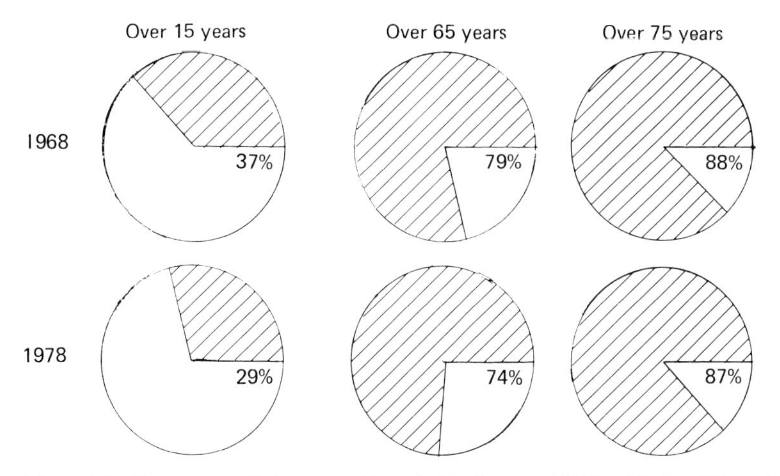


**Figure 1.1** Percentage of edentate subjects in England and Wales. National Dental Survey data for 1968 and 1978. From Heath (1986) with kind permission of the publishers.

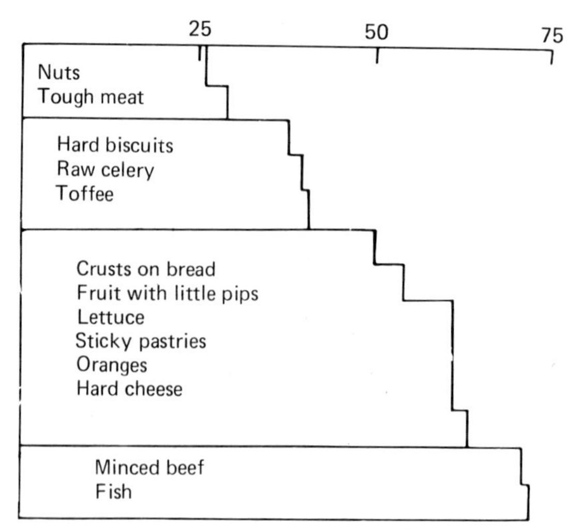


**Figure 1.2** The number of subjects who ever ate each specified food. From Heath (1972) with kind permission of the publishers.

The masticatory effectiveness of elderly denture wearers is only one-sixth that of young dentate adults with natural teeth and their chewing cycle time is twice as long (*Figure 1.3*; Heath 1982). A simple clinical test involving a square of chewing gum (*Figure 1.4*a) shows that a patient without dentures moves the piece ineffectually about the mouth so that it still has the original pattern on it, or it may be merely folded (*b*). Patients chewing with only an upper denture mark the surface but some evidence of the original flattened shape usually remains (*c*). Patients with both dentures often manage to produce a well chewed bolus (*d*)—at best a small soft ball after 20 strokes (*e*). Watching people chewing the test piece of gum gives a picture of the style of chewing.

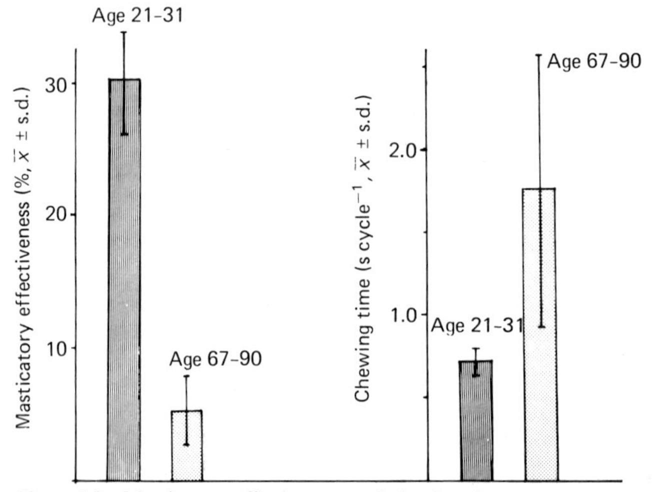


**Figure 1.3** Masticatory effectiveness and chewing time (s/cycle). Means and s.d. for young fit dentate adults ($n = 13$) and housebound edentate subjects ($n = 50$). From Heath (1982) with kind permission of the publishers.

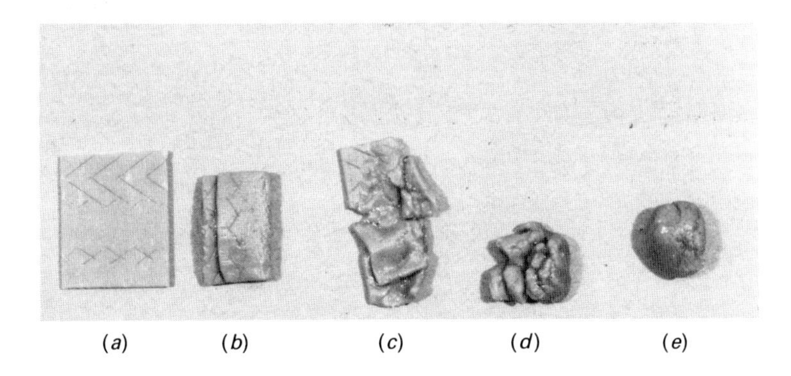


**Figure 1.4** Chewing can be assessed with a small square of chewing gum chewed for 20 strokes. Test pieces (a) unused; (b) sucked and folded by a patient without dentures; (c) marked, but retaining original flat areas, by a patient with both dentures; (d) well marked by a patient with both dentures; (e) a homogeneous soft ball well chewed. From Heath (1982) with kind permission of the publishers.

The ability to chew gum effectively correlates significantly with maximum biting force (*Figure 1.5*; Heath, 1982) but many of these elderly people could exert less than 5 kg on a bite force gauge. To get some concept of the very low forces achieved by some old people, a dentist can use his finger as a biting force gauge; 2 kg barely hurts the finger but produces an indentation (*Figure 1.6*).

The clinical picture is somewhat obscured by man's considerable physiological reserve and his adaptability. Our masticatory system was designed to cope with a wide range of raw foods and the healthy, young digestive system can cope with most cooked foods unchewed (Farrell, 1956) but there are no data to suggest that this remains true with advancing age. Although there is deterioration of mastication with tooth loss, the same pattern of chewing is typically used so that food is less well chewed when swallowed.

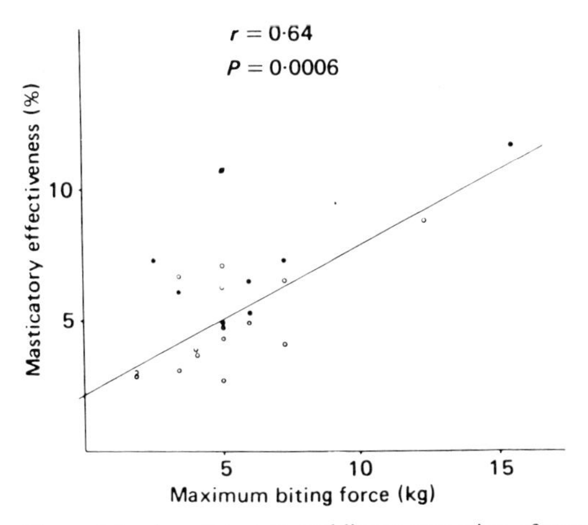


**Figure 1.5** A scattergram and linear regression of masticatory effectiveness on maximum biting force. Those people who ever ate tough meat are shown as solid dots. From Heath (1982) with kind permission of the publishers

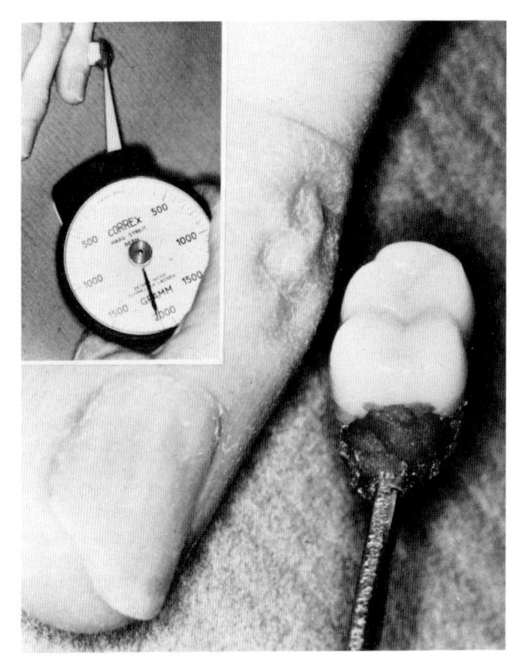


**Figure 1.6** A finger can be cautiously used as a 'clinical' biting force gauge for elderly patients. 2 kg barely hurts but does produce an indentation of the skin. From Heath (1982) with kind permission of the publishers.

These typical patterns are best demonstrated with video records showing the normal coordination and the slowing with age and dentures. It happens that for patterns of mandibular movement giraffes are similar to man though slower, so it is particularly easy to follow the near vertical opening phase followed by lateral shift towards the chewing side and the swing back towards the midline with a pause in or near central tooth contact. Small mammals chew faster, illustrating the fine rapid coordination of mammalian oral physiology. The normal human mouth with natural teeth is equally capable, at an intermediate speed, as can be seen using video radiography to follow the mastication of a radio-opaque biscuit.

This necessarily raises questions about the rheology and fracture toughness of foods which can be eaten by the many whose oral state and adaptation is critically inadequate. The advent of successful osseointegrated dental implants to stabilize dentures provides dramatic restoration of function and comfort but at a price which restricts their availability.

For those less fortunate, dentists carefully minimize the functional loss and counsel on dietary selection of the 'least worst' options, but as is proposed in Chapter 26 the time seems ripe for more interdisciplinary, cooperative research involving the combined efforts of physiologists, clinicians, food scientists and rheologists as well as industry.

## References

FARRELL, J.H. (1956). The effect of mastication on the digestion of food. *British Dental Journal*, **100**, 149–155

GRAY, P.G., TODD, J.E., SLACK, G.L. and BULMAN, J.S. (1970). *Adult Dental Health in England and Wales in 1968*. London, HMSO

HEATH, M.R. (1972). Dietary selection by elderly persons, related to dental state. *British Dental Journal*, **132**, 145–148

HEATH, M.R. (1982). The effect of maximum biting force and bone loss upon masticatory function and dietary selection of the elderly. *International Dental Journal*, **32**, 345–356

HEATH, M.R. (1986). Gerodontics: the role of an MSc course in the United Kingdom. *Gerodontics*, **2**, 239–242

TODD, J.E. and WALKER, A.M. (1980). *Adult Dental Health Vol. 1; England and Wales 1968–1978*, pp. 8–9. London, HMSO

2

# MIXED AND FILLED GELS—MODELS FOR FOODS

G.J. BROWNSEY and V.J. MORRIS
*AFRC Institute of Food Research, Norwich, UK*

## Introduction

The majority of natural and fabricated foods are solids containing substantial quantities of water ($\sim$50–90%). The solid-like consistency can usually be attributed to the presence of certain proteins and polysaccharides. These biopolymers may be natural constituents of the plant or animal tissue or they may be additives incorporated in order to manipulate the mechanical properties and perceived texture of fabricated foods. The gel state provides a good model for many solid-like food systems. However, there exists a hierarchy of gel structures (Morris, 1985; 1986a) which may be used to model the complex molecular networks formed by the protein and/or polysaccharide components within the food system. These structures range from simple *single component* networks used to describe the behaviour of individual biopolymers, through binary networks or *mixed gels* used as the simplest models for multicomponent structures present in real foods, to composite or *filled gels* used to assess the role of particulates such as fat globules, liquid droplets, fibres, gas bubbles, crystallites or cellular components present in food systems. Such simple descriptive models may be of value in at least three areas of research. Physicists and chemists are interested in developing molecular models for gelation. Rheologists are interested in developing mathematical descriptions of the deformation and failure properties of food gels. Finally, sensory scientists and nutritionists are interested in relating perceived texture to measurable physical and mechanical properties, using multicomponent gels as a basis for manipulating the textural characteristics of foods. This chapter will consider how such simple network structures can and have been successful in providing better descriptions of food systems.

## Single component gels

Single component polymeric networks are the simplest molecular models for biopolymer gels. However, they can only be considered as realistic models for very simple food systems such as flans, aspics or table jellies. The advantage of studying single component gels is that they provide ideal models for investigating mechanisms of gelation and also for devising mathematical descriptions of rheological and mechanical behaviour. Such gels thus represent the first step towards a description of more complex multicomponent gels.

The intimate details of most biopolymer gelation processes are still not completely understood. The gelation of food polysaccharides and food proteins has been described in recent review articles (Clark and Lee-Tuffnell, 1986; Ledward, 1986; Morris, 1985, 1986b). Proteins and polysaccharides can be classified as 'structural' or 'storage' biopolymers (Morris, 1986a). Physical and physicochemical studies have established the basic principles of most gelation processes and thus revealed the main factors which control gelation (Clark and Lee-Tuffnell, 1986; Ledward, 1986; Morris, 1985, 1986b; Rees *et al.*, 1982). Such studies explain the cold-setting mechanisms of 'structural' polysaccharides such as pectins or alginate and 'structural' proteins such as gelatin as well as the heat-setting mechanisms of 'storage' proteins such as globular proteins or the 'storage' polysaccharide starch. The effects of external variables such as pH, ionic strength, temperature and the nature of added co-solutes can be explained in terms of molecular models for gelation. Thus the dependence of the melting points of many thermo-reversible gels on ionic strength can often be attributed to the stabilization of charged helical structures. Site binding of specific cations can account for the need for such cations in order to ensure gelation. Similarly marked variations in the mechanical properties and opacity of globular protein gels prepared at different pH or ionic strength are determined by their surface charge. An important feature of such studies on molecular mechanisms has been the identification of structural features of biopolymers which are essential for intermolecular association and gelation. These studies have led to modifications to extraction methods (Rees, 1972) and both chemical (Rees, 1972) and enzymic (McCleary and Neukom, 1982; Skjäk-Braek, 1984) modifications to biopolymers in order to improve functionality. Single component gels also provide good models for examining the effects of co-solutes on gelation although few systematic studies have been reported.

A limitation of physicochemical studies is the tendency to focus attention on the junction zones within the gels and to neglect general descriptions of network structures. However, considerable progress has been made in developing mathematical descriptions of the mechanical properties of polymer gels. The concentration dependence of the elastic moduli may be described by equations developed by Hermans (1965) and based on the Flory–Stockmayer theory of gelation (Flory, 1941a,b,c; Stockmayer, 1943, 1944). The original formulation due to Hermans has been extended (Clark and Lee-Tuffnell, 1986) on the basis of the cascade theory of gelation (Gordon, 1962; Gordon and Ross-Murphy, 1975). Such equations can be used to model the concentration dependence of the shear modulus and have been applied to describe the properties of 'structural' polysaccharide (Clark *et al.*, 1982, 1983), 'structural' protein (Clark *et al.*, 1983) and 'storage' protein (Bikbov *et al.*, 1979; Clark and Lee-Tuffnell, 1986; Clark *et al.*, 1982; Richardson and Ross-Murphy, 1981) gels. The possibility of extending such treatments to include large deformation and failure has been discussed by McEvoy, Ross-Murphy and Clark (1984).

## Two component mixed gels

Binary polymer networks provide the simplest examples of mixed multicomponent gels. It is useful to distinguish (Morris, 1986a) between two types or classes of binary gel. In type I gels only one of the two polymers actively forms part of the polymer network whereas in type II gels both polymers are incorporated into the molecular network.

### *Type I gels*

Type I gels consist of a polymer network entrapping a second soluble polymer (*Figure 2.1*). Let the polymer which forms the network be designated A and the soluble entrapped polymer designated B. The presence of polymer B may influence gelation of polymer A, affect conformational transitions of polymer A and/or swell the polymeric network. Small quantities of dextrans when added to aqueous gelatin solutions have been shown (Tolstoguzov *et al.*, 1974; Tolstoguzov and Braudo, 1983) to increase the rate of helix formation and the rate of gelation. Similarly the excluded volume effect of added polyethylene glycols or dextrans have been shown to favour the compact helical structure of DNA and thus raise the melting point of the helix (Laurent, Preston and Carlsson, 1974). Stainsby (1980) has suggested that the excluded volume effect of entrapped polymers might be put to practical use in an attempt to swell gels and hence combat syneresis. There is evidence (Comper and Laurent, 1978; Hedbys, 1961; Hedbys and Dohlman, 1963) that hyaluronates, sulphated proteoglycans or glycosaminoglycans swell the collagen structures of animal tissue. The removal of such polymers enzymically or otherwise has been shown (Comper and Laurent, 1978) to cause the gel network to shrink. Similar results have been obtained on model systems consisting of polysaccharides inserted into gelatin gels (Comper and Preston, 1974; Meyer, Comper and Preston, 1971). The interest in such effects appears to have been restricted to models for connective tissue. The relative merits of added polymers as alternatives to co-solutes for modifying network–solvent interactions remains to be determined. Enzymic removal of entrapped polymer and the existence of gel shrinkage could provide a mechanism for distinguishing between type I and type II gels.

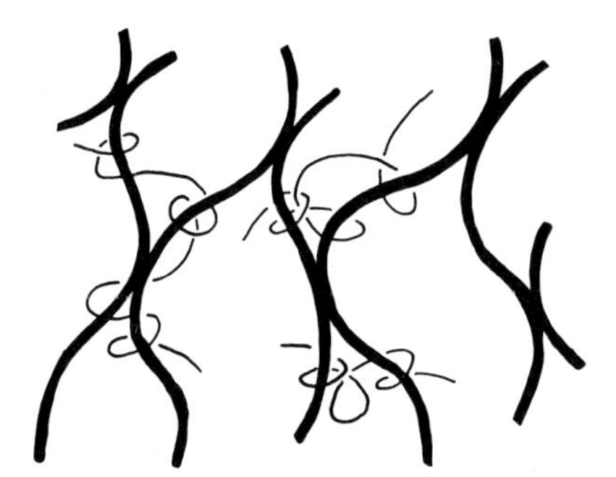

**Figure 2.1** Type I gel network showing a soluble polymer entrapped inside a polymeric network.

### *Type II gels*

Type II binary gels are formed from two polymers under conditions which favour gelation of each component. Let the two polymers be designated as A and B. The type of molecular network formed will depend on the relative magnitudes of A–A, B–B and A–B interactions. The spatial distribution of the two polymers within the formed gel will depend on the extent of mixing of the two components prior to gelation and also on the degree of demixing which occurs during the gelation process. Different networks may be prepared from a given set of ingredients by varying the method of mixing and the method and/or sequence of gelation of each component. Schematic diagrams of three extreme examples of molecular networks which could be formed from two components are shown in *Figure 2.2*. These structures have been called (Morris, 1986a) coupled networks (*Figure 2.2a*), phase-separated networks (*Figure 2.2b*) and interpenetrating networks (Sperling, 1981) (*Figure 2.2c*).

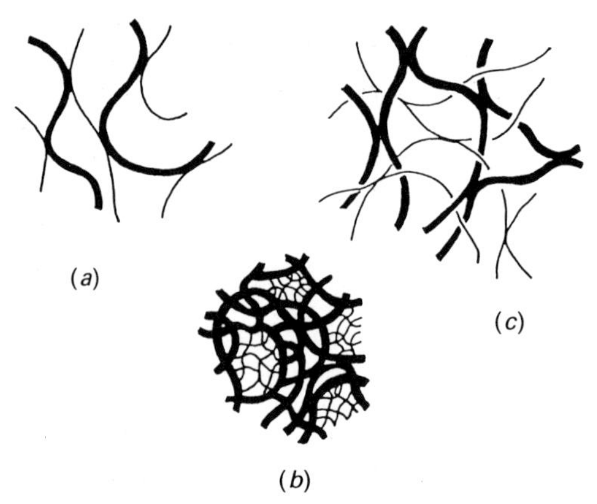


**Figure 2.2** Type II gel networks. Idealized representations of (a) coupled network; (b) phase-separated networks; and (c) interpenetrating networks.

*Coupled networks*

To create a coupled network it is necessary to have a favourable A–B interaction between the two polymers. Coupled networks are attractive commercially because they offer the prospect of developing new mechanical or textural properties. In certain cases it may be possible to take two polymers, neither of which will gel alone, and mix these polymers under conditions which promote intermolecular (A–B) binding and subsequent gelation of the mixture.

The most commonly encountered mixed biopolymer gels are probably 'storage' protein gels. Most globular proteins found or added to foods are mixtures. Examples include soya, liquid or powdered egg, and blood proteins. Even in the case of simple mixtures of globular proteins it has not been possible to establish whether coupled networks or phase-separated networks are formed (Clark and Lee-Tuffnell, 1986). Detailed studies of mixed globular protein gels require the development of techniques for labelling specific biopolymers and visualizing their spatial distribution within the gel.

Chemical cross-linking has been used to try and create coupled networks. Cross-linking of alginate esters with gelatin via the formation of amide bonds has been reported by Stainsby (1980). He reports that starch–alginate ester mixtures may be gelled under mildly acid conditions; binding has been attributed to the formation of transesterification linkages between alginate ester residues and starch hydroxyl residues. Specific ion interactions are said (Snoeren, 1976) to occur between kappa-casein and kappa-carrageenan. Such interactions are claimed (Snoeren, 1976) to result in mixed coupled gel networks responsible for stabilizing milk products.

Favourable interactions between different polysaccharides will require the presence of chemically compatible regions within each polymer chain. There are several examples of synergism in binary polysaccharide systems which have been attributed to intermolecular binding between the two polymers and coupled network formation.

The junctions zones in coupled gels (*Figure 2.2a*) are new ordered structures. Physical and chemical techniques may be employed to test the hypothesis of intermolecular association. The most powerful direct physical technique is X-ray diffraction of oriented fibres prepared from the gels. This technique may be used to test molecular models at atomic resolution and has proved reliable in studies on single component gel systems (Rees *et al.*, 1982). Synergism between alginate and pectin has been reported (Morris and Chilvers, 1984; Thom *et al.*, 1982; Toft, 1982) and the near mirror image similarity between guluronic acid and galacturonic acid suggest a stereochemically possible mechanism for intermolecular binding. Comparative circular dichroism studies on alginate, pectin and pectin–alginate have been cited (Thom *et al.*, 1982) as evidence for intermolecular binding. As yet there are no reported X-ray diffraction studies on alginate–pectin mixed gels. Synergism between plant galactomannans (carob gum, tara gum, enzymically modified guar) and the bacterial polysaccharide xanthan gum or certain algal polysaccharides (kappa-carrageenan, furcellaran and agarose) has been attributed to intermolecular binding (Dea, 1981; Dea and Morrison, 1975; Dea, McKinnon and Rees, 1972; Dea *et al.*, 1977; Morris *et al.*, 1977) between the backbone of the galactomannan and the ordered helical structure of the other polysaccharide (*Figure 2.3a*). Slightly modified versions of this model have also been proposed (McCleary, 1979; Tako and Nakamura, 1985; Tako Asato and Nakamura, 1984).

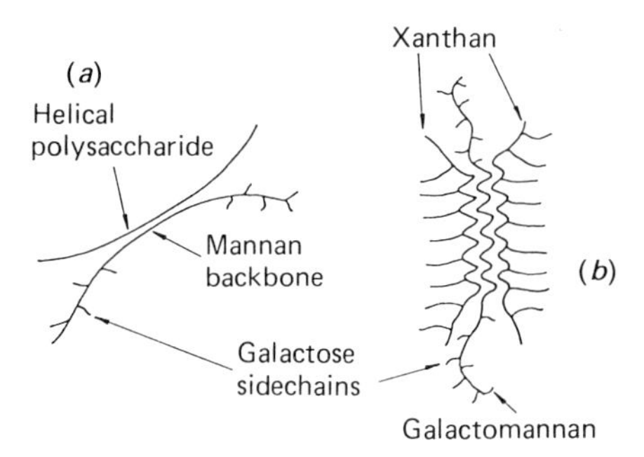


**Figure 2.3** Schematic models for intermolecular binding between two polysaccharides: (a) binding schemes due to Dea and coworkers suggested for galactomannan binding to xanthan, kappa-carrageenan or agarose; (b) binding scheme proposed by Cairns *et al.* (1986b, 1987) for galactomannan–xanthan binding.

The algal polysaccharides kappa-carrageenan, furcellaran and agarose form thermo-reversible gels. Addition of galactomannans (tara, carob) reduces the total polymer concentration required for gelation and modifies the texture of the gel. X-ray diffraction studies of oriented fibres prepared from kappa-carrageenan–carob or kappa-carrageenan–tara mixed gels yield patterns indistinguishable from those obtained for pure kappa-carrageenan gels (Cairns, Miles and Morris, 1986a; Cairns *et al.*, 1986b, 1987; Carroll, Miles and Morris, 1984; Miles, Morris and Carroll, 1984). In similar unpublished studies on furcellaran–carob, furcellaran–tara, agarose–carob and agarose–tara mixed gels it also proved to be impossible to detect the proposed (*Figure 2.3a*) intermolecular binding. In the absence of evidence for intermolecular binding and coupled network formation further experiments were carried out in order to investigate the orientation and spatial location of each polysaccharide within such mixed gels (Cairns *et al.*, 1987). It has been observed that oriented fibres prepared from carob samples are initially poorly crystalline but crystallize on storage (Cairns, Miles and Morris, 1986b; Cairns *et al.*, 1987). When aligned carob–kappa-carrageenan fibres were stored, crystallization of the carob resulted in diffraction rings superimposed upon the oriented fibre pattern characteristic of kappa-carrageenan junction zones (Cairns *et al.*, 1987). These results showed no evidence for cocrystallization of carob and kappa-carrageenan and suggest random orientation of carob even within aligned fibres. Carrageenans are naturally tagged with sulphate residues and galactomannans may be labelled with fluorescein without interfering with gelation (Cairns, Miles and Morris, 1986a; Cairns *et al.*, 1987). These labels have been used to visualize the spatial location of each polymer within a mixed gel. Such studies failed to reveal any evidence for gross phase separation in carob–kappa-carrageenan or tara–kappa-carrageenan mixed gels. The simplest interpretation of the combined experimental data would suggest a type I network (*Figure 2.1*). Enzymic degradation of galactomannans within such gels may provide an experimental method for testing this conclusion. Recent but limited rheological studies of tarra–kappa-carrageenan and carob–kappa-carrageenan gels (Cairns *et al.*, 1986a) suggest that the mechanical properties of these mixed gels may be related to those equivalent properties of the pure kappa-carrageenan gel. Should such studies prove to be widely applicable then this would simplify description of these and similar mixed gels. The absence of evidence for intermolecular binding in these systems is perhaps not too surprising in view of the lack of any stereochemical compatibility between the galactomannan backbone and either the carrageenan, furcellaran or agarose helices.

Mixtures of xanthan gum with certain galactomannans (carob gum or tara gum) are unusual in that gelation occurs over wide composition ranges under conditions for which neither component alone will form gels (Dea and Morrison, 1975; Dea *et al.*, 1977; Morris *et al.*, 1977). Recent X-ray diffraction studies of oriented fibres prepared from carob–xanthan and tara–xanthan mixed gels yield new diffraction patterns characteristic of xanthan–galactomannan binding (Cairns, Miles and Morris, 1986b; Cairns *et al.*, 1987; Morris and Miles, 1986). By varying temperature and ionic strength it is possible to modify the conformation of the xanthan (Lambert, Milas and Rinaudo, 1985). Thus it is possible to mix xanthan and galactomannans with xanthan in either the ordered helical or disordered 'coil' form. Such mixing experiments suggest that intermolecular binding and gelation only occur if the galactomannan is mixed with xanthan in the disordered coil state (Cairns, Miles and Morris, 1986b; Cairns *et al.*, 1987). The mixing data coupled with a qualitative analysis of the X-ray diffraction patterns suggest that intermolecular binding involves a cocrystallization of sections of the denatured xanthan molecules with segments of the galactomannan chain (Cairns, Miles and Morris, 1986b; Cairns *et al.*, 1987). Such an interaction is illustrated schematically in *Figure 2.3b* which is not intended to accurately reflect the true stoichiometry of the interaction, the three dimensional arrangement of the macromolecules or even the need for unsubstituted segments of the galactomannan. *Figures 2.3a* and *2.3b* emphasize that the interaction does not occur with the xanthan helix. The proposed interaction is aesthetically pleasing because of the stereochemical compatibility between the mannan backbone of the galactomannan and the cellulosic backbone of xanthan. The addition of galactose sidechains to the mannan backbone results in a new crystalline structure which may be regarded as a modified mannan crystal lattice (Winter, Chien and Bouckris, 1984). The simplest model for intermolecular binding would be a random cocrystallization of xanthan with the galactomannan resulting in a further perturbation of the mannan lattice (Cairns *et al.*, 1987). The need for stereochemically compatible structures would explain the inability to detect galactomannan binding with kappa-carrageenan, agarose or furcellaran and would account for the marked synergism observed between xanthan and glucomannans such as konjac mannan (Dea, 1981).

*Phase-separated networks*

In the absence of favourable A–B interactions, the mutual incompatibility of unlike biopolymers at the high concentrations required to ensure gelation, suggest that phase-separated networks (*Figure 2.2b*) are the most likely model for the gelation of binary polymer mixtures. Phase separation has been observed in 'structural' protein–'structural' polysaccharide (Clark *et al.*, 1983; Moritaka *et al.*, 1980; Tolstoguzov, 1978; Watase and Nishinari, 1980) and 'storage' protein–'structural' polysaccharide (Clark *et al.*, 1982; Clark and Lee-Tuffnell, 1986) mixed gels. In the case of protein–polysaccharide gels selective staining methods have been used to visualize the phase separation (Clark *et al.*, 1982, 1983; Clark and Lee-Tuffnell, 1986). For a mixed gel formed from two polymers A and B there is a particular ratio (A/B) called the phase inversion point where the system changes from a matrix of gel A containing inclusions of gel B to a matrix of gel B containing inclusions of gel A. At the phase inversion point the system may consist of two continuous interpenetrating networks which will be discussed later in this chapter. Both gelatin and agarose form thermo-reversible gels and phase inversion in gelatin–agarose mixed gels is indicated by a sharp change in melting temperature (Moritaka *et al.*, 1980; Watase and Nishinari, 1980).

Practically, two phase mixtures have limited use only when one phase is the major continuous phase but with truly immiscible components, useful properties are often possible over the entire composition range. For a composite, the dispersed phase can be both high or low in modulus, as well as being polymeric. If a homogeneous single phase structure is produced, then it is true to say that the modulus is roughly intermediate between the two components, but this is not the case for two phase systems. Such systems may be modelled by an upper and lower bound approach (Manson and Sperling, 1976). The upper ($G_U$) and lower ($G_L$) boundaries of the elastic modulus $G$ can be calculated under conditions of isostrain and isostress respectively (*Figure 2.4*). For a simple phase separated system the experimental modulus should

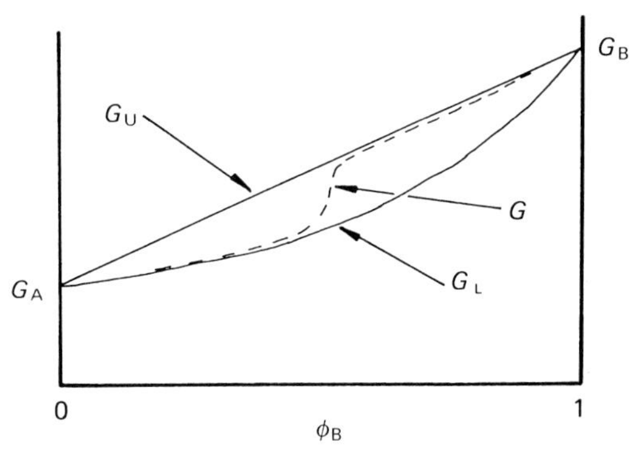


**Figure 2.4** Expected dependence of the shear modulus ($G$) for a simple phase-separated network on the volume fractions of the two components. The upper bound $G_U$ and the lower bound $G_L$ are related to the moduli of the pure components: $G_U = (\phi_A G_A + \phi_B G_B)$ and $G_L = (\phi_A G_A^{-1} + \phi_B G_B^{-1})^{-1}$. $\phi_A$, $\phi_B$, $G_A$, $G_B$ are respectively the volume fractions and moduli of components A and B. $\phi_A + \phi_B = 1$.

follow the lower boundary prior to phase inversion and then invert and follow the upper boundary (*Figure 2.4*). Such an approach has been used to model 'structural' protein–'structural' polysaccharide and 'storage' protein–'structural' polysaccharide gels (Clark and Lee-Tuffnell, 1986; Clark *et al.*, 1982, 1983). Such modelling revealed the need to take into account the effects of redistribution of solvent between the two component phases. This leads to a change in the effective polymer concentration within each phase and thus a change in the effective modulus of each phase. McEvoy, Ross-Murphy and Clark (1984) describe an attempt to extend such modelling of the low deformation elastic modulus to include high deformation and failure properties of phase-separated gel networks.

### *Interpenetrating networks*

An interpenetrating network (IPN) is a combination of two polymeric species in network form, at least one being prepared in the immediate presence of the other (*Figure 2.2c*). There are, in fact, many possible preparative routes to form such structures and it is a consequence of this that a variety of morphologies and structural classifications have arisen (Sperling, 1981). One of the most common types of IPN is where a preformed network is swollen and a second system is crosslinked *in situ*. If there is a high degree of miscibility, then an IPN will result; if not, phase separation is observed with an IPN structure only at the boundaries.

Such structures have been under study for many years with synthetic polymers where the benefits of IPN structures are that they often show improved miscibility over that of the constituent parts. Initially, such dual network formation was considered as a means of suppressing gross phase separation, allowing the formation of composites containing a well dispersed phase but without the need for heavy mixing. Additionally, the degree of interprenetration will reduce the magnitude of structural flaws imposed by the phase boundaries to provide stronger structures.

True IPNs are rare, due to phase separation or the possibility of interactions which may give rise to coupled networks. Nevertheless, it is necessary that a means of identifying a structure as an IPN be available, the most usual entails monitoring the glass transition temperature ($T_g$) which should be sensitive to the local structure within the material. However, the existence of a single $T_g$ is not in itself a sufficient criterion that two materials are miscible since the extent is not yet established to which heterogeneity in local structure influences the transition temperature. It may be that even on a microscopic scale the structure is heterogeneous. The macroscopic properties of these real interpenetrating networks are not simply determined by the usual additivity relations. It appears that the more continuous phase dominates the properties of a real IPN so that it is not only important that the two networks are formed simultaneously but also that they develop at the same rate. Stronger structures are possible if one component develops faster, truly simultaneous interpenetrating networks being weaker. Since most of the more interesting IPNs are phase separated, the term 'interpenetrating phases' has been said to be a better description of the structures. Mixed gels can be treated as a particular case of an IPN (Thomas and Sperling, 1978).

## Filled gels

Filled gels are models which embody a further level of complexity and provide good models for many food systems. The matrix gel may be a single component or multicomponent gel. It is convenient to consider the simplest example of a single component gel matrix. In real food systems the filler particles may be fibres, gas bubbles, liquid droplets, crystals, fat globules or cellular components such as starch granules or seeds. The properties of the matrix and filler may vary during cooking or processing. Indeed, the size, shape and deformability of starch granules change during cooking and baking, and starch based foods retrograde on storage. Processing of dairy products will influence the formation and crystallization of water and fats. The textural changes observed in the fibrous structures of muscle when animals are killed and the meat cooked are well known.

Model composites have been employed to investigate the effects of the mechanical properties of the matrix gel, the volume fraction and rigidity of the filler particle and the filler–matrix interaction. Gelatin has been found to be a convenient matrix material for such studies. Ring and Stainsby (1982) used a variety of fillers to probe the effects of volume of the filler, deformability of the filler and the shear modulus of the matrix on the low deformation shear modulus. The effect of filler shape was studied by Richardson *et al.* (1981) using filled gelatin gels and a general discussion of the effects of filler shape on the mechanical properties of filled polymers is contained in the review article by Chow (1980). Brownsey *et al.* (1986) have extended the studies of Ring and Stainsby (1982) to include large deformation and failure. These authors have considered the effects of volume fraction and rigidity of the filler particles and the matrix–filler interaction. The authors used crosslinked dextran (Sephadex) to fill gelatin gels. *Figure 2.5* shows the effect of increasing the volume fraction of a filler particle of fixed rigidity modulus greater than the modulus of the matrix. At low compression ratios ($\lambda$) the beads enhance the stiffness of the gels. However, they introduce an additional failure mechanism at higher compression ratios due to stress localization at the filler–matrix interface. The stiffness of the filler beads can be varied by varying the crosslink density and this can be monitored by equilibrium swelling experiments. Matching the filler–matrix eliminates this failure mechanism. By using

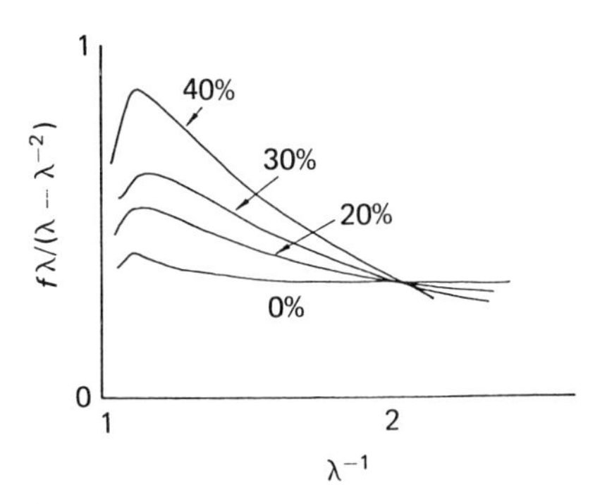


**Figure 2.5** Mooney–Rivlin plots showing the effect of filler particles on gel failure. Matrix gel 6 wt% gelatin. Filler particles Sephadex G50. Volume fractions of filler particles are shown next to each curve. $f$ = compressive force, $\lambda$ is the compressive ratio = ratio of length of specimen to original length. After Brownsey *et al.* (1987). The units of $f$ are $10^4$ N m$^{-2}$.

sulphopropyl-substituted Sephadex and varying the pH, Brownsey *et al.* (1987) have attempted to modify the filler–matrix interaction and note the effects on deformation and failure.

The modulus is only one property of concern. Another and more important as regards large deformation behaviour is that of crack propagation or fracture. Cracks can be arrested by low modulus inclusions and it is often the case that it is the heterogeneities that determine physical properties. Often, the characteristics of the interfacial region govern the ultimate mechanical properties attainable. However, the nature of interactions of fillers leading to the phenomena of reinforcement is still in doubt.

Starch is perhaps the most important gelling agent used in the food industry and starch gels are good examples of filled gels. An important industrial problem with starch-based foods is the 'staling' phenomena which occur on storage. Until recently (Miles, Morris and Ring, 1984, 1985; Miles *et al.*, 1985; Orford *et al.*, 1987; Ring and Stainsby, 1982) the molecular basis of the gelation and retrogradation of starch was poorly understood. Development of a molecular description relied on the recognition (Ring and Stainsby, 1982) that starch may be modelled as a composite, and on comparative studies (Miles, Morris and Ring, 1984, 1985; Miles *et al.*, 1985) on the amylose matrix, on the intact starch gel and on the gelatinized granules. Such studies have revealed that starch gels consist of gelatinized starch granules having an amylopectin skeleton embedded in an interpenetrating amylose matrix. Gelation is dominated by that of the amylose matrix and this step accounts for the opacity of starch gels and the critical starch concentration (~5–6%) necessary for gelation (Miles, Morris and Ring, 1984, 1985; Miles *et al.*, 1985). The amylose network is partially crystalline, the crystallites can only be melted at temperatures above 100 °C (Mestres, 1986). The long term changes in stiffness upon storage are related to crystallization of the branched polymer amylopectin (Miles *et al.*, 1985). The simplest explanation is a crystallization of amylopectin within the granules leading to some degree of reformation of granular structure, stiffening of the granules and enhanced reinforcement of the amylose matrix (Miles *et al.*, 1985). Amylose–amylopectin binding cannot definitely be excluded and the possibility of such binding has been discussed by Mestres (1986). Orford *et al.* (1987) have extended the earlier studies of Miles, Morris and Ring (1984, 1985) and Miles *et al.* (1985) to investigate higher starch concentrations and the effects of botanical source of the starch both on gelation and retrogradation. At higher concentrations, restricted swelling of the granules upon gelatinization and the consequent reduced solubilization of amylose is an important factor. The long term changes on storage, as monitored by the low deformation shear modulus, are sensitive to the botanical source of the starch (Orford *et al.*, 1987) suggesting a subtle interplay between the effects of matrix stiffness and granule rigidity. In recent studies Brownsey *et al.* (1987) have observed significant changes in the high deformation and failure properties of starch gels which depend on the botanical source of the starch.The present studies can be extended to include other components and to provide more realistic models of starch-based foods. Eliasson (1983) has discussed starch–gluten mixtures. Monoglycerides and glycerol monostearate are used as anti-staling or anti-firming agents. Whittam, Ring and Orford (1986) have probed the effects of fatty acids and monoglycerides through comparative studies of amylose and starch gels. The association and crystallization of amylose and the crystallization of amylopectin within gelatinized granules will influence the enzymic degradation of starch-based foods and will be important determinants of 'resistant starch'.

## Sensory properties

The similarity between gels and most foods (Tolstogusov and Braudo, 1983) has often been remarked upon and as a consequence, food gels often serve as good models. They allow studies to be undertaken without the influence and necessary control of factors that would otherwise be detrimental to a more direct food study. As noted earlier, the relation between material structure and physical property is not yet well established. However, as far as the consumer is concerned, it is the sensory properties and not the physical that are of major importance; that of texture being of most direct relevance for structured foods. Examination of the definition of texture, that it is a 'result of the food's underlying microstructure and physical properties . . .' explains why much early effort was directed towards establishing a relationship between subjective textural attributes and some physical measurements (DeMan *et al.*, 1976; Stanley and Tung, 1976). Also, it is clear that the physical properties of the food are important, especially the rheological properties. Since foods must break down and/or flow in a desirable manner, these properties are important and therefore efforts have also been directed towards relating rheological to textural measurements (Bistany and Kokini, 1983). Initially, studies concentrated on the fluid properties, particularly mouthfeel with any textural attributes of the more structured materials being related to failure/fracture related processes. Rheological studies of gels (Mitchell, 1976, 1980; Mitchell and Blanshard, 1976a, 1976b) used both single component structures and mixed systems with other polysaccharides incorporated (Christensen and Trudsoe, 1980; Daget and Collyer, 1984). Gelatin has many desirable sensory characteristics and many attempts have been made to duplicate these attributes but this has been difficult to attain; often a new set of properties has been produced (Harper and Hepworth, 1985). Homogeneous structures are not very abundant (Antonov *et al.*, 1980; Moritaka *et al.*, 1980; Raeuber and Nikolaus, 1980) and most studies have used mixed gels (Lee, Rosenau and Peleg, 1983) acknowledging that the deformability of the structural components as well as the time dependent relaxation effects are important contributory factors. Relatively complex structures such as sponges (Torres *et al.*, 1978) can be made relatively easily using amylose alone, whilst the incorporation of more rigid or fibrous fillers to a gel matrix does produce desirable characteristics (Montejano, Hamann and Lanier, 1983, 1986). Much mechanical information relevant to textural assessments is obtained in the regime of large deformations (Peleg, 1985) with many relevant studies on protein gels (Hamann, 1983). In such systems, phase separation is obviously important and several workers have observed that opaque gels from globular proteins are more fragile and brittle than translucent gels (Nakamura, Fukano and Taniguchi, 1982). A more ordered structure is considered to produce stronger gels (Schmidt, 1981). Matrix–filler interactions are important and studies of the fracture properties of fibrous meat composites (Dransfield, Lockyer and Prabhakaran, 1986) are of particular relevance to the ultimate textural characteristics.

Mechanical failure in food is said to be related to sensory texture during mastication (Montejano, 1983) and correlations have been produced between such measures. However, such correlations as may exist, either between physical and structural parameters or between physical and textural may be suspect (Hermansson, 1982), being dependent on the experimental design chosen. Much work has been purely phenomenological in that it does not claim to offer an explanation for the origin of the observed effect but recent studies (Munoz, Pangborn and Noble, 1986a, 1986b) should help quantify the textural characterization, simple food gels being but a first step in the level of complexity necessary to understand structured foods.

## References

ANTONOV, Y.A., GRINBERG, V.Y. ZHURAVSKAYA, N.A. and TOLSTOGUZOV, V.B. (1980). Liquid two-phase water–protein–polysaccharide systems and their processing into textured protein products. *Journal of Texture Studies*, **11**, 199–215

BIKBOV, T.M., GRINBERG, V.YA., ANTONOV, YU.A., TOLSTOGUZOV, V.B. and SCHMANDKE, H. (1979). On the concentration dependence of the elasticity modulus of soyabean globulin gels. *Polymer Bulletin*, **1**, 865–869

BISTANY, K.L. and KOKINI, J.L. (1983). Dynamic viscoelastic properties of foods in texture control. *Journal of Rheology*, **27**, 605–620

BROWNSEY, G.J., ELLIS, H.S., RIDOUT, M.J. and RING, S.G. (1986). Reinforcement of gels by deformable particles. In *Gums and Stabilisers for the Food Industry 3, 1st Edition*, (Phillips, G.O., Wedlock, D.J. and Williams, P.A., Eds), pp. 525–533. London and New York, Elsevier Applied Science

BROWNSEY, G.J., ELLIS, H.S., RIDOUT, M.J. and RING, S.G. (1987). Elasticity and failure in composite gels. *Journal of Rheology*, (in press)

CAIRNS, P., MILES, M.J. and MORRIS, V.J. (1986a). X-ray fibre diffraction studies of kappa carrageenan–tara gum mixed gels. *International Journal of Biological Macromolecules*, **8**, 124–127

CAIRNS, P., MILES, M.J. and MORRIS, V.J. (1986b). Intermolecular binding of xanthan gum and carob gum. *Nature*, **322**, 89–90

CAIRNS, P., MILES, M.J., MORRIS, V.J. and BROWNSEY, G.J. (1986a). Comparative studies of the mechanical properties of mixed gels formed by kappa carrageenan and tara gum or carob gum. *Food Hydrocolloids*, **1**, 89–93.

CAIRNS, P., MORRIS, V.J., MILES, M.J. and BROWNSEY, G.J. (1986b). Synergistic behaviour in kappa carrageenan–tara gum mixed gels. In *Gums and Stabilisers for the Food Industry 3, 1st Edition*, (Phillips, G.O., Wedlock, D.J. and Williams, P.A., Eds), pp. 597–604. London and New York, Elsevier Applied Science

CAIRNS, P., MILES, M.J., MORRIS, V.J. and BROWNSEY, G.J. (1987). X-ray fibre diffraction studies of synergistic binary polysaccharide gels. *Carbohydrate Research*, **160**, 411–423

CARROLL, V., MILES, M.J. and MORRIS, V.J. (1984) Synergistic interactions between kappa carrageenan and locust bean gum. In *Gums and Stabilisers for the Food Industry 2. Application of Hydrocolloids, 1st Edition*, (Phillips, G.O., Wedlock, D.J. and Williams, P.A., Eds), pp. 501–506. Oxford, Pergamon

CHOW, T.S. (1980). The effect of particle shape on the mechanical properties of filled polymers. *Journal of Material Science*, **15**, 1873–1888

CHRISTENSEN, O. and TRUDSOE, J. (1980). Effect of other hydrocolloids on the texture of kappa carrageenan gels. *Journal of Texture Studies*, **11**, 137–147

CLARK, A.H. and LEE-TUFFNELL, C.D. (1986). Gelation of globular proteins. In *Functional Properties of Food Macromolecules, 1st Edition* (Mitchell, J.R. and Ledward, D.A., Eds), pp. 203–272. London and New York, Elsevier Applied Science

CLARK, A.H., RICHARDSON, R.K., ROBINSON, G., ROSS-MURPHY, S.B. and WEAVER, A.C. (1982). Structure and mechanical properties of agar/BSA co-gels. In *Progress in Food and Nutrition Science, Vol. 6. Gums and Stabilisers for the Food Industry: Interactions of Hydrocolloids, 1st Edition*, (Phillips, G.O., Wedlock, D.J. and Williams, P.A., Eds), pp. 149–160. Oxford, Pergamon

CLARK, A.H., RICHARDSON, R.K., ROSS-MURPHY, S.B. and STUBBS, J.M. (1983). Structural and mechanical properties of agar/gelatin co-gels. Small deformation studies. *Macromolecules*, **16**, 1367–1374

COMPER, W.D. and LAURENT, T.C. (1978). Physiological function of connective tissue polysaccharides. *Physiological Reviews*, **58**, 255–315

COMPER, W.D. and PRESTON, B.N. (1974). Model connective tissue systems. A study of polyion mobile ion and of excluded volume interactions of proteoglycans. *Biochemical Journal*, **143**, 1–9

DAGET, N. and COLLYER, S. (1984). Comparison between quantitative descriptive analysis and physical measurements of gel systems and evaluation of the sensorial method. *Journal of Texture Studies*, **15**, 227–245

DEA, I.C.M. (1981). Specificity of interactions between polysaccharide helices and β-1,4 linked polysaccharides. In *Solution Properties of Polysaccharides, 1st Edition*, (Brant, D.A, Ed.) pp. 439–454. Washington, ACS Symposium Series 150

DEA, I.C.M. and MORRISON, A.A. (1975). Chemistry and interactions of seed galactomannans. *Advances in Carbohydrate Chemistry and Biochemistry*, **31**, 241–312

DEA, I.C.M., MCKINNON, A.A. and REES, D.A. (1972). Tertiary and quaternary structure in aqueous polysaccharide systems which model cell wall cohesion. Reversible changes in conformation and association of agarose, carrageenan and galactomannans. *Journal of Molecular Biology*, **68**, 153–172

DEA, I.C.M., MORRIS, E.R., REES, D.A., WELSH, E.J., BARNES, H.A. and PRICE, J. (1977). Association of like and unlike polysaccharides: mechanism and specificity in galactomannans, interacting bacterial polysaccharides, and related systems. *Carbohydrate Research*, **57**, 249–272

DEMAN, J.M., VOISEY, P.W., RASPER, V.F. and STANLEY, D.W. (1976). *Rheology and Texture in Food Quality*. Westport, Connecticut, AVI Publishers

DRANSFIELD, E., LOCKYER, D.K. and PRABHAKARAN, P. (1986). Changes in extensibility of raw beef muscle during storage. *Meat Science*, **16**, 127–142

ELIASSON, A.C. (1983). Physical properties of starch in concentrated systems such as dough and bread. Doctoral Thesis, University of Lund

FLORY, P.J. (1941a). Molecular size distribution in three-dimensional polymers. I. Gelation. *Journal of American Chemical Society*, **63**, 3083–3090

FLORY, P.J. (1941b). Molecular size distribution in three-dimensional polymers. II. Trifunctional branching units. *Journal of American Chemical Society*, **63**, 3091–3096

FLORY, P.J. (1941c). Molecular size distribution in three-dimensional polymers. III. Tetrafunctional branching units. *Journal of American Chemical Society*, **63**, 3096–3100

GORDON, M. (1962). Good's theory of cascade processes applied to the statistics of polymer distributions. *Proceedings of the Royal Society* **A268**, 240–259

GORDON, M. and ROSS-MURPHY, S.B. (1975). The structure and properties of molecular trees and networks. *Pure and Applied Chemistry*, **43**, 1–26

HAMANN, D.D. (1983). Structural failure in solid foods. In *Physical Properties of Food. 1st Edition*, (Peleg, M. and Bagley, E.B., Eds), pp. 351–383. Westport, Connecticut, AVI Publishers

HARPER, K.A. and HEPWORTH, A. (1985). *Texture Modifying Agents, 1st Edition*. Queensland Agricultural College

HEDBYS, B.O. (1961). The role of polysaccharides in corneal swelling. *Experimental Eye Research*, **1**, 81–91

HEDBYS, B.O. and DOHLMAN, C.H. (1963). A new method for the determination of the swelling pressure of the corneal stroma *in vitro*. *Experimental Eye Research*, **2**, 122–129

HERMANS, J.R. (1965). Investigation of the elastic properties of the particle network in gelled solutions of hydrocolloids. I. Carboxymethyl cellulose. *Journal of Polymer Science, Part A*, **3**, 1859–1868

HERMANSSON, A. (1982). Gel characteristics—compression and penetration of blood plasma gels. *Journal of Food Science*, **47**, 1960–1964

LAMBERT, F., MILAS, M. and RINAUDO, M. (1985). Sodium and calcium counterion activity in the presence of xanthan polysaccharide. *International Journal of Biological Macromolecules*, **7**, 49–52

LAURENT, T.C., PRESTON, B.N. and CARLSSON, B. (1974). Conformational transitions of polynucleotides in polymer media. *European Journal of Biochemistry*, **43**, 231–235

LEDWARD, D.A. (1986) Gelation of gelatin. In *Functional Properties of Food Macromolecules, 1st Edition*, (Mitchell, J.R. and Ledward, D.A., Eds), pp. 171–201. London and New York, Elsevier Applied Science

LEE, Y.C., ROSENAU, J.R. and PELEG, M. (1983). Rheological characterisation of Tofu. *Journal of Texture Studies*, **14**, 143–154

MANSON, J.A. and SPERLING, L.H. (1976). *Polymer Blends and Composites, 1st Edition.* London, Heyden Press

MCCLEARY, B.V. (1979). Enzymic hydrolysis, fine structure, and gelling interaction of legume-seed D-galacto-D-mannans. *Carbohydrate Research*, **71**, 205–230

MCLEARY, B.V. and NEUKOM, H. (1982). Effects of enzymic modifications on galactomannans. In *Progress in Food and Nutrition Science, Vol. 6. Gums and Stabilisers for the Food Industry: Interactions of Hydrocolloids, 1st Edition*, (Phillips, G.O., Wedlock, D.J. and Williams, P.A., Eds), pp. 109–118. Oxford, Pergamon

MCEVOY, H., ROSS-MURPHY, S.B. and CLARK, A.H. (1984). Large deformation and failure properties of biopolymer gels. In *Gums and Stabilisers for the Food Industry 2. Applications of Hydrocolloids, 1st Edition*, (Phillips, G.O, Wedlock, D.J. and Williams, P.A., Eds), pp. 111–122. Oxford, Pergamon

MESTRES, C. (1986). Gelification de l'amidon de mais modifé thermiquement application à la fabrication des pates alimentaires. Doctoral Thesis, University of Nantes

MEYER, F.A., COMPER, W.D. and PRESTON, B.N. (1971). Model connective tissue systems. A physical study of gelatin gels containing proteoglycans. *Biopolymers*, **10**, 1351–1364

MILES, M.J., MORRIS, V.J. and CARROLL, V. (1984). Carob gum–kappa carrageenan mixed gels: Mechanical properties and X-ray fibre diffraction studies. *Macromolecules*, **17**, 2443–2445

MILES, M.J., MORRIS, V.J. and RING, S.G. (1984). Some recent observations on the retrogradation of amylose. *Carbohydrate Polymers*, **4**, 73–77

MILES, M.J., MORRIS, V.J. and RING, S.G. (1985). Gelation of amylose. *Carbohydrate Research*, **135**, 257–269

MILES, M.J., MORRIS, V.J., ORFORD, P.D. and RING, S.G. (1985). The roles of amylose and amylopectin in the gelation and retrogradation of starch. *Carbohydrate Research*, **135**, 271–281

MITCHELL, J.R. (1976). Rheology of gels. *Journal of Texture Studies*, **7**, 313–319

MITCHELL, J.R. (1980). The rheology of gels. *Journal of Texture Studies*, **11**, 315–337

MITCHELL, J.R. and BLANSHARD, J.M.V. (1976a). Rheological properties of alginate gels. *Journal of Texture Studies*, **7**, 219–234

MITCHELL, J.R. and BLANSHARD, J.M.V. (1976b). Rheological properties of pectate gels. *Journal of Texture Studies*, **7**, 341–351

MONTEJANO-GALTON, J.G. (1983). Thermally induced gelation of selected protein systems. Rheological changes during processing, final strengths, texture profile analysis, sensory texture and microstructure. Thesis, North Carolina State University

MONTEJANO, J.G., HAMANN, D.D. and LANIER, T.C. (1983). Final strengths and rheological changes during processing of thermally induced fish muscle gels. *Journal of Rheology*, **27**, 557–579

MONTEJANO, J.G., HAMANN, D.D. and LANIER, T.C. (1986). Comparison of two instrumental methods of sensory texture of protein gels. *Journal of Texture Studies*, **16**, 403–424

MORITAKA, H., NISHINARI, K., HORIUCHI, H. and WATASE, M. (1980). Rheological properties of aqueous–gelatin gels. *Journal of Texture Studies*, **11**, 257–270

MORRIS, E.R., REES, D.A., YOUNG, G., WALKINSHAW, M.D. and DARKE, A. (1977). Order–disorder transition for a bacterial polysaccharide in solution. A role for polysaccharide conformation in recognition between *Xanthomenas* pathogen and its plant host. *Journal of Molecular Biology*, **110**, 1–16

MORRIS, V.J. (1985). Food gels–roles played by polysaccharides. *Chemistry and Industry*, 159–164

MORRIS, V.J. (1986a). Multicomponent gels. In *Gums and Stabilisers for the Food Industry 3, 1st Edition*, (Phillips, G.O., Wedlock, D.J. and Williams, P.A., Eds), pp. 87–99. London and New York, Elsevier Applied Science

MORRIS, V.J. (1986b). Gelation of polysaccharides. In *Functional Properties of Food Macromolecules, 1st Edition*, (Mitchell, J.R. and Ledward, D.A., Eds), pp. 121–170. London and New York, Elsevier Applied Science

MORRIS, V.J. and CHILVERS, G.R. (1984). Cold setting alginate–pectin gels. *Journal of the Science of Food and Agriculture*, **35**, 1370–1376

MORRIS, V.J. and MILES, M.J. (1986). The effect of natural modifications on the functional properties of extracellular bacterial polysaccharides. *International Journal of Biological Macromolecules*, **8**, 342–348.

MUNOZ, A.M., PANGBORN, R.M. and NOBLE, A.C. (1986a). Sensory and mechanical attributes of gel texture I. Effect of gelatin concentration. *Journal of Texture Studies* **17**, 1–16.

MUNOZ, A.M., PANGBORN, R.M. and NOBLE, A.C. (1986b). Sensory and mechanical attributes of gel texture II. Gelatin, sodium alginate and kappa-carrageenan gels. *Journal of Texture Studies*, **17**, 17–36

NAKAMURA, R., FUKANO, T. and TANIGUCHI, M. (1982). Heat induced gelation of hen's egg yolk low density lipoprotein (LDL) dispersion. *Journal of Food Science*, **47**, 1449–1453

ORFORD, P.D., RING, S.G., CARROLL, V., MILES, M.J. and MORRIS, V.J. (1987). The effect of concentration and botanical source on the gelation and retrogradation of starch. *Journal of the Science of Food and Agriculture*, **39**, 169–177

PELEG, M. (1985). A note on the various strain measures at large compressive deformations. *Journal of Texture Studies*, **15**, 317–326

RAEUBER, M.J. and NIKOLAUS, M. (1980). Structure of food. *Journal of Texture Studies*, **11**, 187–198

REES, D.A. (1972). Polysaccharide gels—a molecular view. *Chemistry and Industry*, 630–636

REES, D.A., MORRIS, E.R., THOM, D. and MADDEN, J.K. (1982). Shapes and interactions of carbohydrate chains. In *The Polysaccharides Vol. 1, 1st Edition*, (Aspinall, G.O., Ed.), pp. 195–290. London, Academic Press

RICHARDSON, R.K. and ROSS-MURPHY, S.B. (1981). Mechanical properties of globular protein gels II. Concentration, pH concentration, pH and ionic strength dependence. *British Polymer Journal*, **13**, 11–16

RICHARDSON, R.K., ROBINSON, G., ROSS-MURPHY, S.B. and TODD, S. (1981). Mechanical spectroscopy of filled gelatin gels. *Polymer Bulletin*, **4**, 541–546

RING, S.G. and STAINSBY, G. (1982). Filler reinforcement of gels. In *Progress in Food and Nutrition Science, Vol. 6. Gums and Stabilisers for the Food Industry: Interactions of Hydrocolloids, 1st Edition*, (Phillips, G.O., Wedlock, D.J. and Williams, P.A., Eds), pp. 323–329. Oxford, Pergamon

SCHMIDT, R.H. (1981). Gelation and coagulation. In *Protein Functionality in Foods, 1st Edition*, (J.P. Cherry, Ed.), pp. 131–147. ACS Symposium Series 147

SKJÄK-BRAEK, G. (1984). Enzymic modification of alginate. In *Gums and Stabilisers for the Food Industry 2. Applications of Hydrocolloids, 1st Edition*, (Phillips, G.O., Wedlock, D.J. and Williams, P.A., Eds), pp. 523–533 Oxford, Pergamon

SNOEREN, T.H.M. (1976). *Kappa Carrageenan, 1st Edition*, Monograph 174. Wageningen, H. Veman and B.V. Zonen

SPERLING, L.H. (1981). *Interpenetrating Polymer Networks and Related Materials, 1st Edition*, New York, Plenum Press

STAINSBY, G. (1980). Proteinaceous gelling systems and their complexes with polysaccharides. *Food Chemistry*, **6**, 3–14

STANLEY, D.W. and TUNG, M.A. (1976). Microstructure of food and its relation to texture, In *Rheology and Texture in Food Quality*, (DeMan, J.M., Voisey, P.W., Rasper, V.F. and Stanley, D.W., Eds), Westport, Connecticut, AVI Publishers pp. 28–78.

STOCKMAYER, W.H. (1943). Theory of molecular size distribution and gel formation in branched chain polymers. *Journal of Chemical Physics*, **11**, 45–55

STOCKMAYER, W.H. (1944). Theory of molecular size distribution and gel formation in branched polymers II. General cross-linking. *Journal of Chemical Physics*, **12**, 125–131

TAKO, M. and NAKAMURA, S. (1985). Synergistic interactions between xanthan and guar gum. *Carbohydrate Research*, **138**, 207–213

TAKO, M., ASATO, A. and NAKAMURA, S. (1984). Rheological aspects of the intermolecular interaction between xanthan and locust bean gum in aqueous media. *Agricultural and Biological Chemistry*, **48**, 2995–3000

THOM, D., DEA, I.C.M., MORRIS, E.R. and POWELL, D.A. (1982). Interchain associations of alginate and pectins. In *Progress in Food and Nutrition Science, Vol. 6. Gums and Stabilisers for the Food Industry: Interactions of Hydrocolloids, 1st Edition*, (Phillips, G.O., Wedlock, D.J. and Williams, P.A., Eds), pp. 97–108. Oxford, Pergamon

THOMAS, D.A. and SPERLING, L.H. (1978). Interpenetrating polymer networks. In *Polymer Blends, Vol. 2*, (Paul, D.R. and Newman, S., Eds), pp. 1–33. London, Academic Press

TOFT, K. (1982). Interactions between pectins and alginates. In *Progress in Food and Nutrition Science, Vol. 6. Gums and Stabilisers for the Food Industry: Interactions of Hydrocolloids, 1st Edition*, (Phillips, G.O., Wedlock, D.J. and Williams, P.A., Eds), pp. 89–96. Oxford, Pergamon

TOLSTOGUZOV, V.B. (1978). *Artificial Foodstuffs, 1st Edition*. Moscow, Nauka Publishers

TOLSTOGUZOV, V.B. and BRAUDO, E.E. (1983). Fabricated foodstuffs as multicomponent gels. *Journal of Texture Studies*, **14**, 183–212

TOLSTOGUZOV. V.B., BELKINA, V.P., GULOV, V.JA., GRINBERG, V.YA., TITOVA, E.F. and BELAVZERA, E.M. (1974). Phasenzustand, struktur und mechanische eigenschaften des gelertigen systems wasser-gelatine-dextran. *Die Stärke*, **26**, 130–138

TORRES, A., SCHWARTZBERG, H.G., PELEG, M. and RUFNER, R. (1978). Textural properties of amylose sponges. *Journal of Food Science*, **43**, 1006–1009

WATASE, M. and NISHINARI, K. (1980). Rheological properties of agarose–gelatin gels. *Rheologica Acta,* **19**, 220–225

WHITTAM, M.A., RING, S.G. and ORFORD, P.D. (1986). Starch–lipid interactions: the effect of lipids on starch. In *Gums and Stabilisers for the Food Industry 3, 1st Edition*, (Phillips, G.O., Wedlock, D.J. and Williams, P.A. Eds), pp. 555–563. London and New York, Elsevier Applied Science

WINTER, W.T., CHIEN, Y.Y. and BOUCKRIS, H. (1984). Structural aspects of food galactomannans. In *Gums and Stabilisers for the Food Industry 2. Applications of Hydrocolloids, 1st Edition*, (Phillips, G.O., Wedlock, D.J. and Williams, P.A., Eds), pp. 535–539. Oxford, Pergamon

3

# GEL STRUCTURE OF FOOD BIOPOLYMERS

ANNE-MARIE HERMANSSON

*SIK—The Swedish Food Institute, Göteborg, Sweden*

## Introduction

The three-dimensional network structure is of importance for many properties of food products such as texture, fat- and waterholding, binding, diffusion properties, etc. Reviews of protein and polysaccharide gelation have recently been published by Clark and Lee-Tuffnell (1986) and Morris (1986). In spite of all the work done we still know too little about the microstructure of gels and the mechanisms involved in gel formation, especially on the supermolecular level. Such knowledge is necessary in order to improve our understanding of relationships between the structure and other physical properties of food gels.

Gel formation can be regarded as a special form of phase separation. Biopolymer gels can roughly be divided into two types; one is phase-separated aggregated gels where the specific molecular properties are of minor importance in relation to colloidal properties; the other is gels formed by association of molecules into strands in a more ordered way. As will be shown later one polymer can form both types of gel and the transition from one type to the other can take place within a very narrow range of pH, ionic strength, concentrations of other components, substituents, etc.

In order to increase our understanding of the structure of various types of biopolymer gels information is needed on the following points:

1. Structures of biopolymers in solutions compared with their structure in gels.
2. Mechanisms controlling the formation of strands, superstrands and aggregates.
3. Characteristics of strands, superstrands and aggregates.
4. Mechanisms of network formation and characteristics of junction zones.
5. Distributions of structural domains.
6. Nature of long range density fluctuations.

In this chapter the structures of some pure protein and polysaccharide systems will be discussed with regard to some of the points stated above. The complexity of mixed gels and pitfalls that may arise from work with complex food systems will also be briefly touched upon in order to stress the importance of elucidating the structures we deal with in process and product development work.

## Protein gels

### *Myosin*

Several proteins have the ability to form fine stranded as well as coarsely aggregated gel structures. Myosin is such a protein. It is very suitable for a case study because of its characteristic shape, which makes it easy to identify in solutions as well as in early stages of aggregation prior to gelation. *Figure 3.1* shows myosin molecules from bovine *m. semi-membranosus* (Hermansson, Langton and Olsson, 1987). The molecules are characterized by two globular or pear shaped heads and a tail and have a total length of about 190 nm (Elliott and Offer, 1978; Walker, Knight and Trinick, 1985).

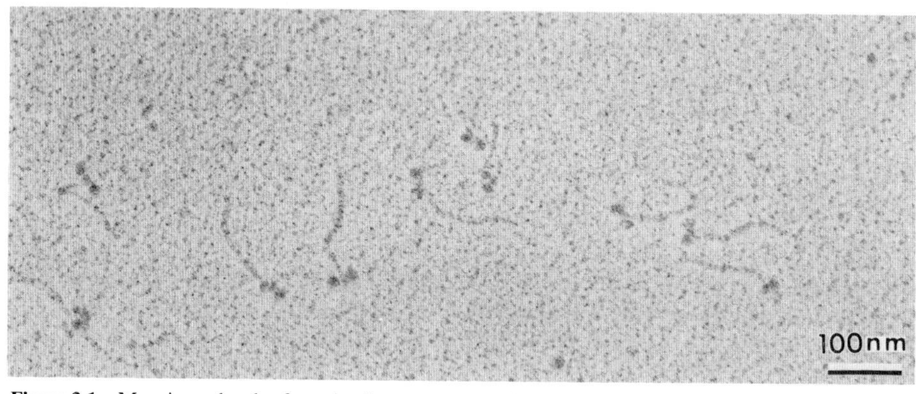


**Figure 3.1** Myosin molecules from bovine *m. semimembranosus* spread on mica and rotary replicated.

Myosin is one of the most important myofibrillar proteins. In the thick filaments in muscle the myosin molecules are packed so that the tails form the backbone and the heads swing out from the surface. The myofibrils dissolve at high salt concentration which makes it possible to extract and fractionate a pure myosin solution. In our study of myosin gels the behaviour of myosin at various combinations of pH and ionic strength was investigated before and after heat treatment (Hermansson, Harbitz and Langton, 1986; Hermansson and Langton, 1987; Hermansson, Langton and Olsson, 1987).

When myosin solutions were heated to 60 °C at pH 5.5–6.0, a common pH range for meat products, two completely different types of gels were formed. A fine stranded gel structure formed at low ionic strength (0.25 M KCl) whereas a gel consisting of globular aggregates formed at high ionic strength (0.6 M KCl) (Hermansson, Harbitz and Langton, 1986). *Figure 3.2* shows a scanning electron micrograph of the fine stranded gel structure formed at pH 5.5 and 0.25 M KCl and *Figure 3.3* shows the aggregated gel structure formed at pH 6.0 and 0.6 M KCl.

An interesting observation was that all fine stranded gels formed from turbid solutions and all aggregated gels formed from clear solutions. The phase diagram in *Figure 3.4* shows the combinations of pH and ionic strengths resulting in clear and turbid solutions respectively. As seen from *Figure 3.4* the turbidity was high at pH 4.0–5.0 in the whole salt concentration range studied. All turbidity measurements

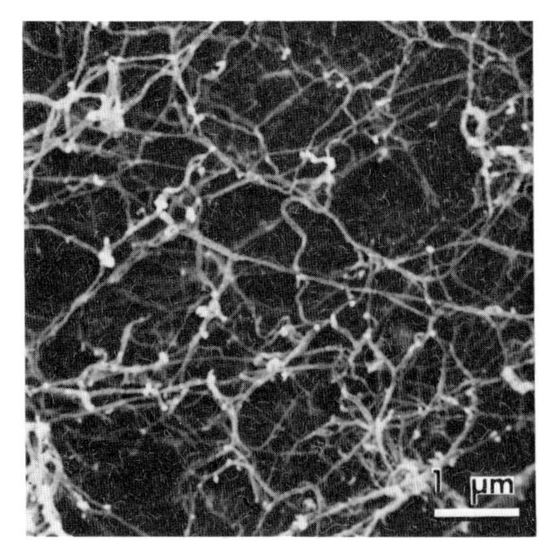


**Figure 3.2** SEM micrograph of a fine stranded myosin gel at pH 5.5 and 0.25 M KCl after heat treatment at 60 °C.

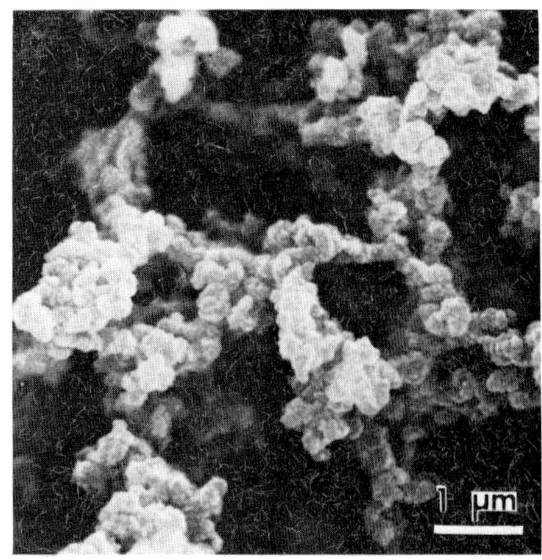


**Figure 3.3** SEM micrograph of an aggregated myosin gel at pH 6.0 and 0.6 M KCl after heat treatment at 60 °C.

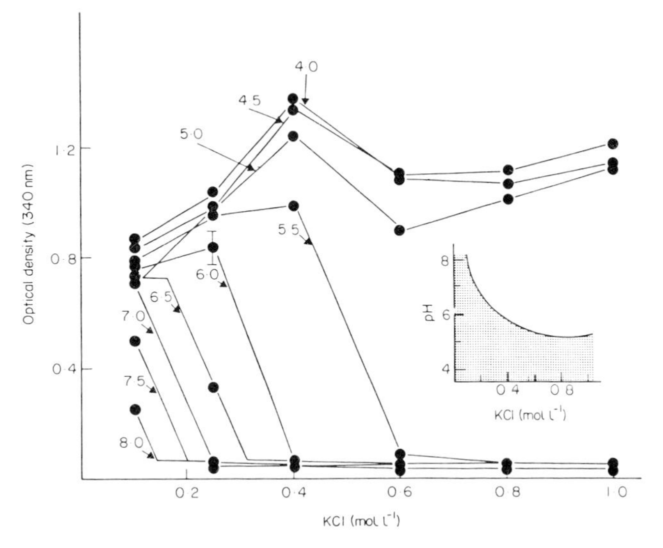


**Figure 3.4** The turbidity of unheated myosin solutions, measured as optical density at 340 nm, as a function of pH and ionic strength. Inset: a phase diagram showing the combinations of pH and ionic strength resulting in clear solutions (optical density of 0.10 at 340 nm) and in turbid suspensions. From Hermansson, Harbitz and Langton (1986).

were made at a protein concentration of 0.1% by weight and all gels formed at a protein concentration of 1.0% by weight. When the pH of the higher concentration solution at 0.6 M KCl was lowered by dialysis to pH 4.0, a strand-like gel structure formed spontaneously. This network structure did not change in character on heating to 60 °C. *Figure 3.5* shows the network structure formed on dialysis at 4 °C and *Figure 3.6* shows the network structure after subsequent heat treatment at 60 °C.

The results shown this far suggest that the conditions required for the formation of

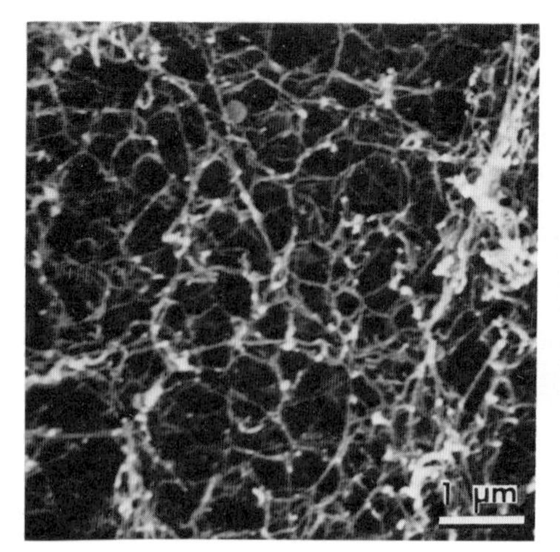


**Figure 3.5** SEM micrograph of a myosin gel formed spontaneously at pH 4.0 and 0.6 M KCl at 4 °C.

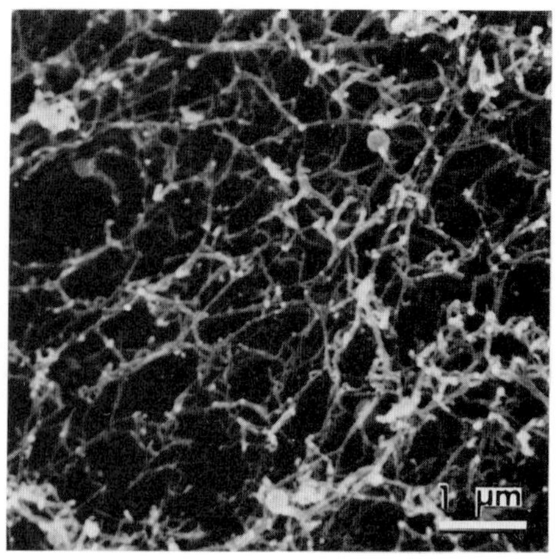


**Figure 3.6** SEM micrograph of a myosin gel formed spontaneously at pH 4.0 and 0.6 M KCl at 4 °C and then heat treated at 60 °C for 30 minutes.

strand-like myosin gels are present in the solution already before heat treatment and gel formation, since all fine stranded gels but not the coarsely aggregated gels originated from turbid solutions, and also that there might be some structural differences between the fine stranded gels formed spontaneously at pH 4.0 and high ionic strength and the heat-induced fine stranded gels formed at pH 5.5 and low ionic strength.

It is well known that myosin can form synthetic filaments at pH 6.0–8.0 if the ionic strength is lowered to 0.1–0.3 by dialysis or dilution which are similar in character to the native thick filaments (Huxley, 1963; Kaminer and Bell, 1966). However, little is known about filament formation below pH 6.0 and the nature of filaments from bovine myosin prepared after the onset of rigor. We were therefore interested in characterizing the structure of bovine myosin in turbid solutions at pH 4.0 and 5.5 to see how that structure was related to the gel structure of heat-induced gels and gels formed spontaneously without any heat treatment. As shown in *Figure 3.7* filaments were indeed formed on dialysis to pH 5.5 and 0.25 M KCl.

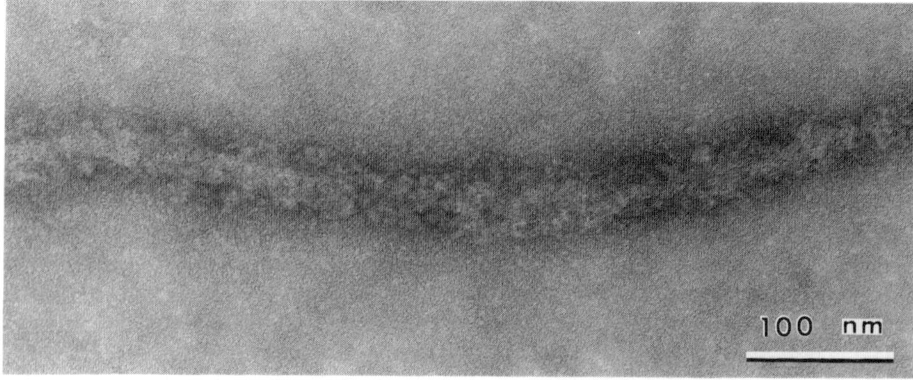


**Figure 3.7** Synthetic filament of myosin formed by dialysis at pH 5.5 and 0.25 M KCl. The filament was prepared for TEM by negative staining.

A number of EM preparation techniques have been used in order to characterize the myosin filament structure at pH 5.5 and 0.25 M KCl (Hermansson and Langton, 1987). The filaments had a backbone of approximately 25 nm in width surrounded by a fringe of globular material, the myosin heads. The total width of the filament is about 45 nm which is somewhat wider than the previously reported synthetic filaments.

The way the filaments interacted before heat treatment has a bearing on the structure of the gel network. Frequently the myosin heads at the end of one filament interacted with the myosin heads on the side of another filament. Filaments were also found to interact side by side forming parallel pairs (Hermansson and Langton, 1987). Some examples are given in *Figures 3.8a* and *3.8b*. Heating to 60 °C results in complete protein denaturation. As shown by *Figure 3.9*, the shape of the filament is maintained but the identity of the individual heads is lost and heads have fused together. Even after heat treatment there was an axial periodicity of about 35 nm. Before heat treatment the axial periodicity of the filament was 43 nm.

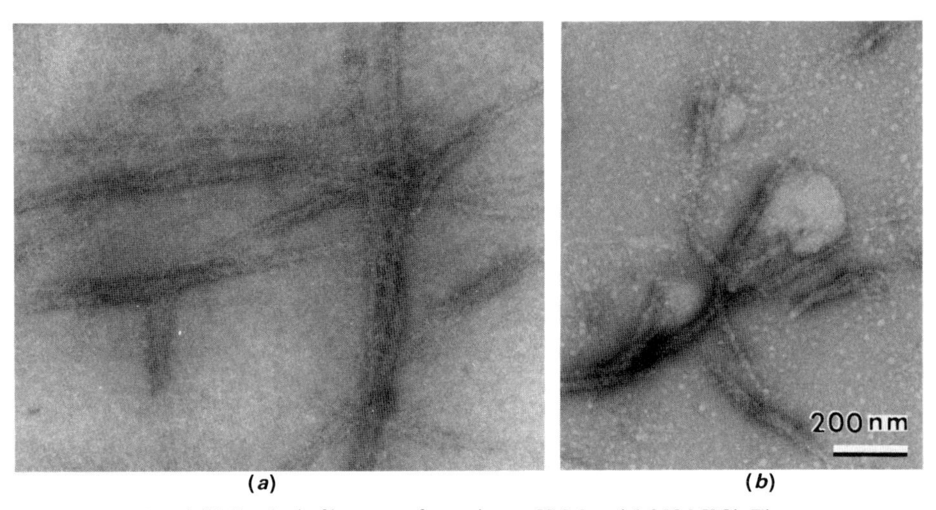


**Figure 3.8** (a) and (b) Synthetic filaments of myosin at pH 5.5 and 0.25 M KCl. The filaments were prepared for TEM by negative staining.

*Figures 3.9* and *3.10* show that the structure of the myosin filaments after heat treatment of a dilute myosin solution and prepared for TEM by negative staining corresponds very well to the strand structure of a myosin gel at a higher concentration after preparation for SEM by chemical fixation, dehydration and critical point drying. Preparation of myosin gels by thin sectioning gave information on the nature of junction zones and the organization of filaments in the network structure (Hermansson and Langton, 1987). Typical for the myosin gel formed at pH 5.5 and 0.25 M KCl are Y-junctions, which are believed to originate from the previously described end to side interactions between filaments. This type of branching does not give rise to a fine network with small pores but a rather open one. Another typical feature of filamentous myosin gels is the pairwise interactions of filaments side by side forming ladders with interactions between the two filaments at regular spacings at 35–40 nm. These ladders are believed to contribute to gel rigidity (Hermansson and Langton, 1987).

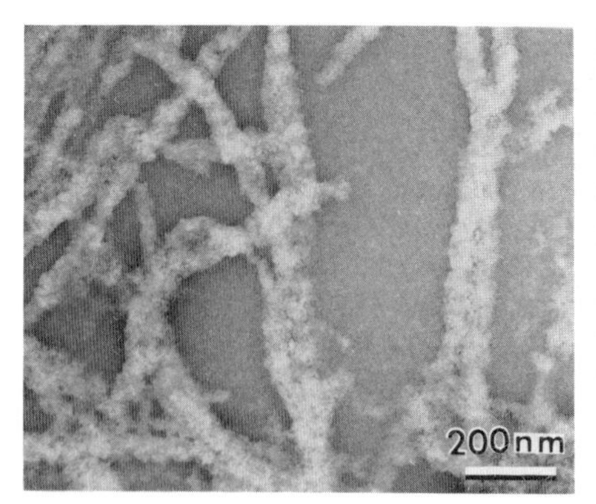


**Figure 3.9** TEM micrograph of a synthetic filament of myosin at pH 5.5 and 0.25 M KCl after heat treatment of a dilute solution at 60 °C and negative staining.

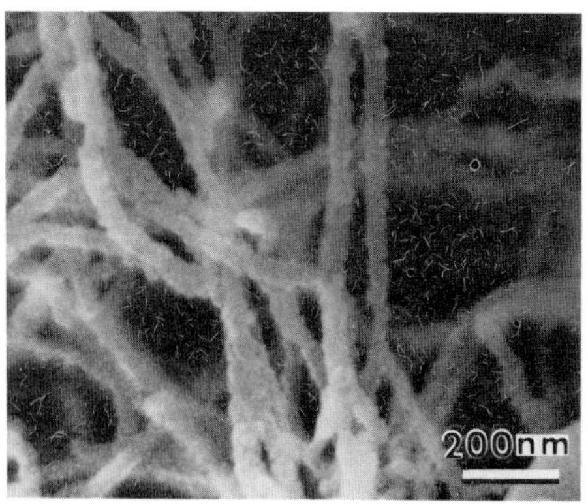


**Figure 3.10** SEM micrograph showing details of the fine stranded gel structure formed at pH 5.5 and 0.25 M KCl after heat treatment at 60 °C.

As was shown earlier, a fine stranded gel structure formed spontaneously without heat treatment when the pH was lowered to 4 at 0.6 M KCl. From differential scanning calorimetry (DSC) measurements Fretheim, Egelandsdal and Harbitz (1985) have concluded that denaturation takes place when the pH of myosin is lowered below pH 5 because of the absence of endothermic peaks. We thought it would be of interest to take a closer look at this type of gel to see how it differed from the fine stranded gel formed at pH 5.5 and low ionic strength.

As can be seen from *Figure 3.11*, filaments formed also at pH 4.0 and 0.6 M KCl but there are some structural differences between these and those formed at pH 5.5 and lower ionic strength. An interesting observation is that even if DSC measurements showed that the internal structure is broken by lowering the pH, the shape of individual myosin heads has remained, which was not the case after heat denaturation. Thus acid and heat denaturation affect the molecular shape and molecular interactions of myosin quite differently. The filament backbone at pH 4.0 has mostly a diameter of approximately 25 nm, the same as at pH 5.5 but the heads are protruding further out from the backbone in a more irregular way than was the case at pH 5.5. Fringes can be seen where the heads extend approximately 50 nm from the backbone but also clusters of myosin molecules protruding far out from the side of the backbone and the filament ends. These clusters seem to play a role in the junctions between filaments. In the gel spontaneously formed at pH 4 the filaments often interact at approximate right angles which results in a fine network different from the more branched structure formed at pH 5.5. The length of the strands between junction zones was in the range 200–600 nm. This is a more efficient network than that formed at pH 5.5 and may be the reason for the spontaneous gel formation at such a low protein concentration as 1%. From thin sections it was seen that the presence of parallel alignments of filaments forming ladders made an important contribution also to the filamentous gel forming at pH 4.0. After heat treatment details of the filaments were lost but the characteristics of the gel network remained intact (Hermansson and Langton, 1987). An important conclusion from this study is that features of filamentous myosin gels are present already before heat treatment and are determined by the conditions during filament formation.

When clear solutions of myosin at high ionic strength were heated to 60 °C an

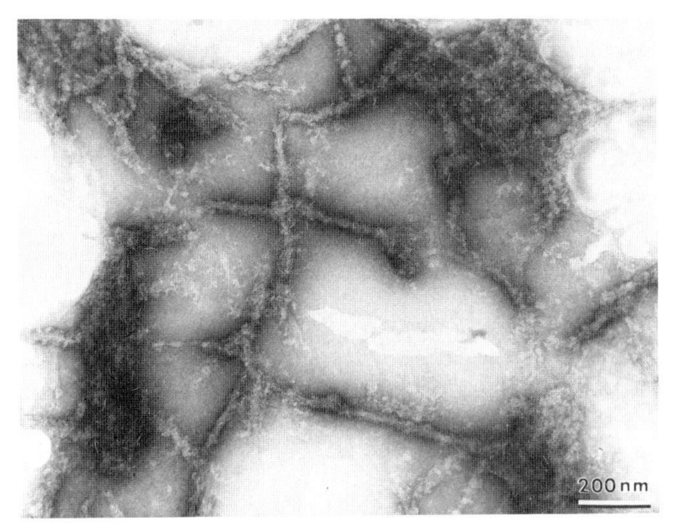


**Figure 3.11** Synthetic filaments of myosin at pH 4.0 and 0.6 M KCl prepared for TEM by negative staining.

aggregated type of gel was found, which was completely different from the strand-like filamentous gel structures previously described. The difference in character between the two types of gels can be seen from *Figures 3.2* and *3.3*. These types of SEM micrographs do not allow any interpretation about interactions on the molecular level. Ishioroshi, Samejima and Yasui (1979) speculated that a similar gel structure formed from rabbit myosin at 60 °C and 0.6 M KCl was due to head to head interactions. However, this interpretation was based on observations of aggregation made at 35 °C, which is below the denaturation temperature of myosin (Kawakami *et al.*, 1971; Yasui, Morita and Takahashi, 1966). It was therefore of interest to characterize structural changes of myosin solutions and gels at high ionic strength in the temperature range 35–60 °C in order to see if changes taking place at 35–45 °C had any bearing on the structure of gels formed at 60 °C.

Rotary shadowing of dilute solutions at 0.6 M KCl heated in the range 35–45 °C revealed that myosin molecules associated in an ordered manner that was completely different from the way myosin associated into filaments at low ionic strength (Hermansson, Langton and Olsson, 1987). Common for these associations were head to head interactions. Spherical micelles were observed where the heads had fused together and the tails were sticking out from the centre. Frequently we also observed an association of heads and tails in a periodic way where a number of myosin molecules were aligned in parallel. An example of a parallel alignment is shown in *Figure 3.12*.

Apart from micelles and parallel alignments, individual myosin molecules as well as big unordered aggregates were seen on heating solutions in the temperature range 35–45 °C. Gels formed already at 45 °C and in these gels the above structures could be seen even if the tendency for phase separation and the formation of big aggregates dominated the gel structure (Hermansson, Langton and Olsson, 1987). DSC measurements have shown that the myosin head portion denatures at 50–53 °C at pH 6–7 and ionic strength 0.6–1.0 (Samejima, Ishioroshi and Yasui, 1983; Wright and Wilding, 1984). At higher temperatures, of significance for meat products, complete phase

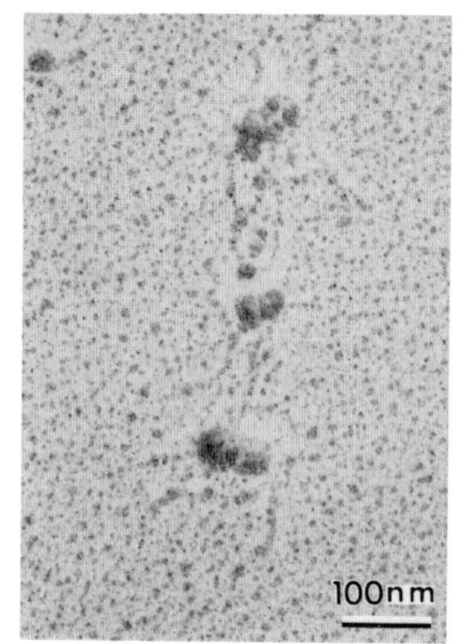


**Figure 3.12** Parallel alignments of a myosin solution at pH 6.0 and 0.6 M KCl after heat treatment at 40 °C. The TEM micrographs were prepared by the mica technique.

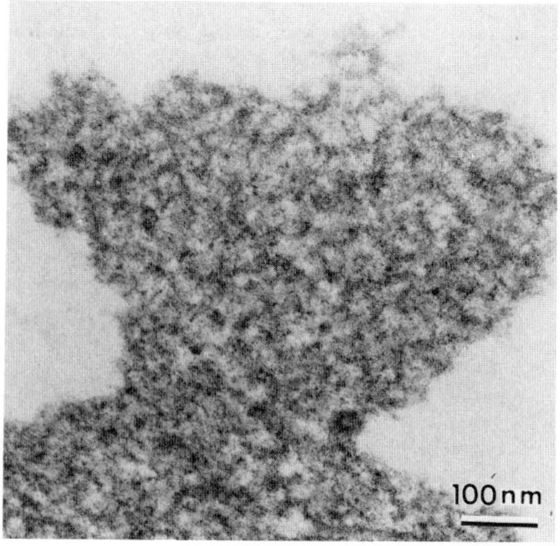

**Figure 3.13** TEM micrograph of a thin section of an aggregate in the coarsely aggregated myosin gel formed at pH 6.0 and 0.6 M KCl after heat treatment at 60 °C.

separation took place, where the identity of the myosin molecules is completely lost. *Figure 3.13* shows a thin section through one of the aggregates of the gel structure of myosin at 0.6 M KCl after heating to 60 °C.

The discrepancies between the results obtained at low and high temperatures of myosin, shown in *Figures 3.12* and *3.13* clearly illustrate the danger of extrapolating results obtained under one set of conditions to explain what takes place under another set of conditions. In this case the ordered associations of myosin molecules formed at 35–45 °C have nothing to do with the coarsely aggregated gel structure formed at higher temperatures.

## Globular proteins

Phase-separated gels consisting of globular aggregates of similar appearance can form from several proteins regardless of their molecular conformations. Apart from myosin this type of structure has been observed for globular proteins such as β-lactoglobulin, serum albumin and ovalbumin. The size of the aggregates can vary considerably between gels depending on environmental conditions and the state of the raw material but the aggregates are relatively uniform in size within each gel structure.

*Figure 3.14* shows the gel structure of β-lactoglobulin at pH 5.5 after heat treatment to 85 °C and *Figure 3.15* shows a similar structure of a crude whey protein concentrate at pH 7.0 after heat treatment at 95 °C (Hermansson, 1986). As can be seen from *Figures 3.3, 3.14* and *3.15*, the gels of myosin, β-lactoglobulin and whey protein concentrate are similar in character but the sizes of the aggregates between the gels are quite different. More knowledge is needed in order to find out the mechanisms

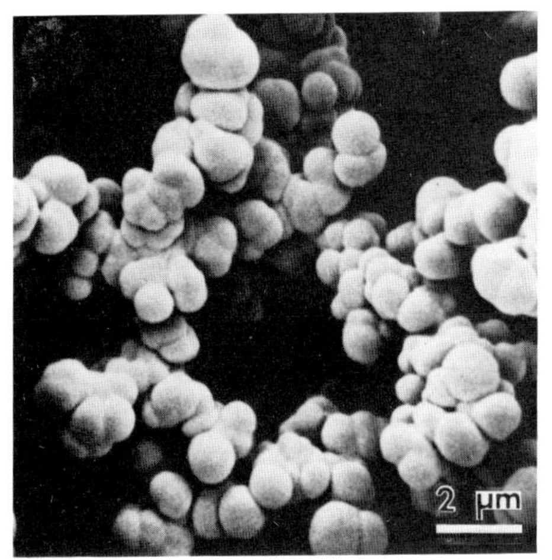


**Figure 3.14** SEM micrograph of a β-lactoglobulin gel at pH 5.5 after heat treatment to 85 °C.

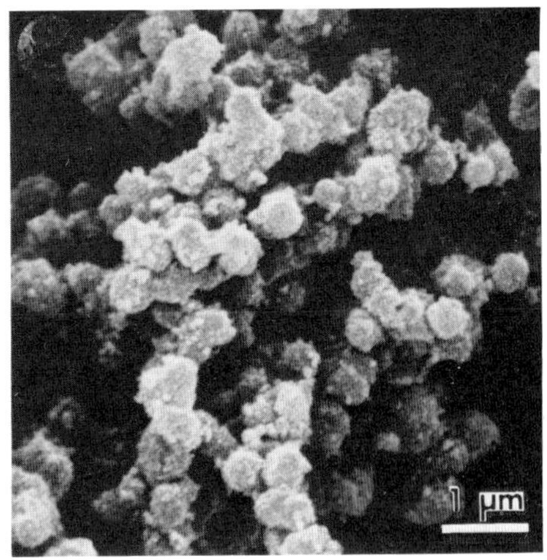


**Figure 3.15** SEM micrograph of a whey protein gel at pH 7.0 after heat treatment to 95 °C.

controlling aggregate size as well as the alignments of aggregates into a three-dimensional network structure.

If the energy barrier against random aggregation is sufficient the molecules can arrange themselves into strands and a fine stranded network structure is obtained. For example, β-lactoglobulin can form a transparent fine stranded gel structure as well as an opaque aggregated structure. If the molecules are rendered more hydrophilic a transparent gel will form and the transition from one type of gel structure to the other can be triggered off by a small shift in pH ($\Delta$pH $\sim$0.1) of the solution before heat treatment. Preliminary results from studies of fine stranded gel structures suggest that the strands are formed by a more complex mode of aggregation than single strands of molecules linked together like a string of beads. High resolution electron microscopy and improved preparation techniques need to be developed in order to establish the exact nature of the very fine strands of transparent β-lactoglobulin gels in the pH range 7–9.

The gel structures of glycinin and conglycinin of soya beans are dependent on the quaternary structures of these storage proteins (Hermansson, 1985). Both proteins form aggregated as well as fine stranded gels. The strands of gels at pH 7 are 10–15 nm in diameter which is far greater than the diameter of individual subunits of the molecules. In the case of glycinin evidence has been given that the strands consist of associated subunits in a circular arrangement. *Figure 3.16a* shows a thin section of a glycinin gel formed in distilled water. The gel is built up by regular strands with an outer diameter of 12–15 nm. Cross sections of strands show that they are built up like hollow tubes. (Some examples are encircled.) *Figure 3.16b, c* and *d* show the circular arrangements of subunits in the strands at higher magnifications. The latter were prepared by ultrarapid freezing and freeze etching (Hermansson, 1985).

The results have shown that proteins can form different types of structures and the transitions from one type of gel structure to another can be triggered off by changes in variables such as pH, ionic strength and heating conditions.

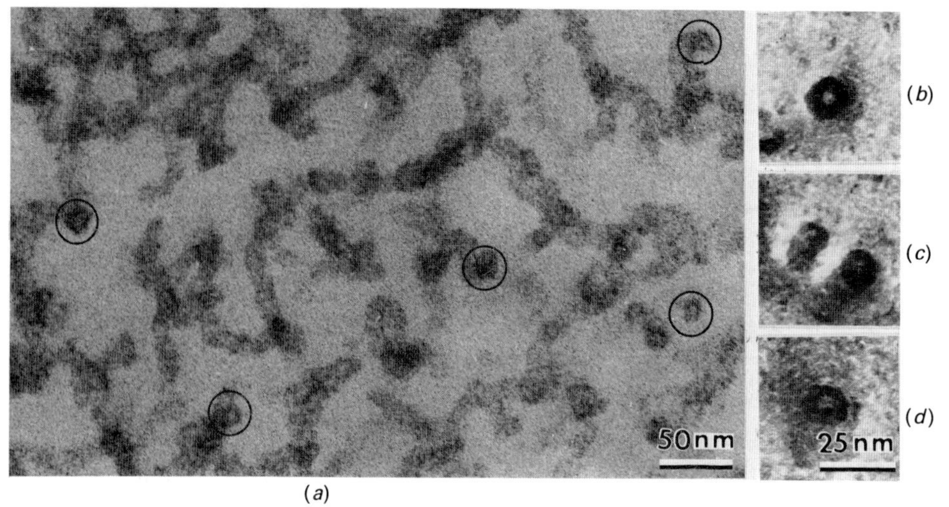


**Figure 3.16** (a) TEM micrograph of a thin section of a glycinin gel made in distilled water at 95 °C. Some cross sections of strands are encircled. (b), (c), (d) Cross sections of glycinin strands prepared by freeze etching and rotary shadowing.

## Polysaccharides

Extensive work has been made in order to elucidate the gelation mechanisms of polysaccharides (Morris, 1986). In spite of this there is very little information available about two- and three-dimensional network structure on the supermolecular level. Ethylhydroxyethylcellulose (EHEC) derivatives will be used as a case to exemplify associations on the supermolecular level of polysaccharides. The studies made on EHEC structures are interesting in comparison to those made on myosin. In the latter case the network structure was changed by a change in ionic strength and/or pH. In the case of EHEC changes in the supermolecular structure are achieved by changes of the molecule itself, i.e. by different degrees of substitution of ethyl and hydroxyethyl groups. An increased substitution of ethyl groups is believed to promote hydrophobic interactions, whereas the substitution of hydroxyethyl will

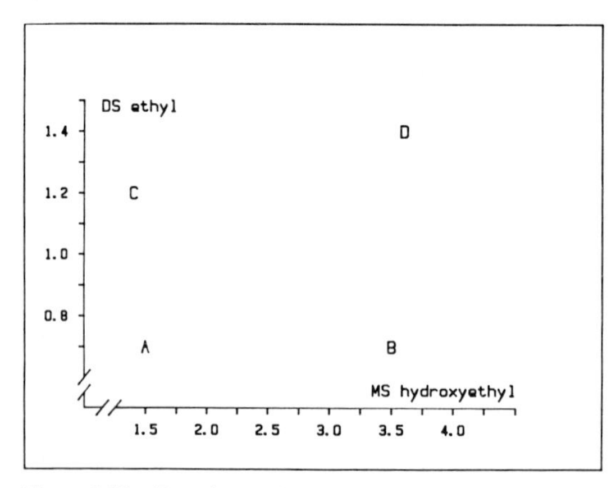


**Figure 3.17** Experimental design of the EHEC study.

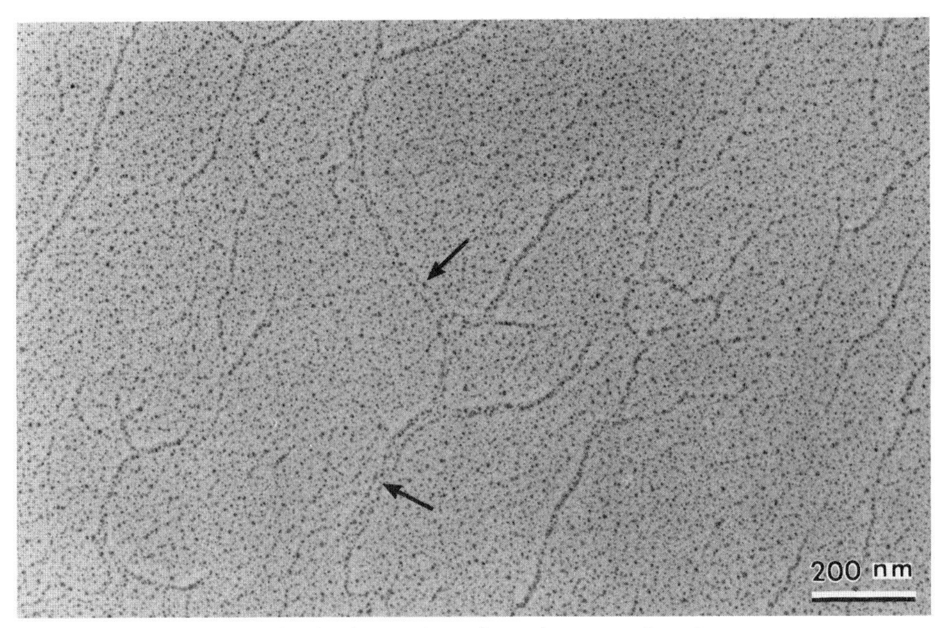


**Figure 3.18** Structure of derivative A spread on mica and rotary replicated.

contribute to the solubility of the biopolymer molecules. EHEC can be used as a stabilizer and binder in foods as well as in paints and other products.

The supermolecular structure of EHEC spread on mica surfaces was investigated at various degrees of substitution of ethyl and hydroxyethyl groups as shown in *Figure 3.17* and the molecular chain length was approximately 200 nm. *Figure 3.18* shows the superstrands of derivative A with a low degree of ethyl as well as hydroxyethyl substitution. Typical of this structure is Y-shaped junctions formed from superstrands containing several molecules. A characteristic feature is also that the superstrands are composed of two strands often aligned in parallel or twisted around each other (see arrows). If the degree of hydroxyethyl substitution is increased and the molecules rendered more soluble a different type of network is formed. *Figure 3.19a* shows such a network structure of derivative B. This is a more flexible network with voids and a different type of branching than that of derivative A. Also in this case the superstrands are composed of several strands. In the centre and in the background of *Figure 3.19a* there is a very fine network which may be difficult to see. *Figure 3.19b* shows a domain of sample B with this type of fine network structure. This structure is also present in the background of sample A and is believed to correspond to the entanglement bulk structure of the solution. The formation of superstrands is believed to be a phase separation process originating from the bulk structure shown in *Figure 3.19b.* At high degrees of ethyl substitution as in derivative C, where the solubility is low and the hydrophobicity high, micelles form as seen in *Figure 3.20.* Also in this case a fine network structure is present in the background.

Similar to myosin, EHEC can form open branched supermolecular structures, flexible network structures and micellular structures. Contrary to myosin the structural changes of EHEC described above were caused by a change of the molecular composition and not by environmental factors such as pH and ionic strength.

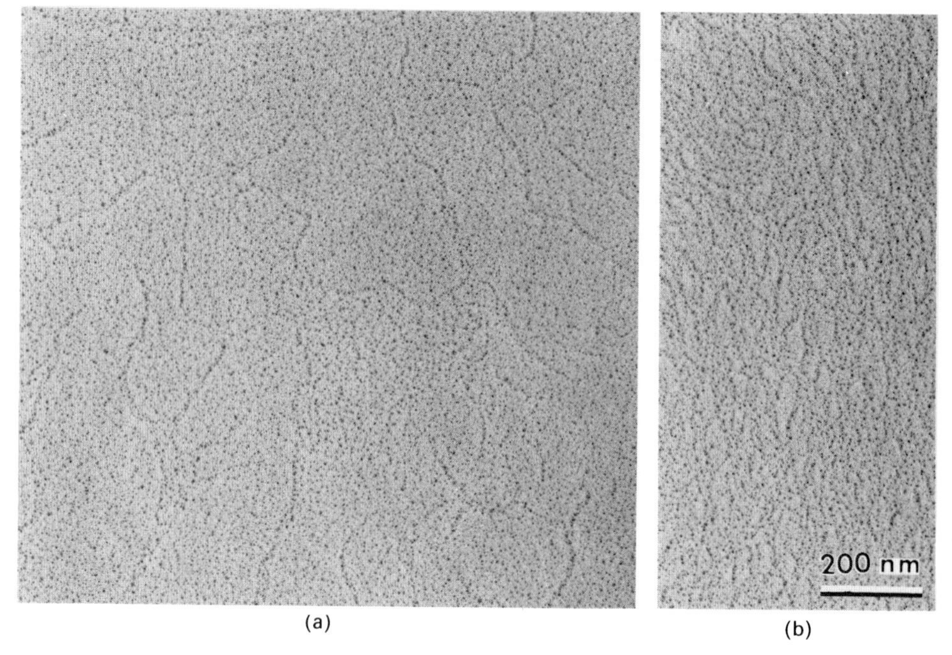


**Figure 3.19** (a) and (b) Structures of derivative B spread on mica and rotary replicated.

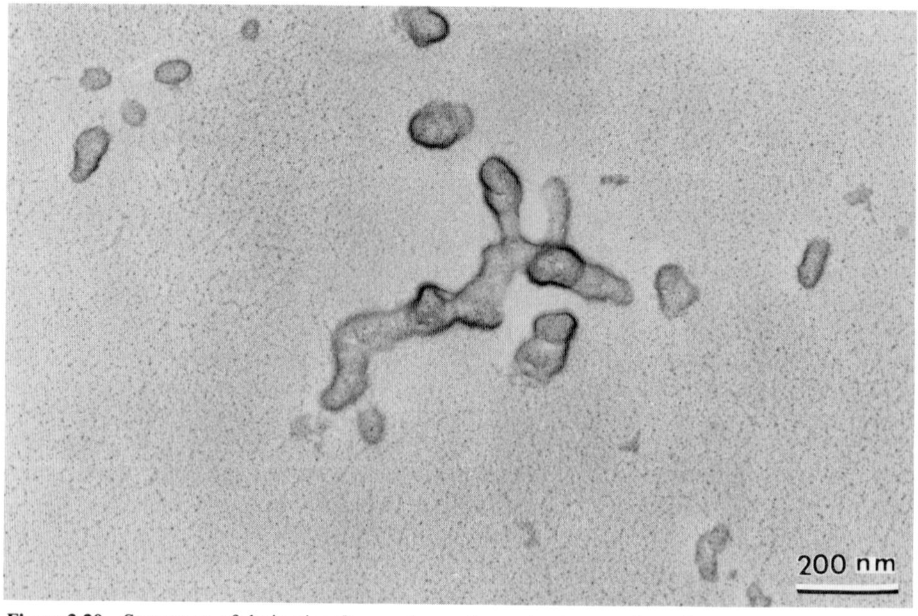


**Figure 3.20** Structures of derivative C spread on mica and rotary replicated.

However, the formation of superstrands of the cellulose derivative is also highly dependent on environmental conditions and the presence of other components. *Figure 3.21* depicts the structure of derivative D with a high degree of substitution of ethyl and hydroxyethyl groups showing the presence of small micelles as well as single superstrands. The addition of sodium chloride promotes the formation of complex superstrands and *Figure 3.22* shows such a superstrand where it is clearly seen how the strand splits up in two and fuses together at two points (see arrows).

The results on EHEC reveal the presence of several structural domains within one preparation. Similar results have been obtained also for other polysaccharide systems such as agarose, kappa- and iota carrageenan gels where superstrand structures as well as entanglement structures are present within one gel structure.

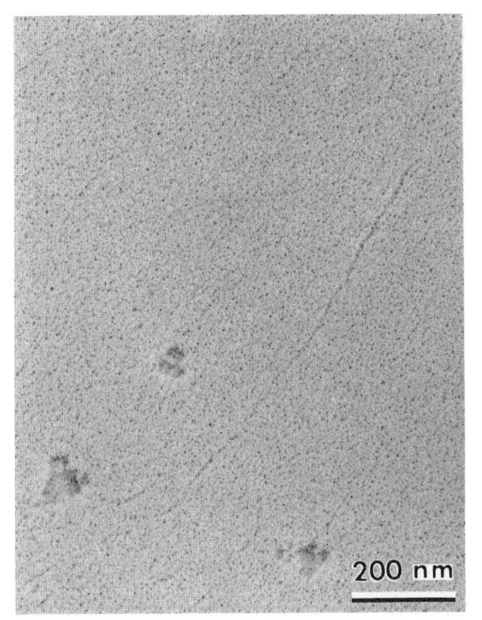


**Figure 3.21** Structure of derivative D spread on mica and rotary replicated.

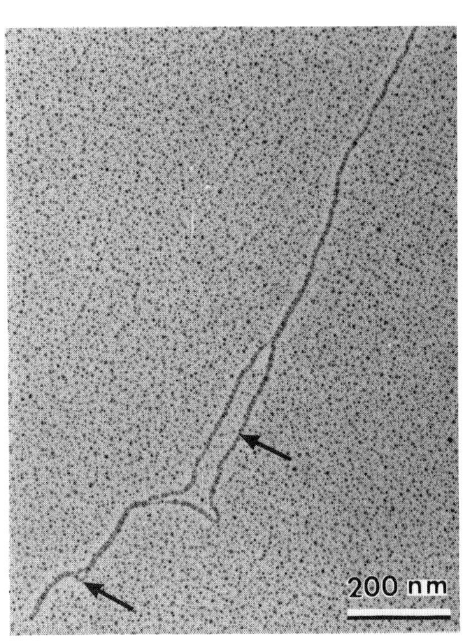


**Figure 3.22** The effect of 0.2 M NaCl on the structure of derivative D after preparation by the mica technique.

## Complex food systems

Even if several structural domains can be found within one polysaccharide system such gels are not what we normally think of when considering mixed gels from a structural point of view, but rather a mixture of fine stranded structures, aggregates, particles, cellular materials etc. In this sense starch gels are more typical. *Figure 3.23* shows an example of a wheat starch gel composed of swollen starch granules embedded in a fine stranded network structure. This gel was prepared for scanning electron microscopy by ultrarapid freezing and the microscope was equipped with a cryostage which allows work with fully hydrated samples.

In a real food product the structure is even more complex, since starch exists together with proteins and other components in a multiphase system. There are many pitfalls that may arise when dealing with such complex materials during food

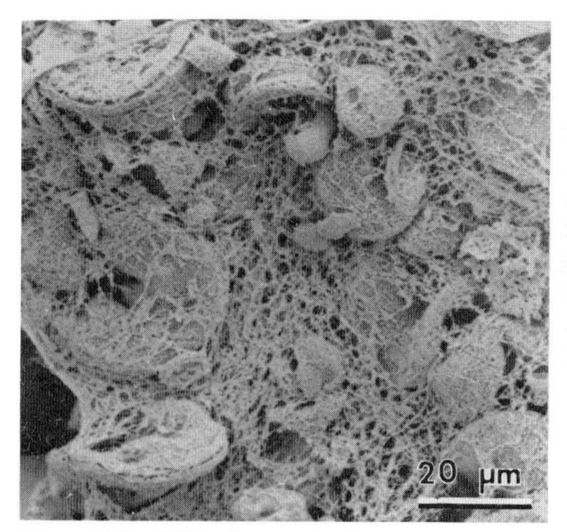


**Figure 3.23** SEM micrograph of a wheat starch gel after heat treatment at 95 °C.

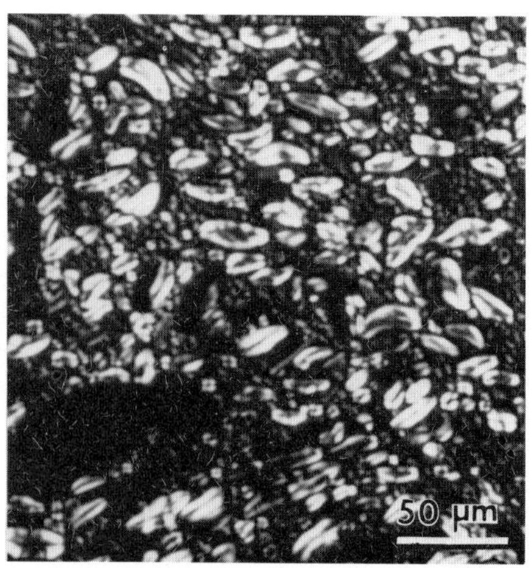


**Figure 3.24** Light micrograph showing semigelatinized starch in the centre of a pasta product photographed with plane polarized light.

processing and several factors need to be taken into account. One of the most important is to know the state of the continuous phase. An example of this complexity will be given from a study of wheatmeal which contains around 80% starch. In traditional products of wheatmeal such as bread, pasta, etc. the gluten protein makes up the continuous and the starch the discontinuous phase both before and after cooking. *Figure 3.24* shows the semigelatinized starch phase in the centre of a pasta product. It is not, up to the present time, generally known that these phases can reverse upon more severe processing such as extrusion cooking or drum drying, where the starch phase melts and the gluten phase remains insoluble. This results in a continuous phase of amorphous starch with dispersed gluten particles. The way the gluten is dispersed depends on the mechanical action during processing.

*Figure 3.25* shows a light micrograph of an extruded particle of wheatmeal. The discontinuous protein phase appears dark as a result of green staining. The gluten particles have a certain orientation due to the mechanical action during extrusion. *Figure 3.26* shows the structure of wheatmeal after drum drying. Also in this case it is quite clear that the darkly stained protein forms the discontinuous phase. The starch phase of these products is very reactive and prone to aggregation and complex formation upon further processing such as reheating in the presence of water. Therefore, the use of extruded or drum dried wheatmeal in a product such as a sauce, will result in a structure that is completely different from that of the original wheat dough. This is just one example of how processing can alter the role of structural components in a mixed gel and how new structural elements of significance for the product properties can form during processing.

We often have rather naive ideas about gel structures, partly due to lack of methods for their characterization. Studies by means of NMR, X-ray, neutron scattering, etc. provide valuable information over small distances but not of orders in the range 10–1000 nm, which are of importance for the network structures of food biopolymer gels. Measurements of gelation by rheology, optical rotation, light scattering, DSC, etc. give general information about phase transitions but not whether a transition applies for the overall structure or is specific for structural domains and no details of the

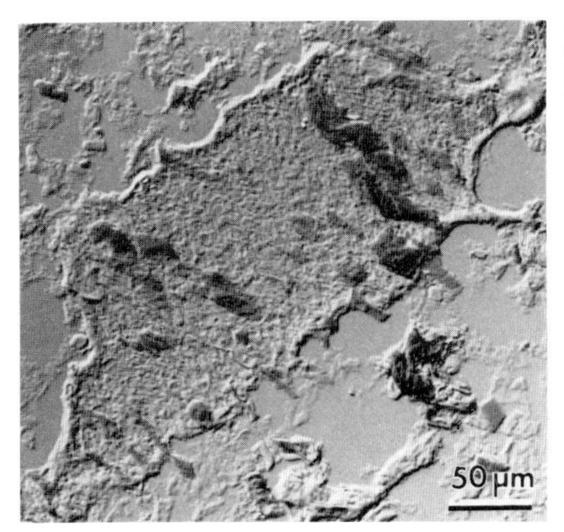


**Figure 3.25** Light micrograph of wheatmeal after extrusion at 148 °C, 600 psi and 10% added water, photographed with differential interference contrast. The protein phase appears dark due to staining with green.

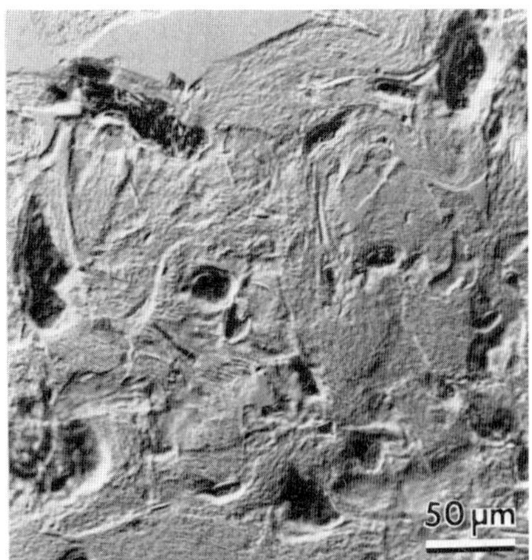


**Figure 3.26** Light micrograph of wheatmeal after drum drying, photographed with differential interference contrast. The protein phase appears dark due to staining with green.

network structure as such. This chapter has illustrated that electron microscopy can be a useful tool in order to gain information about gel structures. More methods are needed for the study of orders over long distances, especially in the range 100–1000 nm. In the future we need to combine methods that provide information on different dimensional levels in order to increase our understanding of food biopolymer gels.

## Acknowledgements

The author would like to thank Maud Langton, Elvy Olsson and Ina Storm for skilful assistance with the electron micrographs and Olle Stenberg at Berol Chemistry for collaboration and permission to publish the work on cellulose derivatives.

## References

CLARK, A.H. and LEE-TUFFNELL, C.D. (1986). Gelation of globular proteins. In *Functional Properties of Food Macromolecules*, (Mitchell, J.R. and Ledward, D.A, Eds), Chapter 5. London and New York, Elsevier Applied Science

ELLIOTT, A. and OFFER, G. (1978). Shape and flexibility of myosin molecules. *Journal of Molecular Biology*, **123**, 505–519

FRETHEIM, K., EGELANDSDAL, B. and HARBITZ, O. (1985). Slow lowering of pH induces gel formation of myosin. *Food Chemistry*, **18**, 169–178

HERMANSSON, A.-M. (1985). Structure of soya, glycinin and conglycinin gels. *Journal of the Science of Food and Agriculture*, **36**, 822–832

HERMANSSON, A.-M. (1986). Water and fatholding. In *Functional Properties of Food Macromolecules*, (Mitchell, J.R. and Ledward, D.A., Eds), pp. 273–314. London and New York, Elsevier Applied Science

HERMANSSON, A.-M. and LANGTON, M. (1987). Filamentous structures of bovine myosin in diluted suspensions and gels. *Journal of the Science of Food and Agriculture* (in press)

HERMANSSON, A.-M., HARBITZ, O. and LANGTON, M. (1986). Formation of two types gels from bovine myosin. *Journal of the Science of Food and Agriculture*, **37**, 69–84

HERMANSSON, A.-M., LANGTON, M. and OLSSON, E. (1987). Gelation and aggregation mechanisms of myosin at high ionic strength. Submitted for publication.

HUXLEY, H.E. (1963). Electron microscope studies on the structure of natural and synthetic protein filaments from striated muscle. *Journal of Molecular Biology*, **7**, 281–308

ISHIOROSHI, M., SAMEJIMA, K. and YASUI, T. (1979). Heat-induced gelation of myosin: factors of pH and salt concentrations. *Journal of Food Science*, **44**, 1280–1284

KAMINER, B. and BELL, A.L. (1966). Myosin filamentogenesis: effects of pH and ionic concentrations. *Journal of Molecular Biology*, **20**, 391–401

KAWAKAMI, H., MORITA, J., TAKAHASHI, K. and YASUI, T. (1971). Thermal denaturation of myosin, heavy meremyosin and subfragment. *Journal of Biochemistry*, **70**, 635–648

MORRIS, V.J. (1986). Gelation of polysaccharides. In *Functional Properties of Food Macromolecules*, (Mitchell, J.R. and Ledward, D.A., Eds), Chapter 3. London and New York, Elsevier Applied Science

SAMEJIMA, K., ISHIOROSHI, M. and YASUI, T. (1983). Scanning calorimetric studies on thermal denaturation of myosin and its subfragments. *Agricultural and Biological Chemistry*, **47**, 2373–2380

WALKER, M., KNIGHT, P. and TRINICK, J. (1985). Negative staining of myosin molecules. *Journal of Molecular Biology*, **184**, 535–542

WRIGHT, D. and WILDING, P. (1984). Differential scanning calorimetric study of muscle and its proteins: myosin and its subfragments. *Journal of the Science of Food and Agriculture*, **35**, 357–372

YASUI, T., MORITA, J. and TAKAHASHI, K. (1966). Inhibition by some neutral organic solutes of the denaturation of myosin A — adenosine triphosphatose. *Journal of Biochemistry*, **60**, 303–316

4

# THE STRUCTURE AND STABILITY OF EMULSIONS

E. DICKINSON
*Procter Department of Food Science, University of Leeds, UK*

## Introduction

'... let me introduce you to that leg of mutton,' said the Red Queen. 'Alice—Mutton: Mutton—Alice.'
[*Through the Looking-Glass*, Lewis Carroll]

An emulsion is a heterogeneous system containing one liquid dispersed in another as droplets of colloidal or microscopic size. One of the liquid phases is normally an aqueous solution. So we can have either an oil-in-water emulsion (O/W) or a water-in-oil emulsion (W/O) depending on which liquid phase makes up the droplets. Unless stated otherwise, it will be assumed here that we are discussing oil-in-water emulsions, since food colloids are most commonly of this type (e.g. milk, mayonnaise, salad dressing, cake batter). Butter and margarine are notable examples of the W/O type. Many meat emulsions, despite the name, do not qualify as *bona fide* emulsions because their fat particles are macroscopic.

When applied to food emulsions, the word 'liquid' is interpreted liberally. Depending on the temperature, the oil phase, and also occasionally the aqueous phase, may be in a state of partial crystallization (as in whipped cream or ice cream); or the aqueous phase may exist in the form of a weak gel (as in a meat paste or a dairy dessert). Crystals in food emulsions can be of various shapes and sizes depending on the temperature history of the ingredients, and the processing and storage conditions. Emulsion droplets are close to spherical in shape, but their sizes are often widely distributed. We say that emulsions are polydisperse. Because they are troublesome to characterize and to make reproducibly, polydisperse systems do not find favour with many colloid scientists, and some have certainly been put off working with emulsions for this very reason. In the field of food colloids, however, we have little choice in the matter, since monodisperse polystyrene latices or silica spheres have rather a limited appeal as food ingredients!

Making emulsions is still regarded as being as much an art as a science. In principle, of course, there is nothing to it: just take a mixture of oil + water + emulsifier, supply a large amount of mechanical energy, and out comes an emulsion. It requires a little care in the choice of ingredients and processing conditions to make an emulsion that will remain 'stable' for several days, but given the right equipment and materials there is usually no great problem. Where difficulties do arise, however, is in ensuring good long-term stability (from weeks to years), and in understanding why apparently small changes in ingredients or processing conditions can lead to a loss of this long-term stability.

The primary process of emulsification is the creation of fresh interface through the disruption of large droplets into smaller ones. This disruption is facilitated by a lowering in the surface free energy as emulsifier adsorbs at the nascent oil–water interface. As well as its surface active role in reducing the energy needed to make new interface, the emulsifier has an important secondary role in protecting newly formed droplets from immediately recoalescing. We can therefore define an 'emulsifier' as a single chemical component, or mixture of components, having the capability for promoting emulsion formation and stabilization by interfacial action. It is useful to distinguish an emulsifier from a so-called 'stabilizer'. A stabilizer is a chemical component, or mixture of components, which can confer long-term stability on an emulsion, possibly by a mechanism involving adsorption, but not necessarily so. Low molecular weight surfactants (lecithin, Spans, Tweens, etc.) are food emulsifiers, whereas polysaccharide hydrocolloids (carrageenan, xanthan gum, etc.) are food emulsion stabilizers. Food proteins like casein or gelatin are particularly convenient in that they are able to fulfil both the emulsifying and the stabilizing roles.

In connection with the term 'emulsifier', two further points are worth mentioning in passing. Firstly, far from enhancing emulsion formation, one of the main uses of low molecular weight emulsifiers in food processing is to induce a partial *de*stabilization of emulsion droplets by displacing adsorbed protein from the oil–water interface, e.g. during whipping or churning (Darling and Birkett, 1987). Secondly, as well as the more precise colloid science usage, one has to recognize that in many areas of the food business the term emulsifier is also applied to materials that more generally promote the shelf life of foods, e.g. by interacting with starch, or by modification of lipid crystallization.

Emulsion science is a big subject, and so it is clear that a short article such as this must be selective in emphasis. For the reader who desires a comprehensive account, there fortunately exist several reviews that are reasonably up-to-date. In separate chapters of a recently published volume, Walstra (1983) and Tadros and Vincent (1983) set out the general principles of emulsion formation and stability in excellent fashion. The physical chemistry approach to food emulsions is described in some detail by Dickinson and Stainsby (1982), and in a more condensed form by Stainsby (1986). Aspects of dairy emulsions can be found in the book by Mulder and Walstra (1974), and more briefly, but with some updating, in Walstra and Jenness (1984). There are also specialized reviews on various topics such as emulsion rheology (Sherman, 1968), food emulsifiers (Krog and Lauridsen, 1976), protein-stabilized systems (Halling, 1981), gums and hydrocolloids (Sharma, 1981), and competitive protein adsorption (Dickinson, 1986). It is the aim of the present chapter to discuss emulsion stability in relation to emulsion structure. The author's viewpoint, for better or worse, is that of a physical chemist, with one eye directed towards fundamental principles and the other towards the requirements of the food technologist. The approach will be illustrated with reference to the creaming instability and the structure of a creamed layer. Before getting too detailed, however, we need carefully to define what we mean by 'stability' and 'structure'.

## Stability

'That's a great deal to make one word mean,' Alice said in a thoughtful tone.

Stability, as applied to colloids, is a kinetic concept and not a thermodynamic one (Dickinson and Stainsby, 1982). In the food area, some emulsions (like cake batter) need only a short lifetime, whereas others (like salad cream) must remain stable for months or even years. Stability implies no tendency towards structural change. A stable emulsion, therefore, is one in which the number and arrangement of droplets changes imperceptibly slowly with time.

Loss of stability has several possible manifestations in an O/W emulsion. One may identify five distinct phenomena:

1. *Creaming.* This is gravitational (or centrifugal) separation of oil droplets into a concentrated, and probably distinct, layer at the top of an emulsion sample, with no associated change in the droplet size distribution.
2. *Flocculation.* This is aggregation of droplets under the influence of attractive forces acting between them, to form small or large flocs with no associated change in the individual droplet sizes.
3. *Coalescence.* This is the coming together of creamed or flocculated droplets to form larger spherical droplets. The limiting situation is complete 'breaking' of the emulsion into two (partly) immiscible liquid phases.
4. *Ostwald ripening.* This is growth of larger droplets at the expense of smaller ones due to mass transport of soluble disperse phase material through the aqueous continuous phase.
5. *Inversion.* This is the abrupt change in state from an O/W emulsion to a W/O emulsion.

Ostwald ripening is usually insignificant in food emulsions because of the extremely low mutual solubilities of triglycerides and water. Emulsion phase inversion can sometimes be important, notably in butter making, but it differs from the other processes in requiring large amounts of mechanical energy, and in being a composite phenomenon, usually involving both flocculation and coalescence, as well as some restabilization through droplet disruption. This leaves us with the three primary instabilities: (1) creaming, (2) flocculation, and (3) coalescence.

Stability *per se* is, of course, not enough. The food technologist certainly aims to produce a stable emulsion—or sometimes, deliberately, a partially destablized one—but the aim also is to give it other essential properties of a chemical, physical, microbiological, nutritional, or organoleptic nature. Certain emulsions (salad dressings, spreads, margarine) are required to have certain textural and rheological properties over a specified temperature range. The volume fraction of the dispersed phase may have to be kept low for reasons of economy (flavour emulsions) or nutritional speciality (low-fat dressings); or high to control texture (mayonnaise) or satisfy legislative contraints (double cream). The aqueous phase may have to be of high salinity (meat emulsions), low pH (fruit juice clouding agents), or alcoholic (cream liqueurs). There may be other dispersed phases present, like protein particles (homogenized milk), starch granules (cake batter), or air bubbles (whipped topping). In addition, there may be extra product requirements, such as a controlled turbidity (soft drinks), or an ability to resist extremes of temperature, low (ice cream) or high (gravies and sauces). Set against this great diversity of product requirements, the search for universal principles of food emulsion stability seems a worthwhile, if somewhat daunting, task.

A food emulsion is a chemically complicated system. As well as water and oil (triglycerides), there may be proteins, polar lipids, carbohydrates, and dissolved salts. The way in which the various chemical components affect the overall stability of a particular food emulsion will, in general, be complex and unknown. Common experience, however, tells us that some components are very much more important than others. Invariably, in O/W emulsions, disorganized protein molecules and low molecular weight surfactant molecules play a crucial role at the oil–water interface, and high molecular weight polysaccharides have a controlling influence on the rheology of the dispersion medium. The chemistry of the dispersed phase is normally unimportant.

Over the years scientists have identified a number of physical factors that appear to relate to emulsion stability. *Table 4.1* lists 12 such factors, together with an assessment, necessarily subjective, of their relative importance in creaming, flocculation and coalescence. The overview is based mainly on experimental observation, buttressed where appropriate by theoretical considerations. Rheology is given its own separate heading in *Table 4.1*, in recognition of the strong links between emulsion stability and emulsion rheology. The source of the simplification, in which a multitude of chemical variables is reduced to a manageable set of physical variables, lies in the fact that the same physical phenomenon can arise from a whole plethora of different chemical compositions and molecular arrangements. *Table 4.1* reminds us what to look out for when a certain type of instability occurs. What it can rarely do, however, without further physicochemical information, is to tell us in detail how a particular emulsion A would be affected by addition of gum X, lipid Y, or salt Z.

There is no shortage of theory in textbooks on surface and colloid science. Nevertheless, at a pragmatic level, the emulsion technologist is justified in asking as to

**Table 4.1** MAIN PHYSICAL FACTORS AFFECTING FOOD EMULSION STABILITY AND RHEOLOGY

| | *Creaming* | *Flocculation* | *Coalescence* | *Rheology* |
|---|---|---|---|---|
| Droplet size | *** | ** | * | * |
| Droplet size distribution | *** | ** | | ** |
| Volume fraction of dispersed phase | *** | *** | *** | *** |
| Density difference between phases | *** | | | |
| Viscosity (rheology) of continuous phase | *** | *** | ** | *** |
| Viscosity (rheology) of dispersed phase | | | | * |
| Viscosity (rheology) of adsorbed layer | | | *** | ** |
| Thickness of adsorbed layer | | ** | *** | ** |
| Electrostatic interaction between droplets | * | ** | ** | * |
| Macromolecular interaction between droplets | | *** | ** | ** |
| Fat crystallization | | | *** | *** |
| Liquid crystalline phases | * | * | ** | ** |

*** = Generally important; ** = often important; * = sometimes important.

how much of the basic theory can in fact be used in practice to predict the behaviour of food systems. Quantitatively speaking, the straight answer must be disappointingly little. Though theories of electrostatic and polymeric stabilization are relatively well developed (Tadros and Vincent, 1983), they cannot be used to make numerical predictions about flocculation in food systems because values of key parameters are either uncertain or completely unknown. For instance, we know far too little about molecular configurations of, and charge distributions on, adsorbed proteins at the oil–water interface, and in mixed proteinaceous systems we usually do not even know the interfacial composition. The situation is no better with regard to coalescence—in fact, if anything, the gap between theory and useful prediction here is even more yawning. This just leaves us with creaming, for which the problems of quantitative prediction are fortunately much less severe. While theories of flocculation and coalescence are merely useful to the food scientist as an aid to understanding and as a guide to general behaviour, the theory of creaming can actually be used as a predictive tool. This is because the creaming of droplets is essentially unaffected by the detailed properties of adsorbed layers, and only weakly so by interactions between droplets (except under conditions of strong flocculation). Herein lies the justification of the emphasis on creaming in the rest of this chapter.

## Structure

> 'When I make a word do a lot of work like that,' said Humpty Dumpty, 'I always pay it extra.'

One need look no further than this volume to see that the concept of structure in foods has connotations that run from the purely chemical to the overtly textural. In the emulsion context, what we mean by structure is (1) the way in which individual droplets are arranged relative to one another, and (2) the way in which emulsifying and stabilizing components (surfactants, macromolecules, crystals, granules, etc.) are arranged in relation to the oil–water interface. Structural concepts (1) and (2) cover a spatial scale from $10^{-9}$ m to, say, $10^{-4}$ m, i.e. corresponding to the whole of the colloidal size range and extending well into the microscopic size range.

Suppose we have an idealized colloidal system in which all the particles are spherical and of the same size. Then, at its simplest, the structure depends on just two factors: the volume fraction of particles, and their degree of aggregation. *Figure 4.1* illustrates the sort of structures that can occur at low and high volume fractions as the degree of flocculation increases in the absence of any significant gravitational separation. The structure of the dilute stable dispersion (*Figure 4.1a*) reminds one of a dilute gas, whereas the concentrated stable dispersion (*Figure 4.1d*) resembles a dense fluid. Under some circumstances, a monodisperse stable system may take the form of an ordered crystal-like structure (Pieranski, 1983). Attractive forces between the particles tend to make the structure less uniform. If the system is dilute and weakly flocculated (*Figure 4.1b*), there coexists, possibly in equilibrium, a mixture of monomer particles and small aggregates (dimers, trimers, etc.). If, on the other hand, the system is concentrated and weakly flocculated (*Figure 4.1e*), the individual aggregates tend to lose their identity; the associated rheological behaviour (usually shear-thinning) is extremely sensitive to the interparticle forces. Under strongly flocculated conditions, large aggregates exist in dilute systems (*Figure 4.1c*), and gel-like network structures at higher particle volume fractions (*Figure 4.1f*). More

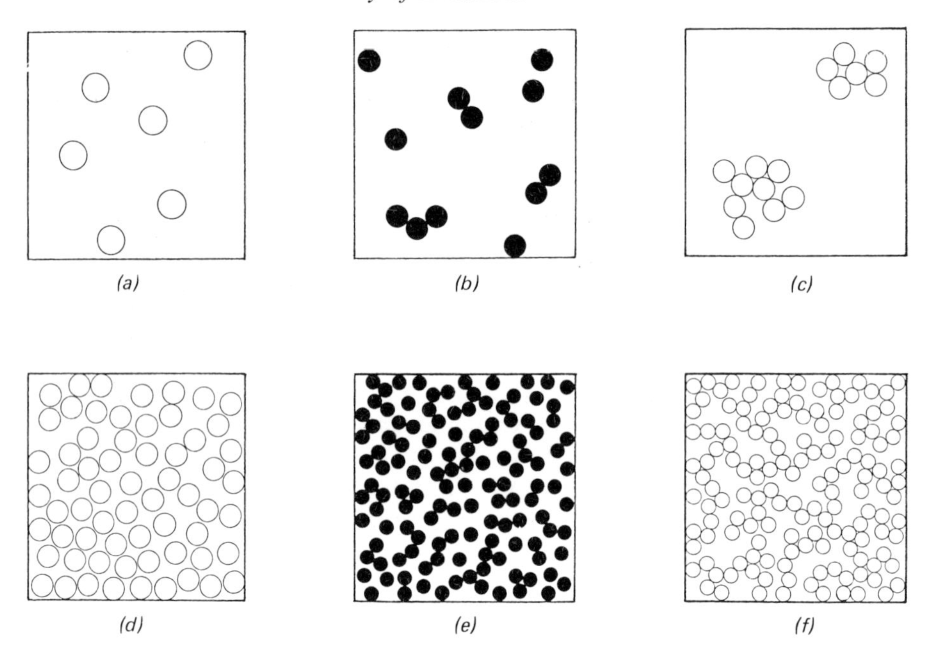


**Figure 4.1** Dependence of colloidal structure on the particle volume fraction and the degree of flocculation: (a) dilute, stable; (b) dilute, weakly flocculated; (c) dilute, strongly flocculated; (d) concentrated, stable; (e) concentrated, weakly flocculated; (f) concentrated, strongly flocculated.

information on the various types of colloidal phases and the possible transitions between them may be found elsewhere (Dickinson, 1983).

Real emulsions differ from these idealized structures for several reasons. For a start, creaming leads to a non-homogeneous distribution of droplets which is continuously changing with time. Droplets are not identical in size, and the geometrical packing of polydisperse spheres differs from that for monodisperse spheres, especially at high volume fractions. Emulsion droplets may be deformed from their normal spherical shape in highly concentrated systems, under strong flow conditions, or where crystals are present. In addition to all this, most food emulsions are complex chemical systems whose overall structure depends on interactions between droplets and other entities: protein particles, gas bubbles, starch granules, fat crystals, ice crystals, and gel-like structures of various types. Covering a wide range of dispersed phase volume fractions, *Table 4.2* shows the diversity of stabilizing structures typically encountered in food emulsions. To some extent, every type of food emulsion has its own particular set of structural features controlling stability.

Various sorts of structure may occur at the oil–water interface: solid particles, monolayers of surfactant or protein, multilayers, and liquid crystalline phases. Few food emulsions are stabilized simply by low molecular weight emulsifiers at monolayer coverage; some flavour emulsions for soft drinks are notable exceptions. As far as liquid crystals are concerned, there is certainly evidence that they can stabilize emulsions (O/W and W/O), and also that stability is correlated with well-defined mesophases in the oil/water/emulsifier phase diagram (Krog, Riisom and Larsson, 1985). But, according to Darling and Birkett (1987), the mechanism probably does

**Table 4.2** STABILIZING STRUCTURES AND DISPERSED PHASE VOLUME FRACTIONS OF SOME FOOD EMULSIONS

| *Food product* | *Volume fraction* | *(Main) stabilizing structure* |
|---|---|---|
| Soft drinks | 0.001–0.002 | Surfactant membrane |
| Homogenized milk | 0.03–0.04 | Protein membrane |
| Butter/margarine | 0.16 (water) | Semi-crystalline continuous fat phase |
| Whipped cream | 0.2–0.4 | Aggregated fat-globule network/air cells |
| Meat emulsions | 0.2–0.5 | Gelled protein matrix |
| Salad dressing | 0.3–0.4 | Lipoprotein membrane/hydrocolloids |
| Mayonnaise | 0.6–0.8 | Lipoprotein membrane/droplets |

not operate in most food systems because the lipid emulsifier concentration is far too low for liquid crystalline phases to develop at the oil–water interface.

Most food emulsions are stabilized by protein at the oil–water interface. Often the protein is aggregated, and droplets are prevented from flocculating and coalescing by the presence of a thick, protective, gelatinous layer at the surface. In dairy emulsions much of the adsorbed protein remains in the form of discrete colloidal particles—casein micelles or fragments thereof (Mulder and Walstra, 1974); and mayonnaise is stabilized by particles of egg yolk (Dickinson and Stainsby, 1982). In addition, composite stabilizing structures of protein + anionic polysaccharide may occur in the form of interfacial complexes or mixed gel layers, especially in meat emulsions where differences between mixed bulk gelation and surface complex formation are rather indistinct (Tolstoguzov, 1986).

Hydrocolloid structures in the aqueous dispersion medium are involved in stabilizing emulsions like desserts and sauces. While strictly outside the scope of the present chapter, the subject of such thickeners does lead us to a couple of matters of direct relevance to colloid stability in general.

In attempting to correlate emulsion stability with the rheological behaviour of the dispersion medium, we need to know the apparent viscosity of the medium at values of the stress and strain rate well below those commonly existing in conventional laboratory rheometers. The existence or otherwise of a very small yield stress can have a crucial influence on stability. In the semi-dilute régime of de Gennes (1979), one may not be sure, close to the polymer–polymer overlap concentration, whether the actual microviscosity experienced by the dispersed particles is the same as that measured in the laboratory on a macroscopic scale, even if the stress and strain rate are of the right order. What experimental results that do exist, e.g. with gelatin solutions (Dickinson, Lam and Stainsby, 1984), would seem to suggest is that the two viscosities are the same in the absence of gelation.

Another complication is that the hydrocolloid may itself be surface active, either because it is a protein, or because it is a polysaccharide complexed with protein or with low molecular weight surfactant. The amount of polymeric material available for thickening or gelation in the bulk phase may be substantially reduced by adsorption, as illustrated schematically in *Figure 4.2*. Such an effect has recently been demonstrated on a quantitative basis in experiments with model gelatin-stabilized emulsions (Dickinson, Stainsby and Wilson, 1985). It is now established that the relationship between polymer adsorption and colloid stability involves a delicate balance of factors. In the past it was always thought that flocculation arises through polymer bridging between particles, but now it is recognized (Napper, 1983) that

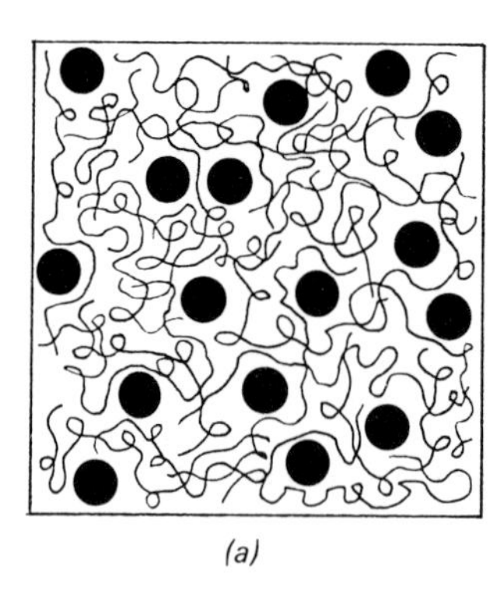

(a)

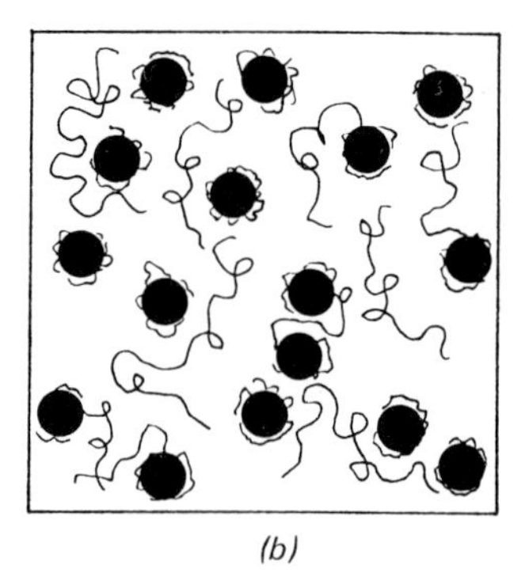

(b)

**Figure 4.2** Schematic representation of different ways of partitioning hydrocolloid between surface and bulk phase: (a) non-adsorbing macromolecules; (b) adsorbing macromolecules.

aggregation may also be induced with strongly *non*-adsorbing macromolecules through a mechanism akin to phase separation of polymer-rich and particle-rich regions.

To focus attention on the relationship between stability and structure, we examine the theory of creaming and the properties of creamed layers. We turn also from the real world of complex food emulsions to the idealized world of the proverbial 'model system'.

## Creaming

> 'I'm afraid I don't quite understand,' said Alice. 'It gets easier further on,' Humpty Dumpty replied.

Consider a single, undeformable, spherical droplet of radius $a$ and density $\rho$. The magnitude of the gravitational force $F_g$ acting on it in a medium of density $\rho_0$ is given by

$$F_g = 4\pi a^3 |\rho - \rho_0| g/3, \tag{1}$$

where $g$ is the acceleration due to gravity. Equilibrium creaming over a vertical distance $h$ is significant when the potential energy $F_g h$ is comparable with the thermal energy $kT$. So, the steady state condition for negligible creaming from Eqn (1) is

$$a \ll [3kT/4\pi |\rho - \rho_0| gh]^{1/3}, \tag{2}$$

where k is Boltzmann's constant and $T$ is the absolute temperature. However, Eqn (2) does not provide a technologically useful criterion for a non-creaming emulsion, because in practice it is sufficient to reduce gravitationally-induced separation to a low rate, as opposed to eliminating it altogether. Due to the effects of mechanical and thermal convention during storage, the equilibrium state is rarely attained in any case.

Moving at a steady speed $v$ in a Newtonian medium of viscosity $\eta_0$, an isolated spherical droplet, rigid and uncharged, experiences a hydrodynamic frictional force $F_f$ of magnitude

$$F_f = 6\pi\eta_0 av. \quad (3)$$

By equating forces $F_g$ and $F_f$, we obtain the well-known Stokes expression for the settling speed $v_s$ of an isolated droplet:

$$v_s = 2a^2|\rho - \rho_o|g/9\eta_0. \quad (4)$$

With appropriate parameter values for micron-sized oil droplets in water, Eqn (4) gives a predicted settling speed of $\sim 1\ \text{mm h}^{-1}$, low enough to prevent observable creaming in milk over a time scale of a few days. Though the theory assumes that the particle is uncharged, Eqn (4) can be used for charged droplets with little error. Surface charge does reduce $v_s$, but only by a few per cent.

The Stokes formula [Eqn (4)] tells us that there are three ways to inhibit creaming in a dilute unaggregated emulsion. The most obvious approach is to make the droplets small (as $v_s \propto a^2$). This is achieved most effectively by high pressure homogenization (Walstra, 1983). The size of the largest droplets is the crucial factor, since all manufactured emulsions are to some extent polydisperse. In practice, even after the most intense homogenization a finite amount of residual creaming is inevitable because it is impossible to ensure that every single large droplet ($\gtrsim 5\ \mu\text{m}$) will have been disrupted. Normally, one is content to think in terms of a mean settling speed. The quantity $a^2$ in Eqn (4) is replaced by

$$< a^2 > = \sum_i n_i a_i^5 / \sum_i n_i a_i^3, \quad (5)$$

where $n_i$ is the number of droplets of radius $a_i$ (as measured, for instance, with a Coulter counter). The significance of the terms $a_i^3$ and $a_i^5$ in Eqn (5) lies in their being proportional, respectively, to the volume and the volume times the speed of the *i*th droplet (Greenwald, 1955).

Theoretically, creaming can be eliminated altogether by matching densities of dispersed and continuous phases. In reality, of course, the food technologist has little room for manoeuvre when confronted by a combination of chemical, legal and toxicological constraints. Nature has fixed the density difference between food oils and water at around $10^2\ \text{kg m}^{-3}$, but variations of up to 50% or so may occur when allowance is taken for factors like fat crystallization, and the presence of added sugar or alcohol. Increasing the oil phase density is one way of producing a density match. Indeed, at one time, in order to stabilize soft drink emulsions, brominated vegetable oils ($1.3 \times 10^3\ \text{kg m}^{-3}$) were mixed with citrus oils ($0.85 \times 10^3\ \text{kg m}^{-3}$); but nowadays the use of brominated oils is prohibited or restricted in most Western countries, and the same is true for many other potential 'weighting agents' (Shankaracharya *et al.*, 1980).

The density $\rho_s$ of the stabilizing layer around a droplet is usually different from the density $\rho_i$ of the internal phase, and for a protein-stabilized oil-in-water system we typically have $\rho_s > \rho_0 > \rho_i$. With small droplets the adsorbed layer density becomes a significant factor in determining the net particle density $\rho$. As adsorbed layer thickness is more or less independent of droplet size, the droplets in a polydisperse emulsion have a distribution of net particle densities as well as a distribution of particle radii. This leads to the situation, with emulsions of high protein load (e.g. homogenized milk), of the smallest droplets being more dense than the dispersion medium, so that they can never be creamed even in a centrifuge (Dalgleish and Robson, 1985). As well as having a positive effect on flocculation and coalesence stability, we note therefore that a high protein load inhibits creaming by reducing the droplet size during emulsification and the density difference after emulsification.

The final way of affecting the creaming rate in Eqn (4) is to adjust the viscosity of the continuous phase. Particle movement during creaming exerts small stresses within the fluid of the dispersion medium; these stresses are largest at the particle surfaces. For an isolated settling droplet, the stress $S$ at the surface is not uniform, but its magnitude is given roughly by

$$S \sim a|\rho - \rho_0|g. \tag{6}$$

Numerically, the value is $\sim$ 10 mPa for a 10 µm diameter oil droplet in water. To stop creaming then, we need to add thickening or gelling agent to the aqueous phase so as to produce a dispersion medium having a Bingham-type yield stress exceeding $\sim$ 10 mPa. If the medium has no real yield point, but is very pseudoplastic, the large effective viscosity at the low stress implied by Eqn (6) may still have the effect, say, of reducing the creaming rate of a 10 µm droplet to that of a 1 µm droplet. This is the physical basis underlying the stabilization of a salad cream emulsion by, for example, xanthan gum (Hibberd *et al.*, 1987).

The differential creaming speeds of droplets of different sizes in a polydisperse system has its effect on the evolving structure of the creamed emulsion. Various types of 'equilibrium' state may be envisaged, as illustrated in *Figure 4.3*. With fine emulsion droplets, where effects of gravity and Brownian diffusion are comparable, we expect a continuous droplet density distribution between top and bottom of the container, droplets at the top being bigger and more numerous than those lower down (*Figure 4.3a*). With coarser droplets, Brownian motion may be swamped by the effect of gravity, in which case a distinct creamed layer is formed with a sharp boundary separating the cream from the continuous phase (*Figure 4.3b*). Often, the situation is intermediate between *Figure 4.3a* and *4.3b*: a visibly perceptible creamed layer is formed, containing mainly the larger droplets, but this has a rather indistinct boundary with the rest of the dispersed system, a dilute emulsion of the smaller (Brownian) droplets. In this lower region, the density of droplets decreases towards the bottom of the container, at equilibrium according to a Boltzmann-type distribution (*Figure 4.3c*).

To be able to predict the rate of creaming in non-dilute polydisperse emulsions, one must look further than the simple Stokes relation [Eqn (4)]. First, then, let us consider how a finite volume fraction of dispersed phase affects the single particle settling speed in the absence of interparticle colloidal interactions.

With rigid spheres at low volume fraction ($\phi \leqslant 0.02$), the mean creaming speed $v$ is less than the Stokes value $v_s$ by an amount which is proportional to $\phi$:

$$v = v_s(1 - 6.55\phi). \tag{7}$$

The derivation of Eqn (7) assumes that the dispersed particles are randomly distributed (Batchelor, 1972). Contributions to the coefficient $-6.55$ from three distinct physical effects are identified: the general backflow and a local pressure gradient arising from the net motion of the other particles (–5.00); liquid entrainment from nearby particles moving in the same direction (+ 0.28); and a passive hydrodynamic slowing down due to the presence of force-free neighbours (–1.83). The third of

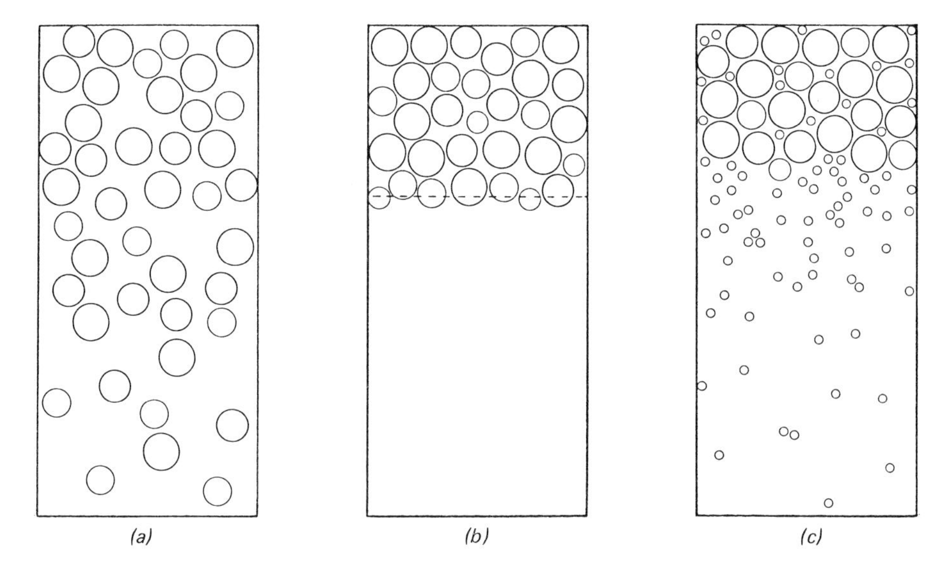


**Figure 4.3** Structural states of creamed emulsions with various particle size distributions: (a) small droplets; (b) large droplets; (c) small + large droplets.

these terms also affects particle diffusion (Dickinson, 1983), but the other two are special to sedimentation (creaming), having their origin in the fact that all the particles are tending simultaneously to move in the same direction, which has obvious consequences for the net motion of the dispersion medium.

There is a temptation in some quarters to interpret the dependence of $v$ on $\varphi$ as being due simply to the increase in effective viscosity of the surrounding fluid caused by the presence of other droplets. Intuitively reasonable though it may seem at first sight, this interpretation is, however, not correct. If it were, then the coefficient of $\varphi$ in Eqn (7) would be just –2.5, since the shear viscosity $\eta$ of a rigid sphere suspension is given by Einstein's classic result:

$$\eta = \eta_0(1 + 2.5\phi). \tag{8}$$

Another plausible argument says that the effect of volume fraction on the mean creaming speed is due, in part, to the effective difference in density $|\rho - \rho_0|$ being reduced by a factor $(1 - \phi)$, as each droplet of density $\rho$ may be thought of as creaming in a medium of effective density $\rho_0' = \rho\phi + \rho_0(1 - \phi)$. When results from the viscosity and density arguments are combined, we get

$$v = v_s(1 - \phi)/(1 + 2.5\phi) = v_s(1 - 3.5\phi), \tag{9}$$

which is still quite different from the exact Batchelor result [Eqn (7)]. In a bounded system, what makes creaming so strongly dependent on volume fraction is that for every droplet that moves up an equivalent volume of continuous phase must move down, thereby retarding the motion of the other droplets.

The implication so far has been that the settling particles have no direct forces acting amongst them, and are therefore randomly distributed. In reality, due to the influence of London dispersion forces, emulsion droplets larger than $\sim 1\ \mu m$ have a tendency to gather together, and so, except under conditions where interparticle electrostatic repulsion is very long ranged (extremely low electrolyte concentration), droplets are closer together on average than is assumed in the simple theory. The net result is a lowering of the numerical value of the coefficient of $\phi$ in Eqn (7) from 6.55 to between 5.5 and 5, based on the refined theory of Batchelor (1982), and supported by experiments on model systems (Buscall *et al.*, 1982).

The above discussion is relevant, with necessary changes, to a polydisperse system of low volume fraction. We simply use the sets of values of $v_s$ corresponding to the particular droplet sizes under consideration, together with the total volume fraction $\phi$. Fortunately, we need not worry that emulsion droplets are not truly rigid spheres: with respect to creaming, they act as if they are (Dickinson and Stainsby, 1982). The oil phase viscosity has no influence on the kinetics of creaming.

At volume fractions above a few per cent, multiparticle hydrodynamic interactions become important, the relationship between $v$ and $\phi$ becomes non-linear, and no rigorous theoretical treatments are available. So, at moderate or high volume fractions, creaming theory is a semi-empirical business. One equation that fits experimental data for a wide range of particulate systems is due to Barnea and Mizrahi (1973):

$$v = v_s(1 - \phi)/(1 + \phi^{1/3}) \exp [5\phi/3(1 - \phi)]. \tag{10}$$

[The term $(1 - \phi)$ in Eqn (10) is the so-called 'density correction' mentioned above; the exponential terms come from the 'viscosity correction'.] Equation (10) tells us that the effect of the other droplets is to retard the creaming speed to about one-third of the Stokes value in, say, a cream liqueur of volume fraction 0.2; to about 10% of the Stokes value in, say, a salad dressing of volume fraction 0.4; and to about 2% of the Stokes value in, say, a mayonnaise of volume fraction 0.6.

Throughout the preceding discussion, the creaming speed $v$ has referred to the mean velocity of a droplet relative to the fluid medium. Sometimes it is more convenient, however, to consider the creaming speed $v_c$ relative to the container wall:

$$v_c = v(1 - \phi). \tag{11}$$

The total volume of droplets crossing unit area per unit time, the flux density $F$, is then given by

$$F = \phi v_c = \phi(1 - \phi)v. \tag{12}$$

Notice that mass transport by creaming is insignificant in very concentrated emu sions ($v \rightarrow 0$) as well as in very dilute ones ($\phi \rightarrow 0$). Mass transport is maximized at a volume fraction $\phi^*$ which satisfies the equation

$$(\mathrm{d}F/\mathrm{d}\phi)_{\phi = \phi^*} = 0. \tag{13}$$

Let us assume *faute de mieux* that the dependence of $v$ on $\phi$ is given by Eqn (10). This gives $\phi^* = 0.17$ (Zimmels, 1983), a prediction which is useful technologically as it tells us how to optimize the efficiency of centrifugal creaming by diluting a concentrated emulsion.

In non-dilute emulsions, the effects of polydispersity can be rather complicated. When the droplet size distribution is known, and a reliable form for $v(\phi)$ is available [e.g. Eqn (10)], one can in principle evaluate numerically how the droplet size distribution as a function of height changes with the time of creaming (Zimmels, 1983). But this type of calculation is rarely done in practice because of its considerable computational complexity; if done properly, the mass transport equation has to include a diffusive term whose magnitude increases towards the lower end of the droplet size distribution. In this connection, we note that the systematic (non-Brownian) motion of small droplets in a concentrated polydisperse emulsion is dominated by the 'tidal' backflow of continuous phase in the opposite direction to the motion of the large droplets. This means that the small droplets, irrespective of their density relative to the fluid medium, may actually be moving in the opposite direction to what one would normally expect! These aspects of polydispersity need to be borne in mind when partial separations of fat phase from non-dilute emulsions are induced by centrifugal creaming.

Flocculation can have a crucial effect on creaming kinetics. In cold fresh milk, creaming is rapid because the fat globules are flocculated by macromolecular material called agglutinin (Walstra and Jenness, 1984). In manufactured emulsions, flocculation may occur for one of several possible reasons—inadequate or excessive homogenization, bridging of droplets by hydrocolloid stabilizer, aggregation of protein initially adsorbed on different droplets, and so on. When flocculation is incipient, it is enhanced by the presence of polydispersity. Differential creaming speeds of small and large droplets cause them to come into close proximity, and therefore to stick together, more often than they would due to Brownian motion alone (Melik and Fogler, 1984). Unfortunately, however, there is no way at present of quantifying reliably the extent of flocculation in food emulsions. One just has passively to note that flocculation in dilute emulsions increases creaming markedly, owing to the fact that aggregates are hydrodynamically bigger than individual droplets. Conversely, in concentrated emulsions, flocculation may inhibit creaming. This happens when a semi-continuous floc structure permeates the whole system. Large flocs are then geometrically unable to pass by one another without extensive rearrangement, and further emulsion concentration can occur only by liquid being exuded from the gaps in the flocculated structure (by a process akin to gel syneresis). At this point the distinction between creamed and uncreamed systems becomes ill-defined, since the structure of a concentrated flocculated emulsion may resemble that of a creamed layer formed from a dilute emulsion.

The ultimate volume fraction of a creamed layer depends on the balance between the interparticle forces (attractive and repulsive) and the external forces (gravitational or centrifugal). For non-deformable monodisperse spheres, the maximum theoretical volume fraction is $\phi = 0.74$ corresponding to a close-packed hexagonal array. Particles of the size of emulsion droplets do not have sufficient thermal energy to form ordered structures. So, in practice, the maximum volume fraction is $\phi = 0.64$ corresponding to random close-packing, although values considerably lower than this often occur in the loose packings produced by sedimentation or creaming. As a creamed layer grows in thickness, it will tend to become more compacted under the compressive influence of other droplets in the layer. At high degrees of polydispersity, high volume fractions ($\phi \geqslant 0.7$) may be achieved, as small droplets fill up the voids amongst the large ones. Flocculation, on the other hand, will have the effect of reducing the cream layer volume fraction. Any sticking together of droplets during creaming, or sticking of freshly creamed droplets to the creamed layer, tends to inhibit the formation of a close-packed structure. Instead, we get a more open gel-like sediment structure like that illustrated in *Figure 4.1e* or *4.1f*.

A concentrated emulsion in which the flocculated droplets are compressed into a coherent structure is called a cohesive cream. Whereas weakly flocculated droplets are redispersed by mild agitation, a cohesive cream is not. In general, creamed layers become more difficult to redisperse the longer they have been formed, due to a gradual strengthening of adhesive contacts between adsorbed layers on droplets in close contact. In most creamed layers formed under normal gravity, the droplets retain their spherical shape. But, with emulsions containing coarse droplets, and with cohesive creams produced by centrifugation, the droplets may be deformed into polyhedral foam-like cells, possibly with some associated partial coalescence.

Theoretical attempts to reconcile the link between sediment structure and the dynamic processes occurring during sedimentation are currently making use of computers to simulate aggregation behaviour. *Figure 4.4* represents part of a computer-generated sediment of volume fraction $\phi = 0.33$ recently simulated in our laboratory (Ansell and Dickinson, 1986). The sediment is produced by a process of irreversible single-particle accretion, and the model includes allowances for the effects of gravity, Brownian motion, and various types of interparticle interactions. What the simulations are telling us is that it is possible to control the volume fraction of cohesive sediments (or cohesive creams) over a wide range (0.1–0.6) by adjusting the strength of the sedimenting field and the nature of the colloidal interactions. As yet, however, the technologically important relationship between sediment structure and sediment rheology is not well understood.

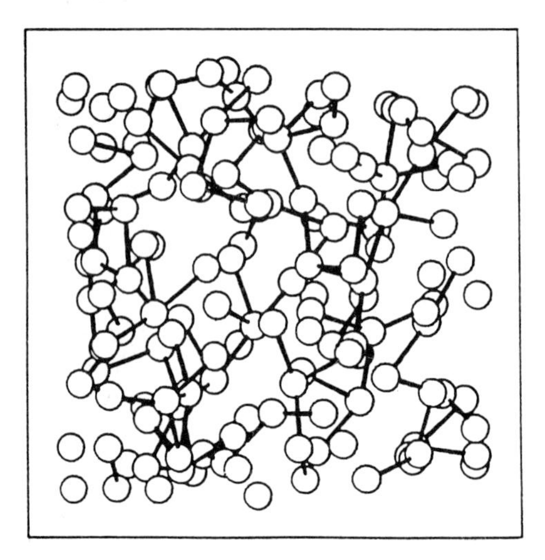

**Figure 4.4** Snapshot of three-dimensional sediment structure ($\phi = 0.33$) formed by Brownian dynamics computer simulation. The viewing box is a cube of side length $12a$, where $a$ is the particle radius. For the sake of clarity, particles are denoted by circles of radius smaller than $a$, with bonds between circles representing points of direct contact between adjacent pairs of particles. Apparently loose particles are, in fact, connected to others outside the viewing box. For further details see Ansell and Dickinson (1986).

## Flocculation and coalescence

'Are five nights warmer than one night, then?' Alice ventured to ask.
'Five times as warm, of course.'
'But they should be five times as *cold*, by the same rule—'

By way of completeness, it seems appropriate to end this chapter by collecting together some heuristic rules of stability with respect to flocculation and coalescence. Setting out the rules is easy; discerning which rules apply in practice to which emulsions is, however, a little tricky!

Whether an emulsion is stabilized by polymer or particles, the thicker the adsorbed film, the greater is the stability with respect to both flocculation and coalescence, but especially the latter. More practically, this means that a high surface excess concentration, and hence a high emulsifier concentration, gives good stability. Most of the film thickness should, however, be located on the continuous phase side of the interface: proteins cannot stabilize water-in-oil emulsions. Weak (so-called secondary-minimum) flocculation is important with large droplets ($\geqslant 2\,\mu m$), but may be absent altogether with small droplets ($\leqslant 0.5\,\mu m$) (Dickinson and Stainsby, 1982). Thick adsorbed layers reduce secondary-minimum flocculation. Electrostatic stabilization is favoured by an aqueous continuous phase of low ionic strength and a pH well away from the isoelectric point of any adsorbed protein. Polymeric stabilization is favoured by good solvent conditions for any adsorbed macromolecules.

Low molecular weight amphiphiles are good for making emulsions; high molecular weight amphiphiles give good long-term stability. Anything forming a 'complex' at the oil–water interface improves stability with respect to coalescence. Combining an oil-soluble surfactant with a water-soluble surfactant is an often used ploy; in food systems, this usually means a mixture of Spans and Tweens, e.g. to stabilize hop oil emulsions for use in beer (Chilton and Laws, 1980). Addition of a low molecular weight surfactant to a protein-stabilized emulsion may cause immediate destabilization as adsorbed protein is displaced from the oil–water interface. Any perturbation to the system (heating, change in pH, addition of calcium ions, etc.) which causes protein to become insoluble or aggregated, is likely also to induce flocculation. Stability with respect to coalescence, however, is greatest when adsorbed protein is near its isoelectric point, since the adsorbed film strength is greatest there (Graham, 1976).

All other things being equal, concentrated emulsions are less stable than dilute ones. At high enough volume fraction, any emulsion is *de facto* flocculated. Coalescence in a concentrated emulsion, or a creamed layer, is greatly increased by fat crystal formation, especially in the presence of agitation (van Boekel, 1980). Freezing an emulsion is one of the most effective ways of breaking it. In emulsions having fat crystals or liquid crystalline phases at the oil–water interface, temperature fluctuations (high or low) are particularly deleterious (Rydhag, 1977). Stability is enhanced by components which induce thickening or gelling in the continuous phase ['self-bodying' mixed surfactants (Barry and Eccleston, 1973), polysaccharide gums, etc.] or viscoelasticity at the oil–water interface (protein films, liquid crystalline mesophases, etc.).

Flocculation can be distinguished from coalescence by its reversibility with respect to dilution, stirring, change of pH, and so on. To avoid changing the state of flocculation, emulsions being prepared for particle sizing or rheological measurement should be diluted with a solvent of the same quality (pH, ionic strength, etc.) as that of the continuous phase. As large flocs are much more easily disrupted than small ones (Dickinson and Stainsby, 1982), even very low shear rates will lead to some deflocculation. Reversal of creaming without significant deflocculation may be achieved, however, by gentle end-over-end rotation of the containing vessel (Tadros and Vincent, 1983).

Finally, we note that Walstra (1987) has designated 'partial coalescence' as a separate category of instability in its own right. When droplets contain fat crystals, as they often do in food emulsions, they tend to aggregate into non-spherical clumps, rather than flowing into larger spherical droplets. By forming non-spherical aggregates and semi-solid network-type structures, partial coalescence is accompanied by large changes in emulsion rheology. Like (full) coalescence, it is enhanced by agitation, but more so; and also by addition of low molecular weight surfactants to a protein-stabilized emulsion. Fats having a broad triglyceride composition produce more stable emulsions than those having a narrow range of co-crystallizing triglycerides (Darling and Birkett, 1987). Mixed crystals produce softer mechanical structures, and so the emulsions are less susceptible to partial coalescence.

## References

ANSELL, G.C. and DICKINSON, E. (1986). Sediment formation by Brownian dynamics simulation: effect of colloidal and hydrodynamic interactions on the sediment structure. *Journal of Chemical Physics*, **85**, 4079–4086

BARNEA, E. and MIZRAHI, J. (1973). A generalized approach to fluid dynamics of particulate systems. Part 1. General correlation for fluidization and sedimentation in solid multiparticle systems. *Chemical Engineering Journal*, **5**, 171–189

BARRY, B.W. and ECCLESTON, G.M. (1973). Influence of gel networks in controlling consistency of O/W emulsions stabilized by mixed emulsifiers. *Journal of Texture Studies*, **4**, 53–81

BATCHELOR, G.K. (1972). Sedimentation in a dilute dispersion of spheres. *Journal of Fluid Mechanics*, **52**, 245–268

BATCHELOR, G.K. (1982). Sedimentation in a dilute polydisperse system of interacting spheres. Part 1. General theory. *Journal of Fluid Mechanics*, **119**, 379–408

BUSCALL, R., GOODWIN, J.W., OTTEWILL, R.H. and TADROS, TH.F. (1982). The settling of particles through Newtonian and non-Newtonian media. *Journal of Colloid and Interface Science*, **85**, 78–86

CHILTON, H.M. and LAWS, D.R.J. (1980). Stability of aqueous emulsions of the essential oil of hops. *Journal of the Institute of Brewing*, **86**, 126–130

DALGLEISH, D.G. and ROBSON, E.W. (1985). Centrifugal fractionation of homogenized milks. *Journal of Dairy Research*, **52**, 539–546

DARLING, D.F. and BIRKETT, R.J. (1987). Food colloids in practice. In *Food Emulsions and Foams*, (Dickinson, E., Ed.), pp. 1–29. London, Royal Society of Chemistry

DE GENNES, P.-G. (1979). *Scaling Concepts in Polymer Physics*. Ithaca, New York, Cornell University Press

DICKINSON, E. (1983). Dispersions of interacting colloidal particles. *Annual Reports C*, pp. 3–37. London, Royal Society of Chemistry

DICKINSON, E. (1986). Mixed proteinaceous emulsifiers: review of competitive protein adsorption and the relationship to food colloid stabilization. *Food Hydrocolloids*, **1**, 3–23

DICKINSON, E. and STAINSBY, G. (1982). *Colloids in Food*. London, Applied Science

DICKINSON, E., LAM, W.L.-K. and STAINSBY, G. (1984). Microviscosity of dilute latex + gelatin dispersions determined by dynamic light-scattering. *Colloid and Polymer Science*, **262**, 51–55

DICKINSON, E., STAINSBY, G. and WILSON, L. (1985). An adsorption effect on the gel strength of dilute gelatin-stabilized oil-in-water emulsions. *Colloid and Polymer Science*, **263**, 933–934

GRAHAM, D.E. (1976). Structure of adsorbed protein films and stability of foams and emulsions. PhD Thesis, Council for National Academic Awards, London

GREENWALD, H.L. (1955). Theory of emulsion stability. *Journal of the Society for Cosmetic Chemists*, **6,** 164–177

HALLING, P.J. (1981). Protein-stabilized foams and emulsions. *CRC Critical Reviews in Food Science and Nutrition*, **15,** 155–203

HIBBERD, D.J., HOWE, A.M., MACKIE, A.R., PURDY, P.W. and ROBINS, M.M. (1987). Measurement of creaming profiles in oil-in-water emulsions. In *Food Emulsions and Foams*, (Dickinson, E., Ed.), pp. 219–229. London, Royal Society of Chemistry

KROG, N. and LAURIDSEN, J.B. (1976). Food emulsifiers and their associations with water. In *Food Emulsions*, (Friberg, S., Ed.), pp. 67–139. New York, Marcel Dekker

KROG, N.J., RIISOM, T.H. and LARSSON, K. (1985). Applications in the food industry. Part 1. In *Encyclopedia of Emulsion Technology. Vol. 2. Applications*, (Becher, P., Ed.), pp. 321–365. New York, Marcel Dekker

MELIK, D.H. and FOGLER, H.S. (1984). Gravity-induced flocculation. *Journal of Colloid and Interface Science*, **101,** 72–83

MULDER, H. and WALSTRA, P. (1974). *The Milk Fat Globule: Emulsion Science as Applied to Milk Products and Comparable Foods*. Wageningen, Pudoc

NAPPER, D.H. (1983). *Polymeric Stabilization of Colloidal Dispersions*. London, Academic Press

PIERANSKI, P. (1983). Colloidal crystals. *Contemporary Physics*, **24,** 25–73

RYDHAG, L. (1977). The effect of temperature and time on emulsion stability. In *Physical, Chemical and Biological Changes in Food Caused by Thermal Processing*, (Høyem, T. and Kvåle, O., Eds), pp. 224–238. London, Applied Science

SHANKARACHARYA, N.B., SHANKARANARAYANA, M.L., LEWIS, Y.S. and NATARAJAN, C.P. (1980). Clouding agents for soft drinks. *Indian Food Packer*, **34,** 39–47

SHARMA, S.C. (1981). Gums and hydrocolloids in oil–water emulsions. *Food Technology*, **35,** 59–67

SHERMAN, P. (1968). Rheology of emulsions. In *Emulsion Science*, (Sherman, P., Ed.), pp. 217–351. New York, Academic Press

STAINSBY, G. (1986). Foaming and emulsification. In *Functional Properties of Food Macromolecules*, (Mitchell, J.R. and Ledward, D.A., Eds), pp. 315–353. London, Elsevier Applied Science

TADROS, Th.F. and VINCENT, B. (1983). Emulsion stability. In *Encyclopedia of Emulsion Technology. Vol. 1. Basic Theory*, (Becher, P., Ed.), pp. 129–285. New York, Marcel Dekker

TOLSTOGUZOV, V.B. (1986). Functional properties of protein-polysaccharide mixtures. In *Functional Properties of Food Macromolecules*, (Mitchell, J.R. and Ledward, D.A., Eds), pp. 385–415. London, Elsevier Applied Science

VAN BOEKEL, M.A.J.S. (1980). Influence of fat crystals in the oil phase on the stability of oil-in-water emulsions. PhD Thesis, University of Wageningen

WALSTRA, P. (1983). Formation of emulsions. In *Encyclopedia of Emulsion Technology. Vol. 1. Basic Theory*, (Becher, P., Ed.), pp. 57–127. New York, Marcel Dekker

WALSTRA, P. (1987). Overview of emulsion and foam stability. In *Food Emulsions and Foams*, (Dickinson, E., Ed.), pp. 242–257. London, Royal Society of Chemistry,

WALSTRA, P. and JENNESS, R. (1984). *Dairy Chemistry and Physics*. New York, Wiley

ZIMMELS, Y. (1983). Theory of hindered sedimentation of polydisperse mixtures. *American Institute of Chemical Engineers Journal*, **29,** 669–676

5

# STRUCTURE AND PROPERTIES OF LIQUID AND SOLID FOAMS

G. JERONIMIDIS
*Department of Engineering, University of Reading, UK*

## Introduction

Liquid and solid foams are found in a variety of natural and manufactured foods. Liquid-filled cellular structures are present in fresh fruit and vegetables; gas-filled ones are obtained by processing methods in bread, cakes, ice-cream, confectionery, etc. Low density in food introduces desirable textural and visual attributes as well as being an economic advantage in terms of cost per unit volume. Texture influences customer acceptance of foods and it is therefore important to be able to relate mechanical behaviour to structure and to properties associated with texture, particularly for solid and semi-solid foamed substances. So, for example, resistance to cutting and flowing, cracking, physiological feel when biting and masticating, depend on foam geometry, viscoelastic response of base materials, rate of application of load, etc. Similar considerations apply when dealing with damage resistance in handling and storage. The complex structure of foams, the variety and combination of loads and deformations which are imposed on them are such that relationships between texture and mechanical properties are difficult to establish. They are, however, crucial for a rational design of processed commercial products in which specific textural characteristics are to be introduced and for a better understanding of the influence of process variables. Modern plant breeding techniques allow some element of design in fruit and vegetables as well. Mechanical and structural features can be selected so as to improve texture or damage resistance; the difficulty is in deciding which characteristic to select.

In this chapter the relationships between mechanical properties of foams and structure will be discussed at a rather general level. Present theories for describing their behaviour will be discussed in relation to elastic, non-elastic response and fracture.

## Classification of foamed structures

Foamed structures are multiphase systems where different structural levels interact to control mechanical behaviour. They can be classified according to a variety of criteria but for the purpose of this chapter three elements are particularly useful for providing the basis for further analysis:

1. The physical state of the base substances.
2. The geometry of the foamed system.
3. The mechanical properties of the base substances and their foamed derivatives.

The first important distinction to make is between liquid and solid foams. A liquid or liquid-like foam can only exist as a two-phase, closed-cell structure since its very existence depends on the equilibrium between the pressure of the gas and the surface tension of the liquid walls. If the gaseous phase is eliminated, through diffusion for example, the foam collapses and reverts to liquid. If the surface tension of the liquid is changed by adsorption of other substances the liquid film may rupture. Furthermore, bubble coalescence is energetically favoured because of the reduction in surface area and hence in surface energy.

Stabilization of liquid foams requires careful control of diffusion processes and bubble–bubble interaction and it is an area of considerable interest to both food and polymer processing industries (Harding, 1967; Prins, 1976). Solid foams, which will be discussed in greater detail further on, often originate from liquid-like foamed structures for which the stability considerations mentioned earlier apply. When polymeric substances with viscoelastic behaviour are involved, elastic and viscous forces need to be considered in addition to surface tension. Solidification of a viscoelastic foam into a mechanically stable structure can be obtained by lowering temperature or by chemical means such as crosslinking. The geometrical details which are frozen in by these methods are responsible for the final density and mechanical properties.

Different considerations apply to the liquid-filled cellular structures of fruits and vegetables. In this case growth processes with continuous addition of material are responsible for the final morphology and, as will be discussed later, the pressure developed inside cells has a stabilizing effect on the structure.

Man-made solid foams can be subdivided into high and low density varieties. The distinction depends on whether solidification occurs before or after isolated gas bubbles present in the liquid have interacted with each other. Perfect spheres can be packed in cubic or hexagonal arrays (*Figure 5.1*) for which the volume fraction occupied by the spheres is about 70% and 90% respectively. This means that when gas bubbles in a foam expand or multiply so as to occupy 60–70% of the total volume or more, they will interact, deforming into thin-walled polyhedral cells. A perfectly closed packed (hexagonal) system of spheres will change into a structure of pentagonal-dodecahedral cells (*Figure 5.2*). In practice such regularity is impossible to achieve and the final geometry of each cell will be more or less irregular with a high proportion of pentagonal and hexagonal cell walls.

A high density foam can then be defined as a structure with a volume fraction of spherical bubbles of less than about 20%. Mechanical properties of these closed-cell systems, such as stiffness, strength and fracture, are in general adequately described using 'rules of mixtures' to modify the corresponding properties of the unfoamed

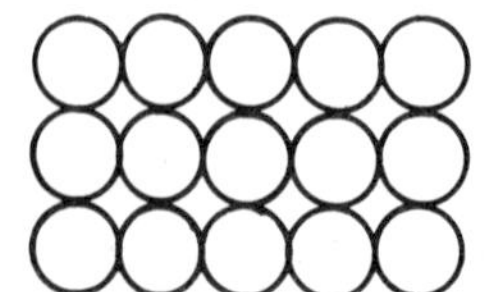

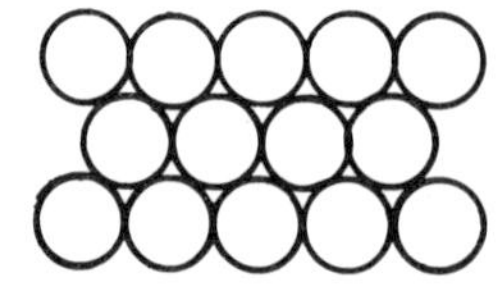

**Figure 5.1** Cubic and hexagonal arrays of spheres.

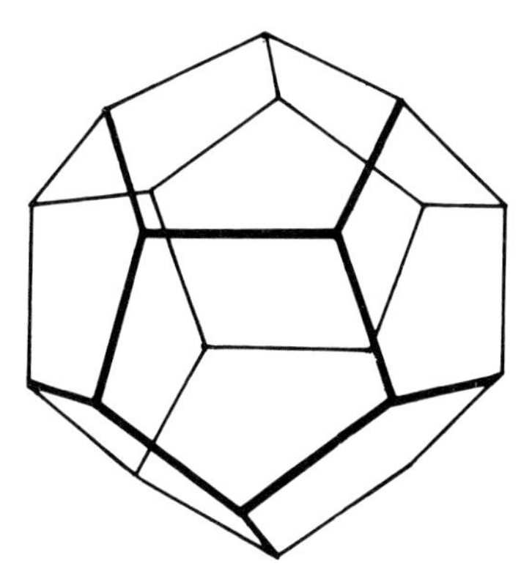

**Figure 5.2** Pentagonal-dodecahedral shape obtained from contacting spheres in a hexagonal, close-packed array.

materials as in composite materials and reinforced rubbers (Guth, 1945; Mullins and Tobin, 1965).

The tensile Young's modulus of a high density rigid foam, for example, may be expressed as:

$$E_f = E_s V_s + E_i V_i \tag{1}$$

where $E_f$, $E_s$, $E_i$ are, respectively, the Young's moduli of the foam, the solid substance and the spherical inclusion. The volume fractions of solid and inclusion are $V_s$ and $V_i$. The overall density of the composite structure is given by:

$$\rho_f = \rho_s V_s + \rho_i V_i \tag{2}$$

where again s refers to solid and i to inclusion.

For a foamed substance $E_s \gg E_i$ (the inclusion being the gas bubble) and $\rho_s \gg \rho_i$. Then:

$$E_f \simeq E_s V_s \tag{3a}$$

$$\rho_f \simeq \rho_s V_s \tag{3b}$$

from which

$$E_f = E_s \left(\frac{\rho_f}{\rho_s}\right) \tag{4}$$

The elastic modulus of high density foams is proportional to the amount of solid substance. This implies that basic deformation mechanisms in the solid are not changed significantly by the inclusion of a gaseous phase. Low density foams differ in several respects from high density ones. The thin cell walls separating adjacent cells of a closed-cell foam may rupture during foaming permitting free communication of gas and if a sufficiently high proportion of walls is broken the structure changes into an open-cell foam. Often the ruptured cell walls are in a liquid-like state and are drawn back by surface tension effects, thickening the edges and vertices of the polyhedra.

**Figure 5.3** Surface tension effects thickening the edges of an open-cell network.

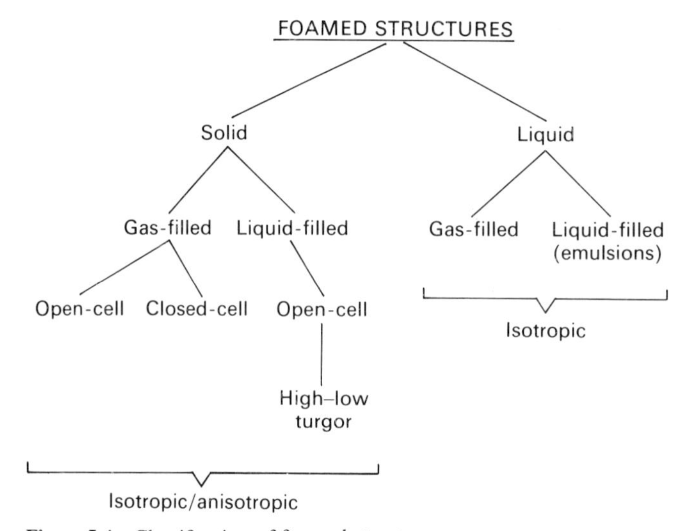


**Figure 5.4** Classification of foamed structures.

The final structure is now a three-dimensional network of randomly orientated struts (*Figure 5.3*).

When loads are applied to open- and closed-cell low density foams, the rod-like edges and thin cell wall plates can deform in ways which are not possible in high density systems and this has a considerable effect on their mechanical properties. *Figure 5.4* shows how foams may be classified in relation to geometry, basic mechanical properties of substances and additional effects such as turgor pressure.

## Elastic properties of solid foams

Depending on foam density, open- and closed-cell geometry and mechanical properties of basic substances, a wide range of elastic responses are possible. Different modes of deformation of the strut and plate elements of the structures are induced by tension, compression and shear. As a consequence, relationships between different elastic properties such as Young's modulus ($E$), shear modulus ($G$), bulk modulus ($K$) and Poisson's ratio ($\nu$) may not always exist. Homogeneous, isotropic solids have the same Young's modulus in tension and compression and two elastic constants are sufficient to characterize their behaviour for any combination of applied stresses. $E$, $G$, $\nu$ and $K$ are related by:

$$G = \frac{E}{2(1+\nu)} \tag{5}$$

$$K = \frac{E}{3(1-2\nu)} \tag{6}$$

In general these equations do not hold for foamed structures, especially those with very low density. Relative density, $\rho_f/\rho_s$ ($\rho_f$ = foam density, $\rho_s$ = density of solid material), plays an important role in determining elastic and strength properties of foams because it is related to geometry and size of wall elements.

Geometrical models with a different degree of complexity can be used to derive load–deformation relationships at the cellular level which are then interpreted in terms of stress–strain behaviour of the foam as a whole. *Figure 5.5* illustrates simple geometries which have been used to model solid foams and associated deformation mechanisms.

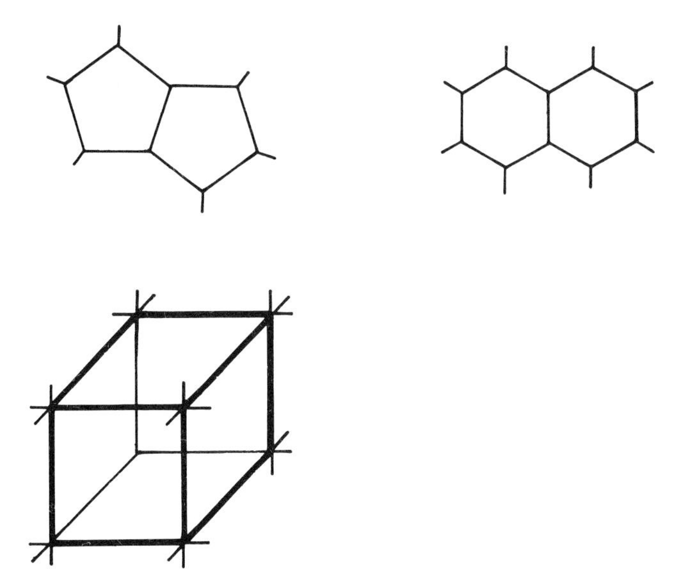

**Figure 5.5** Two- and three-dimensional models for open- and closed-cell foams.

A distinction must be made between rigid and flexible materials because the range of reversible elastic behaviour is very different in the two groups: typically less than 1% for rigid substances and anything up to 100–200% for flexible ones.

### *Elasticity of rigid cellular foams*

Qualitative stress–strain curves in tension, compression and shear are illustrated in *Figure 5.6*. In tension, proportionality between stress and strain holds until plastic deformation or brittle fracture occurs. The linear region in compression is much smaller and is followed by a flat portion where strain increases under practically constant load. The difference in the linear range between tension and compression suggests different modes of deformation of the subunits. Compression stress–strain curves rise steeply when compaction of the foam increases its density to the value of the solid substance.

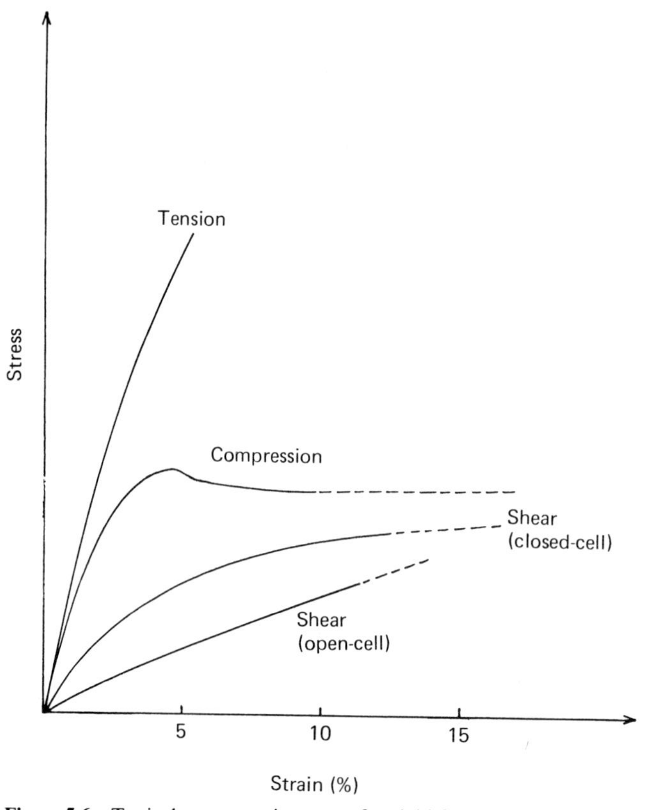


**Figure 5.6** Typical stress–strain curves for rigid foams.

Very little work has been done on shear properties of foams but a good indication of their elastic behaviour can be obtained recalling that a shear stress, $\tau$, can be split into two components, one tensile, one compressive, acting in planes at 45° to the direction of the shear stress (*Figure 5.7*). For open-cell structures, struts parallel to the $x$ or $y$ direction will not contribute significantly because of rotation at the joints. Resistance to deformation is due mainly to the struts at 45° from $x$ or $y$. These will carry direct tension and compression along their lengths and struts at intermediate orientations will carry a combination of tension or compression and bending. On average, shear stress–strain curves will show a small linear portion, limited by the

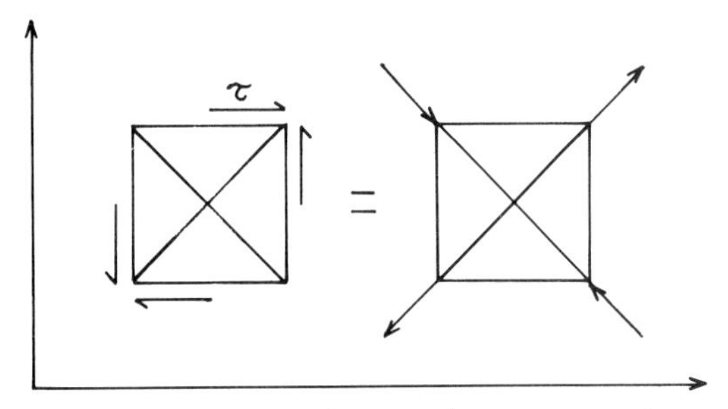


**Figure 5.7** Equivalence between shear stresses and normal stresses.

response of the struts in compression and by the smaller proportion of units at the best orientation (45°) to carry these loads. For closed-cell structures, assuming that the walls are not so thin as to have a negligible effect with respect to the edges, the cell walls will act as shear panels stiffening up the structure. Restrictions are also imposed on the bending of the panels by edge supports and a closed-cell structure will be more resistant to deformation than an equivalent open cell of the same density.

Relationships between moduli of solid substances and their foamed derivatives have been established using different kinds of models. Early work on open-cell polymeric foams was based on the behaviour of a three-dimensional distribution of rods interconnected by rigid nodes. Assuming that, for small strains, bending deformations are neglected, i.e. that only axial tension and compression are carried by the struts, the elastic properties of such a network structure can be related to the Young's modulus of the solid. Calculations made by Cox (1952) on fibrous mats show that the foam Young's modulus $E_f$, shear modulus $G_f$ and Poisson's ratio $\nu_f$ are given by:

$$E_f = \frac{1}{6} E_s V_s \tag{7}$$

$$G_f = \frac{1}{15} E_s V_s \tag{8}$$

$$\nu_f = 0.25 \tag{9}$$

where $V_s$ is the volume fraction of solid base substance and $E_s$ its Young's modulus. In practice:

$$V_s \simeq \frac{\rho_f}{\rho_s} \tag{10}$$

for solid densities between 1 and 2 and volume fractions between 0.2 and 0.8 which cover most food products; hence the elastic properties of the foam can be given in terms of its relative density:

$$E_f = \frac{1}{6} E_s \left(\frac{\rho_f}{\rho_s}\right) \tag{11}$$

$$G_f = \frac{1}{15} E_s \left(\frac{\rho_f}{\rho_s}\right) \tag{12}$$

$$\nu_f = 0.25 \tag{13}$$

According to this model, foam moduli increase linearly with density. The factors 1/6 and 1/15 in Eqns (11) and (12) arise from the three-dimensional statistical distribution. Other authors, by taking detailed geometry into account, have obtained similar results. Gent and Thomas (1963), for example, arrive at the following expression for Young's modulus:

$$E_f = \frac{1}{6} E_s (nAl_0)(1 + \frac{D}{l_0})^2 \tag{14}$$

where $nAl_0$ is the volume fraction of struts with undeformed length $l_0$ and cross-sectional area $A$. $D$ is the diameter of rigid, sphere-like junctions between struts which introduces a small correction factor.

Putting $\beta = D/l_0$, it can be shown (Hilyard, 1982) that:

$$nAl_0 = \frac{3\beta^2}{(1+\beta)^3} \tag{15}$$

$$\frac{\rho_f}{\rho_s} = \frac{3\beta^2}{(1+\beta)^3} \frac{\beta^3}{(1+\beta)^3} \tag{16}$$

For low density foams ($D \ll l_0$)

$$\frac{\rho_f}{\rho_s} \simeq 3\beta^2 \tag{17}$$

and Eqn (14) becomes:

$$E_f \simeq \frac{1}{6} E_s \left(\frac{\rho_f}{\rho_s}\right) \tag{18}$$

which is practically identical to Eqn (11).

*Note*: Eqns (11) and (18) do not yield $E_f = E_s$ when $\rho_f = \rho_s$. This is because the underlying assumptions on which these models are based do not apply at high relative densities and Eqn (4) should be used instead.

Closed-cell systems are more difficult to analyse with these simple assumptions because the plate elements introduce considerable complexity and it may also be necessary to take into account the stabilizing effect of internal gas or liquid pressure. These aspects will be discussed in greater detail in the section on liquid-filled foams.

So far, bending deformations have not been included and this restricts the prediction of elastic properties to very small strains indeed, limiting the value of these models. Corrections due to bending of struts in three-dimensional random distributions have been considered by Menges and Knipschild (1982) who arrived at the following expression for the foam Young's modulus:

$$E_f = K_1 E_s \left(\frac{\rho_f}{\rho_s}\right)^2 \frac{1}{\left(\frac{\rho_f}{\rho_s}\right) + K_2} \qquad (19)$$

where $K_1$ and $K_2$ are constants. Gibson *et al.* (1982), on the other hand, have based their calculations almost entirely on bending of struts and plates in regular geometrical arrangement. Their detailed analysis leads to the result:

$$E_f = K E_s \left(\frac{\rho_f}{\rho_s}\right)^2 \qquad (20)$$

where K is a constant, which shows the same relationship on relative density as Eqn (19).

The important thing to note is that when bending deformations are considered, the Young's modulus of the foam varies as the square of the relative density. This is due to the bending stiffness of rods and plates which depends on the cube of their thickness.

In conclusion, elastic moduli of rigid foams, open- or closed-cell, are controlled by (a) the spatial arrangement of elements, (b) the bending stiffness of rods and plates, (c) the elastic moduli of the base substances.

The models proposed in the literature give, for small strains, the same Young's modulus in tension and compression. The shear modulus, which depends on the same kind of deformation mechanisms operating in tension or compression, is related to foam density in the same way as Young's modulus. In each case, the range of elastic reversible behaviour will depend on the nature of the basic material (ductile, brittle) and on the specific failure modes of the foam as a structure.

### *Elasticity of flexible foams*

When the basic substances which are foamed have rubber-like elasticity, large reversible deformations are possible before fracture. For small strains the elastic moduli of flexible foams can be predicted by the same methods discussed earlier, allowing for the fact that bending will dominate. Typical stress–strain curves are shown in *Figure 5.8*.

There are some differences with rigid foams, especially in tension; the initial linear part, where Young's modulus can be defined, is followed by a region of increasing modulus with strain. This can be explained by the strain-induced reorientation of units in the direction of the applied load (*Figure 5.9*). Elements which started by carrying loads in bending can now carry them in direct tension, increasing the stress which can be carried by the foam more than in proportion to strain.

Reorientation effects can be described by probability density functions which change the probability of finding a segment at a particular angle as a function of strain. Several such functions can be found in the literature: for example, affine deformation functions (Kuhn and Grun, 1942) used in rubber elasticity, and many more (Mardia, 1972). Compressive and shear behaviour are qualitatively similar to those observed in rigid systems. Non-linear deformations are reversible because the bending and buckling of struts and plates which follow the linear response can now be accommodated by the rubber-like substances without fracture or yielding.

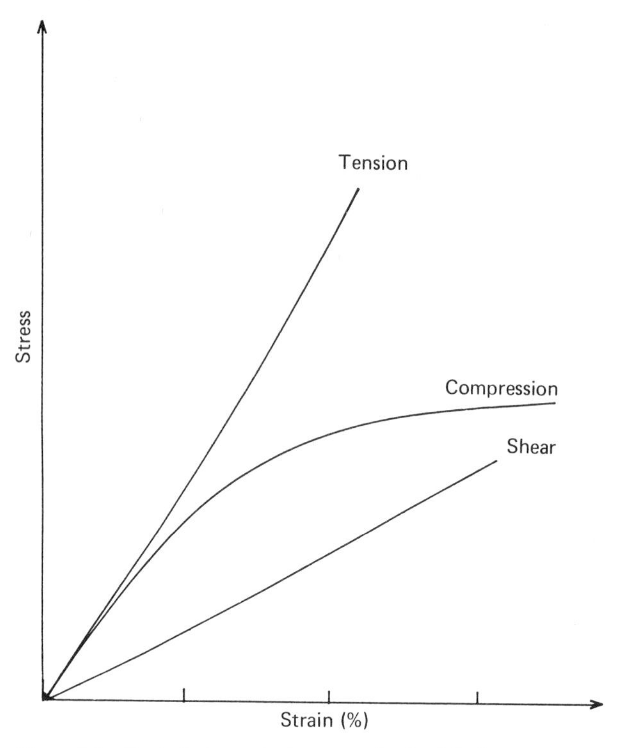


**Figure 5.8** Typical stress–strain curves for flexible foams.

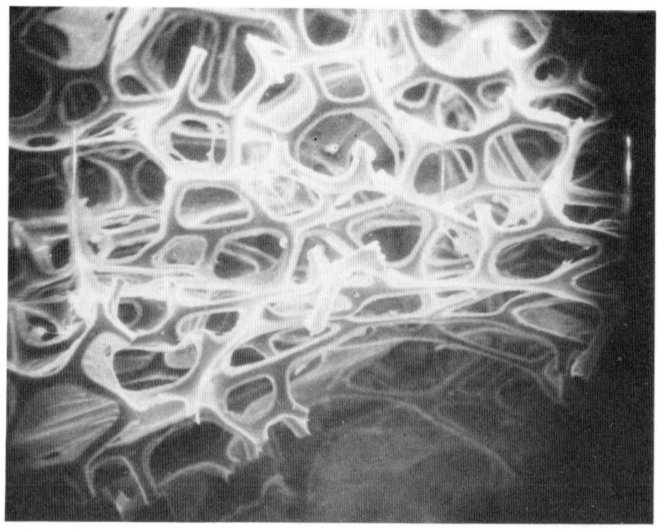

**Figure 5.9** Reorientation effects in tension in a flexible open foam.

## Strength properties of solid foams

When dealing with fracture properties of foams, it is important to distinguish between material failures, structural failures and their relationship; buckling instabilities, for example, may well occur within the elastic range of the base substances. Geometry and strength properties of cellular structures are related to relative density according to the type of loading.

In general the tensile properties of foams are of little importance in both polymers and foods; the only thing which needs to be said is that tensile failure can only occur as a result of fracture in the basic substances. With the help of Eqn (4) it can be shown that a rigid foam made with a brittle material will have a tensile strength $\sigma_f^t$ given by:

$$\sigma_f^t = \sigma_s^t \left(\frac{\rho_f}{\rho_s}\right) \tag{21}$$

where $\sigma_s^t$ is the tensile strength of the solid. Flexible foams will show the same relationship, complicated perhaps by large strains and reorientation effects. In ductile foamed materials post-yielding deformations will also change the geometry very considerably before fracture starts. Detailed analysis of these systems is very difficult but, as a rule, tensile strength will depend on the amount of substance which carries the stress and hence Eqn (21) will hold as a first approximation.

In relation to food texture, compressive and shear strength properties of foamed materials are more relevant from a theoretical and practical point of view. Bending and buckling behaviour of struts and plates are now the critical factors which must be taken into account and which, depending on the properties of base substances, can lead to three types of failures (Ashby, 1983):

1. Elastic instability (elastic buckling).
2. Plastic collapse.
3. Brittle collapse.

Elastic collapse occurs when slender columns and thin plates in compression such as cell edges and walls of cellular foams fail by buckling at stress levels well below the true compressive strength of the materials.

For both struts and plates, the buckling load $P_{crit}$ is given by:

$$P_{crit} = k\frac{E_s I}{l^2} \tag{22}$$

where $E_s$ is the Young's modulus of the solid substance, $I$ the second moment of area of the element in compression, $l$ its length and k a constant which depends on end restraints.

$I = \frac{h^4}{12}$ for struts of square section and thickness $h$

$I = \frac{h^3 w}{12}$ for plates of thickness $h$ and width $w$

Expressions for the second moment of area of other geometries of cross-section can be obtained from first principles or evaluated numerically.

The strong dependence of $I$ on thickness of strut or plate is one of the most important factors in buckling analysis. Examples of buckled struts and plates in foams are given in *Figures 5.10* and *5.11*

Detailed calculations for elastic instabilities in cellular foams have been carried out by Ashby (1983) and Matonis (1964). Apart from minor details, the results obtained by these authors show the same dependence of foam buckling stress, $\sigma_f^{cr}$, on Young's modulus and relative density:

$$\sigma_f^{cr} = K'E_s\left(\frac{\rho_f}{\rho_s}\right)^2 \quad \text{for open-cell foams} \tag{23}$$

$$\sigma_f^{cr} = K''E_s\left(\frac{\rho_f}{\rho_s}\right)^3 \quad \text{for closed-cell foams} \tag{24}$$

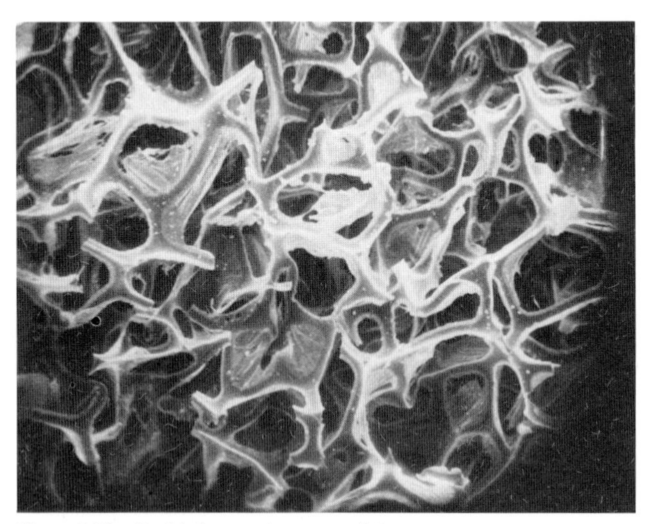

**Figure 5.10** Buckled struts in open-cell foam.

K′ and K″ are constants which depend on the geometrical details of the models chosen.

Experimental results obtained with polymer systems are in good agreement with theory but very little experimental evidence is available for foamed food products.

Post-buckling behaviour, which controls the second stage of compressive failure in low density foams, depends on the mechanical properties of the solid substances. Brittle materials will fracture soon after buckling, ductile ones will deform irreversibly and rubber elastic ones will show high reversible deformations.

Plastic collapse takes place when the materials show ductile behaviour and when bending or buckling deformations produce strains in the struts or plates beyond the elastic limit. In this case it has been shown by Ashby (1983) that the compressive strength of the foam, $\sigma_f^y$, depends on the yield strength of the solid, $\sigma_s^y$, according to:

$$\sigma_f^y = \alpha\sigma_s^y\left(\frac{\rho_f}{\rho_s}\right)^{3/2} \tag{25}$$

where $\alpha$ is a constant.

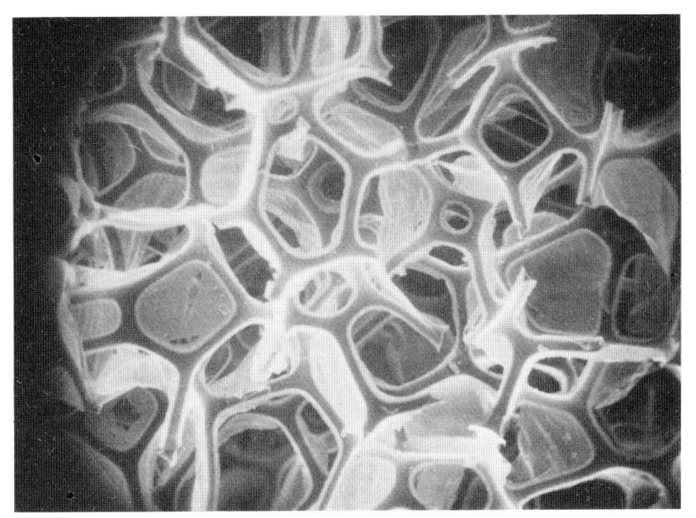

**Figure 5.11** Cell wall buckling in a partially closed foam.

Brittle collapse of failure by crushing occurs in brittle systems when bending or buckling of the foam elements induce tensile stresses comparable to the tensile strength, $\sigma_s^t$, of the base substance. The foam crushing strength, $\sigma_f^c$, is given by

$$\sigma_f^c = \alpha' \sigma_s^t \left( \frac{\rho_f}{\rho_s} \right)^{3/2} \tag{26}$$

where $\alpha'$ is a constant. The similarity between Eqn (25) and Eqn (26) arises from the bending deformations which lead to the same expression for the maximum stress.

Virtually no work has been done on food-based foams to verify the above theoretical predictions with experiment. Materials properties (modulus, strength), process variables (relative density) and failure mechanisms (buckling, yielding) need to be identified and characterized in foam systems relevant to the food industry in order to assess the applicability of the present theories and to relate texture to mechanical behaviour.

## Liquid-filled foams

In broad, general terms the models and theories described earlier will apply to the cellular structures of fruits and vegetables. In this sense they can be considered liquid-filled foams with closed-cell geometry. The cells themselves are either polyhedral, as in apples, tomatoes, potatoes, etc. or elongated cylinders, as in the stalks and stems of many plants.

Water and turgor pressure are the additional factors which need to be taken into account because they have a significant effect on deformation and failure modes at the cell walls. To all intents and purposes water can be considered incompressible and its presence will affect the response of the cellulose skeleton of the foam. Its main effect is the stabilization of the cell walls against buckling; as flat plates or cylindrical shells, their buckling loads are increased by lateral forces acting all along the length on both sides and normal to the surfaces (*Figure 5.12*). The water filling the cells provides this

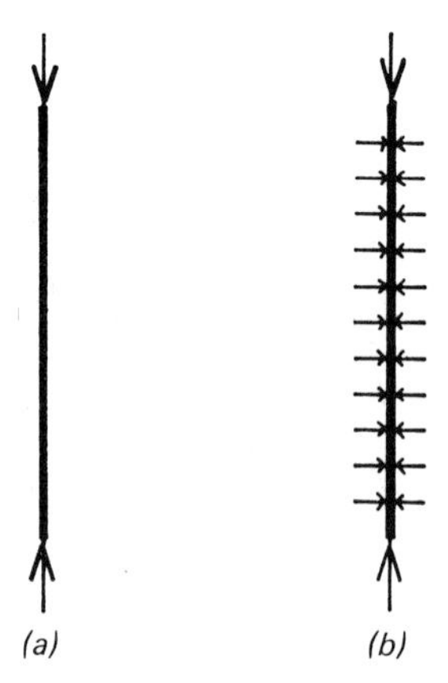


**Figure 5.12** Unsupported and laterally supported struts under axial load. Strut (b) is stabilized by the laterally distributed forces and has a higher buckling load than (a).

supporting effect. This complex phenomenon has been studied in engineering structures (Timoshenko and Gere, 1961) and could provide the basis for modifying the results obtained in gas-filled systems.

The effect of turgor pressure is perhaps even more important in terms of its magnitude. The pressure inside the cell is balanced by tensile stresses in the walls; this is similar to the equilibrium between surface tension and gas pressure in bubbles. The tensile stresses in the cellulosic walls can be very high indeed, depending on the magnitude of the pressure. Direct measurements of turgor pressure are difficult but indirect evidence suggests that values of the order of 1–3 MPa are quite common. The tensile stress in a spherical pressure vessel is given by:

$$\sigma_{max} = \frac{PR}{2t}$$

which is the same as the longitudinal stress in a closed cylinder. $R$ is the radius, $t$ the thickness and $P$ the internal pressure. Assuming an average radius of 50–100 × $10^{-6}$ m and a wall thickness of the order of 5 × $10^{-6}$ m, the calculated tensile stresses in the wall are of the order of 30–60 MPa which are very high indeed. The stresses in the cellulose fibres themselves are probably higher still considering that they represent only a fraction of the hydrated cell wall material. When compressive loads are applied to the foam, buckling of the cell walls will occur at higher loads than in the absence of turgor pressure because the tensile stresses in the wall have to be offset before the walls experience the compressive stresses which lead to instability. However, deformation of the turgid cells will be resisted by the incompressible liquid and this can overstress the walls until rupture in tension occurs, even if the applied load is actually compressive. Tensile crack propagation may also take place in the region between adjacent cell walls (middle lamella), depending on turgor pressure levels and loading geometry.

Depending on the interplay between applied loads, internal turgor pressure and cell geometry, several failure mechanisms are possible in liquid-filled foams: cell wall buckling, cell wall rupture, delamination in the middle lamella. The exact combination of factors which determines actual failure modes and transitions between them is still largely unknown but they are important in relation to textural attributes, such as crispness, and to damage during handling or storage. Some work in this area has been done for apple and potatoes (Holt and Schoorl, 1983) and for unlignified plant material (Jeronimidis and Vincent, unpublished data), but a great deal more research is needed.

## Conclusions

In spite of their complexity the mechanical behaviour of foamed structures can be analysed in terms of deformation and strength characteristics of sub-elements such as cell edges, cell walls, etc. Bulk foam properties can be related to base material properties and to relative density using well established strength of materials and engineering concepts. This approach has been verified for a number of polymer- and metal-based systems and there is no reason why it should not be applicable to the wide range of foams which exist in natural and processed foods. The difficulty lies in the proper measurement of relevant variables and properties. Even simple loading geometries, such as compression between two platens or indentations with a ball, produce complicated states of stress.

Available theories can already improve our knowledge of many foamed food systems if proper experimentation and identification of controlling factors can provide the necessary information.

Other aspects of foam mechanics have hardly been investigated at all, e.g. stability and propagation of buckling deformations, transitions between different failure modes, interactions between fluids and solid structural elements in liquid-filled systems. All these phenomena are relevant to textural attributes which have often been correlated to mechanical measurements but seldom interpreted in terms of detailed mechanical behaviour.

## References

ASHBY, M.F. (1983). The mechanical properties of cellular solids. *Metallurgical Transactions*, **14A,** 1755–1769

COX, H.L. (1952). The elasticity and strength of paper and other fibrous materials. *British Journal of Applied Physics*, **2,** 72–79

GENT, A.N. and THOMAS, A.G. (1963). *Rubber Chemistry and Technology*, **36,** 597

GIBSON, L.J., ASHBY, M.F., SCHAJER, G.S. and ROBINSON, C.I. (1982). The mechanics of two-dimensional cellular materials. *Proceedings of the Royal Society of London*, **A382,** 25–42

GUTH, E. (1945). Theory of filler reinforcement. *Journal of Applied Physics*, **16,** 20–25

HARDING, R.H. (1967). *Morphologies of Cellular Materials, Resinography of Cellular Plastics*, ASTM STP 414. p. 3. American Society for Testing of Materials

HILYARD, N.C. (Ed.) (1982). *Mechanics of Cellular Plastics*, pp. 73–97. London, Applied Science

HOLT, J.E. and SCHOORL, D. (1983). Fracture in potatoes and apples. *Journal of Materials Science*, **18,** 2017–2028

KUHN, W. and GRUN, F. (1942). Beziehungen zwischen elastichen Konstanten und Dehnungsdoppelbrechung hochelasticher Stoffe. *Kolloid Zeitschrift*, **101,** 248–271

MARDIA, K.V. (1972). *Statistics of Directional Data*. London and New York, Academic Press

MATONIS, V.A. (1964). Elastic behaviour of low density rigid foams in structural applications. *Society of Plastics Engineers Journal*, 1024–1030

MENGES, G. and KNIPSCHILD, F. (1982). Stiffness and strength—rigid plastic forms. In *Mechanics of Cellular Plastics*, (Hilyard, N.C., Ed.), pp. 27–72. London, Applied Science

MULLINS, L. and TOBIN, N. (1965). Stress softening in rubber vulcanizates. Part I. Use of strain amplification factor to describe the elastic behaviour of filler-reinforced vulcanized rubber. *Journal of Applied Polymer Science*, **9**, 2993–3009

PRINS, A. (1976). Dynamic surface properties and foaming behaviour of aqueous surfactant solutions. In *Foams*, (Akers, R.J., Ed.), pp. 51–60. London, New York, San Francisco, Academic Press

TIMOSHENKO, S.P. and GERE, J.M. (1961). *Theory of Elastic Stability*. New York, McGraw-Hill

6

# THE POLYMER/WATER RELATIONSHIP—ITS IMPORTANCE FOR FOOD STRUCTURE

P.J. LILLFORD
*Unilever Research, Colworth House, Sharnbrook, Bedfordshire, UK*

## Introduction

Food science and technology contains many references and descriptions of the interaction of water with polymers (and small molecules), and a comprehensive terminology has developed to describe the assorted phenomena. Thus we are concerned with sorption, hydration, plasticization, swelling, gelation, binding and holding of, or by, water at various levels and by a range of structures ranging in size from the molecular to the macroscopic.

Attempts to relate all of these phenomena have not been entirely successful, partly because of the difficulty in measuring the interactions, but mostly because simple relationships between water and polymer are hardly likely to emerge, when theories relating the unique interaction of water molecules with each other have proved so difficult to substantiate[1].

This chapter attempts to distinguish the interactions between water and polymers in large three-dimensional structures at high water content, where microscopic dimensions dominate and the sorption of water at low levels, where molecular structure and conformation play a more significant role.

In the study of water, polymers and food, during the last 15 years various 'experts' have claimed that polymers may bind up to ~ 500 times their own weight of water (gels), or none at all. Such findings certainly generate interest but as far as foods are concerned they are not of themselves important.

Their importance is a consequence of the medical, commercial and economic facts that without appropriate control of the interactions of water with food ingredients in all their complexity, we would die or become seriously ill by the action of microbial spoilage, we would create inedible or unappetizing products and the manufacturing industry could neither effectively operate its equipment nor maintain profitability.

We can only deduce, therefore, that the subject is important, and the science complicated. Otherwise we should have settled things by now!

## Definitions

Some of the complications in the study of polymers and water are self inflicted, particularly with the uses of terminology. Labuza (1985) has pointed out that the terms 'bind' and 'bound' have five definitions in Webster's Dictionary, all of which are recognizable English usage and all, unfortunately, have been used in the scientific description of the behaviour of water. They are:

1. To make secure by tying up.
2. To confine, restrain or restrict as if with bonds.
3. To wrap around or enclose over.
4. To hamper free movement or natural action.
5. To produce cohesion in loosely assembled substances.

It is worth pursuing this literary definition further. For example, we would all agree that a prisoner wearing a strait-jacket and handcuffed to the bars of a cage could be considered 'bound'. If any two of these constraints were removed the prisoner would still be bound, in that free movement would still be restricted. Nonetheless we could all agree that in each case the *mechanism* for restriction was different and the nature of the time-dependent motions of the prisoner would, in each, be different. Unfortunately, the observation techniques applicable to 'water binding' allow us to observe at best only the motion or the transient position of the molecules, or at worst a deviation in behaviour from that of the pure substance. We are then free to deduce the mechanism of binding and it is here where differences and uncertainties arise.

With these ideas in mind, let us now briefly survey the various experimental approaches and results obtained when measuring polymers and water.

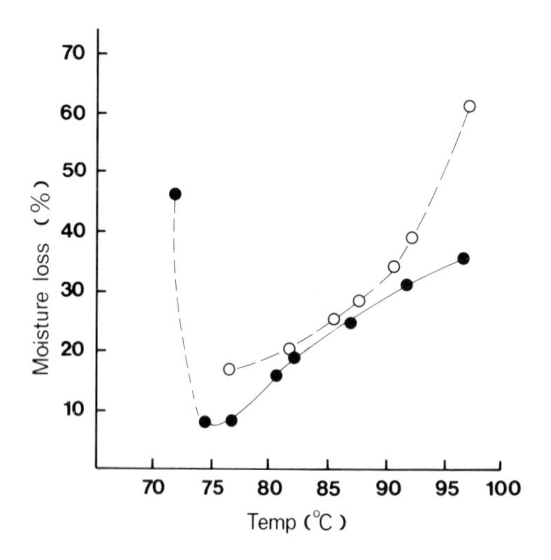


**Figure 6.1** Moisture loss under low speed centrifugation of 5% blood plasma gels. ●——●, pH 9; ○- - -○, pH 7. Redrawn from Hermansson and Lucisano (1982).

## High water content systems e.g. protein and polysaccharide gels and dispersions ($a_w > 0.9–1.00$)

### *Mechanical measurements of 'water holding'*

The bulk properties of the water in these systems appears little changed from its behaviour as a pure material, yet its ability to flow as a liquid is clearly hampered. The simplest explanation is that the water is 'encaged' i.e. free to move over many molecular dimensions but restrained in a structure by capillarity or impermeable barriers. Diffusion measurements of water or dissolved solutes show the latter is not dominant in man-made structures so that capillarity and surface tension appear to be the obvious parameters determining water retention. These forces are summarized in terms of a capillary suction pressure ($P$). For complete wetting, the $\Delta P$ is given by

$$\Delta P = \frac{2\gamma}{r}$$

where $\gamma$ is the surface tension of the liquid phase and $r$ the capillary radius.

Neither of the crucial variables are directly measurable in most food systems but it is self-evident that:

1. Minor contaminants can dramatically influence $\gamma$.
2. $r$ is not a constant but should be represented by a distribution of pore sizes.
3. For pliant systems $r$ is a function of the applied pressure and deformation but the effects may be reversible.
4. For stiff, brittle systems $r$ is less dependent upon the applied pressure, until irreversible fracture occurs.

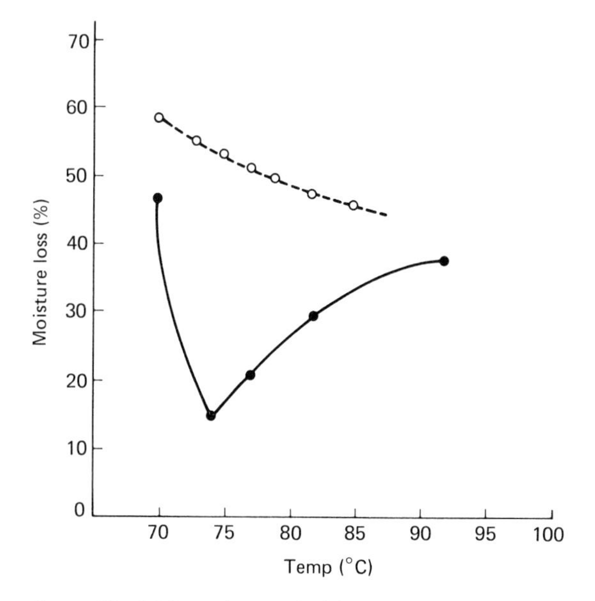


**Figure 6.2** Moisture loss under high speed centrifugation of 5% blood plasma gels. ●——●, pH 9; ○---○, pH 7. Reprinted from Hermansson (1982).

Because of the effects noted above, it is not at all surprising that confusion occurs when different tests involving externally applied pressure are used to compare different materials (Hermansson, 1982; Hermansson and Lucisano, 1982). *Figures 6.1* and *6.2* show the apparently contradictory results of two types of water holding tests on the same material, in this case a heat set gel of blood plasma. At low centrifugal forces (200–800 g) and using a chamber in which the lost moisture is separated from the sample, gels at both pH 7 and pH 9 retain less moisture as the temperature at which they are formed increases (*Figure 6.1*). At high centrifugal forces in chambers where the expelled liquid remains in contact with the sample the results for gels pH 9 are similar but completely different for gels at pH 7.

For the results to be the same from the two tests, the materials must exhibit a capillary pressure which increases very steeply as a function of network concentration and the strain relaxation times must be much longer than the experimental time. Any deviation from this behaviour results in discrepancies between the two methods. *Figures 6.3* and *6.4* show that the stress/strain behaviour of gels is very different as a function of pH and at pH 7 macroscopic failure can occur at deformations as low as 50%. Neither is it surprising that results for such 'water binding' or 'water holding' capacities for three-dimensional swollen structures do not correlate with data derived from absorption isotherms obtained under circumstances where neither the microstructure nor the molecular organization of the substrate is the same.

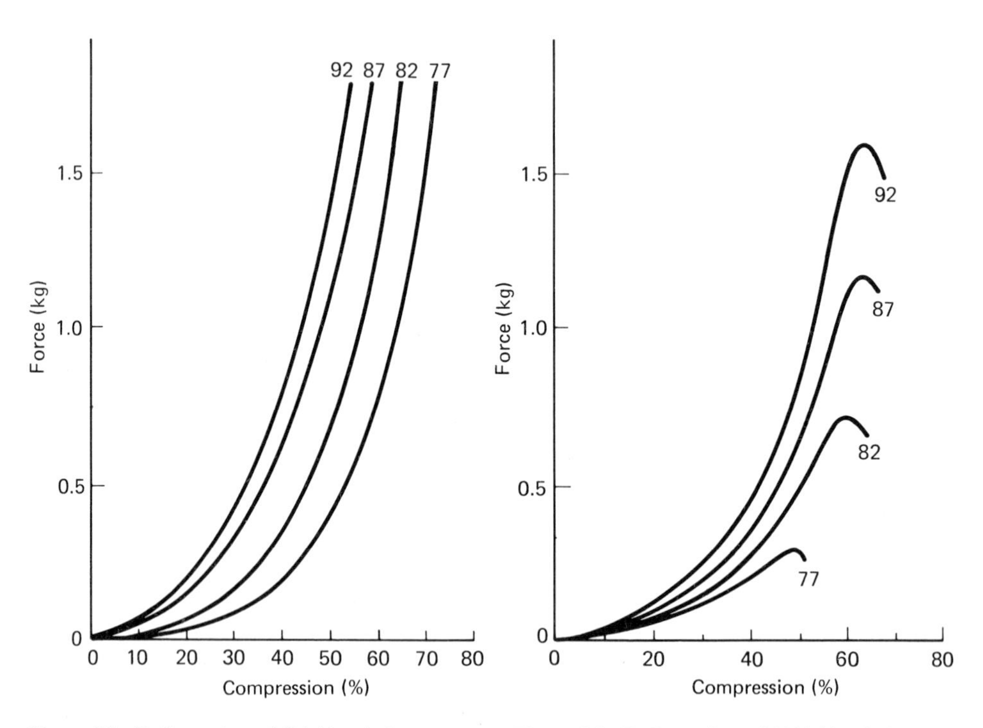


**Figure 6.3** Deformation of 5% blood plasma gels at pH 9.0. Reprinted from Hermansson (1982).

**Figure 6.4** Deformation of 50% blood plasma gels at pH 7.0. Reprinted from Hermansson (1982).

### *Nuclear magnetic resonance of water in gels*

One of the most unfortunate areas of confusion in the concept of polymer/water interactions at high water content has arisen from the uncritical application of nuclear magnetic resonance (NMR). Water protons exhibit two types of relaxation when excited in the presence of a magnetic field, denoted by their spin–spin ($T2$) and spin–lattice ($T_1$) relaxation times. The relaxation processes are influenced by the motion of the molecules but on different time scales, i.e. spin–lattice relaxation processes are sensitive to motions at megahertz frequencies whereas spin–spin processes are sensitive to much slower motions. In pure water all motions are rapid so the two time constants are the same. In the presence of dissolved, dispersed or gelled polymer, *both* relaxation times ($T_1$ and $T_2$) of water protons are reduced. Unfortunately the temptation to equate this effect with restricted motion or binding of the water has been overwhelming and NMR relaxation times and line widths ($\Delta\nu$) have been used to measure and compare water binding for widely different materials. In most cases, the assumptions and therefore the correlation of results is entirely fallacious. Most workers now accept that the observed relaxation times are a weighted average of those of the unmodified 'free' water and a small motionally modified fraction, which are in fast exchange. The observed relaxation rate may be defined as ($1/T_{obs}$). Then

$$\frac{1}{T_{obs}} = \frac{1 - P_b}{T_w} + \frac{P_b}{T_b}$$

where $P_b$ and $T_b$ relate to the motionally modified fraction and $T_w$ to free water.

Now both the quantity ($P_b$) and relaxation times ($T_b$) of the modified fractions will depend on polymer types, molecular weight, etc. so that equivalent relaxation times will only accidentally relate to the capability of the organized structure to physically retain water. The same arguments apply to line width ($\Delta\nu$) measurements since

$$\Delta\nu = \frac{1}{\pi T_2}$$

More recently we have used NMR relaxation phenomena to examine the microstructure of polymer gels and dispersions (Lillford, Clark and Jones, 1980). In heterogeneous systems, where domains greater than 50 μm are present, the water molecule *cannot* average the structural heterogeneity (*Figure 6.5*). The distribution of relaxation times can then be approximated to a distribution of pore sizes for a given substrate. As described above a measurement of pore sizes is useful information to describe the *relative* capillary suction of two structures. However, the comparisons must be restricted to similar substrates, otherwise the same complications of variable $P_b$ and $T_b$ described above can dominate the data. At best this latter use of NMR is useful to describe the undeformed structure of the material and will not necessarily relate to water retention under mechanical deformation.

The method can be used to examine the effects of organization at a microstructural level on the water retention properties of polymer structures. For example, *Figure 6.6* shows the effect of freeze/thaw cycle on spin–spin relaxation of water in a simple agarose gel. Clearly the microheterogeneity has been created by the freezing process but this is easily quantified by curve fitting the decay process. The results show (*Table 6.1*) that ~ 30% of the water is now in much larger pores than the original gel and of

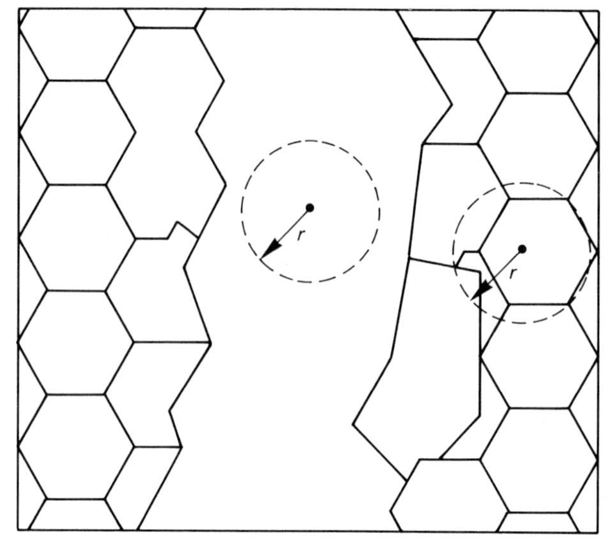


**Figure 6.5** Schematic representation of the effect of microheterogeneity on water proton relaxation.

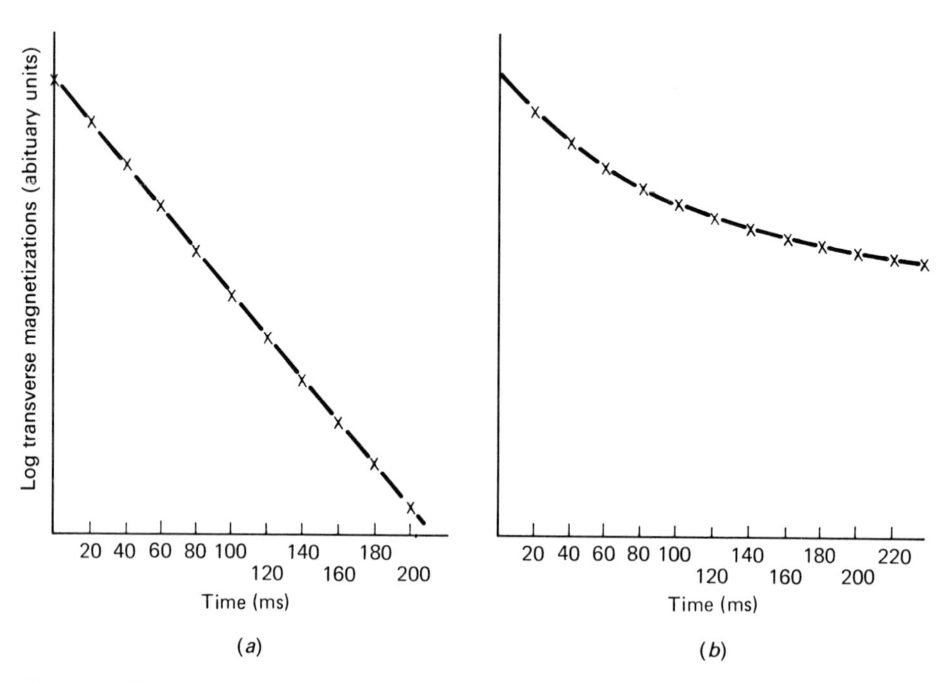


**Figure 6.6** Transverse proton relaxation in a 4.8 wt% agarose gel; (a) homogeneous, (b) freeze/thaw damaged. Reprinted from Lillford, Clark and Jones (1980).

this fraction 10–15% is unaffected by the polymer network and can be expected to synerese or drip from the sample. *Figure 6.7* shows the decay curve for water in a synthetic fibre bundle. Clearly the form of the decay curve can be used to monitor the concentration and packing of the fibres in the sample.

**Table 6.1** MULTIEXPONENTIAL ANALYSIS OF AGAROSE GELS

| *Gel state* | *No. of processes* | *Amplitude* (%) | $T_2$ (ms) |
|---|---|---|---|
| Fresh | 1 | 100 | 46 |
| Frozen/thawed | 3 | 63.3 | 37 |
| | | 22.5 | 157 |
| | | 14.2 | 1140 |
| Frozen/thawed | 4 | 11.5 | 18 |
| | | 60.6 | 47 |
| | | 16.0 | 240 |
| | | 11.9 | 1290 |

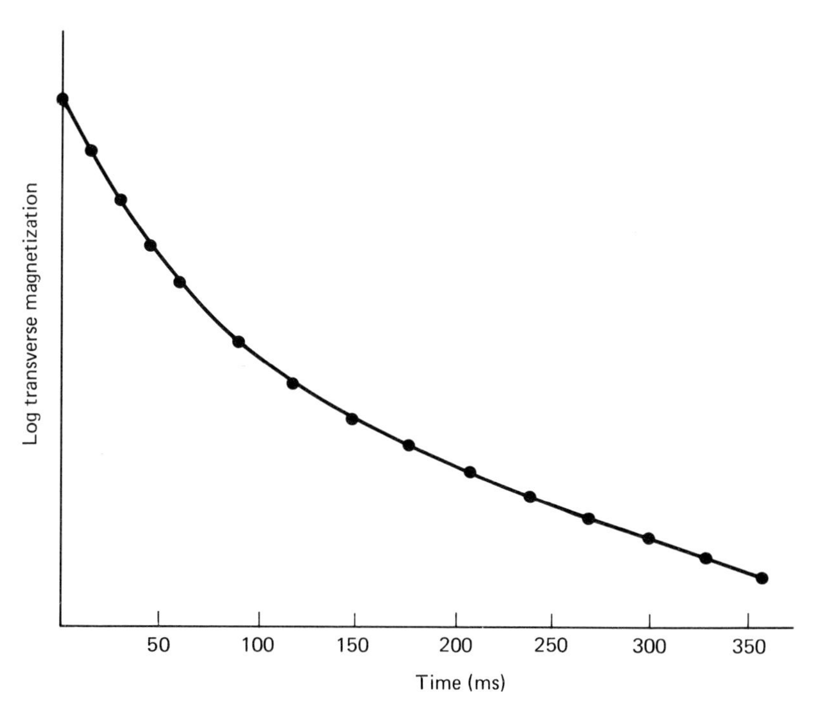


**Figure 6.7** Transverse proton relaxation in a bundle of spun protein fibres. Reprinted from Lillford, Clark and Jones (1980).

In summary, the interaction of polymers with large volumes of water and their ability to retain it must be regarded as a property of the three-dimensional microstructure, not simply of the material itself or of its molecular structure.

## Intermediate/low moisture content systems ($a_w < 0.9$)

In these systems it is less obvious to consider water as encaged in a structure; the analogy usually adopted is of water molecules 'handcuffed' or bound to the macromolecular surfaces via hydrogen bonding, electrostatic interaction, van der Waal forces, etc. Many experimental approaches have been used and provide a significant spread of results. There is no need to expect identical results since the different techniques are capable of, and should be identified as, providing different statements as to the amount and state of this 'hydration' or 'bound' water.

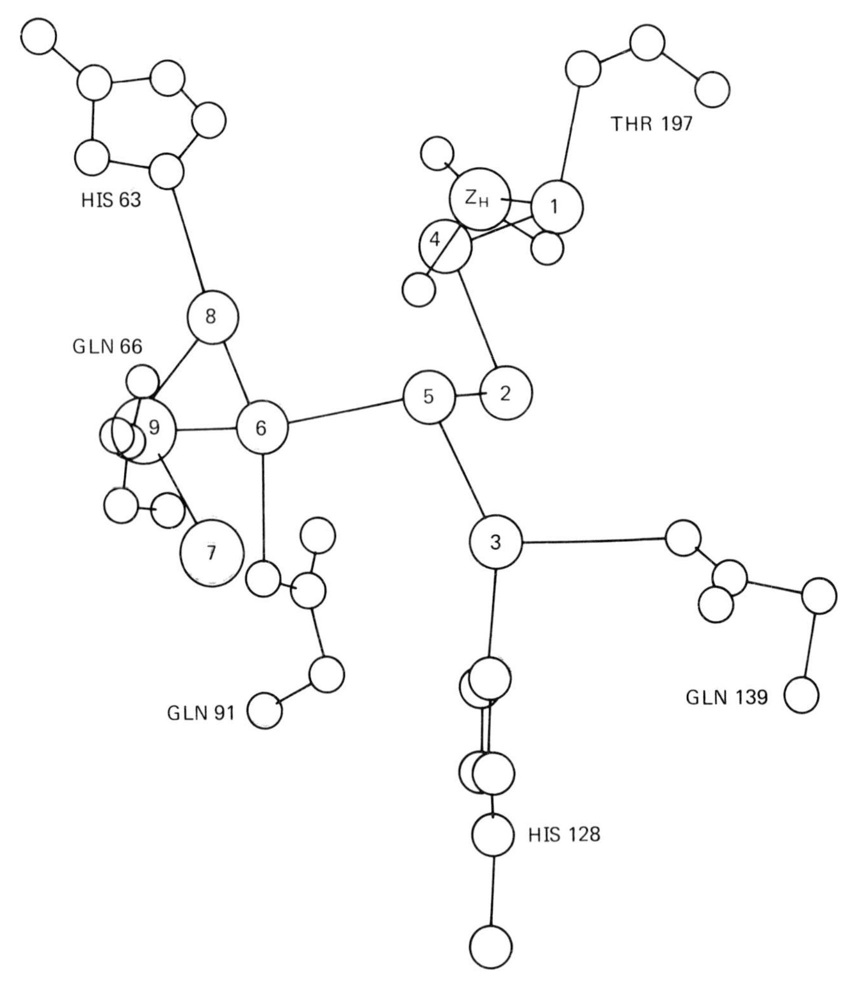


**Figure 6.8** X-ray structure of the active site region of carbonic anhydrase, showing locations of oxygen atoms from nine water molecules. Reprinted from Liljas *et al.* (1972).

Gravimetric methods of controlled adsorption or desorption tell us something of the amount of water associated with the polymer but nothing of its location or state. Calorimetric freezing studies tell us even less, only that some water is 'missing' from the normal behaviour of a simple solution or the pure solvent. Infrared or Raman spectroscopy can identify different states of water, but the presence of a specific frequency identifiable with the absorbed species implies a lifetime only slightly longer than the observation frequency. NMR, on the other hand, can detect changes in molecular motions induced by interaction at frequencies from megahertz downwards, but is incapable of separating the various molecule interactions responsible unless the lifetimes on each site are relatively long ($> 10^{-6}$–$10^{-3}$ s) which is rarely the case.

In special circumstances, structural models can be calculated from NMR relaxation provided short range magnetic fields are 'labelling' surface located water molecules. Only X-ray and neutron diffraction provide primary structural information via some spatially oriented scattering function (Liljas *et al.*, 1972). However, the information relates to a time-averaged location of water molecules, and does not imply a long occupancy of the hydration sites (*Figure 6.8*).

Let us now consider some particular results in some more detail.

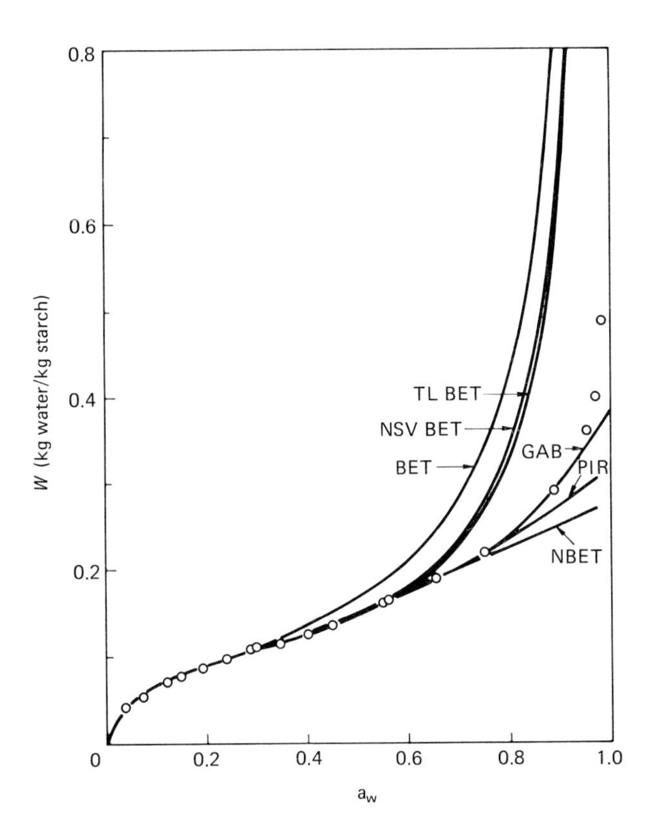


**Figure 6.9** Calculated fitting of the resorption isotherms for native potato starch. Reprinted from Van den Berg (1985).

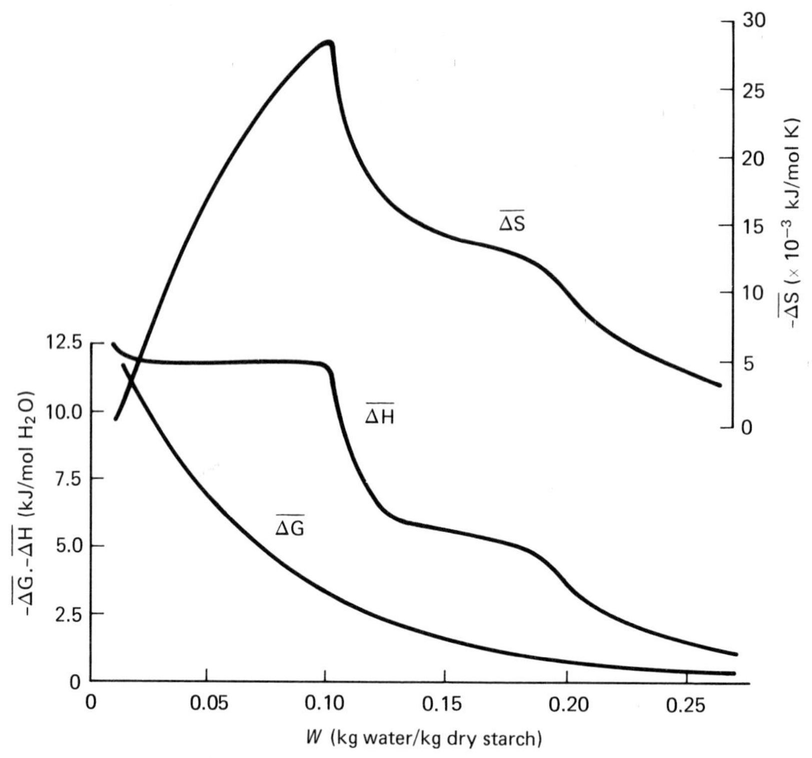


**Figure 6.10** Differential thermodynamic functions for water resorption on native potato starch. Reprinted from Van den Berg (1985).

### *Adsorption studies (Van den Berg, 1981, 1985)*

The models of adsorption are required to explain the sinusoidal form of isotherms commonly observed for polymer/water systems (*Figure 6.9*). The BET equation was the earliest to be employed and has the form

$$\frac{W}{W_B} = \frac{Ca_w}{(1 - a_w)(1 - a_w + Ca_w)}$$

where $W$ is the adsorbed water (g); $W_B$ is the monolayer coverage (g); C is the constant relating to the system and $a_w$ is the water activity. In other words the fit is via two parameters but the form of the equation allows it to describe aspects of both monolayer adsorption and subsequent adsorption on top of the first layer. In practice the experimental data are not fitted at higher values of $a_w$ ($> 0.7$) but the incorporation of another parameter allows the fit to be extended so that the equation becomes (GAB equation):

$$\frac{W}{W_B} = \frac{KCa_w}{(1 - Ka_w(1 - Ka_w + KCa_w)}$$

The incorporation of this third parameter is equivalent to the identification of *either* successive adsorption layers with different polymer affinity *or* an affinity which is itself

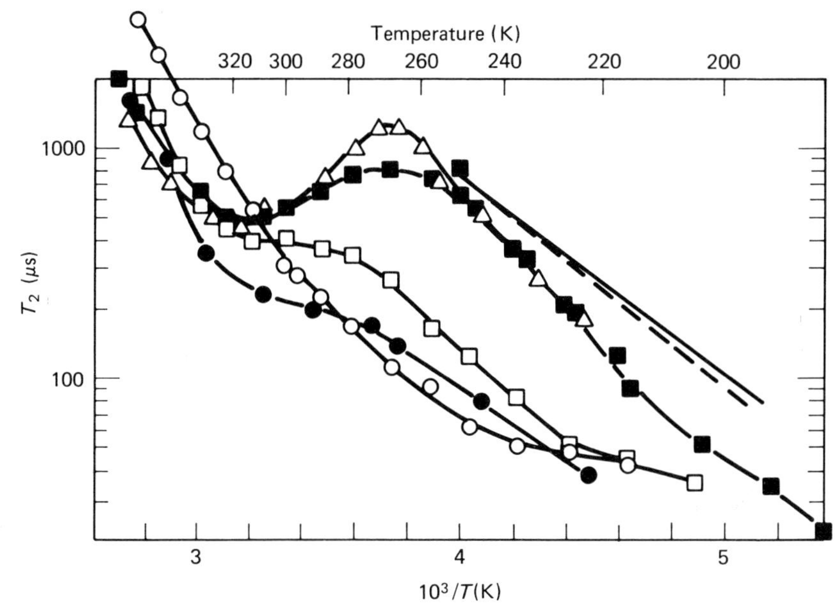


**Figure 6.11** Transverse relaxation times of water protons in agarose at low moisture contents. ○, 14% $H_2O$/agarose powder; ●, 17% $H_2O$/agarose film; □, 24% $H_2O$/agarose film; ■, 29% $H_2O$/agarose film; △, 33% $H_2O$/agarose film; – – –, 38% absorbed $H_2O$/agarose powder; ——, 10% frozen gel.

$a_w$ dependent, i.e. successive modification of the substrate by the adsorption process can be fitted.

Adsorption/desorption studies frequently exhibit hysteresis, demonstrating the presence of slow rate processes within systems of low moisture content. The adsorption arm is usually regarded as being that in thermodynamic equilibrium. Calculation of the partial enthalpies and entropies of water and adsorbant have been carried out, and for a swelling material that is plasticized by the addition of small amounts of water one can see peculiar changes in the *entropies* of substrate and water as hydration develops (*Figure 6.10*). Certainly starch, and presumably many other hydrophilic polymers are far from inert in the adsorption process.

We can measure the motional freedom of water molecules and its relative changes as adsorption proceeds. For instance, using H relaxation measurements we can see the increasing mobility of water on agar (Ablett *et al.*, 1978). Above 10% moisture at least two states of water are detected with exchange lifetimes of milliseconds (*Figure 6.11*). We can also see that this is at much lower levels of water than those where a normal freezing transition is detected. *Figure 6.12* shows that the relaxation rate of all of the non-freezable water varies linearly with water content. This suggests that during sequential attachment of water molecules the whole structure becomes more motionally flexible. Certainly a model in which water is attached sequentially to different sites on an immobile and unaffected substrate is untenable.

Experiments of a similar type on lysozyme are reported (Poole and Finney, 1984) to demonstrate the sequential hydration of particular residues (*Table 6.2*). It is interesting that the enzyme activity is detectable at water contents as low as 20 w/w%. Presumably this corresponds to a level of plasticization of the whole molecule which allows sufficient segmental motion for the catalytic events of substrate binding and

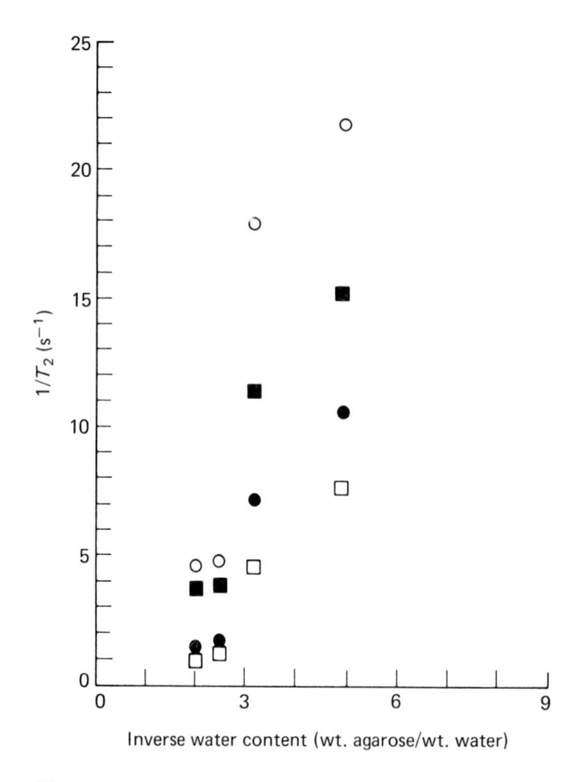


**Figure 6.12** Transverse relaxation rates of water protons in agarose at low moisture contents. ○, $1/T = 4.4$; ■, $1/T = 4.3 \times 10^3$; ●, $1/T = 4.0$; □, $1/T = 3.8$. Reprinted from Ablett *et al.* (1978).

**Table 6.2** EVENTS ACCOMPANYING THE SEQUENTIAL HYDRATION OF LYSOZYME. ADAPTED FROM POOLE AND FINNEY (1984)

| *Degree of hydration* (g water/g protein) | *Water* | *Protein* |
|---|---|---|
| 0 | Hydration of acidic groups | Ionization residues assume normal pK values<br>Protein loosens up |
| 0.1 | Acid residues saturated<br>Side chain polar group saturation | Side chain/backbone conformational shifts |
| 0.2 | Peptide NH saturation | Enzyme activity begins |
| 0.3 | Peptide C=O saturation<br>Polar group monolayer coverage<br>Apolar surface coverage | |

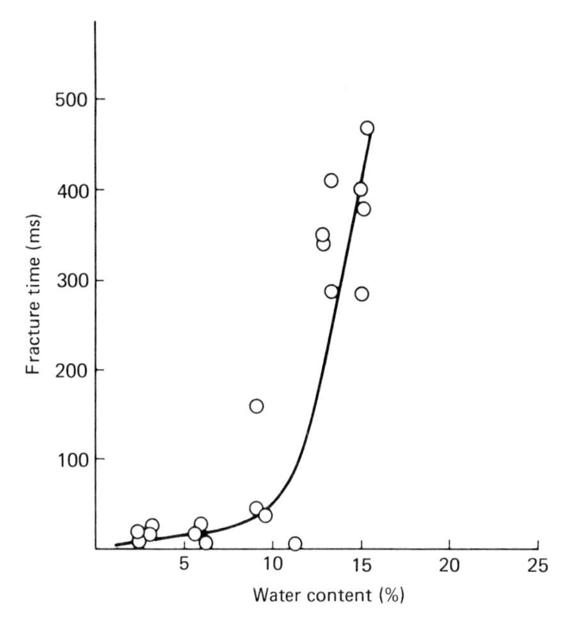


**Figure 6.13** Fracture time in a three-point bend test of a cereal food as a function of moisture content.

turnover. It would be helpful to know whether this increased plasticity is detectable in the bulk properties of the protein powder.

Water motion and bulk mechanical properties can be linked (Ablett, Attenburrow and Lillford, 1986). For example, plasticization of cereal structures coincides with abrupt increases in water molecular mobility (*Figures 6.13* and *6.14*).

The argument still rages concerning the validity of applying thermodynamics and

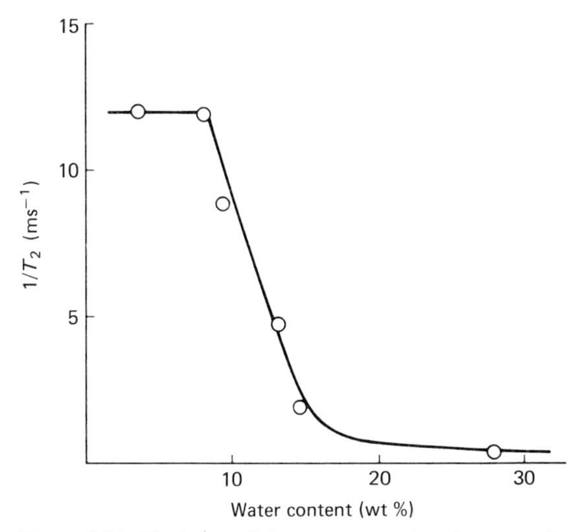


**Figure 6.14** Variation of the transverse relaxation rate of water protons in a cereal food with moisture content.

binding equilibria to adsorption phenomena. Nonetheless, application of increasingly sophisticated spectroscopic techniques such as cross polarization NMR, FT–IR and laser Raman spectroscopies will elucidate greater detail of the organization of polymer–water interaction. Despite all the criticism, the simpler studies of gravimetric analysis still provide convenient numerical classification of the types of materials and their interaction with water.

### *Desorption*

It is perfectly realistic to extend the meaning of the word 'desorption' to embrace any process by which liquid water is removed from a dispersion or solution as well as evaporation and therefore freezing may be included.

Evaporative desorption is usually recognized as being the non-equilibrium process responsible for hysteresis in the desorption/adsorption cycle. When water levels are reduced to very low levels then irreversible damage may occur. This may or may not be true when water activity is maintained at a high value throughout. For example, whereas a lysozyme solution can be dehydrated to 1 g/g water and rehydrated without apparent damage, the same treatment applied to actomyosin results in an insoluble precipitate. Similarly, whereas the sulphated iota carrageenan gels can be reversibly frozen, agarose gels exhibit massive freeze/thaw syneresis.

Because of its importance in the stability of preserved foods, 'desorption' has been very widely studied and the concept of water activity has been heavily used and correlated with mechanical and enzymic properties (Duckworth, 1975; Rockland and Stewart, 1981). The concentration effects of both freezing and drying must favour the formation of new interactions but these are very system-dependent. For example, the insolubilization of gadoid fish proteins is explained via permanent formaldehyde crosslinking which does not occur in other species. Insolubilization of oilseed proteins may involve disulphide exchange or specific $Ca^{2+}$ ion interactions. The agarose example certainly involves enhanced hydrogen bonding but is thermally reversible. It is probable that some of the systems so cherished by the food industry are *usually* in a metastable kinetic state so that dehydration is merely promoting a shift towards the equilibrium minimum free energy state.

Franks (1985a) has recently emphasized the significant difference between the effects of freeze-concentration consequent upon ice-crystal growth and the actual effects of lowering temperature on the shift in the equilibrium electrolyte dissociation constants; and similarly that decreased temperature may differently affect the balance between hydration and hydrophobic interactions in a food product. This aspect of polymer/water interaction must have its parallel in drying experiments, where the extent of deviation from non-ideality increases rapidly as the system is concentrated, but is rarely considered.

More often than not the studies of 'desorption' are of directly observable freezing curves or weight loss curves, and the measurements relate to *kinetically* stabilized states where the interaction of polymers and water cannot be treated in terms of equilibrium thermodynamics. Nonetheless, the amount of water (in g/g solute) which remains unfrozen is remarkably consistent in experimental terms for any given polymer. Such observation led Kuntz and Kauzmann (1974) to propose 'hydration numbers' for proteins and amino acids which produced a reasonably high degree of internal self-consistency (*Tables 6.3* and *6.4*). Furthermore when the proportion and

**Table 6.3** PROPOSED AMINO ACID HYDRATIONS BASED ON NUCLEAR MAGNETIC RESONANCE STUDIES OF POLYPEPTIDES

| *Amino acid residues* | *Hydration*[a] |
|---|---|
| Ionic | |
| Asp⁻ | 6 |
| Glu⁻ | 7 |
| Tyr⁻ | 7 |
| Arg⁺ | 3 |
| His⁺ | 4 |
| Lys⁺ | 4 |
| Polar | |
| Asn | 2 |
| Gln | 2 |
| Pro | 3 |
| Ser, Thr | 2 |
| Trp | 2 |
| Asp | 2 |
| Glu | 2 |
| Tyr | (3) |
| Arg | 3 |
| Lys | 4 |
| Non-polar | |
| Ala | 1 |
| Gly | 1 |
| Phe | (0) |
| Val | 1 |
| Ile, Leu, Met | 1 |

[a] Moles of water per mole of amino acid.

**Table 6.4** PREDICTION OF PROTEIN HYDRATION FROM COMPOSITION AND POLYPEPTIDE RESULTS

| | *Hydration* (g $H_2O$/g protein) | |
|---|---|---|
| *Protein, native* | *Calculated* | *Observed* |
| Lysozyme | 0.36 | 0.34 |
| Myoglobin | 0.45 | 0.42 |
| Chymotrypsinogen | 0.39 | 0.34 |
| Chymotrypsin | 0.36 | 0.33 |
| Ovalbumin | 0.37 | 0.33 |
| Bovine serum albumin (BSA) | 0.45 | 0.40 |
| Haemoglobin (denatured) | 0.42 | 0.42 |
| BSA + urea | 0.45 | 0.44 |
| BSA, pH 3 | 0.32 | 0.30 |

Calculation assumes that all residues are fully hydrated. This is perhaps reasonable for the denatured proteins but leads to a small positive error unless allowance is made for 'buried' groups. This correction was done for lysozyme, yielding a calculated value of 0.335.

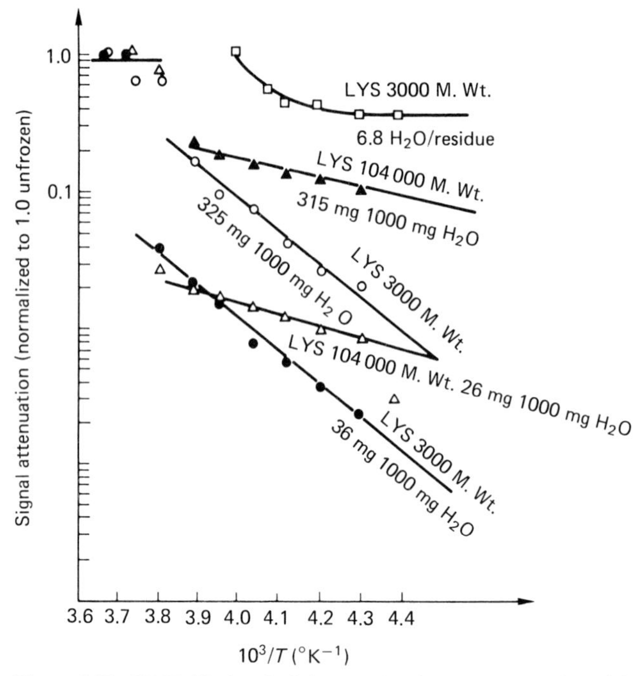


**Figure 6.15** NMR 'H signal of the non-freezing component in polylysine solutions. Reprinted from Woodhouse, Derbyshire and Lillford (1975).

relaxation times of the hydration water are used to calculate the observed, exchange averaged NMR relaxation times of other compositions (e.g. solution states or drying, reduced $a_w$ states) the agreement between that and the observed is not unreasonable. Therefore those proposed metastable states apparently provide a degree of experimental stability.

Subsequent studies have shown the sensitivity of the apparent hydration number to the measurement temperature and the molecular weight of the polymer (Woodhouse, Derbyshire and Lillford, 1975). In particular, the higher the molecular weight the greater the ability to maintain water 'unfrozen' at low temperatures (*Figure 6.15*). It has recently been argued (Franks, 1985b) that the ability of concentrated systems to form glassy states reproducibly on supercooling or during dehydration provides an appropriate explanation, i.e. the equilibrium phase diagram is traversed by a kinetically controlled glassy state which arrests crystallization of both the solvent water and the polymer systems (*Figure 6.16*). This proposal will be developed in greater detail in later chapters.

## Acknowledgements

The author wishes to express his thanks to Professor F. Franks, Professor W. Derbyshire, Dr G. Attenburrow and Mr S. Ablett for considerable patience and experimental support.

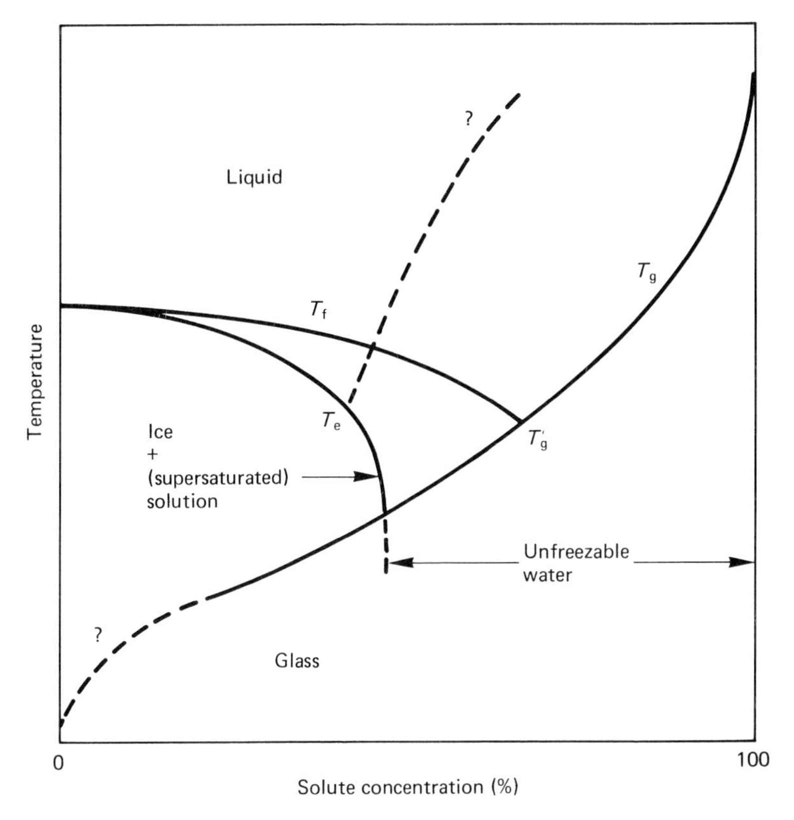


**Figure 6.16** Schematic state diagram for a water/polymer system in which the latter does not readily crystallize. Reprinted from Franks (1985a).

## References

ABLETT, S., ATTENBURROW, G.E. and LILLFORD, P.J. (1986). Significance of water in the baking process. In *Chemistry and Physics of Baking*, (Blanshard, J.M.V., Frazier, P.J. and Galliard, T., Eds), pp. 30–41 Royal Society of Chemistry Special Publication No. 56, London, Royal Society of Chemistry

ABLETT, S., LILLFORD, P.J., BAGHDADI, S.M. and DERBYSHIRE, W. (1978). Nuclear magnetic investigations of polysaccharide films, gels and sols. *Journal of Colloid Interface Science*, **67,** 355–377

DUCKWORTH, R.B. (Ed.) (1975). *Water Relations of Foods*. London, Academic Press

FRANKS, F. (1985a). Complex aqueous systems at low temperatures. In *Properties of Water in Foods*, (Simatos, D. and Multon, J.L., Eds), pp. 497–509. Dordrecht, Martinus Nijhoff

FRANKS, F. (1985b). *Biophysics and Biochemistry at Low Temperatures*. Cambridge, Cambridge University Press

HERMANSSON, A-M. (1982). Gel characteristics: compression and penetration of blood plasma gels. *Journal of Food Science*, **47,** 1960–1964

HERMANSSON, A.M. and LUCISANO, M. (1982). Gel characteristics: water binding properties of blood plasma gels and methodological aspects of water binding of gel systems. *Journal of Food Science*, **47,** 1955–1964

KUNTZ, I.D. Jr and KAUZMANN, W. (1974). Hydration of proteins and polypeptides. *Advances in Protein Chemistry*, **28,** 239–345

LABUZA, T.P. (1985). Water binding of humectants. In *Properties of Water in Foods*, (Simatos, D. and Multon, J.L., Eds), pp. 421–446. Dordrecht, Martinus Nijhoff

LILJAS, A., KANNAN, K.K., BERGSTEN, P-C., WAARA, I., FRIDBORG, K., STRANDBERG, B., CARLBOM, U., JARUP, L., LOVGREN, S. and PETEF, M. (1972). Crystal structure of human carbonic anhydrase C. *Nature, New Biology*, **235,** 131–137

LILLFORD, P.J., CLARK, A.H. and JONES, D.V. (1980). Distribution of water in heterogeneous food and model systems. In *Water in Polymers*, (Rowlands, S.P., Ed.). ACS Symposium Series, **127,** 178

POOLE, P.L. and FINNEY, J.L. (1984). Sequential hydration of dry protein: a direct difference IR investigation of sequence homologs lysozyme and α-lactalbumin. *Biopolymers*, **23,** 1647–1666

ROCKLAND, L.B. and STEWART, G.F. (Eds) (1981). *Water Activity: Influences on Food Quality*. London, Academic Press

VAN DEN BERG, C. (1981). Vapour sorption equilibria and other water–starch interactions: a physicochemical approach. Doctoral Thesis, Agricultural University, Wageningen, The Netherlands.

VAN DEN BERG, C. (1985). Development of BET-like models for sorption of water on foods: theory and relevance. In *Properties of Water in Foods*, (Simatos, D. and Multon, J.L., Eds), pp. 119–131. Dordrecht, Martinus Nijhoff

WOODHOUSE, D.R., DERBYSHIRE, W., and LILLFORD, P.J. (1975). Proton magnetic resonance in aqueous solutions of poly-L-lysine hydrobromide. *Journal of Magnetic Resonance*, **19,** 267–278

7

## POLYMER FRACTURE

D.P. ISHERWOOD
*Department of Mechanical Engineering, Imperial College of Science and Technology, London, UK*

## Introduction

The increase in usage of plastics has been one of the most significant developments in modern industrial society. Polymers, on which plastics are based, have taken the place of traditional metals and ceramics in many applications and they can be found not only in domestic products like utensils, toys and footwear but also in packaging and genuine, high integrity engineering components. Plastics have a number of features which are attractive to manufacturers which explains their widespread use. They are light, relatively cheap, resistant to many chemicals which attack metals and are easy to process directly to an intricate finished product in large quantities. It is probably true that it is this last property—ease of processing—which led to the early widespread introduction of plastic items to the marketplace rather than their mechanical service qualities. Indeed it was found that many plastic products broke after a short time, particularly if subjected to impact or fatigue loading. This led to a fairly general feeling that 'plastic' was synonymous with 'inferior' and the legacy of this is still with us today, at least to some extent.

Whilst a good deal of the problems of failure of plastic parts were attributable to poor design or material selection, the difficulties which were experienced gave an additional impetus to the search for understanding of these failures and there has been extensive research activity into polymer fracture in the last 30 years or so. As a consequence there is a large body of knowledge concerning the failure mechanisms in polymers (as well as failure generally); techniques for the toughening of polymers and for the characterization and assessment of their fracture properties have been developed accordingly.

Research has followed two identifiable, but not mutually exclusive, routes: the microscopic, in which molecular aspects are studied, and the macroscopic in which a continuum approach usually focussing on the techniques of fracture mechanics is taken. Any attempt to present a concise picture of the accumulated research evidence on polymer fracture must, of necessity, be highly selective and tend to demonstrate personal experience and preferences. In this case the relevant background has been in the fracture mechanics of thermoplastics. However, some molecular aspects will be mentioned after a brief description of the viscoelastic nature of polymers which is one property which distinguishes them from other materials. Firstly the basic microstructure is described; this defines polymers and provides a useful method for their classification.

### *Polymer structure*

Polymers are materials which consist of long chain-like molecules whose atoms are held together by primary covalent bonds. The molecules may be linear, branched or of network formation as shown in *Figure 7.1*. The usual, but arbitrary, method of polymer classification is by their physical rather than their chemical structure; more precisely, by the way in which molecules are interconnected. If this system is used, each type of polymer has certain common properties. Thus we have:

1. *Thermoplastic* polymers which are linear or branched polymers with no physical linking between molecules and are held together by relatively weak secondary bonds such as Van de Waal's forces; they melt on heating and freeze on subsequent cooling;
2. *Rubbers* which have light crosslinking and are capable of large elastic deformations; and
3. *Thermosetting* polymers (thermosets) which are heavily crosslinked conferring a rigid structure and which degrade on heating.

Thermoplastics may be of completely random structure (*amorphous* or *glassy*), or contain regions of order (*crystalline* regions) within their structure; such polymers are termed 'semi-crystalline' or simply 'crystalline'. Most fracture studies have been carried out on thermoplastics, particularly the glassy form and attention is focussed on these here.

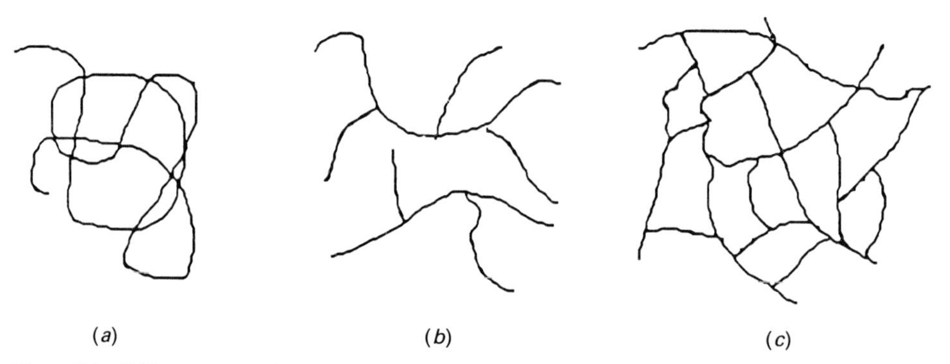


**Figure 7.1** Different types of polymer molecule: (a) linear; (b) branched; (c) network.

### *Viscoelasticity*

When an ideal, elastic solid is subjected to a given stress it suffers an instantaneous, constant deformation which is recovered upon unloading. All energy in deformation is stored and is recoverable. When an ideal, viscous fluid is subjected to stress it deforms at a constant rate and ceases to deform when the stress is removed. Energy is not stored but is irreversibly dissipated as heat. The structure of polymers is such that they have properties of both a solid and a fluid and polymers are said to be viscoelastic. The consequence of this is that polymeric solids have time dependent properties, at least to some extent. For example, when subjected to a constant load a viscoelastic solid suffers an instantaneous deformation followed by further deformation at a rate which decreases with time. This is known as *creep*. When subjected to a constant strain the same solid carries a load which decreases as time passes: this is *relaxation*.

If an oscillating stress is applied to a polymer then the response consists of an elastic component which is in phase with the applied stress and a viscous component which is out of phase with it. The ratio of out of phase to in phase components is called the loss factor, tan $\delta$. It is found that the loss factor varies with temperature and in such a way that peaks in a loss factor against temperature plot are detected. One of these, the $\alpha$ peak, occurs at the glass transition temperature which is the temperature at which a polymer changes abruptly from a glassy to a rubbery form. Peaks occurring at lower temperatures are denoted $\beta$ and $\gamma$ and are associated with local main chain or side group motion. Repeated oscillatory loading has the inevitable consequence that a great deal of energy is dissipated. Since polymers have low thermal conductivity there is an accompanying rise in temperature which may be sufficient to soften the polymer to the extent that it fails. This is *thermal fatigue*. The temperature rise may not be deleterious, however; softening can blunt sharp flaws which can arise in materials. As will be seen later this can increase the resistance to fracture. Although the topic of fatigue, both thermal and dynamic is important, no further mention is made here. Instead the comprehensive work by Hertzberg and Manson (1980) is recommended.

## Molecular aspects of fracture

### *Theoretical and measured strength*

If a polymer is to fracture then there must be a breaking either of the primary or secondary bonds, in which case there would be a 'pull-out' of molecules, or both. Cottrell (1964) showed that the theoretical strength of a solid can be estimated from the variation of interatomic potential energy with separation to give:

$$\sigma_{\text{theo}} = \alpha E \tag{1}$$

where $\sigma_{\text{theo}}$ is the theoretical fracture stress, $\alpha$ is a constant with an approximate value of 0.1 and $E$ is Young's modulus for the solid. In practice, bulk solids fracture at stresses which correspond to an $\alpha$ of the order of 0.001 or less although specially prepared materials such as glass whisker crystals can be produced to give experimental values of about 0.05. Theoretical tensile strengths of polymer molecules have been estimated by Mark (1943) and Vincent (1972) to be in the range of 6000–30 000 $\text{MNm}^{-2}$ whereas actual strengths are about one-two hundredth of these ideal values. A more realistic correlation is obtained if it is assumed that the theoretical strength is reached when the secondary bonds are broken. However, this cannot be the case for thermosetting polymers nor even for high molecular weight (long chain) thermoplastic polymers as it is thought that molecules become entangled so forming anchorage points for effective crosslinks. It is also believed that crystalline regions can act as anchorage points so ideas of fracture being due solely to the breakdown of secondary bonds are untenable despite the improved numerical agreement with experiment.

Vincent (1972) discovered a linear relationship between the measured fracture strength of 13 different polymers and the area density of backbone bonds crossing the fracture plane. In order to ensure experimental consistency, the failure stress was determined at the temperature at which there is a clear transition from ductile to brittle behaviour. This transition is important in polymer fracture and ductile materials can be induced to fail in a brittle manner not only by lowering the temperature but also by increasing the loading rate, introducing cracks and other methods. Vincent's results are reproduced in *Figure 7.2* and the evidence is clear that there is a strong influence of molecular cross-section (i.e. the number of backbone bonds per unit area) on fracture stress. However, the slope of the line is 160 times smaller than the theoretical gradient, which would seem to suggest that molecular factors do not determine failure stress levels. This apparent paradox can be resolved by arguing that flaws are present which intensify stresses locally so that when a molecule is experiencing its failure stress the ambient stress level is much smaller. It will be seen later that flaws can affect local stress levels in this way when the subject of fracture mechanics is considered. In contrast to the comparison between theoretical and experimental failure stresses, the theoretical failure *energies* required to break bonds are much smaller than the corresponding experimental values. This is usually because there is a substantial amount of energy dissipated in plastic or viscoelastic deformations which accompany the fracture process but even if this is disregarded there is still a discrepancy. In tests on highly elastic rubbers for which these dissipative energies are negligible, Lake and Lindley (1965) and Andrews and Fukahori (1977) found that the actual fracture energy was at least an order of magnitude greater than the energy to break a bond. Lake and Thomas (1967) provide an explanation for this by pointing out that a backbone bond can only be broken if *all* the bonds in the chain between crosslinks are deformed to the point of fracture. Since all the energy required to deform these bonds is lost at fracture then the energy needed to break a single bond in a fracture process is sensibly the same as the energy to break all the bonds between crosslinks. Following this argument Lake and Thomas developed a theoretical relationship to predict fracture energy and it has been found to be satisfactory. It is possible that a similar approach can be applied to the results of Vincent's work.

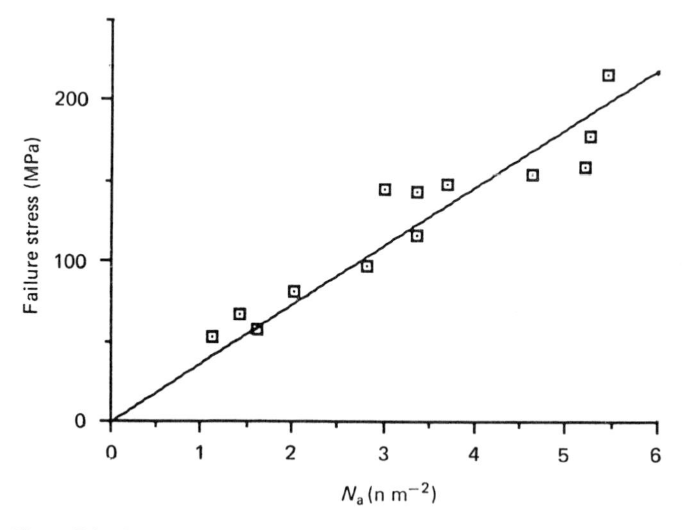


**Figure 7.2** Failure stress at ductile/brittle transition versus number of backbone bonds per unit area. From Vincent (1972).

### *Chain scission, microvoids and crazes*

Evidence that primary bonds break during the overall fracture process is provided by the results obtained using the experimental techniques of electron spin resonance and infrared spectroscopy and have been reviewed by Kausch and DeVries (1975) and Zhurkov and Kusukov (1974) amongst others. The picture that emerges is that, under the influence of stress, molecules uncoil and flow until they are prevented from doing so because of the restrictions imposed by the polymer structure. Such restrictions can be chemical crosslinks in thermosets, crystalline regions which act as anchorage points in semi-crystalline polymers or entanglements in high molecular weight amorphous polymers. Once flow is prevented a small proportion of molecules will be relatively highly loaded and will suffer chain scission, i.e. the breaking of primary bonds.

Chain scission is not coincident with bulk fracture, however, but it can be regarded as an initial stage. Subsequently, under the action of an increasing stress or strain further fractures occur, probably near to molecules already broken, and these accumulate to form *microvoids*. These microvoids can grow and coalesce to form microcracks and/or *crazes*. A craze is formed when the microvoids do not coalesce completely to form a true microcrack and can be considered to be a crack whose opposite faces are linked by ligaments of plastically deformed, oriented polymer and so, unlike a crack, a craze can carry loads. Subsequently a crack will commence and grow in a craze due to breakdown of the linking fibrils. Eventually the crack propagates through the bulk of the material preceded by a craze.

### *Kinetic theory of fracture*

Fracture on the molecular scale has been treated in the Soviet Union by Zhurkov (1965) and Zhurkov and Tomashevsky (1966) using a *kinetic theory of fracture*. In this approach, the activation energy to cause bond rupture is believed to be reduced by the application of stress. Consideration of the kinetics leads to a relationship for the rate of bond rupture which can be expressed as a time to failure of a specimen under the action of a constant applied stress and this is:

$$\ln t_f/c = (G_{AB} - \beta\sigma)/kT \qquad (2)$$

where $t_f$ is time to failure, c is a constant, $G_{AB}$ is the activation energy to cause bond rupture, $\beta$ is an activation volume, $\sigma$ is the applied stress, k is Boltzmann's constant and $T$ is temperature. Therefore a plot of the logarithm of time to failure against applied stress should be linear and experimental results for a number of polymers show that this is so. Moreover, values of the activation energy for fracture are in good agreement with the activation energy for thermal degradation as shown in *Table 7.1*.

The kinetic theory is considered to be useful for explaining crack initiation but is of

**Table 7.1** ACTIVATION ENERGIES (IN kJ/MOL) FOR FRACTURE AND THERMAL DEGRADATION FOR SOME POLYMERS

| *Polymer* | *Fracture* ($G_{AB}$) | *Thermal degradation* |
|---|---|---|
| PVC | 147 | 134 |
| Polystyrene | 227 | 231 |
| PMMA | 227 | 218–222 |
| Polypropylene | 235 | 231–243 |
| PTFE | 315 | 319–336 |
| Nylon 6 | 189 | 181 |

less assistance for dealing with crack propagation, particularly in non-brittle materials which suffer energy losses like plastic deformation, which are not accounted for in the theory. To treat propagation a macroscopic approach is favoured so that the effects of the whole system can be readily included. The central idea is that variations in behaviour at the molecular level will average out on the macroscopic scale and so the material is considered to be a continuum with consistent characterizing bulk properties. Such a treatment is usually termed *fracture mechanics* and the principles of this approach will be briefly presented.

## Fracture mechanics

### *Strain energy release rate and fracture toughness*

The stress raising effect of flaws has been mentioned in connection with the discrepancies found between Vincent's fracture strength measurements and the corresponding theoretical values. The problem of flaws was first addressed by Griffith (1920), who noticed that large glass objects usually had a lower fracture strength than small ones. Griffith proposed that all bodies contain crack-like flaws and that failure occurred at the largest of these. Since larger bodies contain more flaws, they are more likely to contain larger flaws and so their lower strength is a logical consequence. Griffith went on to study a large elastic plate, subjected to a uniaxial tensile stress, $\sigma$, containing a sharp crack of length $2a$ running normal to the line of loading. He found that if the crack were to increase in length by a small amount $2\delta a$ with no external energy input (the loading points are fixed) there is a corresponding *loss* in strain energy per unit thickness of the system equal to:

$$\pi\sigma^2 a \times 2\delta a/E \qquad (3)$$

where $E$ is Young's modulus of the material. Griffith argued that if this energy release is greater than the energy required to set up the surfaces created by the crack growth, $2\gamma$, then the crack would grow and failure of the plate would ensue. Mathematically this condition is:

$$\pi\sigma^2 a/E = 2\gamma \qquad (4)$$

This relationship worked well for the glasses studied by Griffith but was found to be a serious underestimate of the fracture strength of metals and polymers, even glassy ones (Berry, 1961). The main reason for this is that the analysis is for a perfectly elastic material which is rarely found in reality. In practice the stresses at the crack tip are so large that plastic deformation takes place and this dissipates relatively large amounts of energy. It was argued by both Orowan (1952) and Irwin (1948) that provided the plastic region was confined to a small zone at the crack tip then the calculation of the energy released per unit area of crack growth, or *strain energy release rate* (denoted G after Griffith) would still be closely approximated by the elastic analysis. However, the energy required for crack advance must account for the dissipation due to plastic flow near the tip of the flaw. Accordingly, the right hand side of Eqn (4) should be replaced by a new constant which includes both the surface and plastic work (and viscoelastic work, if necessary). This constant, denoted $G_c$, is the critical strain energy release rate, or *fracture toughness*. $G_c$ is considered to be a bulk material property, like modulus or yield stress, which is independent of the mechanical testing situation. The value of G, on the other hand, is governed both by the geometry of the specimen and the loading system. The condition for crack extension is:

$$G = G_c \tag{5}$$

This will generally imply unstable crack propagation for constant $G_c$ since G increases with crack length.

It is instructive to compare the magnitude of $G_c$ with the surface work $\gamma$ for PMMA, which is considered to be a relatively brittle polymer. Strictly $\gamma$ represents the energy to break secondary bonds whereas fracture generally involves the rupturing of primary bonds. The sum of these energies is the intrinsic fracture energy, denoted $G_0$. For PMMA, $2\gamma = 0.078\ \mathrm{Jm^{-2}}$, $G_0 = 0.5\ \mathrm{Jm^{-2}}$ and, typically, $G_c = 500\ \mathrm{Jm^{-2}}$; the overwhelming contribution due to plastic deformation is clear. Because the plastic (and viscoelastic, where relevant) work terms are so much greater than the energies associated with molecular fracture, there appears to be a gap between the macroscopic and microscopic approaches. Andrews (1974) has proposed a generalized theory to bridge this gap. He expresses $G_c$ as the sum of the intrinsic energy, $G_0$ and the other loss terms, collectively denoted $\psi$ and then shows that:

$$G_c = G_0\Phi(\dot{a}, T, e) \tag{6}$$

where $\Phi$ is a loss function. $\dot{a}$ is the rate of crack propagation and $e$ is the strain level. Andrews derives an explicit relationship for $\Phi$ which is test-dependent and has found some success with this approach for crosslinked rubbers, but it is less successful for polyethylene and polycarbonate.

***Fracture toughness measurement***

At present $G_c$ is not deduced from this type of approach but is invariably found directly by testing cracked specimens. If the strain energy release rate, G, is known for a test system, the $G_c$ (a material property) is simply the value of G at which the crack propagates (an initiation value may also be determined). A wide range of test configurations have been used to determine $G_c$, and it is necessary to find the strain energy release rate, G, for each of these as well as for practical and design situations. To treat the general situation, consider a body containing a crack of area A (and crack length $a$ for a planar body), subjected to a load P and suffering a displacement at the loading point and in the loading direction, u, as shown in *Figure 7.3a*. The crack is allowed to increase in area by an amount $\delta A$ and there is a corresponding energy input $\delta U_1$ which is composed of a combination of three parts: $\delta U_2$ (energy irreversibly dissipated); $\delta U_3$ (energy stored); $\delta U_4$ (kinetic energy). Mathematically we have:

$$\delta U_1/\delta A - \delta U_3/\delta A = \delta U_2/\delta A + \delta U_4/\delta A \tag{7}$$

The left hand side of this equation is G, whilst the right hand side has a dissipation rate, $\delta U_2/\delta A$, defined as the fracture resistance R. If R is a constant it becomes $G_c$, the fracture toughness. The remaining term contains the dynamic effects and is usually ignored except in high speed situations; inclusion of this term greatly complicates the analysis but demonstrates some fascinating results as will be shown by Williams (1986). Confining attention to linear elastic fracture mechanics (LEFM) for which the

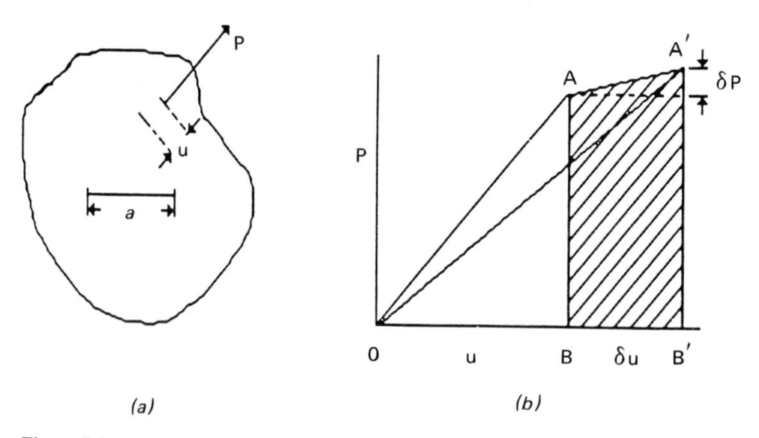


**Figure 7.3** (a) Loading of a cracked body; (b) Load deflection curves for linearly elastic cracked body.

body demonstrates wholly linearly elastic behaviour, typical load displacement curves for the system shown in *Figure 7.3a* are sketched in *Figure 7.3b*. The body is loaded along the line OA; the crack then extends $\delta A$ and the corresponding changes in load and displacement are $\delta P$ and $\delta U$ respectively. If the body were now unloaded, it would do so along the line OA′.

During the incremental crack extension there is an energy input due to the movement of the load (shaded area) equal to $(2P + \delta P)\,\delta u/2$. The stored energy before crack extension is the area of triangle OAB, i.e. $Pu/2$, and after extension the area OA′B′ which is $(P + \delta P)(u + \delta u)/2$. Therefore:

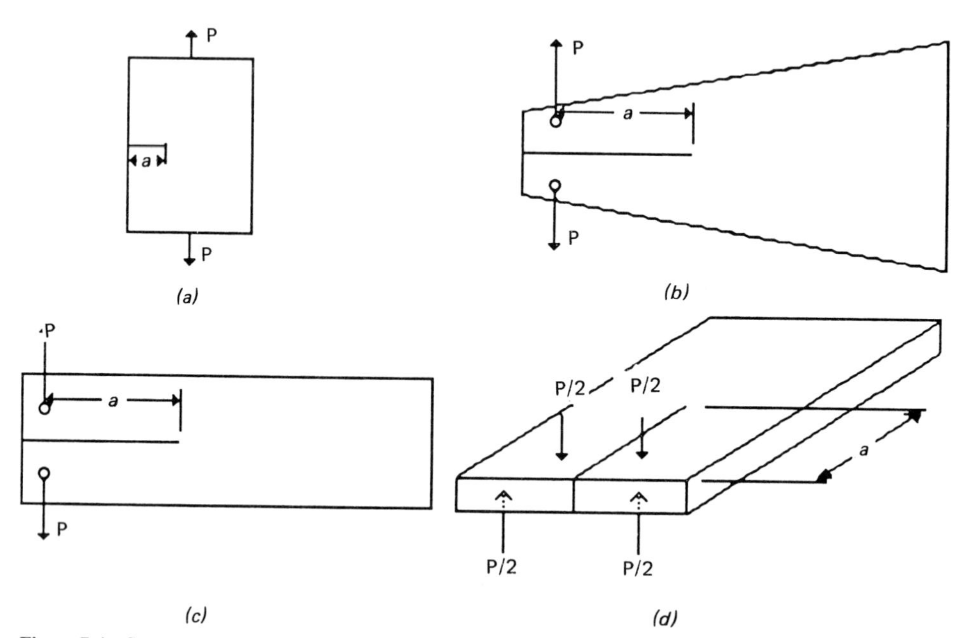


**Figure 7.4** Some common fracture toughness test specimens: (a) single edge notch (SEN); (b) double cantilever; (c) tapered double cantilever; (d) double torsion.

$$G\delta A = (2P + \delta P)\delta u/2 + Pu/2 - (P + \delta P)(u + \delta u)/2 = (P\delta u - u\delta P)/2 \tag{8}$$

This is the area OAA′ and, under fixed grip conditions (no external energy input), is simply the change in stored energy as in the original Griffith condition. Introducing the compliance (reciprocal of stiffness) of the body, C (= u/P), then:

$$G = (P^2/2)dC/dA = (u^2/2C^2)dC/dA \tag{9}$$

The compliance of a cracked specimen can be determined analytically or experimentally and hence G can be computed. Typical test configurations for rigid specimens are shown in *Figure 7.4*. The tapered cleavage (*Figure 7.4c*) and double torsion (*Figure 7.4d*) are of particular importance as they can give constant G irrespective of crack length and so stable crack growth at controlled speeds can be monitored. Some test specimens for flexible polymers are given by Kinloch and Young (1983) and Williams (1984).

### *Non-linear systems*

If the material is non-linear but still elastic then the loading and unloading lines of *Figure 7.3a* will not be straight but the same approach can be followed to give G. For example, if $P = C\,u^n$, and $n$ is constant then:

$$G = ((P^{(1+n)}C^{(1/1+n)})/1 + n)dC/dA \tag{10}$$

which reduces to Eqn (9) for $n = 1$.

Many workers use the symbol J instead of G when considering non-linear systems. This arises from work by Rice (1968) on path independent contour integrals around the crack tip in non-linear materials. Rice showed that such an integral termed the J-integral was the rate of change of energy of the system with crack growth. The use of J has been extended to crack growth in systems where the energy dissipation cannot be considered to be confined to a small region at the crack tip as in large scale yielding which gives rise to a non-linear loading curve. This is not strictly valid because the effect of unloading will be history dependent and, therefore, not unique. However, it has been found to be useful for metals under monotonic loading with the fracture criterion being $J = J_c$ and for small scale yielding $J = G$. This approach has found relatively little usage for polymer fracture but it has been applied on a number of occasions by Williams (1984).

### *Stress intensity factor, K*

An alternative to the energy methods of fracture mechanics is to consider the stress distribution at the tip of a crack. Irwin (1964) has shown that the stresses near the tip of a crack in an elastic solid under any loading system can be expressed in the form (*see Figure 7.5a*):

$$\sigma_{ij} = K(2\pi r)^{-1/2} f(\theta) \tag{11}$$

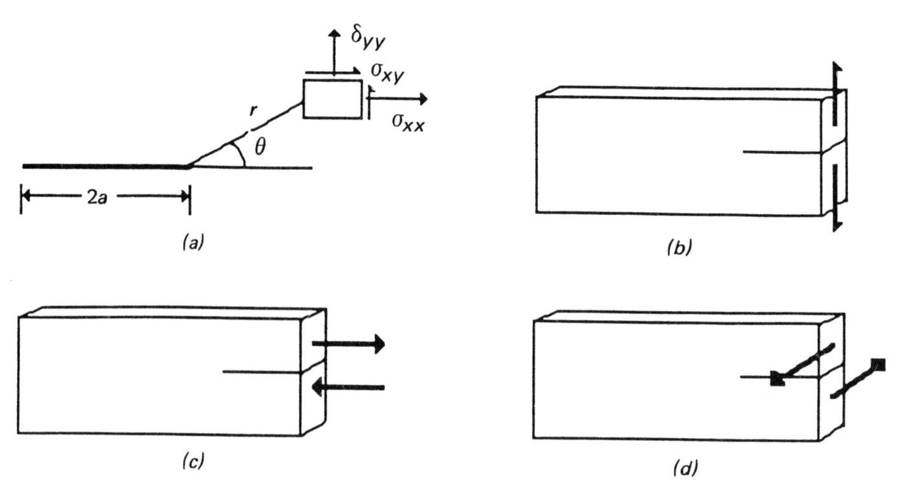


**Figure 7.5** (a) Crack tip coordinates; (b) Mode I or opening mode; (c) Mode II or in-plane shear mode; (d) Mode III or anti-plane shear mode.

where K is the *stress intensity factor* and depends on the geometry and the applied loads. This relationship clearly shows the high stresses near the crack tip and their local nature. Three types of stress intensity factor, denoted $K_I$, $K_{II}$ and $K_{III}$, are identifiable according to the loading modes as shown in *Figures 7.5b, 7.5c* and *7.5d*.

Mode I is by far the most common and if no suffix is given this mode is assumed. The fracture criterion, using this approach is that K reaches a critical value $K_c$, and it can be shown that there is a simple relationship between G and K; for plane stress, $K^2 = EG$ and for plane strain, $K^2 = EG(1 - \nu^2)$ where ν is Poisson's ratio.

## Plastic zones and crazes

### *Yielding and thickness effects*

Yielding marks the onset of plastic deformation in a polymer and is controlled by shear which causes shape changes but no volumetric change. Yielding takes place, therefore, at constant volume. Since plastic flow is a dissipative process then any increase in the extent of yielding represents an increase in the energy requirement for deformation and fracture. If a material fully yields then the yield stress will usually be a reasonable estimate of the fracture stress. On the other hand, if yielding is confined to a small region near the crack tip then the material will tend to be brittle.

The most common yield criterion for use with polymers is the Levy–von Mises criterion which may be expressed in terms of the three, mutually orthogonal principal stresses, $\sigma_1, \sigma_2, \sigma_3$ as:

$$2\sigma_y^2 = (\sigma_1 - \sigma_2)^2 + (\sigma_2 - \sigma_3)^2 + (\sigma_3 - \sigma_1)^2 \tag{12}$$

where $\sigma_y$ is the uniaxial yield stress ($\sigma_2 = \sigma_2 = 0$). The triaxial nature of the condition for yielding has important consequences for the effect of specimen thickness on fracture. A very thick specimen will cause plain strain conditions (no through thickness strain) at the crack tip so that $\sigma_3 > 0$. This means that the applied yielded stresses must be higher to yield a given volume of material. For small scale yielding the size of the plastic zone can be calculated from the elastic field; the yield zone is predicted to be lobar with the plane stress (thin sheet) zone approximately three times the size of that for plane strain. Even in thick specimens the outer edges are in plane stress so the proportion in plane strain will depend on thickness. The consequence of this is that fracture toughness, which depends on the volume of the plastic zone, decreases with increasing thickness (at least above a certain minimum thickness) to the plane strain value. Therefore, standard specimen thicknesses which depend on yield stress have been established to ensure the safest (plane strain) estimate of fracture toughness is made.

In contrast to shear yielding, crazing takes place at increasing volume because of the formation of voids in the crazed region but, like yielding, crazing is a dissipative process. Crazes cannot be modelled in the same way as described above for plastic zones, but it turns out there is a relatively simple craze model which also is applicable to plastic zones. This is the line zone model.

### *The line zone model for plastic zones and crazes*

For most polymers the plastic zones ahead of the crack are not of the lobar shape predicted by the elastic solution but appear as a thickened line ahead of the crack. This is also the form of crazes and both are well described by the line zone model shown in *Figure 7.6* developed independently by Dugdale (1960) and Barenblatt (1962) but usually named after the former. In this model a crack of length $a_0$ is considered to be internally loaded by tensile stresses over the length $(a_0 - a)$. This length is then treated as being the plastic or craze zone with the stresses being the plastic or craze stresses and the actual crack is now length $a$. The attraction of this model is that only the elastic solution for the internally loaded crack is needed to calculate the zone size and displacements. The crack tip opening displacement, $\delta$, is of particular interest as it has been suggested that when it reaches a critical value $\delta_c$, the crack extends. If the model is used to represent a plastic zone, the internal stresses are the yield stress $\sigma_y$ and the crack opening displacement for the crack geometry studied by Griffith is given by:

$$\delta = (8\sigma_y a/\pi E)\ln \sec(\pi\sigma/2\sigma_y) \tag{13}$$

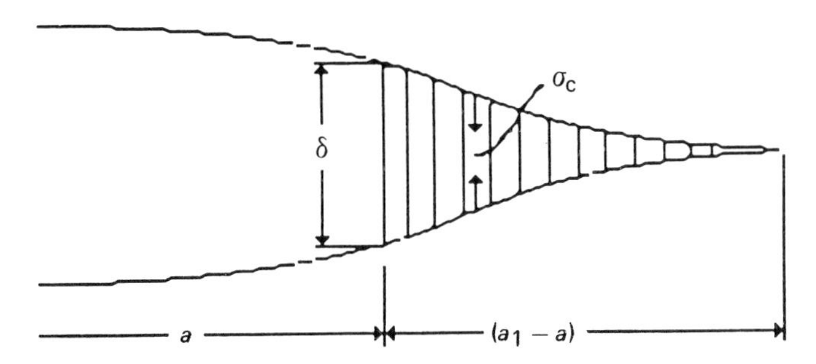


**Figure 7.6** Line zone model of plastic zone or craze.

For small scale yielding $\sigma \ll \sigma_y$ and this reduces to

$$\delta = \pi\sigma^2 a/\sigma_y E \quad (14)$$

or

$$\delta\sigma_y = G = J = K^2/E \quad (15)$$

so there is equivalence of the various fracture criteria for small scale yielding, a result which can be obtained by other methods.

If a craze stress, $\sigma_c$, is substituted for the yield stress, the displacement profile within the craze can be determined. It is found that these are in good agreement with the values measured by Morgan and Ward (1977) and Weidmann and Doll (1978). Therefore, it is concluded that the Dugdale model is a good representation of the craze zone.

## Fracture of glassy thermoplastics—PMMA

Since there are so many polymers, each having its own fracture characteristics, it is necessary to focus attention on a limited range. Historically most work has been performed on poly(methyl methacrylate) (PMMA) whose most common trade name in the UK is 'perspex'. PMMA is a glassy (amorphous) thermoplastic and is well suited to fracture studies being brittle below 80 °C and having consistent fracture properties. It is helpful to use PMMA as the principal model material to develop ideas and describe features of polymer fracture. These will have much in common with other polymers, particularly other glassy thermoplastics but there are variations in behaviour so care should be exercised in extending ideas too far especially when the basic polymer structure differs.

Early work on the fracture of PMMA was carried out by Berry (1961, 1972) who found that LEFM predicted the dependence of failure stress on crack length except for very short cracks when failure stress was crack independent. Berry suggested that the material possessed an 'inherent flaw' which governed fracture under these conditions. It is thought that these flaws are initiated from crazes which grow under the influence of the applied stress. Therefore a study of craze mechanics is important in polymer fracture.

### *Craze growth*

Williams and Marshall (1975) allowed for the viscoelastic nature of polymers by using the Dugdale model with a time dependent craze stress of the form:

$$\sigma_c = At^n \quad (16)$$

This leads to the prediction that the craze length is proportional to time raised to the power of $2n$ and so a plot of the logarithm of craze length against time should be linear. Experimental results for PMMA, rubber modified polystyrene and polycarbonate show excellent linearity and the values for $n$ for these results agree reasonably with other determinations.

At the molecular level, studies of craze growth suggest that, whilst craze initiation is due to the growth and coalescence of microvoids, this is not the usual mechanism for craze growth. Rather, a process known as 'meniscus instability' proposed by Argon and Salama (1977) is favoured.

### *Crack growth*

As the craze length grows, so does the crack opening displacement with a similar dependence on time until a critical value is reached when a crack is initiated. The time for this process to occur is termed the 'incubation time'. Thereafter, the time scale for processes in the craze will be determined by crack speed. Marshall *et al.* (1974), again assuming a critical crack opening displacement, derived a relationship between $K_{Ic}$ and crack speed, $\dot{a}$, of the form:

$$K_{Ic} \propto \dot{a}^n \qquad (17)$$

Using a double torsion test specimen, this relationship was confirmed experimentally for slow crack growth in PMMA over a range of temperatures from $-60\,°C$ to $+80\,°C$ as shown in *Figure 7.7*. It was also confirmed that the critical crack opening displacement was generally constant with temperature by direct measurement (Weidmann and Doll, 1976). This approach also rationalized apparently different results obtained by other workers using different test methods as demonstrated in *Figure 7.8* but it can be seen that at the higher crack speeds there appears to be a mode change, the value of $K_{Ic}$ dropping suddenly. The point at which this occurs in the tests made by Marshall *et al.* (1974) is marked by a cross in *Figure 7.7* so it can be seen that the

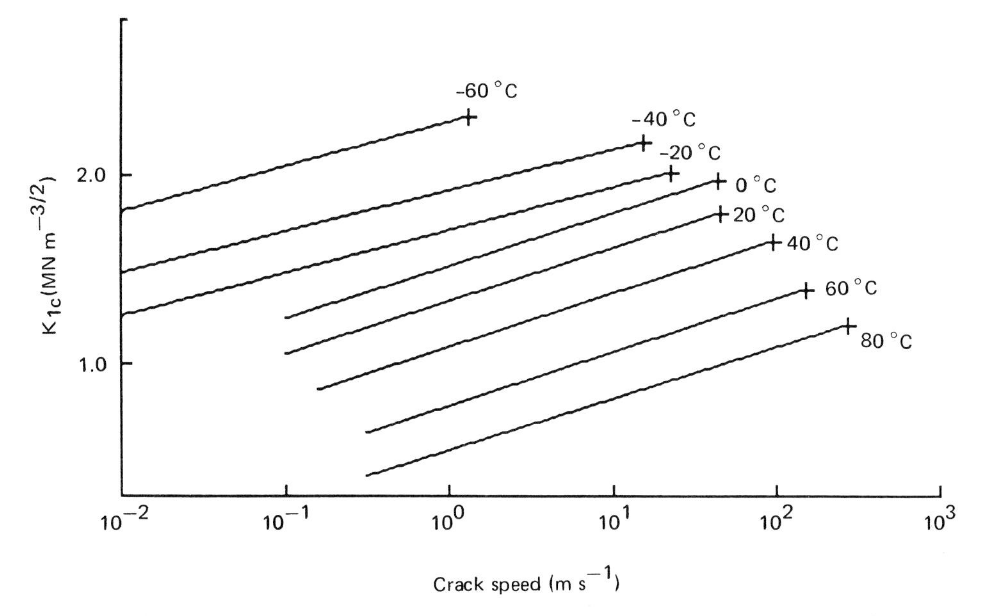


**Figure 7.7** Critical stress intensity factor against crack speed for PMMA for a range of temperatures. The individual data points have been omitted for clarity. From Marshall *et al.* (1974).

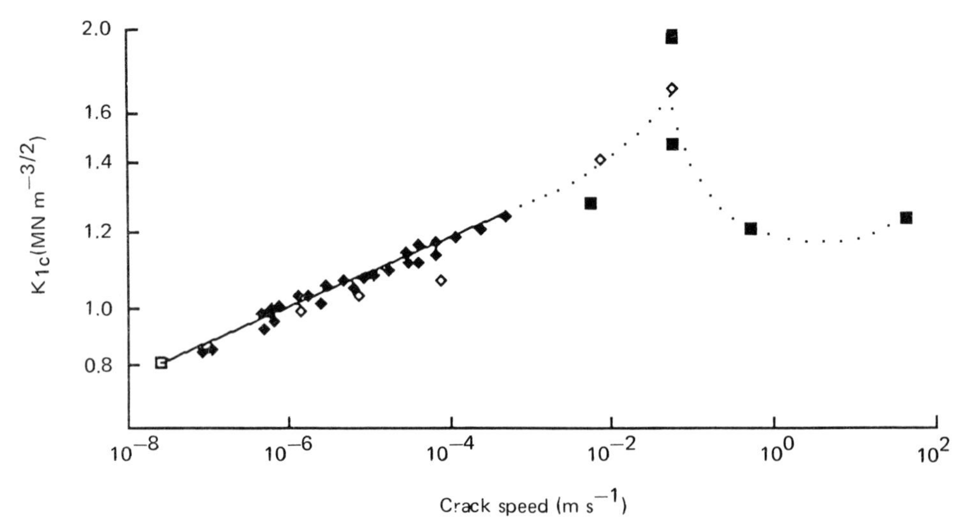


**Figure 7.8** Critical stress intensity factor against crack speed for PMMA. (After: □, Weidmann and Doll (1976); ◆, Beaumont and Young (1975); ◇, Marshall *et al.* (1974); ■, Johnson and Radon (1973).

transition crack speed decreases and the transition critical stress intensity factor increases as the temperature decreases. Cracks can be grown stably at speeds of less than about 50 mm $s^{-1}$; however if attempts are made to steadily increase crack speeds above this value it has been observed that the crack speed jumps to values in excess of 10 m $s^{-1}$ (Johnson and Radon, 1973; Beaumont and Young, 1975; Williams, 1978). This effect has also been reported by Parvin and Williams (1975) for polycarbonate at − 100 °C and by Mai and Atkins (1976) for polystyrene at room temperature. Johnson and Radon (1973) attributed this to β relaxation but the generally accepted mechanism is of an isothermal–adiabatic transition proposed and convincingly argued by Marshall *et al.* (1974). The reasoning is that at low crack speeds the heat generated due to energy dissipation at the crack tip is conducted away to the bulk of the solid but as the speed increases heat generation is generated at a rate much greater than the rate at which it can be conducted away and adiabatic conditions prevail. This leads to local thermal softening and a reduction in fracture toughness. Because polymers have a low thermal conductivity this can only occur at relatively low crack speeds. The rate of heat generation will be related to the rate at which energy is dissipated in crack propagation and this is proportional to the product of $G_{1c}$ and crack speed and Marshall and his co-workers calculated that the increase in temperature of a strip of thickness equal to the crack opening displacement will be proportional to the square root of this product. Williams (1978) extended this approach to show that:

$$\dot{a}K_{1c} = BT_0^4 \tag{18}$$

where $T_0$ is the absolute ambient temperature and B a constant involving thermal properties. This relationship fitted the data for PMMA and B was found to be in reasonable agreement with the calculated value. It also satisfactorily explained the transitions found by Parvin and Williams (1975) and Mai and Atkins (1976).

## Environmental effects

A disconcerting feature of polymers is that they can suffer low stress brittle fracture when loaded in an environment which has no apparent effect on the polymer when it is unstressed. This phenomenon has been the subject of much study and is known as environmental stress cracking (ESC). It can be particularly alarming when the material is tough like polycarbonate (glassy) which fails at low stress after a short exposure to a solution of sodium hydroxide in ethanol or polyethylene (semi-crystalline) which suffers brittle fracture in alcohols or detergent solution. It is believed that the environmental medium penetrates crazes and attacks the craze material because there is a much greater polymer surface area to volume ratio in crazes than in the uncrazed solid. The environment then either weakens the polymer ligaments by a plasticization process or by reducing the surface energy. Williams and Marshall (1974) suggest that the effect of the environment is to reduce the craze stress by some factor and this reduction occurs almost instantaneously on contact. This would lead to an accelerated craze growth by the same relaxation controlled process governing the dry craze previously described. The increased craze growth rate due to reduced craze stress can only be sustained if the rate of penetration of the environmental agent is greater than the accelerated craze growth rate. If this is not the case then the craze growth rate will be controlled by the flow of the environmental fluid through the craze, or *flow controlled craze growth*.

Williams and Marshall used Darcy's law of diffusion in porous media to estimate the flow rate of the environmental fluid and deduced that the craze length, $x_0$, under flow controlled conditions is given by:

$$x_0 = (\mathrm{A}p_0/6\mu)^{1/2}t^{1/2} \tag{19}$$

where A is a constant, $p_0$ is the pressure at the craze mouth, $\mu$ is viscosity (of environmental fluid) and $t$ is time. Experimental results for a number of polymer–liquid combinations show that craze growth increases as the liquid viscosity decreases while tests on PMMA in methanol confirmed the dependence on time predicted by Eqn (20) as do the results found by Miltz *et al.* (1978) for polycarbonate in ethanol. Graham *et al.* (1976) have shown that the craze length at a given time is proportional to K for several systems which suggests that the constant A is proportional to $\mathrm{K}^2$.

There is clear evidence that liquid flow is a controlling mechanism in the environmental cracking of polyethylene, the semi-crystalline polymer for which there is most information on environmental effects (Shanahan and Shultz, 1976; Williams and Marshall, 1975). Williams (1984) argues that flow control marks a transition from crack growth subject to full environmental effects to crack growth at speeds which are too fast for the liquid to affect the process, thus giving the same crack growth rates as in air. He further deduces that the flow controlled transition may be of two types as shown schematically in *Figure 7.9*, where the solid lines are parallel to each other. One form is as for craze growth with the crack speed proportional to $\mathrm{K}^2$ and the other is at a constant crack speed. The results for polyethylene demonstrate the former whilst tests on ABS in butyl oleate made by Kambour and Yee (1981) reveal the latter. In either case the transition crack speed is inversely proportional to the viscosity of the liquid.

Polyethylene is ductile in air but becomes embrittled in an aggressive environment. Glassy thermoplastics, however, can have their fracture toughness increased due to environmental effects. Hakeem and Phillips (1979) demonstrate that this is the case

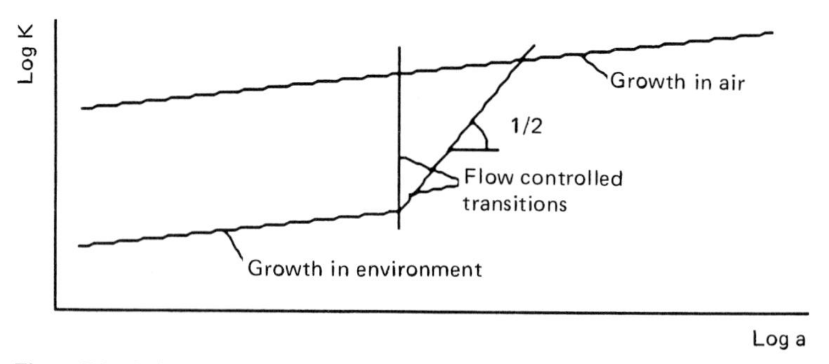


**Figure 7.9** Schematic representation of crack growth in air and environment showing transition region.

for PMMA and that it is due to multiple crazing at the crack tip which effectively blunts the crack. This leads to stick-slip rather than stable crack growth.

## Impact

### *Fracture mechanics approach*

The impact strength is one of the most important mechanical properties of polymeric components but there is some uncertainty about what impact strength is in quantitative terms! Impact tests have been carried out on metals since the last century using conventional Izod and Charpy specimens (*see Figure 7.10*) in swinging pendulum rigs. The energy lost by the pendulum divided by the specimen cross sectional area gives the Izod or Charpy impact strength but, for polymers, different ranking orders of impact strength can be obtained depending on the test used and there is no clear co-relation between them. Fracture mechanics has been a major influence in rationalizing the assessment of a polymer's resistance to impact failure and the approach was developed by Marshall *et al.* (1973) and Brown (1973). Firstly it was recognized that the energy lost by the pendulum, $U_t$, is employed both to fracture the sample, $U_f$, and to protect the broken part(s) with kinetic energy, $U_k$, which may be a significant proportion. Thus:

$$U_t = U_f + U_k \tag{20}$$

For a material with limited crack tip yielding, LEFM applies so that $U_f$ can be related to $G_{Ic}$ if the specimen is assumed to be elastic. This gives:

$$U_f = G_{Ic}BD\Phi \tag{21}$$

where $B$ is specimen thickness, $D$ is width and $\Phi$ is the factor given by:

$$\Phi = C/(\mathrm{d}C/\mathrm{d}(a/D)) \tag{22}$$

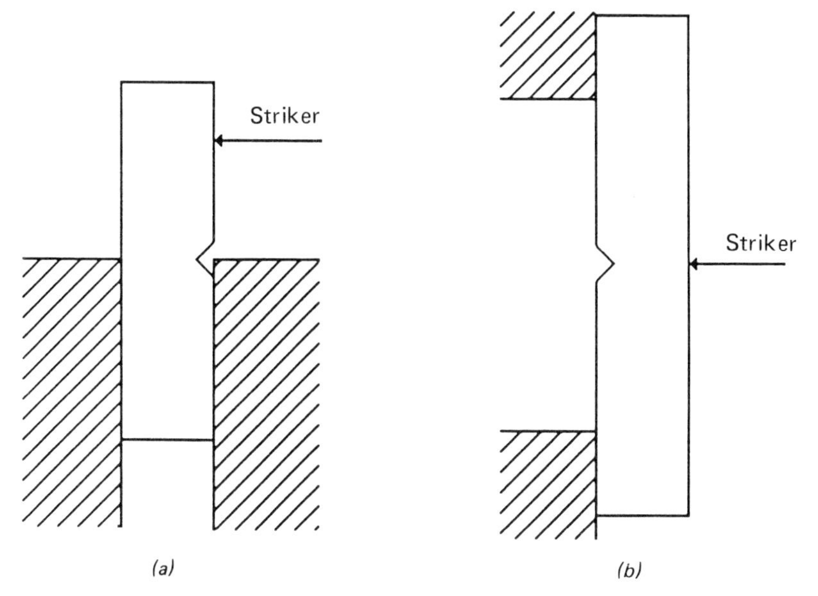


**Figure 7.10** Schematic (a) Izod and (b) Charpy test arrangements.

Compliance, $C$, as a function of crack length, $a$, can be determined experimentally or theoretically for either test geometry. Therefore a plot of energy lost by the striker, $U_t$, against $(BD\Phi)$ should be linear for brittle materials with slope $G_{1c}$ and intercept equal to the kinetic energy. Plati and Williams (1975a) showed that for polyethylene plots of $U_f$ against $(BD\Phi)$, both Izod and Charpy specimens lay on the same line, thus demonstrating the validity of the approach. Plati and Williams (1975b) also tested polycarbonate at different temperatures and their results are reproduced in *Figure 7.11*. The intercept on the energy axis is the kinetic energy loss and is the same for all temperatures as expected.

More ductile materials which exhibit contained but not small scale crack tip yielding can still be assessed using the techniques of LEFM if a plastic zone correction factor is used. This involves modifying the crack length to account for plasticity so that the crack length used in analysis is increased by half the plastic zone size. This simple correction has been used to good effect but gross yielding requires a different approach. If an impact specimen yields fully across its cross-sectional area, $A$, then the J-contour integral approach can be used to give:

$$J_{1c} = 2U_f/A \qquad (23)$$

It is interesting to note that this is just twice the impact energy as defined in the traditional Izod and Charpy tests, but it must be remembered that this approach is applicable to fully yielded test pieces only. This approach has been used by Newmann and Williams (1978) who found it gave a good description of the impact behaviour of ABS.

Returning to the data of Plati and Williams (1975b) as presented in *Figure 7.11*, it can be seen that the slopes differ and show that the fracture toughness in impact *decreases* substantially as the temperature is lowered. This is true of most polymers and contrasts with the low speed fracture toughness which *increases* with reducing

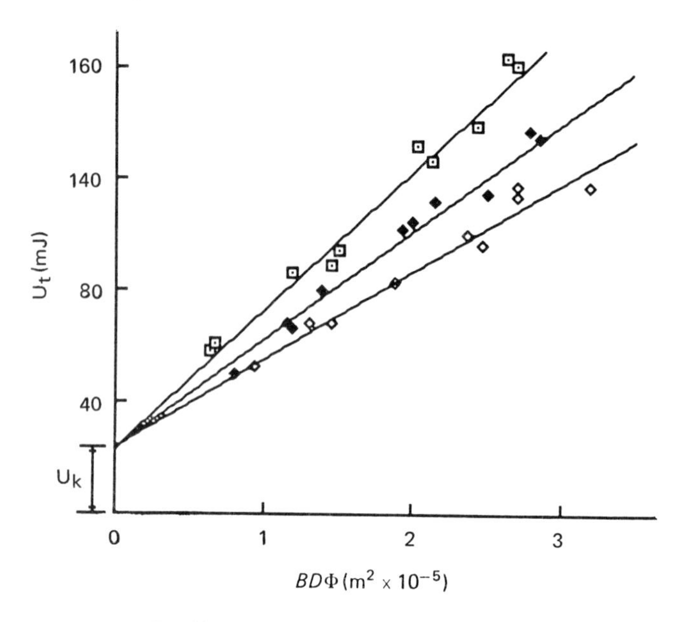


**Figure 7.11** Total impact energy against $BD\Phi$ for polycarbonate at three different temperatures: ,◇ − 100 °C; ,◆ − 40 °C; ⊡, + 20 °C. From Plati and Williams (1975b).

temperature. The variation of fracture toughness with temperature is not monotonic, however, as demonstrated by Kisbenyi *et al.* (1979) who found there were peaks in the impact fracture toughness against temperature relationship for PTFE. These peaks corresponded to the loss peaks suggesting an effect of molecular relaxation on impact.

### *Rate effects*

More recently Williams and Hodgkinson (1981) and Hodgkinson *et al.* (1982, 1983) have examined the effect of impact rate on toughness. If the rate is characterized by a loading time then it was found that $G_{Ic}$ was inversely proportional to the square root of loading time (*see Figure 7.12*). Williams and Hodgkinson propose that an isothermal–adiabatic transition takes place and this leads to a temperature rise and softening of polymer at the crack tip. This causes blunting of the crack which raises the fracture toughness. This may appear to be a contradiction of the explanation that an isothermal–adiabatic transition causes the fall in toughness at a given crack speed described earlier. However, in the previous case the crack was fast running and self sharpening, in the impact test the softening takes place at a *stationary* crack. Williams and Hodgkinson provide an analysis which suitably describes the observed behaviour. Thermal effects are also believed to be responsible for the strong rate effects observed by Isherwood and Younan (1986) in the impact cutting of polymers. In this case the cutting energy increases at a rate greater than that found in conventional impact but at a critical speed the energy ceases to increase and for polymers with high cutting energies the energy decreases with further increase in rate.

Currently Williams (1986) is suggesting that thermal effects are responsible for only some of the effects of rate in impact and favours an explanation based on dynamic behaviour. Williams has shown that the observed behaviour can be predicted using

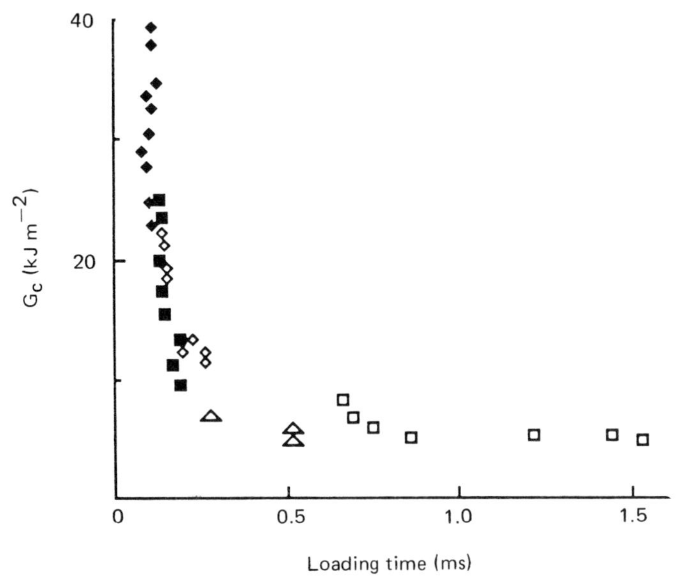


**Figure 7.12** Fracture toughness against loading time for several strain rates: ◆, 573; ◇, 332; ■, 150; □, 87; △, 26 s$^{-1}$. From Williams and Hodkinson (1981).

simple spring-mass models, including contact stiffness effects. Time will tell how this potentially exciting approach develops.

## Toughness and toughening of polymers

The fracture toughness of a polymer depends on its ability to dissipate energy. This can be achieved by either shear yielding or crazing and if either is confined to a small region then the ensuing failure will be brittle. Generally, a polymer which yields will be tougher than one that crazes. Whether a glassy thermoplastic yields or crazes has been shown by Donald and Kramer (1982a, 1982b) to depend on the entanglement structure and so another link has been established between microstructure and gross fracture behaviour. In semi-crystalline polymers matters are more uncertain. Tie molecules between crystals appear to have a rôle akin to entanglements (Peterlin, 1981) and there is evidence that large spherulites tend to cause brittleness (Way *et al.*, 1974). Except at low temperatures, high rates or in aggressive environments, crystalline polymers are more brittle than glassy ones and so the application of LEFM is limited. Consequently techniques based on the generalized theory or J integral and/or critical crack opening criteria are required to cope with extensive energy dissipation.

Strong, tough, crystalline polymers can be produced by orientation of the chains. High strength textile fibres have been produced for some time by solution or solvent spinning and mechanical drawing. Nowadays, strong, light materials for high integrity engineering components are being manufactured.

Glassy polymers may also be toughened by the addition of selected fillers rather than by orientation. Rubber is the most common additive for toughening purposes and particles of about 1 μm in diameter are dispersed in the polymer matrix either by mechanical mixing or in the polymerization process. It was originally believed that these particles acted as crack stoppers but now it is clear that they act as initiation sites for shear yielding or crazing. At first sight it may seem that a system that enhances craze initiation weakens the structure since crazes are the precursors of cracks. It is true that a *single* craze is an initiator for a crack but in rubber-modified systems crazes emanate from individual particles, are stopped by other particles so that large crazes are not grown, and thus demonstrate *multiple* crazing which results in high energy dissipation and toughness is conferred. In order to obtain the desired toughening effect it is necessary to use a filler that is well bonded to the matrix and has appropriate properties since voids do not produce toughness and hard particles have a reduced effect.

The most common rubber-modified plastics are high impact polystyrene (HIPS), which crazes, and acrylo-nitrile butadiene styrene (ABS), which shear yields. In tests on the former Parvin and Williams (1976) found that there was no thinning at the crack tip indicating the increase in volume associated with crazing and there was a strong thickness effect. There appears to be little toughening under plane strain conditions but significant toughening in thin sheets. In ABS, however, thinning does take place due to constant volume shear yielding and there is little thickness effect (Newmann and Williams, 1980). There has been comparatively little work on rigid fillers which seem to be less effective toughening additives but give cheaper materials. As pointed out by Kinloch and Young (1983), there may well be a future for combined rubber and rigid fillers.

## Conclusion

The preceding description only touches on a small fraction of the extensive work that has been undertaken in determining the mechanisms of polymer fracture. It is hoped that the importance of energy dissipation has been emphasized as this is believed to determine the toughness of materials. Important areas have been neglected, however, and it is recognized that the work relating gross properties to microstructure by Donald and Kramer has not received due attention. Similarly fracture of rubbers and thermosets are not mentioned. A comprehensive account of polymer fracture in general is given in the excellent work by Kinloch and Young (1983) which includes these topics. For a fracture mechanics viewpoint Williams (1984) is recommended.

## References

ANDREWS, E.H. (1974). A generalised theory of fracture mechanics. *Journal of Materials Science*, **9**, 887–894

ANDREWS, E.H. AND FUKAHORI, Y. (1977). *Journal of Materials Science*, **12**, 1307

ARGON, A.S. AND SALAMA, M.M. (1977). Growth of crazes in glassy polymers. *Philosophical Magazine*, **35**, 1217–1234

BARENBLATT, G.I. (1962). Mathematical theory of equilibrium cracks. *Advances in Applied Mechanics*, **7**, 56

BEAUMONT, P.W.R. AND YOUNG, R.J. (1975). Failure of brittle polymers by slow crack growth I: Crack propagation in PMMA and time to failure predictions. *Journal of Materials Science*, **10**, 1334–1342

BERRY, J.P. (1961). Fracture processes in polymeric materials I: Surface energy of PMMA. *Journal of Polymer Science*, **10**, 107–115

BERRY, J.P. (1972). Fracture of polymeric glasses. In *Fracture VIII* (H. Leibowitz, Ed.). New York : Academic Press

BROWN, H.R. (1973). A critical evaluation of the impact test for glassy polymers. *Journal of Materials Science*, **8**, 941–948

COTTRELL, A.H. (1964). *Proceedings of the Royal Society of London*, **A292**, 2

DONALD, A.M. AND KRAMER, E.J. (1982a). Effect of molecular entanglements on craze microstructure in glassy polymers. *Journal of Polymer Science*, **20**, 899–909

DONALD, A.M. AND KRAMER, E.J. (1982b). Deformation zones and entanglements in glassy polymers *Polymer*, 23, 1183

DUGDALE, D.S. (1960). Yielding of steel sheets containing slits. *Journal of the Mechanics and Physics of Solids*, **8**, 100–104

GRAHAM, I.D., WILLIAMS, J.G. AND ZICHY, E.L. (1976). Craze kinetics for PMMA in liquids *Polymer*, **17**, 439–442

GRIFFITH, A.A. (1920). The phenomena of rupture and flaws in solids. *Philosophical Transactions of the Royal Society*, **A221**, 163–198

HAKEEM, M.I. AND PHILLIPS, M.G. (1978). *Journal of Materials Science*, **13**, 2284

HODGKINSON, J.M., SAVADORI, A. AND WILLIAMS, J.G. (1983). A fracture mechanics analysis of polypropylene/rubber blends. *Journal of Materials Science*, **16**, 2319–2336

HODGKINSON, M.I., VLACHOS, N.S., WHITELAW, J.H. AND WILLIAMS, J.G. (1982). Drop weight impact tests using laser-Doppler velocimetry. *Proceedings of the Royal Society of London*, **A379**, 133–144

IRWIN, G.R. (1948). Fracture dynamics. In *Fracturing of Metals*, pp 141–166. Cleveland: Americal Society of Metals

ISHERWOOD, D.P. AND YOUNAN, H.R. (1986). An impact cutting test for polymers. *Polymer Testing*, **6**, 267–277

JOHNSON, F.A. AND RADON, F.A. (1873). *Journal of Polymer Science*, **11**, 1995

KAMBOUR, R.P. AND YEE, A.F. (1981). *Polymer Engineering and Science*, **21**, 218–222

KAUSCH, H.H. AND DeVRIES, K.L. (1975). *International Journal of Fracture*, **11**, 727

KINLOCK, A.J. AND YOUNG, R.J. (1983). *Fracture Behaviour of Polymers*. London: Elsevier

KISBENYI, M., BIRCH, M.W. HODGKINSON, J.M. AND WILLIAMS, J.G. (1979). Correlation of impact fracture toughness with loss peaks in PTFE. *Polymer*, **20**, 1289–1297

LAKE, G.L. AND LINDLEY, P.E. (1965). The mechanical fatigue limit for rubber. *Journal of Applied Polymer Science*, **9**, 1233–1251

LAKE, G.J. AND THOMAS, A.G. (1967). The strength of highly elastic materials. *Proceedings of the Royal Society of London*, **A300**, 108–119

MARK, H. (1943). *Cellulose and its Derivatives* (E.Ott, Ed), p. 1001. New York : Wiley Interscience

MAI, Y.W. AND ATKINS, A.G. (1976). *Journal of Materials Science*, **11**, 677

MARSHALL, G.P., COUTTS, L.H. AND WILLIAMS, J.G. (1974). Temperature effects in the fracture of PMMA. *Journal of Materials Science*, **9**, 1409–1419

MARSHALL, G.P., WILLIAMS, J.G. AND TURNER, C.E. (1973). Fracture toughness and absorbed energy measurements in impact tests on brittle materials. *Journal of Materials Science*, **8**, 949–956

MILTZ, J., DIBENEDETTO, A.T. AND PETRIE, S. (1978). The environmental stress crazing of polycarbonate. *Journal of Materials Science*, **13**, 1427–1437

MORGAN, G.P. AND WARD, I.M. (1977) Temperature dependance of craze shape and fracture in poly(methyl methacrylate). *Polymer*, **18**, 87–91

NEWMANN, L.V. AND WILLIAMS, J.G. (1978). The impact behaviour of ABS over a range of temperatures. *Polymer Engineering and Science*, **18**, 893–899

NEWMANN, L.V. AND WILLIAMS, J.G. (1980). Craze growth and fracture in ABS polymers. *Journal of Materials Science*, **15**, 773–780

OROWAN, E. (1952). Fundamentals of brittle behaviour in metals. In *Fatigue and Fracture of Metals* (W.M. Murray, Ed) pp 139–167. New York : Wiley

PARVIN, M. AND WILLIAMS, J.G. (1975). The effect of temperature on the fracture of polycarbonate. *Journal of Materials Science*, **10**, 1883–1888

PETERLIN, A. (1981). *Journal of Macromolecular Science—Physics*, **B19**, 401

PLATI, E. AND WILLIAMS, J.G. (1975a). The determination of the fracture parameters for polymers in impact. *Polymer Engineering and Science*, **15**, 470–477

PLATI, E. AND WILLIAMS, J.G. (1975b) Effect of temperature on the impact fracture toughness of polymers. *Polymer*, **16**, 915–920

SHANAHAN, M.E.R. AND SHULTZ, J. (1976). *Journal of Polymer Science*, **17**, 705–710

VINCENT, P.I. (1972). *Polymer*, **13**, 558

WAY, J.L., ATKINSON, J.R. AND NUTTING, J. (1974). The effect of spherulite size on the fracture morphology of polypropylene. *Journal of Materials Science*, **9**, 293–299

WEIDMAN, G.W. AND DOLL, W. (1978). Some results of optical interference measurements of critical displacements at the crack tip. *International Journal of Fracture*, **14**, R189–194

WILLIAMS, J.G. (1978). Applications of linear fracture mechanics. *Advances in Polymer Science*, **27**, 67–120

WILLIAMS, J.G. (1984). *Fracture Mechanics of Polymers.* Chichester: Ellis Horwood

WILLIAMS, J.G. (1986). *International Journal of Fracture Mechanics* (in press)

WILLIAMS, J.G. AND HODGKINSON, J.M. (1981). Crack blunting mechanisms in tests on polymers. *Proceedings of the Royal Society of London*, **A375**, 231–248

WILLIAMS, J.G. AND MARSHALL, G.P. (1975). Environmental crack and craze phenomena in polymers. *Proceedings of the Royal Society of London*, **A342**, 55–77

YOUNG, R.J. (1981). *Introduction to Polymers.* London: Chapman and Hall

ZHURKOV, S.N. (1965). Kinetic concept of the strength of solids. *International Journal of Fracture Mechanics*, **1**, 311–323

ZHURKOV, S.N. AND KUSUKOV, V.E. (1974). *Journal of Polymer Science*, **12**, 385

ZHURKOV, S.N. AND TOMASHEVSKY, E.E. (1966). *Physical basis of yields and fracture*, p. 200. London: Institute of Physics

# 8

## STRUCTURAL STABILITY OF INTERMEDIATE MOISTURE FOODS— A NEW UNDERSTANDING?

LOUISE SLADE AND H. LEVINE
*General Foods Corporation, New York, USA**

## Introduction

'Water is the most ubiquitous plasticizer in our world' (Sears and Darby, 1982). It has become well established that plasticization by water affects the glass transition temperatures ($T_g$) of many synthetic and natural amorphous polymers (particularly at lower moisture contents), and that depression of $T_g$ can be advantageous or disadvantageous to material properties, processing, and stability (Rowland, 1980). Eisenberg (1984) has stated that 'the glass transition is perhaps the most important single parameter which one needs to know before one can decide on the application of the many non-crystalline (synthetic) polymers that are now available.' Karel (1985) has noted that 'water is the most important ... plasticizer for hydrophilic food components.' The physicochemical effect of water, as a plasticizer, on the $T_g$ of starch and other amorphous or partially crystalline polymeric food materials has been increasingly discussed in several recent reviews and reports (Ablett, Attenburrow and Lillford, 1986; Biliaderis, Page and Maurice, 1986; Biliaderis *et al.*, 1986; Blanshard, 1986; van den Berg, 1985, 1986), dating back to the pioneering doctoral research of van den Berg (van den Berg, 1981; van den Berg and Bruin, 1981).

The critical role of water as a plasticizer of amorphous materials (both water-soluble and water-sensitive ones) has been a focal point of our research, and has developed into a central theme in an active industrial programme in food polymer science of six years' duration. Recently reported studies from our laboratories were based on thermal and thermomechanical analysis methods which were used to illustrate and characterize the polymer physicochemical properties of various food ingredients and products, described as systems of completely amorphous or partially crystalline polymers, oligomers, and/or monomers, soluble in and/or plasticized by water (Levine and Slade, 1987a). Our polymer science approach to understanding structure/property relations of food 'polymers' (a term we use generically to refer to homologous polymer families, including oligomers and monomers) emphasizes the generic similarities between synthetic polymers and food molecules such as starch, gelatin, and sugars (Slade and Levine, 1987a). From a theoretical basis of established structural principles from the field of synthetic polymer science, functional properties of food materials during processing and storage can be explained and often predicted.

This food polymer science approach has developed to unify structural aspects of food materials [conceptualized as completely amorphous or partially crystalline polymer systems, the latter typically based on the classic 'fringed micelle' morphologi-

* Present address: Nabisco Brands Inc., Corporate Technology Group, P.O. 1943, East Hanover, New Jersey 07936–1943, USA

cal model (Billmeyer, 1984; Flory, 1953)] with functional aspects described in terms of water dynamics and glass dynamics (Slade and Levine, 1987b; Slade, Levine and Franks, 1987). These integrated concepts focus on the non-equilibrium nature of all 'real world' food products and processes, and stress the importance to ultimate product quality and stability of maintenance of food systems in kinetically-metastable 'states' (as opposed to equilibrium thermodynamic phases), which are always subject to potentially detrimental plasticization by water. Through this unification, the kinetically-controlled behaviour of polymeric food materials may be described by a map [derived from a solute–solvent state diagram (Franks *et al.*, 1977; MacKenzie, 1977)], in terms of the critical variables of moisture content, temperature ($T$), and time (Levine and Slade, 1986). The map domains of moisture content and temperature, traditionally described using concepts such as '$A_w$', 'bound water', cryoprotection, water vapour sorption isotherms, and sorption hysteresis, can be treated as aspects of water dynamics (Slade, Levine and Franks, 1987). The concept of water dynamics has been used to explain moisture management and structural stabilization of intermediate moisture food (IMF) systems (Levine and Slade, 1987a; Slade and Levine, 1987b; Slade, Levine and Franks, 1987), as reviewed here, and cryostabilization of frozen, freezer-stored, or freeze-dried aqueous glass-forming materials (Levine and Slade, 1987b).

The concept of glass dynamics focuses on the temperature dependence of relationships among composition, structure, thermomechanical properties, and functional behaviour, and has been used to describe a unifying concept for interpreting 'collapse' phenomena (*see* Chapter 9), such as caking during storage of low moisture, amorphous or partially crystalline food powders (Levine and Slade, 1986). Glass dynamics has also proved useful for elucidating physicochemical mechanisms of structural changes involved in various melting and (re) crystallization phenomena which are relevant to many partially crystalline food polymers and processing/storage situations, including, for example, gelatinization and retrogradation of starch (Slade and Levine, 1987a) and gelation of gelatin (Slade and Levine, 1987c). Glass dynamics has also been used to describe amorphous polymeric behaviour of proteins such as wheat gluten and elastin at low moisture (Levine and Slade, 1987a).

The key to our new perspective on water-plasticized polymers and the resulting insights into structural stability of IMFs relates to recognition of the fundamental importance of the above-mentioned dynamic map. Results of our research programme in food polymer science (Levine and Slade, 1987a) have demonstrated that the critical feature of this map is identification of the glass transition as a reference surface which serves as a basis for describing non-equilibrium behaviour of polymeric materials, in response to changes in moisture content, temperature, and time. Kinetics of all diffusion-controlled relaxation processes (including structural collapse phenomena), which are governed by mobility of the water-plasticized polymer matrix, vary from Arrhenius to Williams–Landel–Ferry (WLF) between distinctive temperature/structural domains, which are divided by this glass transition. The viscoelastic, rubbery fluid state, for which WLF kinetics apply (Ferry, 1980), represents the most significant domain for study of water dynamics. One particular location on the reference surface results from behaviour of water as a crystallizing plasticizer and corresponds to an invariant point [i.e. $T'_g$, the $T_g$ of a maximally freeze-concentrated solute/water matrix surrounding the ice crystals in a frozen solution (Franks, 1982a, 1985a,b; Levine and Slade, 1986)] on a state diagram for any particular solute. This location represents, in practice, the glass with maximum moisture content [i.e. $W'_g$ expressed as w% water or g unfrozen water/g solute (Levine and Slade, 1987b)] as a kinetically-metastable, dynamically-constrained solid which is pivotal to characterization of structure and function of amorphous and partially-crystalline polymeric food materials. From a theoretical basis provided by the integrated concepts of water and glass dynamics, a new experimental approach has been suggested for predicting technological performance, product quality, and stability of many polymeric food systems (Levine and Slade, 1987a; Slade and Levine, 1985; Slade, Levine and Franks, 1987), and examples which illustrate the utility of this approach, versus others based on the traditional concept of '$A_w$', will be reviewed here.

To complete this introduction, we should recall that Franks (1985a) has emphasized that 'low', with respect to temperature, is a relative concept which refers to a range of some 800 degrees when interests of both metallurgists and quantum physicists are considered. For the context of this review, based on food polymer science, it is equally necessary to emphasize that 'low', with respect to moisture contents, is a relative concept to refer to everything which is 'not very dilute' (Slade and Levine, 1985). For example, a 38 w% fructose solution at 20 °C (typical of many IMF systems) already exhibits the symptoms of a 'low-moisture' situation (Slade, Levine and Franks, 1987). This insight was derived from a groundbreaking study by Soesanto and Williams (1981) on viscosities ($\eta$) of concentrated aqueous sugar solutions. They demonstrated that, for such glass-forming liquids, in their rubbery state ($\eta > 10$ Pa s) at $20 < T < 80$ °C, the WLF equation (derived from free volume theory for amorphous polymers (Ferry, 1980)) characterizes $\eta(T)$ extremely well. Arrhenius kinetics would not be applicable to describe behaviour of such 'polymers at low moisture' (Slade, Levine and Franks, 1987).

## Theoretical background

### *Polymer structure/property principles*

For 'dry' polymers which are solids at room temperature, two possible structural forms are described as completely amorphous ('glassy') and partially crystalline. The latter term, rather than semi-crystalline, is preferred to describe polymers of relatively low percentage crystallinity, such as native starches [$\simeq$ 15–39% crystallinity (Biliaderis *et al.*, 1986; Blanshard, 1986)] and gelatin [$\simeq$ 20% crystallinity (Jolley, 1970)]. The term 'semi-crystalline' is the preferred usage to describe polymers of $\gtrsim$ 50% crystallinity (Flory, 1953; Wunderlich, 1973), so that it may be conceptually misleading when it is used (Biliaderis *et al.*, 1986) to describe granular starches. Completely amorphous homogeneous polymers manifest a single, 'quasi-second order', kinetic state transition from metastable amorphous solid to unstable amorphous liquid at a characteristic $T_g$. Partially crystalline polymers show two types of characteristic transitions: 1) a $T_g$ of the amorphous component, and 2) at a crystalline melting temperature, $T_m$, which is always at a higher temperature than $T_g$ for homopolymers, a first order, 'equilibrium phase' transition from crystalline solid to amorphous liquid (Fuzek, 1980). The same is true for partially crystalline oligomers and monomers (Soesanto and Williams, 1981; Slade, Levine and Franks, 1987). Both $T_g$ and $T_m$ are measurable as thermomechanical transitions by various instrumental methods, including differential scanning calorimetry (DSC), differential thermal analysis, thermomechanical analysis, and dynamic mechanical analysis (Fuzek, 1980).

***Structural models for partially crystalline polymers***

The 'fringed micelle' model, used classically to describe morphology of partially crystalline synthetic polymers (Billmeyer, 1984; Flory, 1953; Wunderlich, 1973, 1976), is illustrated in *Figure 8.1*. It is particularly useful for conceptualizing a three-dimensional network composed of microcrystallites crosslinking amorphous regions of randomly-coiled chain segments (Jolley, 1970). The model is especially applicable to polymers which crystallize from an undercooled melt or concentrated solution, to produce a metastable network of relatively low percentage crystallinity, containing small crystalline regions of only $\simeq$ 100 Å dimensions (Jolley, 1970; Slade and Levine, 1987c; Wunderlich, 1973). Thus, it has been used to describe the partially crystalline structure of aqueous gels of food polymers such as gelatin (Borchard, Bergmann and Rehage, 1976; Borchard, Bremer and Keese, 1980; Borchard *et al.*, 1976; Djabourov and Papon, 1983; Godard *et al.*, 1978; Jolley, 1970; Slade and Levine, 1984a,b, 1987c; Wunderlich, 1973) and starch (Maurice *et al.*, 1985; Slade, 1984; Slade and Levine, 1984a,b, 1987a,b), in which amorphous regions contain plasticizing water, and microcrystalline regions (which act as physical junction zones) are crystalline hydrates. We have also used this model to conceptualize the partially crystalline morphology of lower molecular weight (MW) carbohydrate systems (Levine and Slade, 1987a) and frozen aqueous solutions of water-compatible, non-crystallizing materials (Slade and Levine, 1985, 1987b; Slade, Levine and Franks, 1987). In the latter case, ice crystals represent the 'micelles' dispersed in a continuous amorphous matrix of unfrozen water and solute (Franks, 1982a, 1985a; Levine and Slade, 1986, 1987b). An important feature of the 'fringed micelle' model, as applied to high MW polymers including gelatin and starch, concerns the interconnections between crystalline and amorphous regions. A single long polymer chain can have helical (or other ordered) segments located within one or more microcrystallites and random-coil segments in one or more amorphous regions (Jolley, 1970; Slade and Levine, 1984a,b; Wunderlich, 1973). Moreover, in amorphous regions, chain segments may experience

**Figure 8.1** 'Fringed micelle' model of the crystalline/amorphous structure of partially crystalline polymers. From Slade and Levine (1987c).

random intermolecular 'entanglement coupling' (Ferry, 1980; Levine and Slade, 1986). So, in terms of their thermomechanical behaviour in response to plasticization by water and/or heat, crystalline and amorphous regions are certainly not independent phases (Slade and Levine, 1984a,b), as described by Biliaderis *et al.* (1986).

In recent years, an extension of the simple 'fringed micelle' model, originally proposed in 1930 (Wunderlich, 1976), has been described to explain further the thermomechanical behaviour of various semi-crystalline synthetic homopolymers (Cheng, Cao and Wunderlich, 1986; Jin, Ellis and Karasz, 1984; Menczel and Wunderlich, 1986; Wissler and Crist, 1980). This 'three-microphase' model incorporates two distinct types of completely amorphous domains, a bulk mobile amorphous fraction and an interfacial (i.e. intercrystalline regions within chain-folded spherulites or between crystallites in stressed systems) rigid amorphous fraction, each capable of manifesting a separate $T_g$, plus a third rigid crystalline component (B. Wunderlich, 1985, personal communication). The sequence of thermal transitions predicted by this model is $T_g$ (mobile amorphous) < $T_g$ (rigid amorphous) < $T_m$ (rigid crystalline), although the magnitude of the change in heat capacity for the middle transition may become vanishingly small due to steric restraints on chain-segmental mobility (Cheng, Cao and Wunderlich, 1986; Jin, Ellis and Karasz, 1984; Menczel and Wunderlich, 1986; Wissler and Crist, 1980; B. Wunderlich, 1985, personal communication). However, it should still be possible to detect this $T_g$ by dynamic mechanical measurements. Recently, the 'three-microphase' model has been postulated to explain multiple thermal transitions observed during non-equilibrium melting of native granular starch (Biliaderis *et al.*, 1986). However, while the model has proved applicable to several linear homopolymers (Cheng, Cao and Wunderlich, 1986; Jin, Ellis and Karasz, 1984; Menczel and Wunderlich, 1986; Wissler and Crist, 1980), its application to normal starches, while well-intentioned, is not necessary and may even be inappropriate (Slade and Levine, 1987a). Normal starch is not a homopolymer, but a mixture of two glucose polymers, linear amylose (of MW $10^5$–$10^6$) and highly-branched amylopectin (of MW $10^8$–$10^9$) (Whistler and Daniel, 1984). Even 'waxy' starch, which contains only amylopectin, is not best described as a homopolymer of glucose, but as a special type of block copolymer in which backbone segments and branch points exist in amorphous domains and crystallizable branches exist in microcrystalline domains (French, 1984; Whistler and Daniel, 1984; Wunderlich, 1980). In a native granule of normal starch, each of these polymers may be partially crystalline (French, 1984), and their amorphous components may each manifest a distinct $T_g$ (at a temperature dependent on MW (Billmeyer, 1984) as well as local moisture content) characteristic of a predominant mobile amorphous domain (Levine and Slade, 1987a).

### *Crystallization mechanism and kinetics for partially crystalline polymers*

A classical three-step crystallization mechanism has been widely used to describe partially crystalline synthetic polymers crystallized from the melt or concentrated solution, by undercooling from $T > T_m$ to $T_g < T < T_m$ (Jolley, 1970; Wunderlich, 1976). The mechanism is conceptually compatible with both the generic 'fringed micelle' model (Hiltner and Baer, 1986) and a specific 'three-microphase' model, where for the latter, the operative $T_g$ would be that of the predominant mobile amorphous phase. It involves the following sequential steps (Wunderlich, 1976):

1. Nucleation (homogeneous)—formation of critical nuclei by initiation of ordered chain segments intramolecularly;
2. Propagation—growth of crystals from nuclei by intermolecular aggregation of ordered segments; and
3. Maturation—crystal perfection (by annealing of metastable microcrystallites) and/or continued slow growth (via Ostwald ripening).

The thermoreversible gelation from concentrated solution of a number of crystallizable synthetic homopolymers and copolymers, with flexible chains of high MW which may be linear or highly branched (Mandelkern, 1986), has recently been reported to occur by the above crystallization mechanism (Boyer, Baer and Hiltner, 1985; Domszy *et al.*, 1986; Hiltner and Baer, 1986; Mandelkern, 1986). This gelation-via-crystallization process [described as a nucleation-controlled growth process (Domszy *et al.*, 1986)] produces a metastable three-dimensional network (Domszy *et al.*, 1986; Mandelkern, 1986) crosslinked by 'fringed micellar' (Hiltner and Baer, 1986) or chain-folded lamellar (Domszy *et al.*, 1986) microcrystalline junction zones composed of intermolecularly-associated helical chain segments (Boyer, Baer and Hiltner, 1985). Such partially crystalline gels may also contain random interchain entanglements in their amorphous regions (Boyer, Baer and Hiltner, 1985; Domszy *et al.*, 1986). The non-equilibrium nature of the process (Domszy *et al.*, 1986) is manifested by 'well known aging phenomena' (Hiltner and Baer, 1986) [i.e. maturation, which can involve polymorphic crystalline forms (Domszy *et al.*, 1986)], attributed to time-dependent crystallization processes which occur after initial gelation. Thermoreversibility of such gels is explained in terms of a crystallization (on undercooling)/melting (on heating to $T > T_m$) process (Domszy *et al.*, 1986; Mandelkern, 1986). Only recently has it been recognized that such synthetic homopolymer–organic diluent gels are not glasses (Blum and Nagara, 1986) ['gelation is not the glass transition of highly plasticized polymer' (Hiltner and Baer, 1986)] but partially crystalline rubbers (Boyer, Baer and Hiltner, 1985), in which diluent mobility (in terms of rotational and translational motion) is not significantly restricted by gel structure (Blum and Nagara, 1986). The temperature of gelation ($T_{gel}$) is above $T_g$ (Blum and Nagara, 1986) in the rubbery fluid range up to $T_g + 100\,°C$, and is related to $T_f$ observed in flow relaxation of rigid amorphous entangled polymers and to $T_m$ observed in melts of partially crystalline polymers (Boyer, Baer, Hiltner, 1985; Hiltner and Baer, 1986). The MW dependence of $T_{gel}$ has been identified (Boyer, Baer and Hiltner, 1985) in terms of an isoviscous state (which may include existence of interchain entanglements) of $\eta_{gel}/\eta_g = 10^5/10^{12} = 1/10^7$, where $\eta_g$ at $T_g \simeq 10^{12}$ Pa s (Franks, 1985a; Soesanto and Williams, 1981).

The distinction among these transition temperatures becomes especially important for elucidating how morphology and structure of food systems relate to their thermal and mechanical behaviour, particularly when experimental methods involve very different timeframes (mechanical measurements during compression tests or over prolonged storage; relaxation times from experiments at acoustic, microwave, or NMR frequencies; thermal analysis at scanning rates varying over four orders of magnitude) and sample preparation histories (time, temperature, concentration). In the case of morphologically homogeneous, molecularly amorphous solids, $T_g$ corresponds to the limiting relaxation temperature for mobile polymer segments. In the case of morphologically heterogeneous, supramolecular networks, the effective network $T_g$ corresponds to the $T_f$ transition above $T_g$ for flow relaxation (Boyer, Baer and Hiltner, 1985) of the network. The ratio $T_f/T_g$ varies from 1.02–1.20 for polystyrene above the entanglement MW (Keinath and Boyer, 1981). $T_f$ defines an isoviscous state of $10^5$ Pa s for entanglement networks (corresponding to $T_{gel}$ for partially crystalline networks) (Boyer, Baer and Hiltner, 1985). $T_{gel}$ of a partially crystalline network would always be observed at or above $T_f$ (network $T_g$) of an entanglement network; both transitions occur above $T_g$, with analogous influences of MW and plasticizing water. As an example, the effective network $T_g$ responsible for mechanical firmness of freshly-baked bread would be near room temperature for low extents of network formation, well above room temperature for mature networks, and equivalent to $T_{gel}$ near 60 °C for stale bread, even though the underlying $T_g$ for segmental motion, responsible for the predominant second-order thermal transition, remains below 0 °C at $T'_g$ (Slade and Levine, 1987a).

Curiously, it has been well-established for a much longer time (Domszy *et al.*, 1986) that the same three-step polymer crystallization mechanism describes the gelation mechanism for the classical gelling system, gelatin–water (Borchard, Bergmann and Rehage, 1976; Borchard, Bremer and Keese, 1980; Durand, Emery and Chatellier, 1985; Godard *et al.*, 1978; Jolley, 1970; Reutner, Luft and Borchard, 1985; Wunderlich, 1980). The fact that the resulting partially-crystalline aqueous gels (Marshall and Petrie, 1980) can be modelled by the 'fringed micelle' structure is also widely recognized (Borchard, Bergmann and Rehage, 1976; Borchard, Bremer and Keese, 1980; Borchard *et al.*, 1976; Djabourov and Papon, 1983; Godard *et al.*, 1978; Jolley, 1970; Slade and Levine, 1984a,b, 1987c; Wunderlich, 1976). However, while the same facts are true with regard to aqueous gelation of starch [i.e. retrogradation, a gelation via crystallization process that follows gelatinization and 'pasting' of partially crystalline native granular starch–water mixtures (French, 1984; Zobel, 1984)], recognition of starch retrogradation as a polymer crystallization process has been much more recent and less widespread (Ablett, Attenburrow and Lillford, 1986; Biliaderis, Page and Maurice, 1986; Biliaderis *et al.*, 1986; Blanshard, 1986). Many of the persuasive early insights in this area resulted from the food polymer science approach of Slade and her various coworkers (Biliaderis, Page and Maurice, 1986; Biliaderis *et al.*, 1985, 1986; Maurice *et al.*, 1985; Slade, 1984; Slade and Levine, 1984a,b, 1985, 1987a).

Slade (1984) and Slade and Levine (1984a,b) were the first to stress the importance of treating gelatinization as a non-equilibrium melting process of partially crystalline, kinetically metastable granular starch in the presence of plasticizing water. In this process, melting of microcrystallites is controlled by prerequisite plasticization ('softening') of random-coil chain segments in interconnected amorphous regions of the 'fringed micelle' network (Slade and Levine, 1984a,b). Slade and coworkers (Maurice *et al.*, 1985; Slade, 1984; Slade and Levine, 1984a,b) recognized that previous attempts (e.g. Biliaderis, Maurice and Vose, 1980; Donovan, 1979) to interpret the effect of water content on observed $T_m$ of starch by Flory–Huggins thermodynamic treatment (Flory, 1953) were inappropriate and had failed because Flory–Huggins theory only applies to equilibrium melting of a partially crystalline polymer with diluent. Slade described retrogradation as a non-equilibrium recrystallization process (involving amylopectin) in completely amorphous (in the case of waxy starches) melts of starch–water (Slade, 1984; Slade and Levine, 1984a,b). She noted that amylopectin recrystallization is a nucleation-controlled process that occurs at $T > T_g$, in the mobile, viscoelastic 'fringed micelle' network plasticized by water (Slade, 1984; Slade and Levine, 1984a,b). As for the aging effects observed in starch gels, Slade (1984) reported that 'analysis of results (of measurements of extent of recrystallization vs. time after gelatinization) by the classical Avrami equation may provide a convenient means to represent empirical data from retrogradation experiments, but published theoretical interpretations (e.g. Kulp and Ponte, 1981) have been misleading'. Complications, due to the non-equilibrium nature of starch recrystallization via the three-step mechanism, limit theoretical utility of the Avrami parameters (Slade, 1984), which were originally derived to define the mechanism of crystallization under conditions of thermodynamic equilibrium (Wunderlich, 1976).

It should be noted that the three-step polymer crystallization mechanism also applies to concentrated aqueous solutions and melts of monomers and oligomers such as low MW carbohydrates (Levine and Slade, 1986, 1987a; Slade and Levine, 1985, 1987b; Slade, Levine and Franks, 1987) and to recrystallization processes in frozen systems of water-compatible materials (Levine and Slade, 1986, 1987a,b).

The classical theory of crystallization kinetics developed for synthetic partially crystalline polymers (Wunderlich, 1976), which is illustrated in *Figure 8.2* [adapted from Jolley (1970)], has also been shown to describe the kinetics of gelatin gelation (Domszy *et al.*, 1986; Durand, Emery and Chatellier, 1985; Jolley, 1970) and starch retrogradation (Slade, 1984; Slade and Levine, 1987a). *Figure 8.2* shows the dependence of crystallization rate on temperature within the range $T_g < T < T_m$, and emphasizes the fact that gelation via crystallization can only occur in the rubbery (undercooled liquid) state (Djabourov and Papon, 1983; Hayashi and Oh, 1983; Reutner, Luft and Borchard, 1985), between temperature limits defined by $T_g$ and $T_m$. These limits, for gelatin (high MW) solutions of concentrations up to $\simeq 65$ w% gelatin, are $\simeq -12\,°C$ and 37 °C, respectively (Reutner, Luft and Borchard, 1985; Slade and Levine, 1984a,b) while for homogeneous and completely amorphous sols or pastes of gelatinized B-type starch containing > 27 w% water, they are $\simeq -5\,°C$ and 60 °C, respectively (Slade and Levine, 1984a,b). The rate of crystallization would be essentially zero at $T < T_g$, because nucleation is a liquid state phenomenon which requires orientational mobility, and such mobility is virtually disallowed (i.e. over realistic times) in a mechanically solid glass of $\eta \gtrsim 10^{12}$ Pa s (Franks, 1982a).

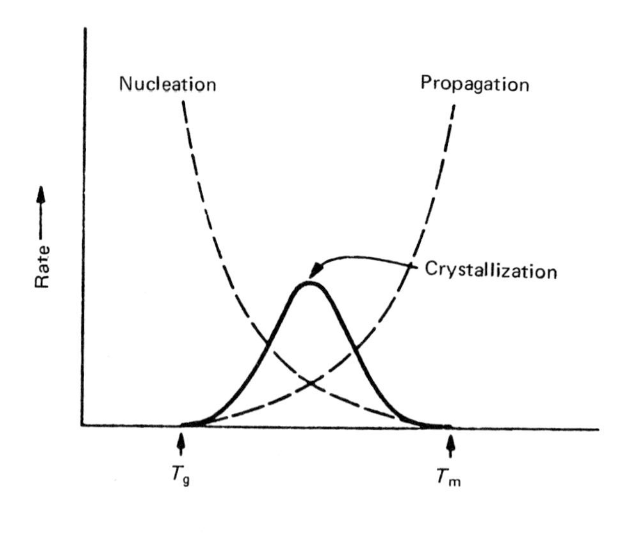


**Figure 8.2** Crystallization kinetics of partially crystalline polymers. From Slade and Levine (1987c).

Likewise, the rate of propagation goes essentially to zero below $T_g$ because propagation is a diffusion-controlled process which also requires the liquid state. At $T > T_m$, the rate of crystallization also goes to zero because, intuitively, one realizes that crystals can neither nucleate nor propagate at any temperature at which they would be melted instantaneously.

As illustrated in *Figure 8.2*, the mechanistic steps of nucleation and propagation each manifest an exponential dependence of rate on temperature, such that nucleation rate increases exponentially with decreasing temperature, down to $T = T_g$, while propagation rate increases exponentially with increasing $T$ up to $T = T_m$ (Jolley, 1970; Wunderlich, 1976). The rate of maturation for non-equilibrium crystallization processes also increases with increasing $T$, up to a maximum $T_m$ of most mature crystals (Slade, 1984). The overall rate of crystallization (i.e. nucleation and propagation), allowing sufficient time at a single temperature, can be maximized at a temperature about midway between $T_g$ and $T_m$, e.g. about room temperature for starch. However, when overall crystallization time is insufficient, the necessity that nucleation precedes propagation and the negative temperature coefficient for nucleation explains the frequently reported results (Wynne-Jones and Blanshard, 1986) for maximum rate of retrogradation at a single temperature below room temperature and negative temperature coefficient for retrogradation of starch. Moreover, Ferry (1948) showed that the rate of gelation of gelatin can be increased further, while the phenomenon of steadily increasing gel maturation over extended storage time can be eliminated, by a two-step temperature cycling protocol for gelation that capitalizes on the crystallization kinetics defined in *Figure 8.2*. Ferry showed that a short period of nucleation at $T$ just above $T_g$, followed by another short period for crystal growth at $T$ just below $T_m$, produced a gelatin gel of maximum and unchanging gel strength in the shortest possible overall time. Recently, Slade has shown that a similar $T$-cycling protocol can be used to maximize starch recrystallization rate in freshly-gelatinized starch–water mixtures with at least 27 w% water (Sladc and Levine, 1987a).

## *Viscoelastic properties of amorphous and partially crystalline polymers*

In the absence of plasticizer, the viscoelastic properties of glassy and partially crystalline polymers depend critically on $T$, relative to $T_g$ of the undiluted polymer. These properties include, for example, polymer specific volume, $V$, as illustrated in *Figure 8.3* (adapted from Ferry, 1980; Cakebread, 1969; Cowie, 1973) for glassy, partially crystalline, and crystalline polymers. From free volume theory, $T_g$ is defined as the temperature at which the slope changes (due to a discontinuity in thermal expansion coefficient) in the $V$ versus $T$ plot for a glass in *Figure 8.3* (Ferry, 1980), while at $T_m$, $V$ shows a characteristic discontinuity, typically increasing $\simeq 15\%$ for crystalline polymers (Wunderlich, 1980). Glassy and partially crystalline polymers also manifest $\eta$ versus $T$ behaviour as illustrated in *Figure 8.4* (adapted from Franks, 1982a), which provides a working definition of a glass (as a mechanical solid capable of supporting its own weight against flow) in terms of its $\eta$. $\eta \simeq 10^{11}$–$10^{14}$ Pa s at $T_g$ (Downton, Flores-Luna and King, 1982; Franks, 1982a; Soesanto and Williams, 1981), which represents the intersection of the curve in *Figure 8.4* with the boundary between glassy and rubbery fluid states. For many typical partially crystalline synthetic polymers, $T_g = 0.5$–$0.8$ of $T_m$ in °K (Batzer and Kreibich, 1981; Franks, 1982a; Wunderlich, 1980). For highly symmetrical pure polymers, $T_g/T_m$ is $< 0.5$,

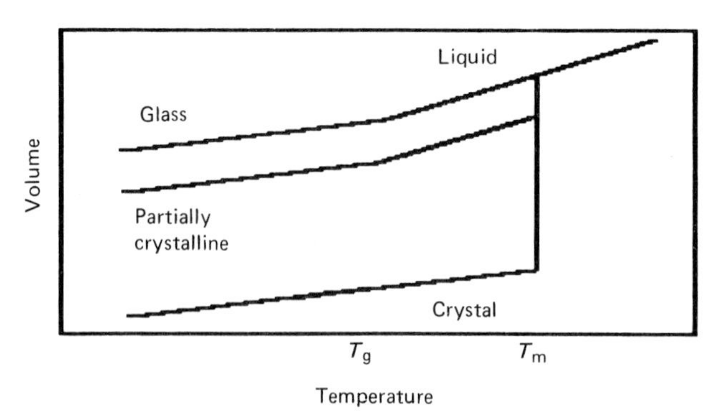


**Figure 8.3** Specific volume as a function of temperature for glassy, crystalline, and partially crystalline polymers. From Levine and Slade (1987a).

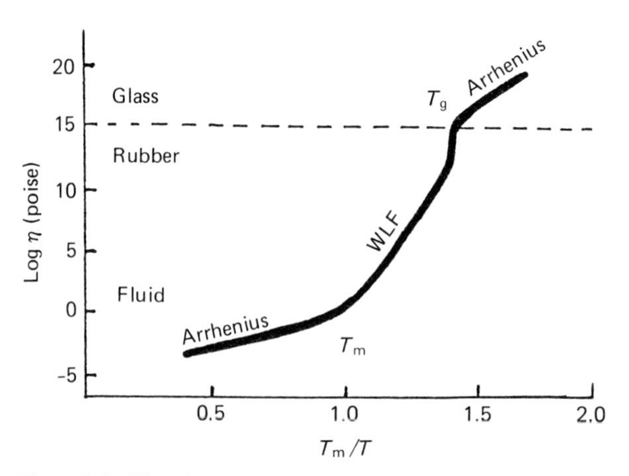


**Figure 8.4** Viscosity as a function of reduced temperature ($T_m/T$) for glassy and partially crystalline polymers. From Levine and Slade (1987a).

while for highly unsymmetrical ones the ratio is > 0.8 (Brydson, 1972; Wunderlich, 1980). $T_g$ is also known to vary with polymer MW in a characteristic manner. For a homologous series of amorphous linear polymers, $T_g$ increases with increasing number-average MW ($\bar{M}n$), due to decreasing free volume, up to the plateau limit for the region of 'entanglement coupling' in rubber-like viscoelastic random networks (typically at $\bar{M}n = 1.25 \times 10^3$–$10^5$ daltons), then levels off with further increases in $\bar{M}n$ (Billmeyer, 1984; Ferry, 1980; Graessley, 1984). For polymers with constant values of $\bar{M}n$, $T_g$ increases with increasing weight-average MW ($\bar{M}w$), due to an increasing local effective η. This contribution of local effective η is reported to be especially important when comparing different MWs in the range of low MWs (Ferry, 1980). We have found that anomalous ratios of $T_m/T_g$ correlate with anomalous relaxation behaviour due to contributions of excess free volume or decreased local effective η (Slade and Levine, 1987a; Slade, Levine and Franks, 1987).

***Effects of water as a plasticizer on thermomechanical properties of solid polymers***

Water as a plasticizer affects both $T_g$ and $T_m$ of polymers. The direct plasticizing effect of increasing moisture content at constant temperature, which is equivalent to the effect of increasing temperature at constant percentage moisture, leads to increased segmental mobility of chains in amorphous regions of glassy and partially crystalline polymers, which in turn produces a primary structural relaxation transition, $T_g$, at decreased temperature (Flink, 1983; Kellaway, Marriott and Robinson, 1978; Starkweather, 1980). The state diagram in *Figure 8.5* (van den Berg, 1986) illustrates the extent of this $T_g$ depressing effect of water for starch. *Figure 8.5* demonstrates that freshly gelatinized starch exemplifies a typical water-compatible, completely amorphous polymer, which manifests a smooth $T_g$ curve from $\simeq$ 125 °C for 'dry' starch to $\simeq -134$ °C, the value commonly stated (but never yet measured) for glassy water (Franks, 1982a). The dramatic effect of water on $T_g$ is seen at low moisture, such that for starch $T_g$ decreases $\simeq$ 6 °C/w% water for the first 10 w% moisture. Levine and Slade (1987a) and other workers (Bair *et al.*, 1981; Batzer and Kreibich, 1981; Franks, 1982a) have found this same extent of plasticization at low moisture (5–10 °C/w% water) to apply widely to completely amorphous and partially crystalline, water-compatible polymers, as well as to many monomeric and oligomeric carbohydrates (Levine and Slade, 1987a; Slade and Levine, 1987a).

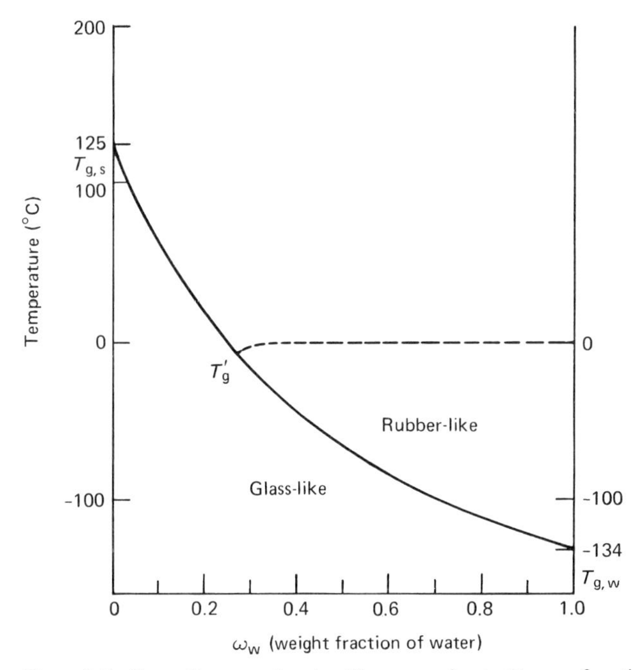


**Figure 8.5** State diagram, showing 'the approximate $T_g$s as a function of mass fraction, for the starch–water system'. Reproduced with permission from van den Berg (1986).

***Mechanism of plasticization***

According to the view still prevalent in much of the literature on polymers (synthetic and natural) (e.g. Carfagna, Apicella and Nicolais, 1982; Moy and Karasz, 1980), the overall mechanism of plasticization of water-compatible glassy polymers by water may have two major components. The first derives from free volume theory, which provides the general concept that low MW plasticizing diluents have high free volume, so that molecular level, water–polymer interactions lead to increased free volume, which allows increased backbone chain segmental mobility, which in turn results in decreased $T_g$ (Bone and Pethig, 1982; Carfagna, Apicella and Nicolais, 1982; Moy and Karasz, 1980; Nakamura, Hatakeyama and Hatakeyama, 1981; Sears and Darby, 1982). [In fact, it is well known that ability of a diluent to depress $T_g$ decreases linearly with increasing diluent MW, albeit with scatter (Boyer, Baer and Hiltner, 1985), as predicted by free volume theory.] This free volume increase alone accounts completely for the plasticizing effect of water on $T_g$ of polystyrene (Nakamura, Hatakeyama and Hatakeyama, 1981). However, for polymers capable of hydrogen bonding with water, it has been claimed that the free volume contribution to the plasticization mechanism can be augmented by a contribution due to the extent of polymer–water hydrogen bonding (Carfagna, Apicella and Nicolais, 1982; Moy and Karasz, 1980). The particular nature of such molecular level, water–polymer interactions has been suggested to be site-specific hydrogen bonding between water and polar groups on a polymer, and extent of this interaction has been said to increase with increasing polymer hydrophilicity (Batzer and Kreibich, 1981; Bone and Pethig, 1982; Carfagna, Apicella and Nicolais, 1982; Hoeve and Hoeve, 1978; Moy and Karasz, 1980; Nakamura, Hatakeyama and Hatakeyama, 1981; Olson and Webb, 1978; Scandola, Ceccorulli and Pizzoli, 1981; To and Flink, 1978). Furthermore, it has been suggested that when a glassy polymer network structure involves interpolymer hydrogen bonds in addition to any covalent crosslinks, the plasticization mechanism may be augmented further by breakage of these interpolymer hydrogen bonds by water, and their replacement by labile water–polymer hydrogen bonds, which would lead to greatly increased molecular mobility and result in an even greater $T_g$ depression (Carfagna, Apicella and Nicolais, 1982; Hoeve and Hoeve, 1978; Moy and Karasz, 1980; Olson and Webb, 1978; Scandola, Ceccorulli and Pizzoli, 1981; To and Flink, 1978). However, recent publications (Ellis, Jin and Karasz, 1984; Jin, Ellis and Karasz, 1984) have expressed strong favour for the free volume concept, contending that the effectiveness of water as a plasticizer of synthetic polymers merely reflects, in part, its low molar mass (Jin, Ellis and Karasz, 1984). Our studies of water-compatible food polymers have led to the same conclusion (Levine and Slade, 1987a). Karasz and coworkers (Ellis, Jin and Karasz, 1984; Jin, Ellis and Karasz, 1984) have come to discount concepts of specific interactions, such as disruptive hydrogen bonding in polymer hydrogen bonded networks and plasticizing molecules becoming 'firmly bound' to polar sites along a polymer chain, in explaining water's plasticizing ability. To negate the prevailing argument for site-binding, they have cited (Jin, Ellis and Karasz, 1984) NMR results which clearly indicate that water molecules in polymers with polar sites have a large degree of mobility. Levine and Slade (1986, 1987b and Chapter 9 of this volume) and Slade and Levine (1987a) have followed Franks' lead (1982a, 1985a,b, 1986) in taking a similar position and using similar evidence to try to dispel popular myths about 'bound water' and 'water-binding capacity' in food polymers such as starch and other aqueous glass-forming materials, including sugars and polyols. For example, NMR was used to test accessibility of water with reduced mobility in retrograded starch gels, and results showed that all of the water in the gel could be freely exchanged with deuterium oxide (Wynne-Jones and Blanshard, 1986).

***Behaviour of partially crystalline polymers***

In partially crystalline polymers, the fact that water plasticization occurs only in amorphous regions (Gaeta, Apicella and Hopfenberg, 1982; Jin, Ellis and Karasz, 1984; Mohajer *et al.*, 1980; Starkweather, 1980; van den Berg, 1981) both further complicates the effect of crystalline regions on the thermomechanical behaviour of amorphous regions and explains how amorphous regions cause the non-equilibrium melting behaviour of crystalline regions. In partially crystalline polymers with anhydrous crystalline regions and a relatively low capacity for water in amorphous regions, the percentage crystallinity affects $T_g$, such that increasing percentage crystallinity generally lends to increasing $I_g$ (Jin, Ellis and Karasz, 1984; Levine and Slade, 1987a), due to two factors. One is the stiffening or 'antiplasticizing' effect of dispersed microcrystalline crosslinks, which leads to decreased mobility of chain segments in interconnected amorphous regions (Gaeta, Apicella and Hopfenberg, 1982; Jin, Ellis and Karasz, 1984). The same effect is produced by covalent crosslinks (Jin, Ellis and Karasz, 1984; ten Brinke, Karasz and Ellis, 1983) which, when produced by radiation, occur only in amorphous regions (Ellis, Jin and Karasz, 1984; Jin, Ellis and Karasz, 1984). [Note the obvious connection between this stiffening effect in partially crystalline polymers modelled by the 'fringed micelle' (whereby $T_g$ of the homogeneous amorphous 'fringe' is increased), and the previously mentioned effect of the rigid crystalline phase on segmental mobility in the rigid amorphous phase (resulting in a higher $T_g$ than that of the bulk amorphous phase) of polymers modelled by the 'three-microphase' structure.] The second factor is the hydrophobicity of crystalline regions relative to the hydrophilicity of the glassy matrix, which is another way of saying that polymer–polymer contacts are much preferred over polymer–water contacts in crystals (Starkweather, 1980). It has been pointed out that in such polymers, only amorphous regions are accessible to penetration and thus plasticization by moisture (Ellis, Jin and Karasz, 1984; Jin, Ellis and Karasz, 1984). For such polymers, analysed based on a 'three-microphase' model, it has been proposed (Ellis, Jin and Karasz, 1984) that a concentration gradient of plasticizing water may be present. Experimental evidence has suggested that moisture content is zero in the rigid crystalline phase, low in the rigid amorphous phase, and highest (i.e. the majority of water in the polymer) in the bulk amorphous phase (Ellis, Jin and Karasz, 1984; Jin, Ellis and Karasz, 1984).

In native starches composed of partially crystalline polymers with hydrated crystalline regions, it is known that hydrolysis by aqueous acid ('acid-etching') or enzymes at $T < T_m$ can occur initially only in amorphous regions (French, 1984). Similarly, acid-etching of retrograded starch progresses in amorphous regions, leading to increased relative crystallinity (or even increased absolute crystallinity, via crystal growth) of the residue (French, 1984). Dehumidification of granular starch proceeds most readily from initially mobile amorphous regions, leading to non-uniform moisture distribution (Slade and Levine, 1987a). In native starch, the initial $T_g$ which determines $T$ of gelatinization ($T_{gelat}$) depends on extent and type (B versus A versus V polymorphs) of crystallinity of the granule (but not on amylose content), and on total moisture and moisture distribution (Levine and Slade, 1987a; Slade, 1984). $T_{gelat}$ increases with increasing percentage crystallinity, an indirect effect due to disproportionation of mobile short branches of amylopectin from amorphous regions to crystalline micelles (Slade and Levine, 1987a). Two other related phenomena are observed in situations of overall low moisture content for partially crystalline polymers with hydrated crystalline regions, including starch (Biliaderis, Page and

Maurice, 1986; Maurice *et al.*, 1985; Slade and Levine, 1984a,b) and gelatin (Marshall and Petrie, 1980; Reutner, Luft and Borchard, 1985; Slade and Levine, 1984a,b): atypically high (Soesanto and Williams, 1981; Wunderlich, 1980) $T_g/T_m$ ratios $\gg 0.80$; and a pronounced apparent depressing effect of water on $T_m$ as well as $T_g$, such that $T_m$ decreases with increasing percentage water.

Previous attempts to analyse the latter phenomenon, using Flory–Huggins thermodynamic treatment for equilibrium melting of a partially crystalline polymer with diluent (Flory, 1953) had failed (Biliaderis, Maurice and Vose, 1980; Donovan, 1979). We reported (Slade, 1984; Slade and Levine, 1984a,b) for starch and gelatin that the correct explanation lies in the *non*-equilibrium nature of the melting process, and the indirect effect of plasticizing water on $T_m$. Melting of native granular starches or gelatin gels is kinetically controlled, due to existence of contiguous microcrystalline and amorphous regions in the 'fringed micelle' network. Relative dehydration of amorphous regions at an initially low overall moisture content leads to a kinetically metastable condition where the effective $T_g$ is higher than the equilibrium $T_m$ of the crystalline regions. Consequently, the apparent $T_m$ is elevated and observed only after softening of amorphous regions at $T_g$. Added plasticizing water acts directly on the continuous glassy regions, depressing their $T_g$ and thus allowing sufficient mobility and swelling for the interconnected microcrystallites, embedded in the 'fringed micelle' matrix (wherein the 'fringe' is an unstable rubber above $T_g$) to melt (by dissociation and concomitant volume expansion) on heating to a less kinetically-constrained $T_m$ only slightly above the depressed $T_g$. In contrast to the case of limited moisture, in an excess moisture situation (e.g. a retrograded wheat starch gel with $\geqslant 27$ w% water or a gelatin gel with $\geqslant 35$ w% water) where the amorphous matrix would be fully plasticized and $T$ would be $> T_g$ ($\simeq -5$ °C and $-12$ °C, respectively, for starch and gelatin gels), the fully hydrated and matured crystalline junctions in such a gel would show the actual (lower) equilibrium $T_m$ ($\simeq 60$ °C and 37 °C, respectively, for retrograded B-type starch and gelatin) (Levine and Slade, 1987a; Maurice *et al.*, 1985; Slade and Levine, 1984a,b, 1987a). Once starch gelatinization was identified (Slade, 1984) as such a non-equilibrium melting process [for which the Flory–Huggins treatment has no theoretical basis (Alfonso and Russell, 1986)], others who had previously tried to treat starch melting data by Flory–Huggins theory (Biliaderis, Maurice and Vose, 1980) subsequently began to recognize and describe results for non-equilibrium melting of granular starches and amylose–lipid crystalline complexes (Biliaderis, Page and Maurice, 1986; Biliaderis *et al.*, 1986).

***Effects of water as a plasticizer on properties of polymers in the rubbery state—the domain of WLF kinetics***

At $T > T_g$, plasticization by water affects the viscoelastic, thermomechanical, electrical and gas permeability properties of completely amorphous and partially crystalline polymers by means of the effect on $T_g$ (Rowland, 1980). The dependence of polymer properties on $T$ in the rubbery range above $T_g$ (typically at least 100 °C above $T_g$), is successfully predicted (Cowie, 1973) by the WLF equation derived from free volume theory (Ferry, 1980; Williams, Landel and Ferry, 1955). The WLF equation (Soesanto and Williams, 1981; Williams, Landel and Ferry, 1955) is shown in Eqn (1):

$$\log\left(\frac{\eta}{\rho T}\Big/\frac{\eta_g}{\rho_g T_g}\right) = -\frac{C1(T-T_g)}{C2+(T-T_g)} \qquad (1)$$

where η is viscosity or some other diffusion-controlled relaxation process, ρ is density, and C1 and C2 are 'universal constants' [17.44 and 51.6 respectively, as extracted from data on numerous polymers and molten glucose (Soesanto and Williams, 1981; Williams, Landel and Ferry, 1955)]. Eqn (1) describes the kinetic nature of the glass transition (Ferry, 1980), and is universally applicable to any glass-forming polymer, oligomer, or monomer (Ferry, 1980; Soesanto and Williams, 1981). The equation defines the exponential $T$ dependence of any diffusion-controlled relaxation process occurring at $T$, versus the rate of the relaxation at a reference $T$, namely $T_g$ below $T$, in terms of log η proportional to $\Delta T$, where $\Delta T = T - T_g$. The WLF equation is required in the $T$ range of the rubbery or undercooled liquid state above $T_g$, and is based on the $T$ dependence of free volume (i.e. the $T$ dependence of segmental mobility), as illustrated in *Figure 8.3*. It is not required much below $T_g$ (i.e. in the glassy solid state) or in the very low η liquid state [< 10 Pa s (Soesanto and Williams, 1981)] $\gtrsim$ 100 °C above $T_g$, where Arrhenius kinetics apply (Ferry, 1980; Williams, Landel and Ferry, 1955). The WLF equation depends critically on the appropriate reference $T_g$ for any particular glass-forming polymer [of any MW and extent of plasticization (Soesanto and Williams, 1981; Williams, Landel and Ferry, 1955)], where $T_g$ is defined as an iso-free volume state of limiting free volume for the liquid, and also approximately as an iso-η state in the range $10^{11} - 10^{14}$ Pa s (Ferry, 1980; Franks, 1982; Soesanto and Williams, 1981; Williams, Landel and Ferry, 1955).

The impact of WLF behaviour on kinetics of diffusion-controlled relaxation processes in water-plasticized polymers at low moisture can be illustrated as follows. For example, relative relaxation rates versus $\Delta T$ calculated from Eqn (1) demonstrate the exponential relationship: for $\Delta T$s of 0, 3, 7, 11 and 21 °C, corresponding rates would be 1, 10, 100, 1000 and 100 000, respectively. Such rates are dramatically different from those defined by the familiar Q10 rule of Arrhenius kinetics for dilute solutions. Another example has already been illustrated by *Figure 8.2*. The propagation step in the mechanism of recrystallization of an amorphous but crystallizable polymer (e.g. freshly-gelatinized starch or molten glucose), initially quenched from the melt to a kinetically metastable solid state, reflects an essentially zero rate at $T < T_g$. Due to immobility in the glass, migratory diffusion of large main-chain segments required for crystal growth would be inhibited over realistic times. However, propagation rate increases exponentially with increasing $\Delta T$ above $T_g$ (up to $T_m$) due to the mobility allowed in the rubbery state (LaBarre and Turner, 1982; Nakazawa *et al.*, 1985).

The critical message to be distilled from the preceding material in this section is that structure/property relations of water-compatible polymers such as starch, at low moisture, are dictated by a moisture/temperature/time superposition (Flink, 1983; Starkweather, 1980). Referring to the starch–water state diagram in *Figure 8.5* as a conceptual 'map', one sees that the $T_g$ curve represents a boundary between physical states in which various diffusion-controlled processes (e.g. collapse and recrystallization) either can (at $T > T_g$, the domain of 'water dynamics') or cannot (at $T < T_g$, the 'glass dynamics' domain) occur over realistic times (Neogi, 1983). The WLF equation defines kinetics of molecular level relaxation processes that can occur above $T_g$ in terms of a non-Arrhenius, non-linear, non-exponential function of $\Delta T$ above this boundary condition.

## Structural stability of intermediate moisture (IM) foods

Dilute aqueous systems can exist as equilibrium systems, governed by energetics and appropriately described by thermodynamic treatments. For such systems measurement of water activity, in terms of an equilibrium vapour pressure relative to the partial pressure of pure water, may have some relevance. This legitimate use of $A_w$ as a true thermodynamic property for equilibrium systems which are composed of a fixed solute or ratio of solutes with varying water content may be predictive of technological performance and physiological viability (Franks, 1982b; Slade and Levine, 1985; Slade, Levine and Franks, 1987; van den Berg, 1981, 1985, 1986; van den Berg and Bruin, 1981). However, specific physicochemical and biochemical contributions from different solutes may prevent intersystem predictions (Slade, Levine and Franks, 1987).

In contrast, IM systems are not equilibrium systems, are not governed by energetics, and cannot be described by thermodynamic treatments (Levine and Slade, 1987a; Slade and Levine, 1985; Slade, Levine and Franks, 1987). In particular, measured vapour pressure for these non-equilibrium IM systems is not a function of $A_w$, and $A_w$ as a thermodynamic property is not relevant to the description of their behaviour. This is true even for *in vivo* biological systems under extreme stress of drought, freezing, or salinity. At best, the measured value of the relative vapour pressure (RVP) of these systems represents a stationary state vapour pressure. More typically, measured RVP is an instantaneous value for a dynamic property of a system under kinetic control. Use of RVPs as a measurement of 'water availability' to predict product quality and biological viability in IM systems would require a detailed physicochemical description of their behaviour. The spectacular lack of predictive utility (van den Berg, 1985) of RVP (so-called $A_w$), in the absence of such a description, is demonstrated by, for example, hysteresis of sorption–desorption isotherms (van den Berg, 1986). Recent discussions (Slade and Levine, 1985; Slade, Levine and Franks, 1987; van den Berg, 1985) about the use of $A_w$ as a credible measure of technological performance and physiological viability have resulted in new conclusions and guidelines for more credible criteria of IMF product quality and biological viability to replace the currently popular usage of '$A_w$' (Slade, Levine and Franks, 1987).

IM systems should be treated as 'low moisture' systems according to the previously defined context of this review. Hence, we have described a set of fundamental principles for structural stabilization and moisture management of IMFs, from the perspective of our concept of water dynamics (Levine and Slade, 1987a; Slade and Levine, 1985). These principles have been derived from structure/property relations for such non-equilibrium IM systems, viewed generically as completely amorphous or partially crystalline polymer systems, consisting of homologous series from monomers to oligomers to high polymers with their solvents and plasticizers. As described earlier, behaviour of these systems is controlled by kinetics, rather than energetics, and this kinetic control (i.e. the rate of approach to equilibrium) varies according to distinctive temperature and percentage moisture domains. On a 'dynamics map' (e.g. the state diagram in *Figure 8.5*), the domain of water dynamics corresponds to the region of $T > T_g$ and $W > W_g$ (Levine and Slade, 1987a; Slade and Levine, 1985; Slade, Levine and Franks, 1987).

The first question to ask in order to determine the contribution of water 'availability' to quality, stability, and viability of IM systems should be 'is water required *per se* or is the requirement simply for mobility accrued by the presence of a low MW species (plasticizer)?' Other molecules (e.g. sugars and sugar alcohols, which act as cryoprotectants in nature) can replace water in maintaining biopolymers (e.g. proteins) in a viable but inactive native state, even under conditions of extreme dehydration (Franks, 1985a). If there is no chemical requirement for water molecules, then it is appropriate to consider the factors that govern mobility (Slade and Levine, 1985).

Mobility of an IM system is governed by temperature and composition, the latter expressed in terms of $\bar{M}w$ of the total system, i.e. polymer(s) and plasticizer(s) (Slade and Levine, 1985). At constant concentration, which defines $\bar{M}w$ of a system, a change in temperature changes mobility. At constant temperature, a change in concentration changes mobility. The extent of mobility can be measured in terms of $\eta$, diffusion, and rates of relaxation or reaction. The temperature domains which influence mobility and kinetic control are determined by absolute temperature with respect to the two characteristic transition temperatures of partially crystalline systems, $T_g$ and $T_m$. As previously illustrated in *Figure 8.4*, $T_g$ is typically 0.5–0.8 (°K) of $T_m$ (Wunderlich, 1980), and only at $T < T_g$ and $> T_m$ are Arrhenius kinetics observed (Slade and Levine, 1985). It bears repeating here that, in contrast, WLF kinetics are observed at $T_g < T < T_m$, which represents the most significant $T$ region (within the domain of water dynamics) for study of water availability (Slade and Levine, 1985). While Arrhenius kinetics typically involve a doubling of rate for an increase in temperature of 10 °K (without identifying any particular reference temperature), WLF kinetics typically involve an order of magnitude increase in rate (equivalent to an order of magnitude increase in diffusion or decrease in $\eta$) for each 3–5 °K increase in temperature above $T_g$ as a reference temperature (Ferry, 1980).

$T_g$ depends on MW (i.e. linear degree of polymerization, DP) for a single substance within a homologous series and on $\bar{M}w$ for a mixture (Billmeyer, 1984). Thus, a 'glass curve' with constant $\eta$ can be defined in terms of temperature and composition; the high temperature tie-point is determined by $T_g$ of the pure high MW component and the low temperature tie-point by that of the low MW component, which serves as a plasticizer (as shown earlier in *Figure 8.5* for the water–starch system) (Slade and Levine, 1985). Other typical solute–water glass curves are highlighted in *Figure 8.6*, where schematic state diagrams, which illustrate the relationship between $T_g$ and $T_m$, are represented for the common sugars sucrose, glucose, and fructose (Levine and Slade, 1987a; Slade and Levine, 1985). Another interesting illustration of what can be gleaned from an analysis of glass curves for complex aqueous mixtures is shown in *Figure 8.7* (Levine and Slade, 1987a; Slade and Levine, 1985). In this artist's rendering of a three-dimensional state diagram for a hypothetical three-component system, both solutes (e.g. a polymer, 2, and its monomer, 1) are non-crystallizing, interacting (i.e. compatible), and plasticized by water, which is the crystallizing solvent, 3. The diagram reveals the postulated origin of a sigmoidal curve of $T'_g$ versus w% solute composition, the $T'_g$ glass curve ABCDEF (Levine and Slade, 1987a; Slade and Levine, 1985). In fact, similar sigmoidal curves of collapse temperature, $T_c$ versus w% concentration for collapse during freeze-drying of analogous three-component aqueous systems have been reported by MacKenzie (1975, p. 288), but 'the basis for (this non-linearity) has not yet been determined'.

As demonstrated in Chapter 9 of this volume, $T_g$ is a useful diagnostic tool. It is a reference temperature which defines a $\eta$ of $\simeq 10^{14}$ Pa s, for systems with a typical ratio of $T_m/T_g \simeq 1.25$–2 (Brydson, 1972; Soesanto and Williams, 1981; Wunderlich, 1980). It is manifested by a dramatic change in $\eta$, diffusion, and heat capacity over a small temperature interval near $T_g$ (Ferry, 1980). Iso-$\eta$ curves differing by an order of magnitude can be drawn as contour lines on a water dynamics map, such as the state diagram for gelatinized starch in *Figure 8.5*, at about 3–5 °K intervals in the region between $T_g$ and $T_m$. These contour lines represent the projection of the time axis

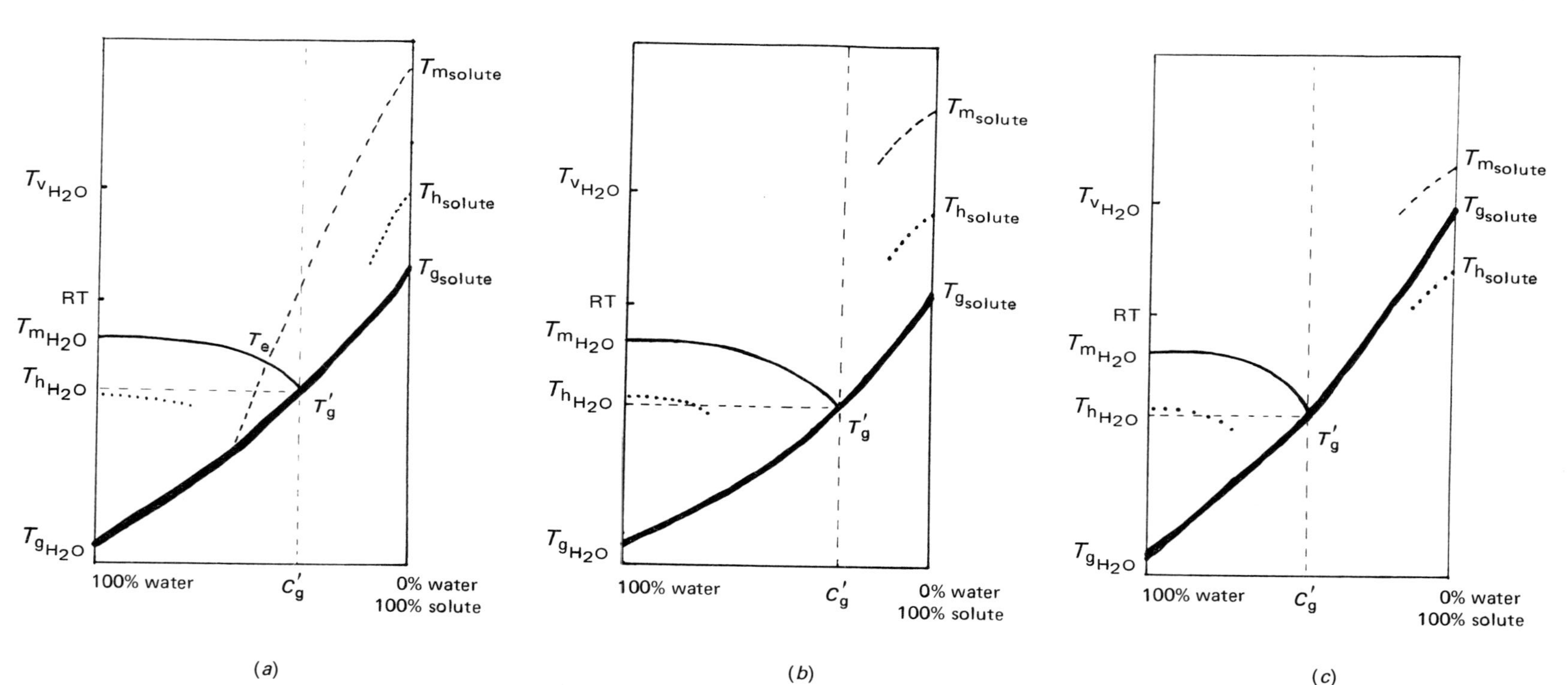


**Figure 8.6** Schematic state diagrams for (a) sucrose, (b) glucose, and (c) fructose, which emphasize the solute–water glass curve, and the relationship between $T_m$, $T_g$, and $T_h$, the estimated homogeneous nucleation temperature, for each pure solute. From Levine and Slade (1987a).

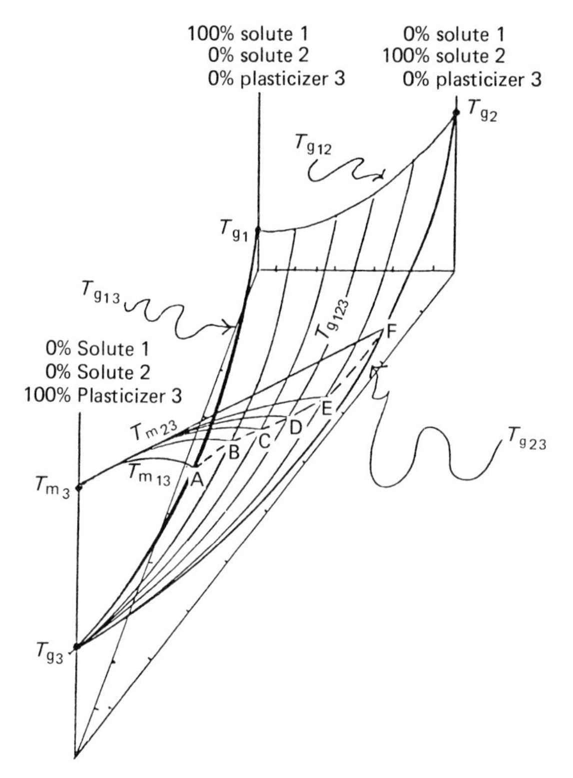


**Figure 8.7** A schematic three-dimensional state diagram for a hypothetical three-component aqueous system. The two solutes (e.g. polymer + monomer) are both non-crystallizing, interacting, and plasticized by water, which is the crystallizing solvent. The diagram illustrates the postulated origin of a sigmoidal curve of $T'_g$ versus w% solute composition. From Levine and Slade (1987a).

(manifested as isograms for relaxation times, viscosity, or relative vapour pressure) from the three-dimensional map onto a two-dimensional map of temperature versus moisture (Slade, Levine and Franks, 1987). While this rubbery domain is typically about 100 °K for 'well-behaved' systems (e.g. pure high polymers) with typical ratios of $T_m/T_g$, there are some notable exceptions and qualifications to this situation. For example, in frozen aqueous solutions the magnitude of the WLF region, in terms of $\Delta T$ between $T'_g$ and $T_m$ of ice, can be as small as 4 °C for a 0.5-DE dextrin (Levine and Slade, 1986; Slade and Levine, 1985). A synthetic polymer with an exceptionally low value of 1.18 for $T_m/T_g$, reportedly due to anomalously large free volume at $T_g$, is bisphenol polycarbonate (Brydson, 1972). Fructose is another important exception (Slade and Levine, 1985).

In a completely amorphous, quench-cooled melt of pure crystalline β-D-fructose, we have observed (by DSC) two separate glass transitions (Levine and Slade, 1987a). The first $T_g$ occurs near 30 °C, which is approximately the same temperature reported for $T_g$ of other monosaccharides such as glucose and mannose (Slade and Levine, 1985; Soesanto and Williams, 1981). However, the second $T_g$ for fructose occurs at 100 °C, only 26 °C below its $T_m$ (Slade and Levine, 1985; Soesanto and Williams, 1981). The observation of two $T_g$s may indicate the presence of two different major anomeric forms of fructose immobilized in the quenched melt (Levine and Slade, 1987a). Similar findings of multiple anomeric forms (which is a common situation in aqueous sugar solutions) in quenched melts have been reported from DSC analyses of other low MW crystalline carbohydrates (Shafizadeh *et al.*, 1971). The higher value $T_g$ of fructose would translate to a $T_m/T_g$ ratio of 1.06, much lower even than that for bisphenol polycarbonate (Slade and Levine, 1985). Since mobility of a glass or high-η rubber depends on the $T_m/T_g$ ratio, which is a predictor of relative free volume for pure substances (Ferry, 1980), the exceptionally low $T_m/T_g$ ratio for fructose would be indicative of high mobility and free volume but low η in a fructose glass at its $T_g$ (Levine and Slade, 1987a). Thus, instead of the typical value of $10^{14}$ Pa s at $\eta_g$ mentioned above, a value of only $10^{11}$ Pa s has been suggested (Slade and Levine, 1985; Soesanto and Williams, 1981) for $\eta_g$ of amorphous fructose. This explains why we and others (Soesanto and Williams, 1981) usually cite a range of three orders of magnitude for $\eta_g$ at $T_g$. We believe that it is important to recognize that this anomalously low $\eta_g$ may account for much of the unusual behaviour of fructose, compared to other more typical monomeric sugars, often observed in IM systems (Chirife, Favetto and Fontan, 1982; Loncin, 1975). For example, the relative instability of fructose systems compared to glucose or mannose systems would be predictable because of greater mobility in a fructose glass or rubber than in a glucose ($T_m/T_g = 1.42$) or mannose ($T_m/T_g = 1.36$) glass or rubber (Levine and Slade, 1987a; Slade and Levine, 1985). As illustrated by results in *Table 8.1* (described below), at the same or lower RVP, fructose affords less microbiological stability than glucose or mannose (Levine and Slade, 1987a; Slade and Levine, 1985).

The ubiquitous application of our food polymer science approach to understanding of aqueous sugar glasses and concentrated solutions can be seen in the analogous relative influence of these same sugars on gelatinization and recrystallization behaviour of starch (Slade and Levine, 1987a) and in dielectric relaxation times of concentrated sugar solutions (Suggett and Clark, 1976).

As the universal plasticizer for biological and food systems, water is a 'mobility enhancer' (Slade, Levine and Franks, 1987). Its low MW leads to a large increase in mobility as moisture content is increased from that of a dry solute to a solution. However, since water is a crystallizing plasticizer, ice formation leads to freeze-concentration of a solution, until $T$ is depressed to $T_g$, where the limited rate of diffusion prevents further crystallization. As described earlier, upon slow freezing, composition changes initially along the equilibrium liquidus curve, and subsequent maximum freeze-concentration by non-equilibrium freezing is observed at $T'_g$ (Franks, 1982a). For each solute or mixture of solutes, $W'_g$ is the characteristic maximum water content of the aqueous glass at $T'_g$ (Levine and Slade, 1986; Slade and Levine, 1985). Even though this water is kinetically metastable and does not freeze within a practical timeframe, it is not bound at subzero temperatures and is not energetically stable (Slade, Levine and Franks, 1987; *see* also Chapter 9 of this volume). Similarly, the apparent anomalous depression of vapour pressure ('water activity') and freezing point are kinetic phenomena, and, like 'bound water', are symptoms of the non-equilibrium rubbery state. The aqueous glass identified by $T'_g$ and $W'_g$ is as homogeneous as the same solution at room temperature. These parameters serve as a third point on the diagnostic glass curve, so that the shape as well as the end-points can be described for each system. The state diagrams in *Figure 8.6* illustrate glass curves for sucrose, glucose and fructose (Levine and Slade, 1987a; Slade and Levine, 1985). Comparison of these diagrams reveals differences in the kinetically metastable domains between $T_m$ and $T_g$, with respect to estimated homogeneous nucleation temperatures ($T_h$) of these sugars. In each case, $T_h$ was estimated from the ratio of $T_h/T_m$ (°K) which, for many partially crystalline polymers, is typically 0.8, with a reported range of 0.78–0.85 (Walton, 1969; Wunderlich, 1976). The relationship between $T_h$ and $T_g$ allows prediction of stability towards recrystallization of concentrated and supersaturated aqueous solutions (Levine and Slade, 1987a; Slade and Levine, 1985). For sucrose and glucose which are known to crystallize readily by undercooling such solutions, $T_h > T_g$, so homogeneous nucleation can occur before vitrification on cooling from $T > T_m$. In contrast, for fructose which is impossible to crystallize (in a realistic time) by the same mechanism, $T_g > T_h$ so, on cooling, vitrification occurs first, thus immobilizing the system and preventing the possibility of homogeneous nucleation (Levine and Slade, 1987a).

The following experimental approach, based on WLF kinetics of partially crystalline polymer systems, has been suggested for investigating water dynamics as a predictive parameter for quality and stability of IM systems (Levine and Slade, 1987a; Slade, Levine and Franks, 1987):

Measure $T_g$ of anhydrous solute.
Measure $T'_g$ and $W'_g$ of freeze-concentrated glass.
Use literature estimate (Franks, 1982a) for $T_g$ of water.
Construct $T_g(c)$ curve to define reference state for kinetic metastability ($\eta \simeq 10^{14}$ Pa s).
Measure $T_m/T_g$ to estimate departure from reference of $\eta \simeq 10^{14}$ Pa s for typical glass behaviour.

Then shelf life is determined by a combination of $\Delta T$ and $\Delta W$, where

| | | |
|---|---|---|
| $\Delta T = T - T_g$ | for constant water content | at $W < W'_g$; |
| $\Delta W = W - W_g$ | for constant $T$ | at $W < W'_g$ and any $T$; |
| $\Delta W = W - W_g$ | for constant $T$ | at $W > W'_g$ and $T > T'_g$; |
| $\Delta T = T - T'_g$ | for | $W > W'_g$; and |
| $\Delta W = W - W'_g$ | for constant $T$ | at $W > W'_g$ and $T < T'_g$. |

For a convenient estimation of relative shelf life, a vector can be constructed from $\Delta T = T - T'_g$ and $\Delta W = W - W'_g$ (Slade and Levine, 1985).

Two examples of use of this approach to describe water availability and product quality are shown in *Figure 8.8* (Levine and Slade, 1987b) for a low-temperature system and *Table 8.1* (Levine and Slade, 1987a; Slade and Levine, 1985) for a system near room temperature. At low temperature, development of iciness (due to grain growth of ice) during freezer storage of ice cream is detrimental to product quality. Rate of development of iciness, which limits useful shelf life, depends on $\Delta T$ of $T$ freezer above $T'_g$ (Levine and Slade, 1986), since this recrystallization process is an example of WLF-governed collapse phenomena (*see* Chapter 9). Because typical ice cream products have $T'_g$ values in the range −30 to −43 °C, they would exist as rubbery fluids (with embedded ice and fat crystals) under typical freezer storage at −18 °C (Cole *et al.*, 1983, 1984), and WLF kinetics would describe the rate of ice crystal growth. *Figure 8.8* contains a WLF plot of log rate of iciness development versus $\Delta T$, which shows that iciness increases with increasing $\Delta T$ with a linear regression coefficient of 0.68. Considering that iciness scores were obtained by sensory evaluation, we believe this unprecedented experimental demonstration of WLF behaviour in a frozen system is remarkable. However, we were amazed to find that this non-Arrhenius behaviour of freezer-stored ice cream is apparently recognized, at least empirically, by the British frozen foods industry. We have seen on ice cream

**Table 8.1** GERMINATION OF MOULD SPORES OF *ASPERGILLUS PARASITICUS* IN INTERMEDIATE MOISTURE (IM) SYSTEMS

| *Design parameters* | | | | | | *IM sample* | | *Days required to germinate at 30 °C* |
|---|---|---|---|---|---|---|---|---|
| *RVP*[a] (30 °C) | $T'_g$ (°C) | $W'_g$[b] (w% $H_2O$) | $T_g$ (°K) | $T_m$ (°K) | $T_m/T_g$ (°K) | *Concentration* (w% $H_2O$) | *Solute type* | |
| *Controls* | | | | | | | | |
| 1.0 | | | | | | 100 | None | 1 |
| ~ 1 | | | | | | 99 | Glucose (α-D) | 1 |
| ~ 1 | | | | | | 99 | Fructose (β-D) | 1 |
| ~ 1 | | | | | | 99 | PVP-40 | 1 |
| ~ 1 | | | | | | 99 | Glycerol | 2 |
| 0.92 | − 21.5 | 35 | 373 | — | | 50 | PVP-40 | 21 |
| 0.92 | − 45.5 | 49.5 | 302 | 444.5 | 1.47 | 60 | α-methyl glucoside[c] | 1 |
| 0.83 | − 42 | 49 | 373 | 397 | 1.06 | 50 | Fructose | 2 |
| 0.83 | − 65 | 46 | 180 | 291 | 1.62 | 60 | Glycerol | 11 |
| 0.99 | − 29.5 | 20 | 316 | 402 | 1.27 | 60 | Maltose | 2 |
| 0.97 | − 32 | 36 | 325 | 465 | 1.43 | 60 | Sucrose | 4 |
| 0.95 | − 23 | 31 | 349 | 406.5 | 1.16 | 50 | Maltotriose | 8 |
| 0.93 | − 41 | 26 | 303 | 412.5 | 1.36 | 50 | Mannose | 4 |
| 0.95 | − 23 | 31 | 349 | 406.5 | 1.16 | 50 | Maltotriose | 8 |
| 0.92 | − 21.5 | 35 | 373 | — | | 50 | PVP-40 | 21 |

| | | | | | | | | |
|---|---|---|---|---|---|---|---|---|
| 0.93 | − 41 | 26 | 303 | 412.5 | 1.36 | 50 | Mannose | 4 |
| 0.87 | − 42 | 49 | 373 | 397 | 1.06 | 54 | Fructose | 2 |
| 0.92 | − 45.5 | 49.5 | 302 | 444.5 | 1.47 | 60 | α-methyl glucoside | 1 |
| 0.87 | − 42 | 49 | 373 | 397 | 1.06 | 54 | Fructose | 2 |
| 0.92 | − 45.5 | 49.5 | 302 | 444.5 | 1.47 | 60 | α-methyl glucoside | 1 |
| 0.70 | − 42 | 49 | 373 | 397 | 1.06 | 30 | Fructose | 2 |
| 0.85 | − 43 | 29 | 304 | 431 | 1.42 | 50 | Glucose | 6 |
| 0.83 | − 42 | 49 | 373 | 397 | 1.06 | 50 | Fructose | 2 |
| 0.82 | − 42.5 | 48 | 293 | — | | 40 | 1/1 Fructose/Glucose | 5 |
| 0.98 | − 26 | 36 | 339 | — | | 50 | PVP-10 | 11 |
| 0.98 | − 42 | 49 | 373 | 397 | 1.06 | 60 | Fructose | 2 |
| 0.93 | − 26 | 36 | 339 | — | | 40 | PVP-10 | 11 |
| 0.95 | − 21.5 | 35 | 373 | — | | 60 | PVP-40 | 9 |
| 0.99 | − 26 | 36 | 339 | — | | 60 | PVP-10 | 11 |
| 0.99 | − 29.5 | 20 | 316 | 402 | 1.27 | 60 | Maltose | 2 |

[a] RVP measured after seven days ‘equilibration’ at 30 °C.
[b] $W'_g$ expressed here in terms of w% water, for ease of comparison with IM sample concentration.
[c] Commercial sample from Staley—technical grade, used as received.

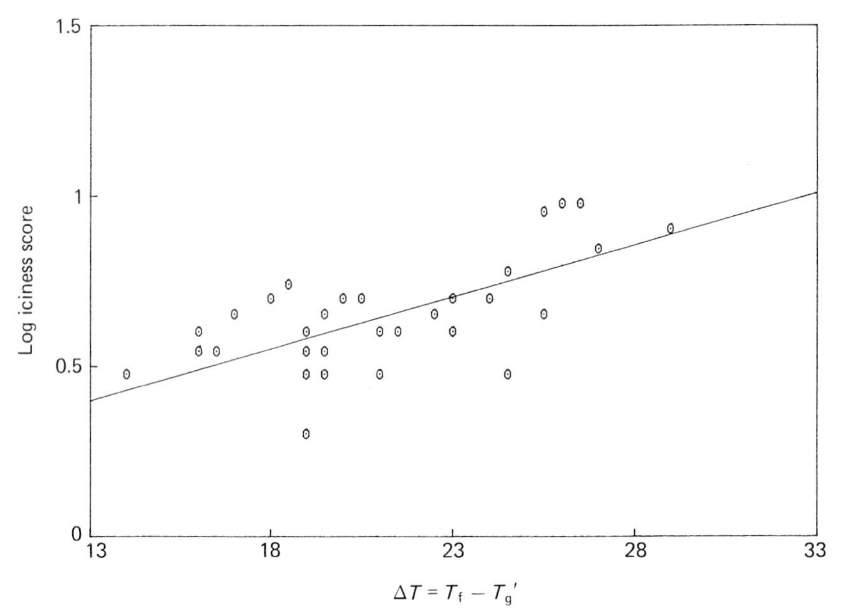


**Figure 8.8** Log iciness score (determined organoleptically, on a 0–10 point scale) as a function of $\Delta T$ ($= T_f - T'_g$), for experimental ice cream products, after two weeks of deliberately abusive (temperature-cycled) frozen storage in a so-called 'Brazilian Ice Box'. From Levine and Slade (1987a).

packages in the UK, the following shelf-life code: in a one-star home freezer (15 °F), storage life = 1 day; ** freezer (0 °F), 1 week; *** (−10 °F), 1 month.

Near room temperature, initial germination of mould spores of an *Aspergillus* depends only on water availability, not on the presence of nutrients (Lang, 1981). The observed rates of germination at 30 °C could not be predicted by measured RVPs, but increased understanding was gained from the suggested approach based on state diagrams to describe kinetics of these partially crystalline polymer systems. The illuminating results shown in *Table 8.1* (Levine and Slade, 1987a; Slade and Levine, 1985) represent a dramatic experimental demonstration of the failure of '$A_w$' to predict relative usefulness of additives for antimicrobial stabilization. The experimental protocol, adapted from a microbiological assay used by Lang (1981), compared inhibitory effects on conidia germination for a series of IM solutions of selected glass formers. Germination is essentially an all-or-nothing process, with massive appearance of short hyphae surrounding the previously bare spores occurring within 24 h at 30 °C in pure water or dilute solution (RVP $\simeq$ 1.0). Various glass formers were assayed in pairs, deliberately matched as to individual parameters of approximately equal RVP (at 30 °C), solute concentration, $\bar{M}$w, $T'_g$ and/or $W'_g$.

For fructose versus glucose (at equal solute concentration, $\bar{M}$w, and $T'_g$), fructose produced a less stable system (i.e. faster germination), even at slightly lower RVP. Likewise for the fructose versus glycerol, maltose versus sucrose, and mannose versus fructose pairs, the solute system with a lower ratio of $T_m/T_g$ was faster to show germination, regardless of RVP values. Thus, apparently due to higher free volume, mobility, and lower $\eta$ in the rubber (as governed by $T_m/T_g$), water availability was greater for fructose ($T_m/T_g = 1.06$) > mannose ($T_m/T_g = 1.36$) > glucose ($T_m/$

$T_g = 1.42$) > glycerol ($T_m/T_g = 1.62$), and maltose ($T_m/T_g = 1.27$) > sucrose ($T_m/T_g = 1.43$), so that greater antimicrobial stabilization was observed for glycerol > glucose > mannose > fructose, and sucrose > maltose. The extraordinary mobility and water availability of IM fructose rubbers was manifested by the same fast germination time observed for solutions of 40–70 w% fructose and corresponding RVPs of 0.98–0.70. Other noteworthy results in *Table 8.1* involved the poly(vinyl pyrrolidone) PVP-40 versus methyl glucoside, maltotriose versus mannose, and PVP-10 versus fructose pairs, for which solute MW, as reflected by $T'_g$, appeared to be the critical variable and determinant of functionality. In each case, the solute of higher MW and $T'_g$ manifested lower water availability in its IM rubber (regardless of RVP values), and thus greater stabilization against germination, in direct analogy with structure/property principles described previously for cryo- and thermomechanical stabilization in the glass dynamics domain (Levine and Slade, 1986).

The definitiveness of these results contrasts sharply with the controversial, contradictory and confusing state of the IMF literature on use of fructose versus glucose to lower '$A_w$' in moisture management applications (Chirife, Favetto and Fontan, 1982; Loncin, 1975). Since this subject is one of great practical and topical interest and importance to the food industry, it represents a good example of the potential utility of our concept of water dynamics, in place of '$A_w$', for predicting functional behaviour from structural properties (Levine and Slade, 1987a; Slade and Levine, 1985; Slade, Levine and Franks, 1987). The current IMF literature (e.g. Chirife, Favetto and Fontan, 1982; Loncin, 1975) can be summarized by the following paradoxical 'truths'. On an equal weight basis, replacement of glucose by fructose results, for many IMF products, in a lower RVP and increased shelf stability. However, if a product has been formulated with glucose to achieve a certain RVP, which has been found to result in satisfactory microbiological stability and organoleptic quality, then reformulation with fructose to the same RVP often has not produced as stable a product. This is so because less fructose than glucose is required to achieve the same RVP, while, as has been shown by our model system studies, at the same RVP fructose solutions are less stable than glucose solutions. [Unfortunately, it has not even been possible for agreement to be reached in the IMF literature on correct RVP values for 50 and 60 w% fructose solutions (Chirife, Favetto and Fontan, 1982; Loncin, 1975).] We feel that the proposed concept of water dynamics can explain such apparently contradictory results and will allow the IMF field to advance beyond its current state of affairs (Slade, Levine and Franks, 1987).

Another aspect of our experimental approach to understanding water dynamics in IM systems has involved investigating the possible correlation between RVP and $W'_g$. RVP was measured, after nine days 'equilibration' at 30 °C, for a series of 67.2 w% solids solutions, and plotted against $W'_g$ for maximally frozen 20 w% solutions of the same solids, as shown in *Figure 8.9* (Levine and Slade, 1987a; Slade and Levine, 1985). The samples represented a quasi-homologous family of sugar syrups, including high fructose corn, ordinary corn, sucrose, and invert syrups, all of which are commonly used ingredients in IMF products. *Figure 8.9* illustrates the fair linear correlation ($r = -0.71$) between decreasing content of unfrozen water in the glass at $T'_g$ and increasing RVP, which is usually assumed to be an indicator of free water content in IM systems at room temperature. The obvious scatter of these data prohibited any fundamental insights into the question of water availability in such systems. This was not unexpected since many of these samples represented non-homologous, polydisperse mixtures of polymeric solutes of unknown $\bar{M}w$, MW distribution and $T_m/T_g$.

A general set of structure/property relationships for IM systems has begun to

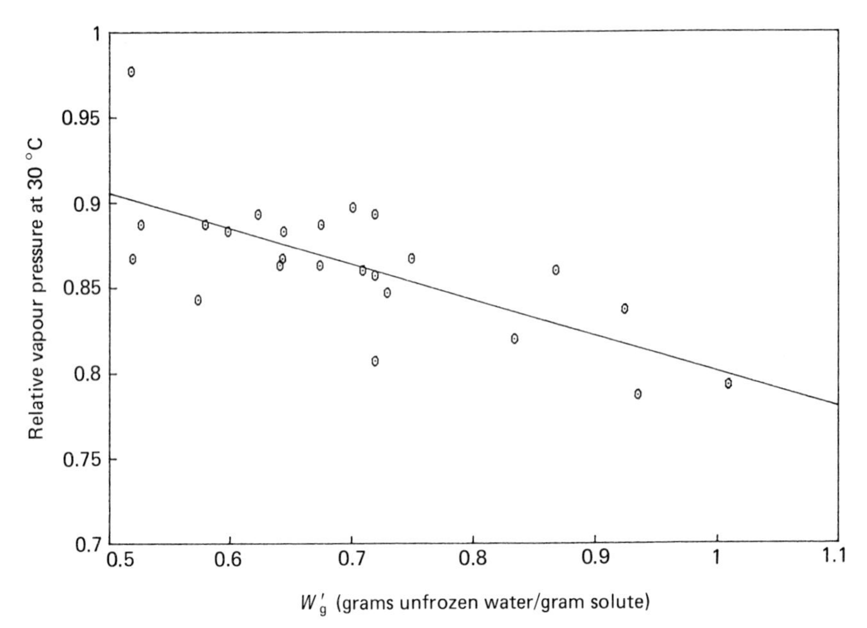


**Figure 8.9** Variation of relative vapour pressure (measured for 67.2 w% solutions of various corn, sucrose, and invert syrup solids, after nine days at 30 °C) against $W'_g$, the composition of the glass at $T'_g$ in g unfrozen water/g solute, for maximally frozen 20 w% solutions of the same syrup solids. From Levine and Slade (1987a).

emerge from experimental results such as those in *Table 8.1*, considered from the perspective of water dynamics in low moisture polymeric systems (Slade, Levine and Franks, 1987). Other results, such as those in *Table 8.2* (Levine and Slade, 1987a; Slade and Levine, 1985), have also been integrated in the development of these new guidelines for more credible criteria (with predictive capability) of quality, stability, and viability in IM systems to replace '$A_w$' or RVP. *Table 8.2* contains literature data (Thill, Schirman and Appleby, 1979; Wiggans and Gardner, 1959) for relative germination times of plant seeds in IM systems at equivalent osmotic pressures (analogous to equivalent RVPs) and temperatures near room temperature. Such results suggest, at best, a rough correlation between increasing solute MW and increasing germination time, but have left unanswered questions about observed differences between glucose, sucrose and mannitol. The emerging guidelines (Levine and Slade, 1987a; Slade and Levine, 1985; Slade, Levine and Franks, 1987) are as follows:

1. For solutes of equal solids content at $T_g$, greater stability is achieved for one with a higher ratio of $T_m/T_g$;
2. For solutes with equal ratios of $T_m/T_g$, greater stability at $T_g$ is achieved by higher solids content;
3. At a given ambient temperature, greater stability results for a smaller $\Delta T$ above $T_g$; and
4. For reasons of efficacy and cost effectiveness of a stabilizing agent for IM systems, choose a material for which vitrification is achieved with low solids content (implying high $W'_g$), $\Delta T$ is small (implying high $T_g$), and $T_m/T_g$ ratio is high (implying low free volume in the glass).

**Table 8.2** RELATIVE GERMINATION TIMES OF SEEDS IN INTERMEDIATE MOISTURE (IM) SYSTEMS AT EQUIVALENT OSMOTIC PRESSURES

| *IM system* | *Relative germination times at 20 °C* |
|---|---|
| *A. Winter wheat seeds* | |
| Water control | 1 |
| Mannitol | 2.75 |
| Poly(ethylene glycol) (MW 20 000) | 5.5 |
| | *Relative germination times at 24 °C* |
| *B. Scarlet globe radish seeds* | |
| Water control | 1 |
| Mannitol | 1.3 |
| Sucrose | 1.5 |
| Glucose | 1.8 |
| PVP (MW not specified) | 9 |

## Conclusion

As emphasized in this review, we have seen, especially since 1980, a growing awareness among a small but increasing number of food scientists of the value of a polymer science approach to the study of food materials and systems. In this respect, food science has followed the compelling lead of the synthetic polymers field. During this decade there has also been an increased recognition of two key elements of this research approach:

1. The critical role of water as a plasticizer of amorphous and partially crystalline materials; and
2. The importance of the glass transition as a physicochemical parameter which can govern product properties, processing, and stability.

Today, the first question we are asked most frequently is 'so what is the $T_g$ of glucose (or starch, gelatin, gluten, etc.)?' The necessary and appropriate response to this question is another question 'at what moisture content ($W_g$) and distribution?' With the answer to the second question (plus information on $W'_g$ relative to $W_g$ for the specific material and situation), one can begin to answer intelligently the original question. In this review, we have tried to illustrate, with examples from our experiences with food materials, how such questions and answers can allow one to understand and explain complex behaviour, design processes, and predict product quality and storage stability, all based on fundamental structure/property relations defined by studies which employ a polymer science approach to low moisture polymeric systems plasticized by water. In future years we can expect to see much progress reported from this emerging, cross-disciplinary research area, especially with regard to the important industrial problem of structural stabilization of intermediate moisture foods.

## Acknowledgements

We thank General Foods Corporation for permission to publish; our former colleague at General Foods, Terry Maurice, for his contributions to our research programme; Cornelis van den Berg and John Blanshard for encouragement of our work; and especially our consultant, Prof. Felix Franks of the University of Cambridge, for invaluable suggestions, discussions, and much encouragement and support over the years.

## References

ABLETT, S., ATTENBURROW, G.E. and LILLFORD, P.J. (1986). The significance of water in the baking process. In *Chemistry and Physics of Baking*, (Blanshard, J.M.V., Frazier, P.J., and Galliard, T., Eds.), pp. 30–41. London, Royal Society of Chemistry

ALFONSO, G.C. and RUSSELL, T.P. (1986). Kinetics of crystallization in semicrystalline/amorphous polymer mixtures. *Macromolecules*, **19,** 1143–1152

BAIR, H.E., JOHNSON, G.E., ANDERSON, E.W. and MATSUOKA, S. (1981). Non equilibrium annealing behavior of poly(vinyl acetate). *Polymer Engineering and Science*, **21,** 930–935

BATZER, H. and KREIBICH, U.T. (1981). Influence of water on thermal transitions in natural polymers and synthetic polyamides. *Polymer Bulletin*, **5,** 585–590

BILIADERIS, C.G., MAURICE, T.J. and VOSE J.R. (1980). Starch gelatinization phenomena studied by DSC. *Journal of Food Science*, **45,** 1669–1680

BILIADERIS, C.G., PAGE, C.M., SLADE, L. and SIRETT, R.R. (1985). Thermal behavior of amylose–lipid complexes, *Carbohydrate Polymers*, **5,** 367–389

BILIADERIS, C.G., PAGE, C.M. and MAURICE, T.J. (1986). Non-equilibrium melting of amylose–V complexes. *Carbohydrate Polymers*, **6,** 269–288

BILIADERIS, C.G., PAGE, C.M., MAURICE, T.J. and JULIANO, B.O. (1986). Thermal characterization of rice starches: a polymeric approach to phase transitions of granular starch. *Journal of Agricultural and Food Chemistry*, **34,** 6–14

BILLMEYER, F.W. (1984). *Textbook of Polymer Science*, 3rd Edition. New York, Wiley-Interscience

BLANSHARD, J.M.V. (1986). The significance of the structure and function of the starch granule in baked products. In *Chemistry and Physics of Baking*, (Blanshard, J.M.V., Frazier, P.J. and Galliard, T., Eds), pp. 1–13. London, Royal Society of Chemistry

BLUM, F.D. and NAGARA, B. (1986). Solvent mobility in gels of atactic polystyrene. *Polymer Preprints*, **27**(1), 211–212

BONE, S. and PETHIG, R. (1982). Dielectric studies of the binding of water to lysozyme. *Journal of Molecular Biology*, **157,** 571–575

BORCHARD, W., BERGMANN, K. and REHAGE, G. (1976). Investigations of gelation phenomena in aqueous gelatin solutions. In *Photographic Gelatin II* (Cox, R.J., Ed.), pp. 57–71. London, Academic Press

BORCHARD, W., BREMER, W. and KEESE, A. (1980). State diagram of the water–gelatin system. *Colloid and Polymer Science*, **258,** 516–526

BORCHARD, W., BERGMANN, K., EMBERGER, A. and REHAGE, G. (1976). Thermodynamische eigenschaften des systems gelatine–water. *Progress in Colloid and Polymer Science*, **60,** 120–129

BOYER, R.F., BAER, E. and HILTNER, A. (1985). Concerning gelation effects in atactic polystyrene solutions. *Macromolecules*, **18**, 427–434

BRYDSON, J.A. (1972). The glass transition, melting point and structure. In *Polymer Science*, (Jenkins, A.D., Ed.), pp. 194–249. Amsterdam, North Holland

CAKEBREAD, S.H. (1969). Factors affecting the shelf life of high boilings. *Manufacturing Confectioner*, **49**, 41–44

CARFAGNA, C., APICELLA, A. and NICOLAIS, L. (1982). The effect of the prepolymer composition of amino-hardened epoxy resins on the water sorption behavior and plasticization. *Journal of Applied Polymer Science*, **27**, 105–112

CHENG, S.Z.D., CAO, M.Y. and WUNDERLICH, B. (1986). Glass transition and melting behavior of poly(oxy-1, 4-phenyleneoxy-1, 4-phenylene carbonyl-1,4-phenylene). *Macromolecules*, **19**, 1868–1876

CHIRIFE, J., FAVETTO, G. and FONTAN, C. (1982). The water activity of fructose solutions in the intermediate moisture range. *Lebensmittel Wissenschaft und Technologie.*, **15**, 159–160

COLE, B.A., LEVINE, H.I., MCGUIRE, M.T., NELSON, K.J. and SLADE, L. (1983). Soft, Frozen Dessert Formulation. US Patent 4 374154

COLE, B.A., LEVINE, H.I., MCGUIRE, M.T., NELSON, K.J. and SLADE, L. (1984). Soft, Frozen Dessert Formulation. US Patent 4 452824

COWIE, J.M.G. (1973). *Polymers: Chemistry and Physics of Modern Materials*. New York, Intertext

DJABOUROV, M. and PAPON, P. (1983). Influence of thermal treatments on structure and stability of gelatin gels. *Polymer*, **24**, 537–542

DOMSZY, R.C., ALAMO, R., EDWARDS, C.O. and MANDELKERN, L. (1986). Thermoreversible gelation and crystallization of homopolymers and copolymers. *Macromolecules*, **19**, 310–325

DONOVAN, J.W. (1979). Phase transitions of the starch–water system. *Biopolymers*, **18**, 263–275

DOWNTON, G.E., FLORES-LUNA, J.L. and KING, C.J. (1982). Mechanism of stickiness in hygroscopic, amorphous powders. *Industrial Engineering Chemistry Fundamentals*, **21**, 447–451

DURAND, D., EMERY, J.R. and CHATELLIER, J.Y. (1985). Investigation of renaturation in gelatin gels. *International Journal of Biological Macromolecules*, **7**, 315–319

EISENBERG, A. (1984). The glassy state and the glass transition. In *Physical Properties of Polymers*, (Mark, J.E., Eisenberg, A., Graessley, W.W., Mandelkern, L. and Koenig, J.L., Eds), pp. 55–95. Washington, American Chemical Society

ELLIS, T.S., JIN, X. and KARASZ, E.E. (1984). The water induced plasticization behavior of semi-crystalline polyamides. *Polymer Preprints*, **25**(2), 197–198

FERRY, J.D. (1948). Mechanical properties of substances of high molecular weight. *Journal of the American Chemical Society*, **70**, 2244–2249

FERRY, J.D. (1980). *Viscoelastic Properties of Polymers*, 3rd Edition. New York, John Wiley & Sons

FLINK, J.M. (1983). Structure and structure transitions in dried carbohydrate materials. In *Physical Properties of Foods*, (Peleg, M. and Bagley, E.B., Eds), pp. 473–521. Westport, AVI Publishers

FLORY, P.J. (1953). *Principles of Polymer Chemistry*. Ithaca, Cornell University Press

FRANKS, F. (1982a). The properties of aqueous solutions at subzero temperatures. In *Water: A Comprehensive Treatise*, Vol. 7, (Franks, F., Ed.), pp. 215–338. New York, Plenum Press

FRANKS, F. (1982b). Water activity as a measure of biological viability and quality control. *Cereal Foods World*, **27**, 403–407

FRANKS, F. (1985a). *Biophysics and Biochemistry at Low Temperatures*. Cambridge, Cambridge University Press

FRANKS, F. (1985b). Complex aqueous systems at subzero temperatures. In *Properties of Water in Foods*, (Simatos, D. and Multon, J.L., Eds), pp. 497–509. Dordrecht, Martinus Nijhoff

FRANKS, F. (1986). Unfrozen water: yes; unfreezable water: hardly; bound water: certainly not. *Cryo-Letters*, **7,** 207

FRANKS, F., ASQUITH, M.H., HAMMOND, C.C., SKAER, H.B. and ECHLIN, P. (1977). Polymeric cryoprotectants in the preservation of biological ultrastructure. I. *Journal of Microscopy*, **110,** 223–238

FRENCH, D. (1984). Organization of starch granules. In *Starch: Chemistry and Technology*, 2nd Edition, (Whistler, R.L., Bemiller, J.N. and Paschall, E.F., Eds), pp. 183–247. Orlando, Academic Press

FUZEK, J.F. (1980). Glass transition temperature of wet fibers: its measurement and significance. In *Water in Polymers*, (Rowland, S.P., Ed.), ACS Symposium Series 127, pp. 515–530. Washington, American Chemical Society

GAETA, S., APICELLA, A. and HOPFENBERG, H.B. (1982). Kinetics and equilibria associated with the absorption and desorption of water and lithium chloride in an ethylene–vinyl alcohol copolymer. *Journal of Membrane Science*, **12,** 195–205

GODARD, P., BIEBUYCK, J.J., DAUMERIE, M., NAVEAU, H. and MERCIER, J.P. (1978). Crystallization and melting of aqueous gelatin. *Journal of Polymer Science: Polymer Physics Edition*, **16,** 1817–1828

GRAESSLEY, W.W. (1984). Viscoelasticity and flow in polymer melts and concentrated solutions. In *Physical Properties of Polymers*, (Mark, J.E., Eisenberg, A., Graessley, W.W., Mandelkern, L. and Koenig, J.L., Eds), pp. 97–153. Washington, American Chemical Society.

HAYASHI, A. and OH, S.C. (1983). Gelation of gelatin solution. *Agricultural and Biological Chemistry*, **47,** 1711–1716

HILTNER, A. and BAER, E. (1986). Reversible gelation of macromolecular systems. *Polymer Preprints*, **27,** 207

HOEVE, C.A.J. and HOEVE, M.B.J.A. (1978). The glass point of elastin as a function of diluent concentration. *Organic Coatings and Plastics Chemistry*, **39,** 441–443

JIN, X., ELLIS, T.S. and KARASZ, F.E. (1984). The effect of crystallinity and crosslinking on the depression of the glass transition temperature in nylon 6 by water. *Journal of Polymer Science: Polymer Physics Edition*, **22,** 1701–1717

JOLLEY, J.E. (1970). The microstructure of photographic gelatin binders. *Photographic Science and Engineering*, **14,** 169–177

KAREL, M. (1985). Effects of water activity and water content on mobility of food components, and their effects on phase transitions in food systems. In *Properties of Water in Foods*, (Simatos, D. and Multon, J.L., Eds), pp. 153–169. Dordrecht, Martinus Nijhoff

KEINATH, S.E. and BOYER, R.F. (1981). Thermomechanical analysis of $T_g$ and $T > T_g$ transitions in polystyrene. *Journal of Applied Polymer Science*, **26,** 2077–2085

KELLAWAY, I.W., MARRIOTT, C. and ROBINSON, J.A.J. (1978). The mechanical properties of gelatin films. I. *Canadian Journal of Pharmaceutical Science*, **13,** 83–86

KULP, K. and PONTE, J.G. (1981). Staling of white pan bread: fundamental causes. *CRC Critical Reviews in Food Science and Nutrition*, **15,** 1–48

LABARRE, E.E. and TURNER, D.T. (1982). Increased water sorption of poly(methyl methacrylate) after removal of methanol. *Journal of Polymer Science: Polymer Physics Edition*, **20,** 557–560

LANG, K.W. (1981). Physical, chemical and microbiological characterization of polymer and solute bound water. Doctoral Thesis, University of Illinois

LEVINE, H. and SLADE, L. (1986). A polymer physico-chemical approach to the study of commercial starch hydrolysis products. *Carbohydrate Polymers*, **6**, 213–244

LEVINE, H. and SLADE, L. (1987a). Water as a plasticizer: physico-chemical aspects of low-moisture polymeric systems. In *Water Science Reviews*, Volume 3 (Franks, F., Ed.). Cambridge, Cambridge University Press (in press)

LEVINE, H. and SLADE, L. (1987b). Thermomechanical behavior of sugar–water glasses and rubbers. *Transactions of the Faraday Society* (in press)

LONCIN, M. (1975). Basic principles of moisture equilibria. In *Freeze Drying and Advanced Food Technology*, (Goldlith, S.A., Rey, L. and Rothmayr, W.W., Eds), pp. 599–617. New York, Academic Press

MACKENZIE, A.P. (1975). Collapse during freeze drying—qualitative and quantitative aspects. In *Freeze Drying and Advanced Food Technology*, (Goldlith, S.A., Rey, L. and Rothmayr, W.W., Eds), pp. 277–307. New York, Academic Press

MACKENZIE, A.P. (1977). Non-equilibrium freezing behavior of aqueous systems. *Philosophical Transactions of the Royal Society of London*, **B278,** 167–189

MANDELKERN, L. (1986). Thermoreversible gelation and crystallization from solution. *Polymer Preprints*, **27,** 206

MARSHALL, A.S. and PETRIE, S.E.B. (1980). Thermal transitions in gelatin and aqueous gelatin solutions. *Journal of Photographic Science*, **28,** 128–134

MAURICE, T.J., SLADE, L., SIRETT, R.R. and PAGE, C.M. (1985). Polysaccharide–water interactions—thermal behavior of rice starch. In *Properties of Water in Foods*, (Simatos, D. and Multon, J.L., Eds), pp. 211–227. Dordrecht, Martinus Nijhoff

MENCZEL, J. and WUNDERLICH, B. (1986). Glass transition of semicrystalline macromolecules. *Polymer Preprints*, **27,** 255–256

MOHAJER, Y., WILKES, G.L., GIA, H.B. and MCGRATH, J.E. (1980). Influence of tacticity and sorbed water on material properties of poly-(N,N′-dimethyl acrylamide). *Polymer Preprints*, **21,** 229–230

MOY, P. and KARASZ, F.E. (1980). The interactions of water with epoxy resins. In *Water in Polymers*, (Rowland, S.P., Ed.), ACS Symposium Series 127, pp. 505–513. Washington, American Chemical Society

NAKAMURA, K., HATAKEYAMA, T. and HATAKEYAMA, H. (1981). DSC studies on the glass transition temperature of polyhydroxystyrene derivatives containing sorbed water. *Polymer*, **22,** 473–476

NAKAZAWA, F., NOGUCHI, S., TAKAHASHI, J. and TAKADA, M. (1985). Retrogradation of gelatinized potato starch studied by DSC. *Agricultural and Biological Chemistry*, **49,** 953–957

NEOGI, P. (1983). Anomalous diffusion of vapors through solid polymers. *American Institute of Chemical Engineering Journal*, **29,** 829–839

OLSON, D.R. and WEBB, K.K. (1978). The effect of humidity on the glass transition temperature. *Organic Coatings and Plastics Chemistry*, **39,** 518–523

REUTNER, P., LUFT, B. and BORCHARD, W. (1985). Compound formation and glassy solidification in the system gelatin–water. *Colloid and Polymer Science*, **263,** 519–529

ROWLAND, S.P. (Ed.) (1980). *Water in Polymers*, ACS Symposium Series 127. Washington, American Chemical Society

SCANDOLA, M., CECCORULLI, G. and PIZZOLI, M. (1981). Water clusters in elastin. *International Journal of Biological Macromolecules*, **3,** 147–149

SEARS, J.K. and DARBY, J.R. (1982). *The Technology of Plasticizers.* New York, Wiley-Interscience

SHAFIZADEH, F., MCGINNIS, G.D., SUSOTT, R.A. and TATTON, H.W. (1971). Thermal reactions of alpha-D-xylopyranose and beta-D-xylopyranosides. *Journal of Organic Chemistry*, **36,** 2813–2818

SLADE, L. (1984). Staling of starch based products. American Association of Cereal Chemists 69th Annual Meeting, Minneapolis, MN, Abstract No. 112

SLADE, L. and LEVINE, H. (1984a). Thermal analysis of starch and gelatin. American Chemical Society NERM 14, Fairfield, CT, Abstract No. 152

SLADE, L. and LEVINE, H. (1984b). Thermal analysis of starch and gelatin. In *Proceedings of the 13th NATAS Conference*, Philadelphia, PA. (McGhie, A.R., Ed.), p. 64

SLADE, L. and LEVINE, H. (1985). Intermediate moisture systems. Presented at Faraday Division, Royal Society of Chemistry, Industrial Physical Chemistry Group, Discussion Conference on Concept of Water Activity, July 1–3, Girton College, Cambridge

SLADE, L. and LEVINE, H. (1987a). Recent advances in starch retrogradation. In *Recent Developments in Industrial Polysaccharides*, (Stivala, S.S., Crescenzi, V. and Dea, I.C.M., Eds), New York, Gordon and Breach Science (in press)

SLADE, L. and LEVINE, H. (1987b). Non-equilibrium behavior of small carbohydrate–water systems. *Pure and Applied Chemistry*, (in press)

SLADE, L. and LEVINE, H. (1987c). Polymer–chemical properties of gelatin in foods. In *Advances in Meat Research*, Volume 4, (Pearson, A.M., Dutson, T.R. and Bailey, A., Eds), pp. 251–266. Westport, AVI Publishers

SLADE, L., LEVINE, H. and FRANKS, F. (1987). Beyond water activity: recent advances in the assessment of food safety and quality. *CRC Critical Reviews in Food Science and Nutrition*, (in press)

SOESANTO, T. and WILLIAMS, M.C. (1981). Volumetric interpretation of viscosity for concentrated and dilute sugar solutions. *Journal of Physical Chemistry*, **85,** 3338–3341

STARKWEATHER, H.W. (1980). *Water in nylon.* In *Water in Polymers*, (Rowland, S.P., Ed.), ACS Symposium Series 127, pp. 433–440. Washington, American Chemical Society

SUGGETT, A. and CLARK, A.H. (1976). Molecular motion and interactions in aqueous carbohydrate solutions. I. Dielectric–relaxation studies. *Journal of Solution Chemistry*, **5,** 1–15

TEN BRINKE, G., KARASZ, F.E. and ELLIS, T.S. (1983). Depression of glass transition temperatures of polymer networks by diluents. *Macromolecules*, **16,** 244–249

THILL, D.C., SCHIRMAN, R.D. and APPLEBY, A.P. (1979). Germination studies on winter wheat seeds. *Agronomy Journal*, **71,** 105–108

TO, E.C. and FLINK, J.M. (1978). 'Collapse', a structural transition in freeze dried carbohydrates. I.–III. *Journal of Food Technology*, **13,** 551–594

VAN DEN BERG, C. (1981). Vapour sorption equilibria and other water–starch interactions; a physico-chemical approach. Doctoral Thesis, Agricultural University, Wageningen

VAN DEN BERG, C. (1985). On the significance of water activity in low moisture systems. Presented at Faraday Division, Royal Society of Chemistry, Industrial Physical Chemistry Group, Discussion Conference on Concept of Water Activity, July 1–3, Girton College, Cambridge

VAN DEN BERG, C. (1986). Water activity. In *Concentration and Drying of Foods*, (MacCarthy, D., Ed.), London, pp. 11–36. Elsevier Applied Science

VAN DEN BERG, C. and BRUIN, S. (1981). Water activity and its estimation in food systems: theoretical aspects. In *Water Activity: Influences on Food Quality*, (Rockland, L.B. and Stewart, G.F., Eds), pp. 1–61. New York, Academic Press

WALTON, A.G. (1969). Nucleation in liquids and solutions. In *Nucleation*, (Zettlemoyer, A.C., Ed.), p. 225. New York, Marcel Dekker

WHISTLER, R.L. and DANIEL, J.R. (1984). Molecular structure of starch. In *Starch: Chemistry and Technology*, 2nd Edition, (Whistler, R.L., Bemiller, J.N. and Paschall, E.F., Eds), pp. 153–182. Orlando, Academic Press

WIGGANS, S.C. and GARDNER, F.P. (1959). Effect of solutes on radish seed germination times. *Agronomy Journal*, **51,** 315–318

WILLIAMS, M.L., LANDEL, R.F. and FERRY, J.D. (1955). Temperature dependence of relaxation mechanisms in amorphous polymers and other glass-forming liquids. *Journal of the American Chemical Society*, **77,** 3701–3706

WISSLER, G.E. and CRIST, B. (1980). Glass transition in semicrystalline polycarbonate. *Journal of Polymer Science: Polymer Physics Edition*, **18,** 1257–1270

WUNDERLICH, B. (1973). *Macromolecular Physics, Volume 1—Crystal Structure, Morphology, Defects*. New York, Academic Press

WUNDERLICH, B. (1976). *Macromolecular Physics, Volume 2—Crystal Nucleation, Growth, Annealing*. New York, Academic Press

WUNDERLICH, B. (1980). *Macromolecular Physics, Volume 3—Crystal Melting*. New York, Academic Press

WYNNE-JONES, S. and BLANSHARD, J.M.V. (1986). Hydration studies of wheat starch, amylopectin, amylose gels and bread by proton magnetic resonance. *Carbohydrate Polymers*, **6.** 289–306

ZOBEL, H.F. (1984). Gelatinization of starch and mechanical properties of starch pastes. In *Starch: Chemistry and Technology*, 2nd Edition, (Whistler, R.L., Bemiller, J.N., and Paschall, E.F., Eds), pp. 285–309. Orlando, Academic Press

9

## 'COLLAPSE' PHENOMENA—A UNIFYING CONCEPT FOR INTERPRETING THE BEHAVIOUR OF LOW MOISTURE FOODS

H. LEVINE AND LOUISE SLADE
*General Foods Corporation, New York, USA**

# Introduction

The extensive recent literature on caking and other so-called 'collapse'-related phenomena in amorphous or partially crystalline food powders (reviewed by Flink, 1983; Karel, 1985; Karel and Flink, 1983) supports the conclusion that such collapse phenomena are consequences of a material-specific structural relaxation process. Our premise is that these consequences represent the microscopic and macroscopic manifestations of an underlying molecular 'state' transformation from kinetically metastable amorphous solid to unstable amorphous liquid, which occurs at $T_g$.

This thesis has developed during six years of active industrial research in the area of food polymer science. Recently reported studies from our laboratories were based on thermal and thermomechanical analysis methods used to illustrate and characterize the polymer physicochemical properties of various food ingredients and products, e.g. rice and starch (Biliaderis *et al.*, 1985; Maurice *et al.*, 1985; Slade, 1984; Slade and Levine, 1984); gelatin (Levine and Slade, 1984; Slade and Levine, 1984, 1987); frozen aqueous solutions of sugars and starch hydrolysis products (SHPs) (Levine and Slade, 1984, 1986; Schenz *et al.*, 1984); and 'intermediate moisture' carbohydrate systems (Levine and Slade, 1984; Slade and Levine, 1985), all of which behave as systems of completely amorphous or partially crystalline polymers, oligomers, and monomers, soluble in and/or plasticized by water.

Plasticization by water, which has been a focal point of the above research, is also a critical element of our unifying concept for collapse phenomena. Others have already noted that 'water is the most ubiquitous plasticizer in our world' (Sears and Darby, 1982), and 'the most important ... plasticizer for hydrophilic food components' (Karel, 1985). It has become well established that plasticization by water affects the $T_g$ of many synthetic and natural amorphous polymers (particularly at low moisture contents), and that $T_g$ depression can be advantageous or disadvantageous to material properties, processing and stability (Rowland, 1980). In this context (especially from the synthetic polymer literature), we have postulated a generalized physicochemical mechanism for collapse, derived from Williams–Landel–Ferry (WLF) theory (Ferry, 1980). This mechanism can be described as follows. As the ambient temperature rises above $T_g$ or as $T_g$ falls below the ambient temperature due to plasticization by water, polymer free volume increases, leading to increased

*Present address: Nabisco Brands Inc., Corporate Technology Group, P.O. 1943, East Hanover, New Jersey 07936–1943, USA

segmental mobility of the polymer chains. Consequently, the viscosity, $\eta$, of the dynamically constrained solid falls below the characteristic $\eta_g$ at $T_g$, thus allowing the glass-to-rubber transition to occur, and permitting viscous liquid flow. In this liquid state, translational diffusion is no longer inhibited, and diffusion-controlled relaxations (including structural collapse) are free to proceed with rates defined by the WLF equation, i.e. rates which increase exponentially with increasing $\Delta T$ above $T_g$ (in K).

To illustrate our concept of collapse phenomena, this report reviews and updates our findings (Levine and Slade, 1984, 1986; Schenz *et al.*, 1984) on the structure/property relationships for two extensive series of food carbohydrates: 1) sugars, glycosides, and polyhydric alcohols, and 2) commercially available SHPs. The properties of SHPs (e.g. dextrins, maltodextrins, corn syrup solids, corn syrups) represent an important, but sparsely researched, subject within the food industry (Murray and Luft, 1973). In contrast, sugars and polyols have been studied extensively, and limited compilations of their characteristic transition temperatures for structural collapse ($T_c$) are available (Franks, 1982). Much can be learned about the functional attributes of SHPs as ingredients in fabricated foods, from a polymer physicochemical approach to studies of the thermomechanical properties of these amorphous, water-soluble polymers of glucose. For example, To and Flink (1978) demonstrated a correlation between increasing $T_c$ and increasing number-average degree of polymerization, $\overline{DP}n$, for a series of SHPs of $2 \leqslant \overline{DP}n \leqslant 16$ (calculated dextrose equivalent (DE) = 52.6–6.9).* The same approach to a systematic study of lower MW saccharide monomers and oligomers can likewise provide new information which is useful in predicting the functional attributes of these common ingredients in both fabricated and natural foods.

In this report, we review our differential scanning calorimetry (DSC) results for the $T'_g$ values of 80 SHPs (DE values of 0.3–100) and 60 polyhydroxy compounds. [As defined by Franks (1982, 1985), $T'_g$ is the particular $T_g$ of the maximally freeze-concentrated solute/water matrix surrounding the ice crystals in a frozen solution.] For the SHPs, this analysis yielded a linear correlation between decreasing DE and increasing $T'_g$, from which we constructed a calibration curve used to predict DE values for other SHPs of unknown DE (Levine and Slade, 1986). The same DE versus $T'_g$, data were also used to construct a predictive map of functional attributes for SHPs, based on a demonstration of their classical $T'$ versus $\bar{M}n$ behaviour as a homologous series of amorphous polymers. Our studies have covered a more extensive range of SHPs and provided a theoretical basis for interpreting the results of To and Flink (1978). For the 60 polyhydroxy compounds, a linear correlation between increasing $T'_g$ and decreasing value of 1/MW is demonstrated here for the first time, and augments the most recent literature (Franks, 1985). This correlation is not quite as good as the one for the SHPs, as one would expect, since these 60 compounds do not represent a single homologous family of monomers and oligomers.

The possibility that SHP functional behaviour can be predicted from the correlation between DE (or $\overline{DP}n$ or $\bar{M}n$) and $T_g$, and that of sugars and polyols from their $T_g$ versus 1/MW relationship, has important implications for a better understanding of the mechanism of collapse.

For the food industry, such predictive capabilities are valuable because various non-equilibrium collapse phenomena affect the processing and storage stability of

*DE is defined by the equation DE = $100/(\bar{M}n/180.16)$, since the reducing sugar content (in terms of the number of reducing end groups) of a known weight of sample is compared to an equal weight of glucose of DE 100 and $\bar{M}n$ 180.16.

many fabricated and natural foods, including frozen products, amorphous dry powders, and candy glasses. We shall discuss the potential (and frequently demonstrated) utility of SHPs in preventing structural collapse, in the context of collapse processes which are often promoted by the heavy use of low MW saccharides, and shall relate these insights to our unifying concept of collapse phenomena.

## Materials and methods

The SHPs used in this study are listed in *Table 9.1*, along with their manufacturers and vegetable sources, wherever known. The majority are typical commercial SHPs, readily available and widely used throughout the food industry, either now or in the past. These materials were analysed as received, and their DE values are those specified by the manufacturers. The low MW polyhydroxy compounds studied are listed in *Table 9.2*. All of the sugars and polyols were reagent grade chemicals, many of them from Sigma. Most of the glycosides were synthesized and purified in our laboratories.

**Table 9.1** $T'_g$ VALUES FOR COMMERCIAL SHPs

| *SHP* | *Manufacturer* | *Starch source* | *DE* | $T'_g$ (°C) |
|---|---|---|---|---|
| AB 7436 | Anheuser Busch | Waxy maize | 0.5 | −4 |
| Paselli SA-2 | AVEBE (1984) | Potato (Ap) | 2 | −4.5 |
| Stadex 9 | Staley | Dent corn | 3.4 | −4.5 |
| 78NN128 | Staley | Potato | 0.6 | −5 |
| 78NN122 | Staley | Potato | 2 | −5 |
| V-O Starch | National | Waxy maize | ? | −5.5 |
| N-Oil | National | Tapioca | ? | −5.5 |
| ARD 2326 | Amaizo | Dent corn | 0.4 | −5.5 |
| Paselli SA-2 | AVEBE (1986) | Potato (Ap) | 2 | −5.5 |
| ARD 2308 | Amaizo | Dent corn | 0.3 | −6 |
| AB 7435 | Anheuser Busch | Waxy/dent blend | 0.5 | −6 |
| Star Dri 1 | Staley (1984) | Dent corn | 1 | −6 |
| Crystal Gum | National | Tapioca | 5 | −6 |
| Maltrin M050 | GPC | Dent corn | 6 | −6 |
| Star Dri 1 | Staley (1986) | Waxy maize | 1 | −6.5 |
| Paselli MD-6 | AVEBE | Potato | 6 | −6.5 |
| Dextrin 11 | Staley | Tapioca | 1 | −7.5 |
| MD-6-12 | V-Labs | | 2.8 | −7.5 |
| Stadex 27 | Staley | Dent corn | 10 | −7.5 |
| MD-6-40 | V-Labs | | 0.7 | −8 |
| Star Dri 5 | Staley (1984) | Dent corn | 5 | −8 |
| Star Dri 5 | Staley (1986) | Waxy maize | 5.5 | −8 |
| Paselli MD-10 | AVEBE | Potato | 10 | −8 |
| Paselli SA-6 | AVEBE | Potato (Ap) | 6 | −8.5 |
| α-Cyclodextrin | Pfanstiehl | | | −9 |
| Capsul | National | Waxy maize | 5 | −9 |
| Lodex Light V | Amaizo | Waxy maize | 7 | −9 |
| Paselli SA-10 | AVEBE | Potato (Ap) | 10 | −9.5 |
| Morrex 1910 | CPC | Dent corn | 10 | −9.5 |
| Star Dri 10 | Staley (1984) | Dent corn | 10 | −10 |
| Maltrin M040 | GPC | Dent corn | 5 | −10.5 |
| Frodex 5 | Amaizo | Waxy maize | 5 | −11 |

**Table 9.1** $T'_g$ VALUES FOR COMMERCIAL SHPs—*continued*

| *SHP* | *Manufacturer* | *Starch source* | *DE* | $T_{g'}$ (°C) |
|---|---|---|---|---|
| Star Dri 10 | Staley (1986) | Waxy maize | 10.5 | −11 |
| Lodex 10 | Amaizo (1986) | Waxy maize | 11 | −11.5 |
| Lodex Light X | Amaizo | Waxy maize | 12 | −11.5 |
| Morrex 1918 | CPC | Waxy maize | 10 | −11.5 |
| Mira-Cap | Staley | Waxy maize | ? | −11.5 |
| Maltrin M100 | GPC | Dent corn | 10 | −11.5 |
| Lodex 5 | Amaizo | Waxy maize | 7 | −12 |
| Maltrin M500 | GPC | Dent corn | 10 | −12.5 |
| Lodex 10 | Amaizo (1982) | Waxy maize | 12 | −12.5 |
| Star Dri 15 | Staley (1986) | Waxy maize | 15.5 | −12.5 |
| MD-6 | V-Labs | | ? | −12.5 |
| Maltrin M150 | GPC | Dent corn | 15 | −13.5 |
| Maltoheptaose | Sigma | | 15.6 | −13.5 |
| MD-6-1 | V-Labs | | 20.5 | −13.5 |
| Star Dri 20 | Staley (1986) | Waxy maize | 21.5 | −13.5 |
| Maltodextrin Syrup | GPC | Dent corn | 17.5 | −14 |
| Frodex 15 | Amaizo | Waxy maize | 18 | −14 |
| Maltohexaose | Sigma | | 18.2 | −14.5 |
| Frodex 10 | Amaizo | Waxy maize | 10 | −15.5 |
| Lodex 15 | Amaizo | Waxy maize | 18 | −15.5 |
| Maltohexaose | V-Labs | | 18.2 | −15.5 |
| Maltrin M200 | GPC | Dent corn | 20 | −15.5 |
| Maltopentaose | Sigma | | 21.7 | −16.5 |
| Maltrin M250 | GPC | Dent corn | 25 | −17.5 |
| N-Lok | National | Blend | ? | −17.5 |
| Staley 200 | Staley | Corn | 26 | −19.5 |
| Maltotetraose | Sigma | | 27 | −19.5 |
| Frodex 24 | Amaizo | Waxy maize | 28 | −20.5 |
| Frodex 36 | Amaizo | Waxy maize | 36 | −21.5 |
| DriSweet 36 | Hubinger | Corn | 36 | −22 |
| Maltrin M365 | GPC | Dent corn | 36 | −22.5 |
| Staley 300 | Staley | Corn | 35 | −23.5 |
| Globe 1052 | CPC | Corn | 37 | −23.5 |
| Maltotriose | V Labs | | 35.7 | −23.5 |
| Frodex 42 | Amaizo | Waxy maize | 42 | −25.5 |
| Neto 7300 | Staley | Corn | 42 | −26.5 |
| Globe 1132 | CPC | Corn | 43 | −27.5 |
| Staley 1300 | Staley | Corn | 43 | −27.5 |
| Neto 7350 | Staley | Corn | 50 | −27.5 |
| Maltose | Sigma | | 52.6 | −29.5 |
| Globe 1232 | CPC | Corn | 54.5 | −30.5 |
| Staley 2300 | Staley | Corn | 54 | −31 |
| Sweetose 4400 | Staley | Corn | 64 | −33.5 |
| Sweetose 4300 | Staley | Corn | 64 | −34 |
| Globe 1642 | CPC | Corn | 63 | −35 |
| Globe 1632 | CPC | Corn | 64 | −35 |
| Royal 2626 | CPC | Corn | 95 | −42 |
| Glucose | Sigma | Corn | 100 | −43 |

**Table 9.2** $T'_g$ VALUES FOR VARIOUS SUGARS, GLYCOSIDES, AND POLYOLS[a]

| *Sugar or polyol* | *MW* | $T'_g$ (°C) | $W'_g$ (g UFW/g) |
|---|---|---|---|
| Ethylene glycol | 62.1 | − 85 | 1.90 |
| Propylene glycol | 76.1 | − 67.5 | 1.28 |
| 1,3-Butanediol | 90.1 | − 63.5 | 1.41 |
| Glycerol | 92.1 | − 65 | 0.85 |
| Erythrose | 120.1 | − 50 | 1.39 |
| Erythritol | 122.1 | − 53.5 | (eutectic) |
| Thyminose (deoxyribose) | 134.1 | − 52 | 1.32 |
| Xylose | 150.1 | − 48 | 0.45 |
| Arabinose | 150.1 | − 47.5 | 1.23 |
| Ribose | 150.1 | − 47 | 0.49 |
| Arabitol | 152.1 | − 47 | 0.89 |
| Ribitol | 152.1 | − 47 | 0.82 |
| Xylitol | 152.1 | − 46.5 | |
| Methyl riboside | 164.2 | − 53 | 0.96 |
| Methyl xyloside | 164.2 | − 49 | 1.01 |
| Quinovose (deoxyglucose) | 164.2 | − 43.5 | 1.11 |
| Fucose (deoxygalactose) | 164.2 | − 43 | 1.11 |
| Rhamnose (deoxymannose) | 164.2 | − 43 | 0.90 |
| Glucose | 180.2 | − 43 | 0.41 |
| Fructose | 180.2 | − 42 | 0.96 |
| Galactose | 180.2 | − 41.5 | 0.77 |
| Allose | 180.2 | − 41.5 | 0.56 |
| Sorbose | 180.2 | − 41 | 0.45 |
| Mannose | 180.2 | − 41 | 0.35 |
| Tagatose | 180.2 | − 40.5 | 1.33 |
| Inositol | 180.2 | − 35.5 | 0.30 |
| Mannitol | 182.2 | − 40 | (eutectic) |
| Galactitol | 182.2 | − 39 | (eutectic) |
| Sorbitol | 182.2 | − 43.5 | 0.23 |
| 2-*o*-Methyl fructoside | 194.2 | − 51.5 | 1.61 |
| β-1-*o*-Methyl glucoside | 194.2 | − 47 | 1.29 |
| 3-*o*-Methyl glucoside | 194.2 | − 45.5 | 1.34 |
| 6-*o*-Methyl galactoside | 194.2 | − 45.5 | 0.98 |
| α-1-*o*-Methyl glucoside | 194.2 | − 44.5 | 1.32 |
| 1-*o*-Methyl galactoside | 194.2 | − 44.5 | 0.86 |
| 1-*o*-Methyl mannoside | 194.2 | − 43.5 | 1.43 |
| 1-*o*-Ethyl glucoside | 208.2 | − 46.5 | 1.35 |
| 2-*o*-Ethyl fructoside | 208.2 | − 46.5 | 1.15 |
| 1-*o*-Ethyl galactoside | 208.2 | − 45 | 1.26 |
| 1-*o*-Ethyl mannoside | 208.2 | − 43.5 | 1.21 |
| Heptulose | 210.2 | − 36.5 | 0.77 |
| 1-*o*-Propyl glucoside | 222.2 | − 43 | 1.22 |
| 1-*o*-Propyl galactoside | 222.2 | − 42 | 1.05 |
| 1-*o*-Propyl mannoside | 222.2 | − 40.5 | 0.95 |
| 2,3,4,6-*o*-Methyl glucoside | 236.2 | − 45.5 | 1.41 |
| Isomaltulose | 342.3 | − 35.5 | |
| Cellobiulose | 342.3 | − 32.5 | |
| Isomaltose | 342.3 | − 32.5 | 0.70 |
| Sucrose | 342.3 | − 32 | 0.56 |
| Gentiobiose | 342.3 | − 31.5 | 0.26 |
| Turanose | 342.3 | − 31 | 0.64 |
| Mannobiose | 342.3 | − 30.5 | 0.91 |
| Lactulose | 342.3 | − 30 | 0.72 |
| Maltose | 342.3 | − 29.5 | 0.25 |
| Maltulose | 342.3 | − 29.5 | |

**Table 9.2** $T'_g$ VALUES FOR VARIOUS SUGARS, GLYCOSIDES, AND POLYOLS[a]—*continued*

| *Sugar or polyol* | *MW* | $T'_g$ (°C) | $W'_g$ (g UFW/g) |
|---|---|---|---|
| Trehalose | 342.3 | −29.5 | 0.20 |
| Lactose | 342.3 | −28 | 0.69 |
| Maltitol | 344.3 | −34.5 | 0.59 |
| Isomaltotriose | 504.5 | −30.5 | 0.50 |
| Panose | 504.5 | −28 | 0.59 |
| Raffinose | 504.5 | −26.5 | 0.70 |
| Maltotriose | 504.5 | −23.5 | 0.45 |
| Stachyose | 666.6 | −23.5 | 1.12 |
| Maltotetraose | 666.6 | −19.5 | 0.55 |
| Maltopentaose | 828.9 | −16.5 | 0.47 |
| Maltohexaose | 990.9 | −14.5 | 0.50 |
| Maltoheptaose | 1153.0 | −13.5 | 0.27 |

[a] Some of our experimental results for $T'_g$ and $W'_g$ of various sugars and polyols were first published in Franks (1985).

All solutions for $T'_g$ determination were 20% w/w solids basis (i.e. 20.0 g solid solute/100.0 g solution), in distilled deionized water. Samples were prepared by mechanical stirring, with gentle heating when necessary, to produce clear solutions (or homogeneous sols for the SHPs of lowest DE).

DSC measurements were performed with a DuPont 990 Thermal Analyzer combined with a Model 910 Differential Scanning Calorimeter equipped with a liquid nitrogen quench-cooling accessory capable of sample cooling at about 50 °C $min^{-1}$. The analogue derivative function on the DuPont 990 allowed the precise determination of transition temperatures, with a reproducibility (for duplicate samples) of ± 0.5 °C for $T'_g$. In practice, 20–30 mg of solution were hermetically sealed in an aluminium sample pan and scanned (against an empty reference pan), at a heating rate of 5 °C $min^{-1}$, from a temperature at least 10 °C below $T'_g$ to + 25 °C. In all cases, initial cooling to well below $T'_g$ ensured maximal freeze concentration, and thus maximally frozen samples.

## Results

*Figure 9.1* shows two typical low temperature DSC thermograms for 20 w% solutions. In each case, the heat flow curve begins at the top (endothermic down), and the analogue derivative trace (zeroed to the temperature axis) at the bottom. For both thermograms, instrumental amplification and sensitivity settings were identical, and sample weights comparable. As illustrated by *Figure 9.1*, the derivative feature of the DuPont 990 greatly facilitates the identification of sequential thermal transitions, assignment of precise transition temperatures, and thus overall interpretation of thermal behaviour, especially for such frozen aqueous solutions exemplified by *Figure 9.1(a)*. Surprisingly, we have found no other reported use of derivative thermograms, in the many DSC studies of such systems, to sort out the small endothermic and exothermic changes in heat flow that occur typically below 0 °C (*see* Franks (1982) for extensive bibliography).

Despite the handicap of such instrumental shortcomings in the past, the theoretical

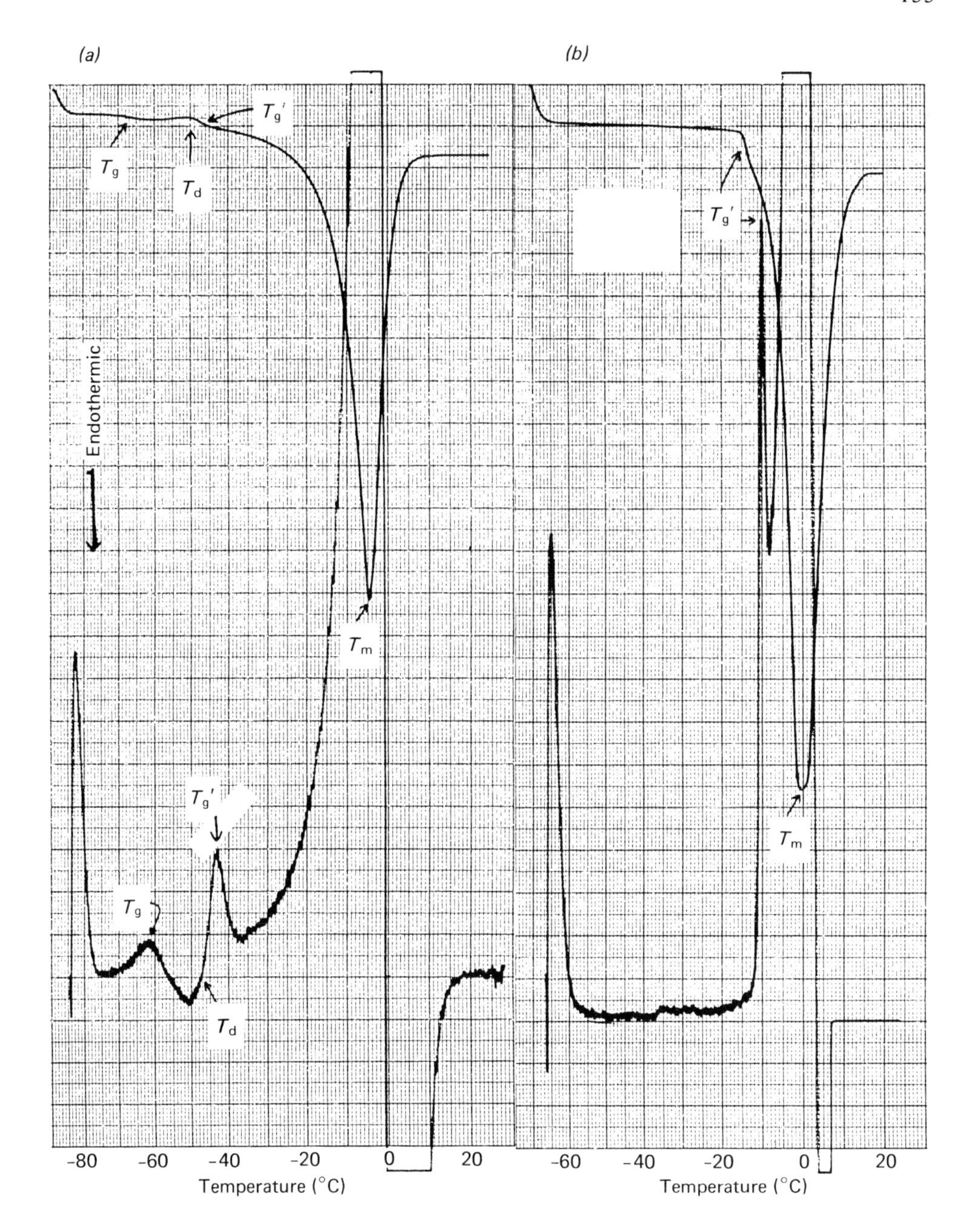


**Figure 9.1** Typical DSC thermograms for 20 w% solutions of a) glucose, and b) Star Dri 10 10-DE maltodextrin (Staley, 1984). In each, the heat flow curve begins at the top (endothermic down), and the derivative trace (zeroed to the temperature axis) at the bottom.

basis for the thermal properties manifested by aqueous solutions at subzero temperatures has become well understood, as described in detail (along with voluminous corroborative experimental results) by Franks (1982, 1985), MacKenzie (1977, 1981), and coworkers (Franks *et al.*, 1977; Luyet, 1960; MacKenzie and Rasmussen, 1972; Rasmussen and Luyet, 1969). Hence, only the new and salient features of the present results will be described. In *Figure 9.1(a)*, after rapid cooling of the glucose solution to below − 80 °C, slow heating revealed a minor $T_g$ at − 61.5 °C, followed by an exothermic devitrification (a crystallization of some of the previously unfrozen water) at − 47.5 °C, followed by another (major) $T_g$, namely $T'_g$, at − 43 °C, and then finally the 'equilibrium' melting of ice at $T_m$. In *Figure 9.1(b)*, the maltodextrin solution thermogram shows only the obvious $T'_g$ at − 10 °C, in addition to $T_m$. These assignments of transitions and temperatures can be reconciled definitively with the appropriate solid–liquid 'state' diagrams, as reported by MacKenzie and Rasmussen (1972) and Franks (1982). In such state diagrams (e.g. *Figure 9.8*), the different cooling/heating paths which can be followed by solutions of monomeric and polymeric solutes are revealed. For the former (e.g. glucose), partial vitrification of the original solution can occur, apparently because the cooling rate is high relative to the ice crystallization rate; whereas for the latter (e.g. a maltodextrin), the cooling rate appears to be low relative to the freezing rate. However, as demonstrated by the thermograms in *Figure 9.1*, in both cases rewarming forces the system through a glass transition at $T'_g$. [Note: in many earlier DSC studies (e.g. MacKenzie, 1977, 1981; Maltini, 1977), performed without benefit of derivative thermograms, a pair of transition temperatures (each independent of initial concentration), called $T$ antemelting ($T_{am}$) and $T$ incipient melting ($T_{im}$), have been reported in place of a single $T'_g$. In fact, for the many cases that we have studied, the reported values of $T_{am}$ and $T_{im}$ bracket that of $T'_g$ (as we measure it), leading us to surmise that $T_{am}$ and $T_{im}$ actually represent the temperatures of onset and completion of the single thermal event (a glass transition) that must occur at $T'_g$, as defined by the state diagram.]

The point of greatest interest in the thermograms in *Figure 9.1* involves $T'_g$. The matrix surrounding the ice crystals in a maximally frozen solution is a supersaturated solution of all the solute in the fraction of water remaining unfrozen. This matrix exists as a kinetically metastable* amorphous solid (a glass of constant composition) at any temperature below $T'_g$, but as a viscoelastic liquid (a rubbery fluid) at any temperature between $T'_g$ and the $T_m$ of ice. Again with regard to a state diagram for a typical solute that does not readily undergo eutectic crystallization [*see* Franks (1982) or *Figure 9.8*], $T'_g$ corresponds to the intersection of the thermodynamically defined liquidus curve and the kinetically determined supersaturated glass curve. As such, Franks (1982) described $T'_g$ as having the appearance of a 'metastable eutectic', in that it represents a quasi-invariant point in the state diagram, invariant in both its characteristic temperature ($T'_g$) and composition (i.e. $C'_g$, expressed as w% solute, or $W'_g$, expressed as g unfrozen water/g solute) for a particular solute. However, 'eutectic' in this usage does not imply a phase separation. This glass which forms, for example, on slow cooling to $T'_g$, acts as a kinetic barrier to further ice formation (within the experimental time frame), despite the continued presence of unfrozen

*The glass at $T'_g$, with invariant composition, has been described in previous literature as 'metastable' (Franks, 1982). We suggest a more discriminating terminology, such as 'kinetically metastable' or 'dynamically constrained'.

water at all temperatures below $T'_g$, as well as to any other diffusion-controlled process. Recognizing this, one begins to appreciate why the temperature of this glass transition is important in several aspects of frozen food technology, e.g. freezer storage stability, freeze concentration, and freeze drying (Franks, 1982, 1985), which can be subject to various recrystallization and 'collapse' phenomena, as will be described later.

The measured $T'_g$ values for the 80 SHPs and 60 polyhydroxy compounds are listed in *Tables 9.1* and *9.2*, respectively. The $T'_g$ for glucose of − 43 °C is midway between reported values for $T_{am}$ and $T_{im}$ (Rasmussen and Luyet, 1969), and within a few degrees of various values for $T_c$ and the transition temperature for recrystallization ($T_r$) (*see Table 9.3*). The same is true of our previously reported $T'_g$ for sucrose of − 32 °C (Schenz *et al.*, 1984). As shown in *Table 9.3*, literature values for $T_c$ and/or $T_r$ (both always independent of initial concentration) for soluble starch of − 5 or − 6 °C

**Table 9.3** COMPARISON OF $T'_g$ VALUES AND LITERATURE VALUES FOR OTHER 'COLLAPSE' TRANSITION TEMPERATURES

| *Substance* | $T_r$[a] (°C) | $T_c$[b] (°C) | $T'_g$ (°C) |
|---|---|---|---|
| Ethylene glycol | − 70[c] | | − 85 |
| Glycerol | − 58, − 65[c] | | − 65 |
| Ribose | − 43 | | − 47 |
| Glucose | − 41, − 38 | − 40 | − 43 |
| Fructose | − 48 | − 48 | − 42 |
| Sucrose | − 32, − 30.5 | − 32 | − 32 |
| Maltose | | − 32 | − 29.5 |
| Lactose | | − 32 | − 28 |
| Raffinose | − 27, − 25.4 | − 26 | − 26.5 |
| Inositol | | − 27 | − 35.5 |
| Sorbitol | | − 45 | − 43.5 |
| Glutamic acid, sodium salt | | − 50 | − 46 |
| Gelatin | − 11 | − 8 | |
| Gelatin (300 Bloom) | | | − 9.5 |
| Gelatin (250 Bloom) | | | − 10.5 |
| Gelatin (175 Bloom) | | | − 11.5 |
| Gelatin (50 Bloom) | | | − 12.5 |
| Bovine serum albumin | − 5.3 | | − 13 |
| Dextran | | − 9 | |
| Dextran (MW 9400) | | | − 13.5 |
| Soluble starch | − 5, − 6[c] | | |
| Soluble potato starch | | | − 3.5 |
| Hydroxyethyl starch | − 21 | | − 6.5 |
| PVP | − 22, − 21, − 14.5 | − 23 | |
| PVP-10 | | | − 26 |
| PVP-40 | | | − 20.5 |
| PVP-44 | | | − 21.5 |
| PEG | − 65, − 43 | − 13 | |
| PEG (MW 200) | | | − 65.5 |
| PEG (MW 300) | | | − 63.5 |
| PEG (MW 400) | | | − 61 |

[a] Recrystallization temperatures (Franks, 1982, p. 297)
[b] Collapse temperatures during freeze drying (Franks, 1982, p. 313)
[c] Luyet (1960), p. 564

(Franks, 1982; Luyet, 1960), and for dextrin of $-9\,°C$ (Luyet, 1960), are also comparable to our $T'_g$ values for similar materials.

*Figure 9.2* shows $T'_g$ plotted versus DE for all the SHPs with specified DE values. There is a linear correlation between increasing $T'_g$ and decreasing DE (regression coefficient $r = -0.98$). Since, as previously defined, DE is inversely proportional to $\overline{DP}n$ and $\bar{M}n$ for this series of SHPs, the results in *Figure 9.2* demonstrate that $T'_g$ increases with increasing $\bar{M}n$. Such a correlation between $T_g$ and $\bar{M}n$ is the general rule for any homologous family of glass-forming monomers, oligomers and polymers (Billmeyer, 1984). The equation describing the regression line in *Figure 9.2* is DE $= -2.2\,(T'_g, °C) - 12.8$. We have shown (Levine and Slade, 1986) that *Figure 9.2* can be used as a calibration curve for interpolating DE values of new or 'unknown' SHPs, in preference to the time-consuming classical methods for DE determination (Murray and Luft, 1973).

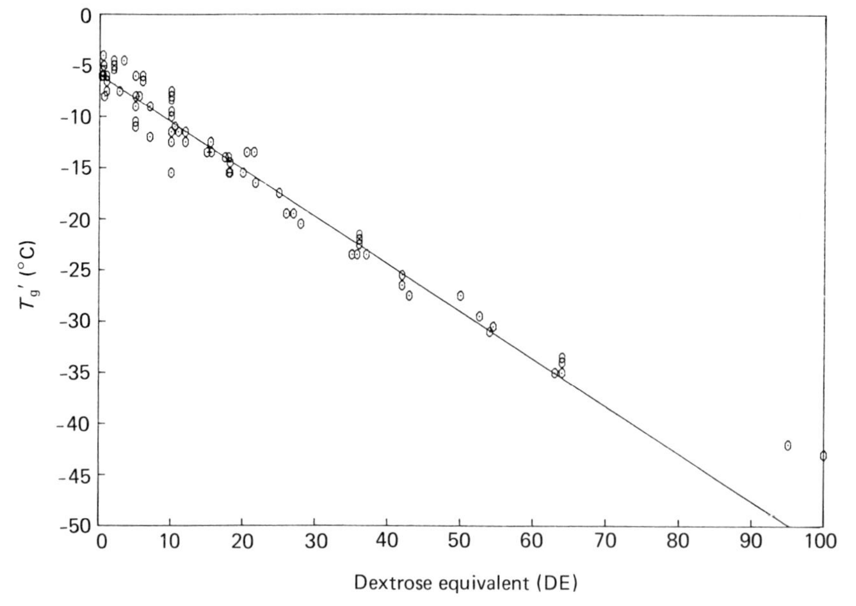


**Figure 9.2** Variation of the glass transition temperature, $T'_g$, for maximally frozen 20 w% solutions against DE value for the commercial SHPs in *Table 9.1*.

For the 60 polyhydroxy compounds, the corresponding plot of $T'_g$ versus 1/MW is shown in the inset of *Figure 9.3*. In this case, the regression line has an $r$ value of $-0.94$, which is slightly lower than the one for the homologous series of SHPs. The major contributor to the scatter in this plot is the series of chemically different glycosides, which do not constitute a homologous family of glass formers.

*Figure 9.4* shows $T'_g$ plotted versus $W'_g$ for 13 of the corn syrups (of DE 26–95) listed in *Table 9.1*. The composition of the glass at $T'_g$ was calculated from the thermogram, specifically from measurement of the area (enthalpy) under the ice melting endotherm. By calibration with pure water, this measurement yields a maximum weight of ice in the frozen sample, and by difference from the known weight of total water in the initial solution, a weight of unfrozen water, per unit weight of solute, in the glass at $T'_g$. A few in the food industry will recognize this procedure as one of several routine

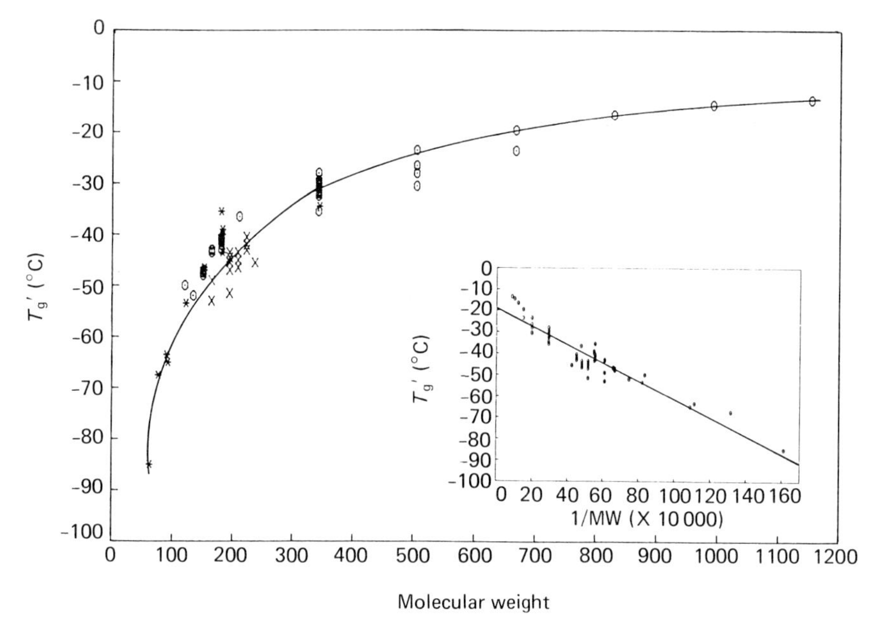


**Figure 9.3** Variation of the glass transition temperature, $T_g'$, for maximally frozen 20 w% solutions against molecular weight for the sugars (o), glycosides (x), and polyhydric alcohols (*) in *Table 9.2*. Inset: a plot of $T_g'$ versus $1/\text{MW} \times 10^4$, illustrating the theoretically predicted linear dependence.

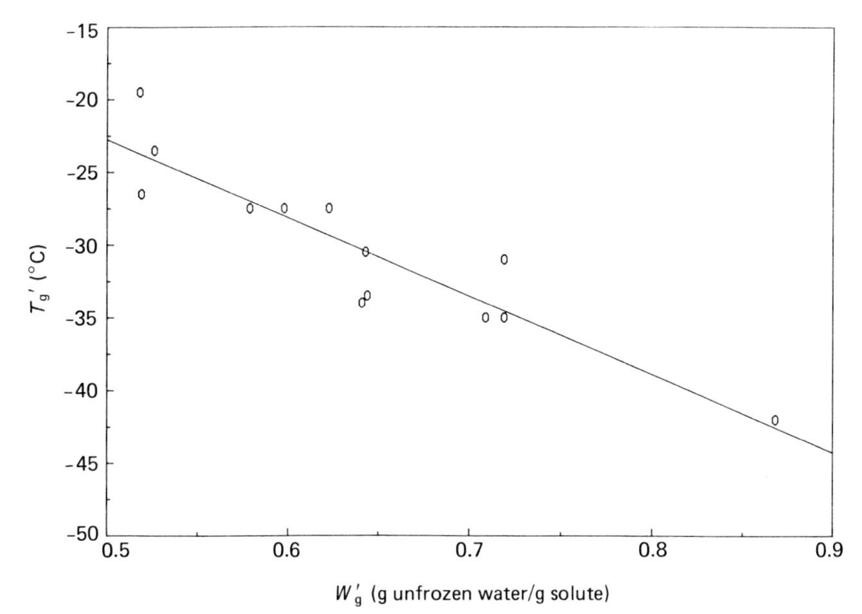


**Figure 9.4** Variation of the glass transition temperature, $T_g'$, for maximally frozen 20 w% solutions against $W_g'$, the composition of the glass at $T_g'$, in g unfrozen water/g solute, for 13 commercial corn syrups from *Table 9.1*.

methods for determining what is referred to as the 'water binding capacity' of a solute. Franks (1982, 1985) has reviewed this subject, and taken pains to point out that this so-called 'bound' water is not truly bound in any energetic sense. It is subject to rapid exchange, has thermally-labile hydrogen bonds, shows cooperative molecular mobility, has a heat capacity, Cp, approximately equal to that of liquid water rather than ice, and has some capability to dissove salts (Levine and Slade, 1984). Furthermore, it has been demonstrated, for water-soluble polymers and monomers alike, that such unfreezability is not due to tight binding by solute, but to kinetic (non-equilibrium) retardation effects on the diffusion of water and solute molecules at the low temperatures approaching the vitrification $T_g$ of the solute/unfrozen water mixture (Franks, 1982).

As shown by the results in *Figure 9.4*, $W'_g$ decreases with increasing $T'_g$ for this homologous series of corn syrup solids solutions. The regression coefficient is − 0.91. In other words, as the average $\bar{M}n$ of the solute(s) increases, the fraction of the total water unfrozen in the glass at $T'_g$ generally decreases. This fact is also illustrated dramatically by the thermograms in *Figure 9.1*. For comparable amounts of total water, the area under the ice melting peak for the glucose solution is much smaller than that for the maltodextrin solution. Once again, in the context of a typical state diagram (e.g. *Figure 9.8*), the above results show that as $\bar{M}n$ of the solute (or mixture of homologous solutes) in an aqueous system increases, the $T'_g/C'_g$ point generally moves up the temperature axis toward 0 °C and to the right along the composition axis toward 100 w% solute. The criticality of this fact will become clear later when we describe SHP functional behaviour vis-à-vis $T'_g$ and the possibilities of inhibiting collapse by formulating a fabricated food product with the intent of elevating $T'_g$.

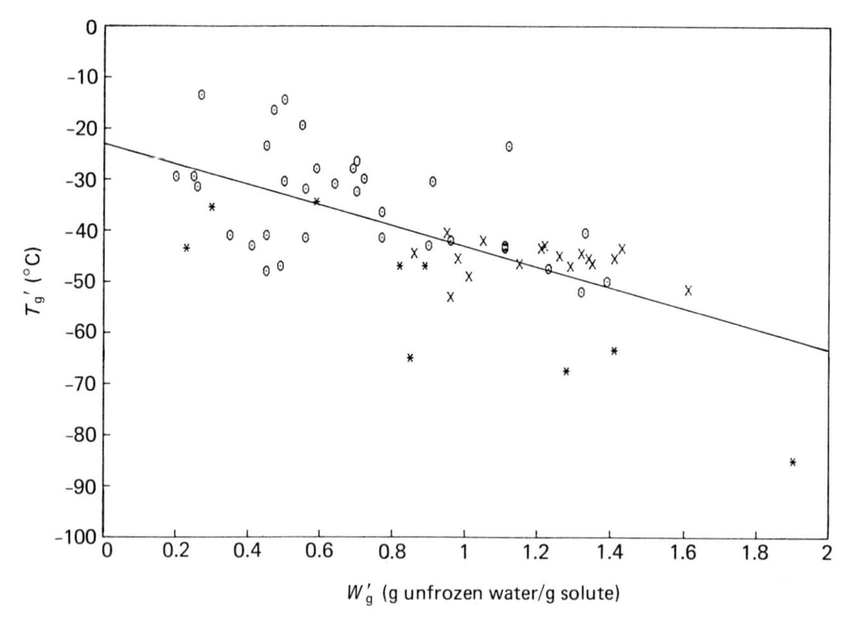


**Figure 9.5** Variation of the glass transition temperature, $T'_g$, for maximally frozen 20 w% solutions against $W'_g$, the composition of the glass at $T'_g$, in g unfrozen water/g solute, for the sugars (o), glycosides (x), and polyhydric alcohols (*) in *Table 9.2*.

In contrast to the results in *Figure 9.4* for a homologous series of mixed glucose monomer and oligomers, the results in *Figure 9.5*, of $T'_g$ versus $W'_g$ for the diverse polyhydroxy compounds listed in *Table 9.2*, yield a regression coefficient of only −0.64. Thus, when Franks (1985) notes that, among the (non-homologous) sugars and polyols most widely used as 'water binders' in fabricated foods, 'the amount of unfreezable water does not show a simple dependence on the MW of the solute', one is wise to pay heed and proceed cautiously. In fact, when the $W'_g$ data in *Table 9.2* are plotted against 1/MW (plot not shown), $r = 0.47$. One must conclude that the plot in *Figure 9.5* obviously cannot be used for predictive purposes, so the safest approach is to rely on measured $W'_g$ values for each potential 'water binding' candidate.* Franks suggests that investigations of $T_m$ and $\eta$ as functions of solute concentration, and of the liquidus curve as a function of solute structure, would be particularly worthwhile.

## Discussion

### *Structure/property relationships for SHPs and polyhydroxy compounds*

The straightforward presentation of the DE versus $T'_g$ data in *Figure 9.2* is not the most rigorous theoretical treatment. Yet, the linear correlation of DE with $T'_g$ and the convenience for practical application in the estimation of DE to characterize SHP samples justify its potential use. The rigorous theoretical dependence of DE on $T'_g$ stems from the respective dependence of each of these parameters on the degree of linear polymerization and MW within a series of monodisperse (i.e. MW = $\bar{M}n$ = weight-average MW, $\bar{M}w$) homopolymers. High polymers can be distinguished from oligomers because of their capacity for molecular chain 'entanglement coupling', resulting in the formation of rubber-like viscoelastic random networks—often called gels, in accord with Flory's (1953, 1974) nomenclature for disordered three-dimensional networks formed by physical aggregation—above a critical polymer concentration (Ferry, 1980). As summarized by Mitchell (1980), 'entanglement coupling is seen in most high MW polymer systems. Entanglements (in gels) behave as crosslinks with short lifetimes. They are believed to be topological in origin rather than involving chemical bonds'. For linear homopolymers (either amorphous or partially crystalline, and not necessarily monodisperse) with $\bar{M}n$ values below the entanglement limit, $T_g$ decreases linearly with increasing $1/\bar{M}n$ (Billmeyer, 1984). The onset of entanglement corresponds to a plateau region in which further increases in MW have little or no effect on $T_g$ (Billmeyer, 1984). There may, however, be a dramatic effect on the viscoelastic properties of the network, resulting, for example, in increased gel strength at constant temperature (Ferry 1980). The conventional presentation of such experimental data is simply $T'_g$ versus $\bar{M}w$ (Billmeyer, 1984), which conveniently displays the plateau region. Two typical

*It is interesting to note the new $W'_g$ results for the series of monomeric glycosides, in terms of a possible relationship between glycoside structure (e.g. size of the hydrophobic aglycone, absent in the parent sugar) and the functionality reflected by $W'_g$. Clearly, the $W'_g$ values for all the methyl, ethyl and propyl derivatives are much greater than those for the corresponding parent sugars. However, $W'_g$ values appear consistently to be maximized for the methyl or ethyl derivatives, but somewhat decreased for the propyl derivatives. These results could indicate that increasing hydrophobicity (of the aglycone) leads to both decreasing $W'_g$ and the demonstrated tendency toward increasing insolubility of propyl and larger glycosides in water.

examples are shown in *Figure 9.6*. The main plot describes the behaviour of a homologous series of amorphous linear poly(vinyl acetate) samples, commercially available as Vinnapas PVAc's from Wacker-Chemie. The plateau region, which includes high polymer samples that demonstrate viscoelastic rheological properties, is clearly observed (Levine and Slade, 1984). In the inset of *Figure 9.6* is shown an idealized figure [from a Perkin-Elmer DSC user's manual (Brennan, 1973)] illustrating the generalized relationship between $T_g$ and $\bar{M}w$. As exemplified in *Figure 9.6*, $T_g$ increases monotonically with increasing $\bar{M}w$ up to the plateau limit for the region of entanglement coupling—typically at $\bar{M}w$ somewhere in the range $10^4$–$10^5$ daltons (Billmeyer, 1984)—then levels off.

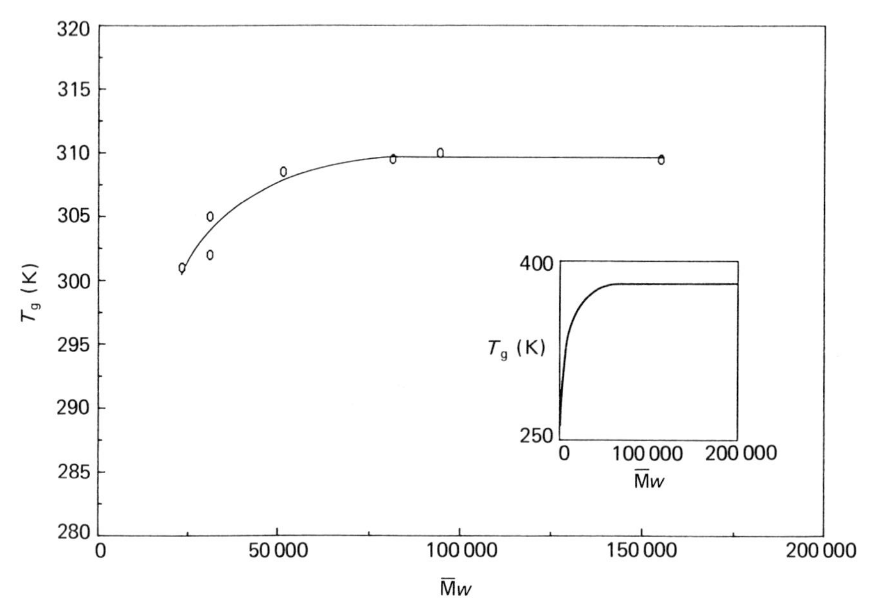


**Figure 9.6** Variation of the glass transition temperature, $T_g$, against $\bar{M}w$ for a series of commercial poly(vinyl acetate) polymers (Levine and Slade, 1984). Inset: an idealized plot of $T_g$ versus $\bar{M}w$ (Brennan, 1973).

To a first approximation, DE has the simple inverse dependence on $\bar{M}n$ defined earlier. Expressing that equation in the form $\bar{M}n = 18016/\text{DE}$ and using the conventional presentation to explore the behaviour of $T'_g$ with MW, we show, in the main plot of *Figure 9.7*, the $T'_g$ results for the SHPs in *Table 9.1*. After a steeply rising portion, a plateau region is reached for SHPs with DE $\leqslant 6$ *and* $T'_g \geqslant -8$ °C. The most likely explanation for this previously-unreported behaviour is that such SHPs experience molecular entanglement in the freeze-concentrated glass that exists at $T'_g$ and $C'_g$. Consequently, SHPs with DE $\leqslant 6$ and $T'_g \geqslant -8$ °C should be capable of forming gel networks (via entanglement), above a critical polymer concentration (which would be related to $C'_g$). Braudo, Plashchina and Tolstoguzov (1984), in their reports of the viscoelastic properties of thermoreversible maltodextrin gels (at $T > 0$ °C), also implicated entanglement coupling above a critical polymer concentration. They concluded that the non-cooperative gelation behaviour shown by maltodextrins is characteristic of semi-rigid chain polymers. This is consistent with Ferry's

(1980) observation that 'molecules which are relatively stiff and extended (in concentrated solution) exhibit the effects of entanglement coupling even more prominently than do highly flexible polymers'. Additional information about thermoreversible maltodextrin (5–8 DE) gels comes from Bulpin, Cutler and Dea (1984), who reported that such gels are composed of a network of high MW (> 10 000) branched molecules derived from amylopectin. These branched molecules represent the structural elements, which are aggregated with, and further stabilized by interactions with, short linear chains (MW < 10 000) derived from amylose.

In the inset of *Figure 9.7* is shown the linear relationship between decreasing $T'_g$ and increasing $1/\bar{M}n$, for SHPs with $\bar{M}n$ values below the entanglement limit. To and Flink's (1978) $T_c$/MW data (in their *Figure 4*, p. 574) showed the same correlation. In fact, the theoretical treatment of the data in the inset is simply a modified version of the presentation in *Figure 9.2*, with the same coefficient $r$ of $-0.98$.

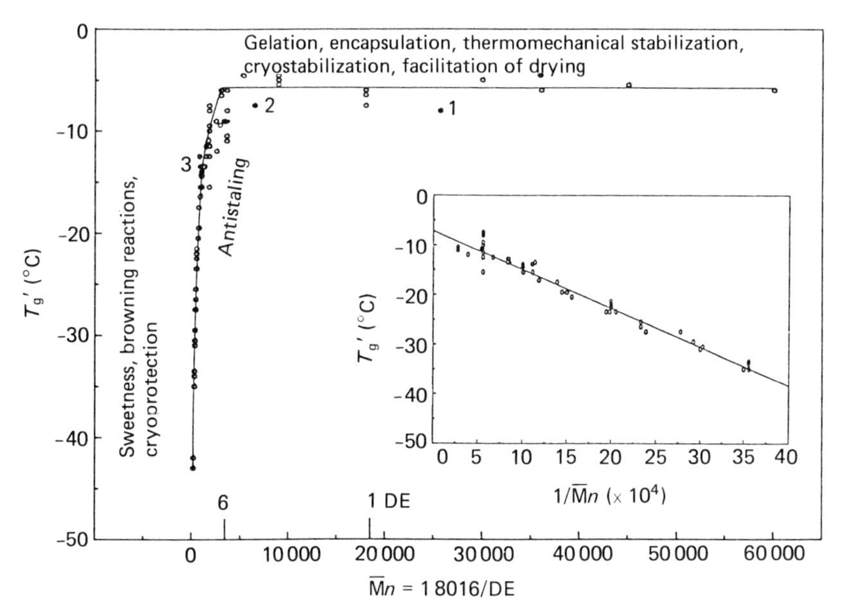


**Figure 9.7** Variation of the glass transition temperature, $T'_g$, against $\bar{M}n$ (expressed as a function of DE) for the commercial SHPs in *Table 9.1*. DE values are indicated by the numbers marked above the $x$-axis. Data points for the maltodextrin MW standards are numbered 1, 2, and 3 to provide MW markers. Areas of specific functional attributes, corresponding to three regions of the diagram, are labelled. Inset: a plot of $T'_g$ versus $1/\bar{M}n \times 10^4$ for SHPs with $\bar{M}n$ values below the entanglement limit, illustrating the theoretically predicted linear dependence.

The possible implications of our new finding and the conclusions we draw from it may explain previously observed but poorly understood aspects of SHP functional behaviour in various food-related applications. The SHPs which fall on the plateau region in *Figure 9.7* have DEs from 6 to 0.3. These DEs correspond to $\overline{DP}n$ values in the range 18 to 370, respectively, and $\bar{M}n$ values between 3000 and 60 000. The data points for the three maltodextrin MW standards are numbered in *Figure 9.7* to provide MW markers. The points for nos. (1) MD-6-40 ($\bar{M}n$ = 27 200; $\bar{M}w$ = 39 300) and (2) MD-6-12 ($\bar{M}n$ = 6500; $\bar{M}w$ = 13 000) fall on the plateau, while no. (3) MD-6-1

($\bar{M}n = 880$; $\bar{M}w = 1030$) is below the entanglement limit. Within this series of SHPs, the minimum linear chain length apparently required for intermolecular entanglement corresponds to $\overline{DP}n \simeq 18$ and $\bar{M}n \simeq 3000$. This fact explains why there is no plateau region in To and Flink's (1978) plot of $T_c$ versus $\overline{DP}n$ for SHPs of $\overline{DP}n \leqslant 16$ and DE $\geqslant 6.9$. Their *Figure 3* (p. 573) and the portion of our *Figure 9.7* for DE $\geqslant 7$ are similar in appearance, in that both show a steeply rising portion for DE $\geqslant 20$, followed by a less steeply rising portion for $20 \geqslant$ DE $\geqslant 7$. Importantly, the entanglement capability evidenced by just such SHPs of DE $\leqslant 6$ (materials not previously studied by the polymer characterization method described herein) may underlie various aspects of their functional behaviour. For example, as described by Slade (1984), sufficiently long linear chain lengths ($\overline{DP}n \gtrsim 18$) of SHPs have been correlated with intermolecular network formation and thermoreversible gelation, and with SHP and starch (re)crystallization by a chain-folding mechanism. It may be that in a partially crystalline SHP gel network, the existence of random interchain entanglements in the amorphous regions and chain-folded microcrystalline junction zones each represents a manifestation of sufficiently long chain length. This suggestion is supported by recent work (Ellis and Ring, 1985; Miles, Morris and Ring, 1985) which showed that amylose gels, which are found to be partially crystalline, are formed by cooling solutions of entangled chains. Miles, Morris and Ring (1985) stated that amylose gelation requires network formation, and this network formation requires entanglement, and they concluded that 'polymer entanglement is important in understanding the gelation of amylose'.

The excellent fit of the experimental data in *Figure 9.7* to the conventional presentation of the behaviour expected for such a family of oligomers and high polymers is gratifying, especially considering the numerous caveats that one must mention about commercial SHPs. For example, in *Figure 9.7*, we used $\bar{M}n$ (and implicitly $\overline{DP}n$) values calculated from DE, while in the conventional form (*Figure 9.6*), $\bar{M}w$ is used as a basis for specifying a typical MW range for the entanglement limit. Furthermore, for highly polydisperse solutes such as commercial SHPs (for which MW distribution, MWD = $\bar{M}w/\bar{M}n$, is frequently a variable), $T'_g$ is rigorously dependent on the average $\bar{M}w$ of a mixture of solutes (Franks, 1982). Despite these facts, the entanglement limit of $\bar{M}n \simeq 3000$ for the SHPs in *Figure 9.7* is within the characteristic range of 1250–19 000 for the minimum entanglement MWs of many typical synthetic linear high polymers (Graessley, 1984). This result for the SHPs in *Figure 9.7* contrasts with the behaviour manifested by the polyhydroxy compounds listed in *Table 9.2*, as illustrated by the $T'_g$ versus MW data in the main plot in *Figure 9.3*. It is clear, from the shape of the curve in *Figure 9.3* (and the linearity of the $T'_g$ versus 1/MW plot in the inset), that the monodisperse sugars, glycosides, and polyols represented do not show evidence of entanglement coupling. For these saccharide oligomers, none of which is larger than a heptamer of MW 1153, the entanglement plateau has not been reached, a result which agrees with the MW range of entanglement limits cited above (Graessley, 1984).

The variable polydispersity of commercial SHPs was mentioned above. Other largely uncontrollable potential variables within the series of SHPs include:

1. Significant lot-to-lot variability of solids composition (i.e. saccharides distribution) for a single SHP, which would affect the reproducibility of $T'_g$; and
2. 'As is' moisture contents (in the generally specified range of 5–10%) for different solid SHPs, which would not affect measured $T'_g$ (since, for example, 15, 20 and 25 w% solutions would all freeze-concentrate to the same invariant $T'_g$ point on the

state diagram), but would affect the calculated $W'_g$. Hence, the $W'_g$ data in *Figure 9.4* were only those for some corn syrups, whose moisture contents are generally more tightly specified.

Another origin of variability among different SHPs of nominally comparable DE concerns the method of production, i.e. hydrolysis by acid, enzyme, or acid plus enzyme (Medcalf, 1985; Murray and Luft, 1973). Especially with regard to enzymatic hydrolysates, each particular enzyme produces a different set of characteristic breakdown products with a unique MWD (Krusi and Neukom, 1984). Still another major variable among SHPs (and even for a single SHP) from different vegetable starch sources involves the original amylose/amylopectin ratio for a starch, and the consequent ratio of linear to branched polymer chains in an SHP (Medcalf, 1985). The influence of this variable can be particularly pronounced among a set of low DE maltodextrins (which would contain higher DP fractions), some from linear amylose-containing dent corn and some from all amylopectin (branched) waxy maize. The consequent range of $T'_g$ values can be quite broad, since, as a generally observed rule, linear chains give rise to higher $T'_g$ than branched chains of the same $\bar{M}w$. This observation is illustrated by several pairs of SHPs in *Table 9.1*. For each pair, of the same DE from the same manufacturer [e.g. Star Dri 1 and Star Dri 10 (1984) versus (1986), Paselli MD-6 versus SA-6 and MD-10 versus SA-10, Morrex 1910 versus 1918], the hydrolysate from amylose-free starch has a lower $T'_g$, than the correspond ing one from a starch containing amylose. This type of behaviour is also exemplified by the $T'_g$ data for the 13 10-DE maltodextrins listed in *Table 9.1*, where $T'_g$ ranges from $-7.5\,°C$ for Stadex 27 from dent corn to $-15.5\,°C$ for Frodex 10 from waxy maize, a $\Delta T$ of 8 °C. The obvious conclusion regarding a suitable maltodextrin for a specific application is that one SHP is not necessarily interchangeable with another of the same nominal DE, but from a different commercial source (Medcalf, 1985). Basic characterization of structure/property relations, e.g. in terms of $T'_g$ (rather than DE, which can be a less significant, and even misleading quantity), is often advisable before one selects such food ingredients.

The fact that linear chains give rise to higher $T'_g$ than branched chains of the same $\bar{M}w$ is also illustrated by the $T'_g$ results for some of the glucose oligomers in *Table 9.2*. Those results demonstrate that, within such a homologous series, $T'_g$ appears to depend most rigorously on the *linear* $\overline{DP}w$ of the solute. From comparisons of the significant $T'_g$ differences among maltose (1→4-linked glucose dimer), gentiobiose (1→6-linked), and isomaltose (1→6-linked); and among maltotriose (1→4-linked trimer), panose (1→4, 1→6-linked), and isomaltotriose (1→6, 1→6-linked); one may conclude that 1→4-linked (linear amylose-like) glucose oligomers manifest greater 'effective' linear chain lengths in aqueous solution (and, consequently, greater hydrodynamic volumes) than oligomers of the same MW which contain 1→6 (branched amylopectin-like) links.

### *Predicted functional attributes of SHPs and polyhydroxy compounds*

Further insights into structure/function relationships for SHPs may be gleaned if one considers *Figure 9.7* as a predictive map of regions of functional behaviour for SHP samples. For example, SHPs which fall on the entanglement plateau demonstrate certain functional attributes, some of which have been reported in the past, but not quantitatively explained from the theoretical basis of the entanglement capability revealed by our studies. Thus, it appears (as described below) that the plateau region defines the useful range of gelation, encapsulation, cryostabilization, thermomechanical stabilization, and facilitation of drying processes. The lower end of the $\bar{M}n$ range corresponds to the region of sweetness, undesirable browning reactions, and cryoprotection. The intermediate region at the upper end of the steeply rising portion represents the area of anti-staling ingredients. The map (labelled as in *Figure 9.7*) can be used to choose individual SHPs or mixtures of SHPs and other carbohydrates (e.g. targeted to a particular $T'_g$ value) to achieve desired complex functional behaviour for specific product applications. Especially for applications involving such mixtures, use can also be made of the data for the polyhydroxy compounds in *Figure 9.3* in combination with *Figure 9.7*. One will recognize that the area represented by the left-hand third of *Figure 9.3* (and the low MW sugars and polyols included therein) corresponds to the sweetness/browning/cryoprotection region of *Figure 9.7*. Likewise, the tri- through hepta-saccharides which occupy the right-hand portion of *Figure 9.3* would be predicted to function similarly to the SHPs in the anti-staling region of *Figure 9.7*.

As a specific example, the synthesis of SHPs capable of gelation from solution should be designed to yield materials of DE $\leqslant 6$ *and* $T'_g \geqslant -8$ °C. This prediction agrees with the results of Richter *et al.* (1976a,b; Braudo *et al.*, 1979), who reported that 25 w% solutions of potato starch maltodextrins of 5–8 DE produce thermoreversible, fat-mimetic gels, and with those of Lenchin, Trubiano and Hoffman (1985), who patented tapioca SHPs of DE $< 5$ which also form fat-mimetic gels from solution. Maltodextrins to be used for encapsulation of volatile flavours/aromas and lipids should likewise be capable of entanglement and network formation ($T'_g \geqslant -8$ °C). As reported by Flink (1983), effectiveness of encapsulation increases with increasing $T_c$, which in turn increases with increasing $\overline{DP}n$ within a series of SHPs, although 'a quantitative relationship between $T_c$ and MW has not been established' (To and Flink, 1978). Maltodextrins of DE $\leqslant 10$ have been used as amorphous coatings for the encapsulation of crystalline salt-substitute particles (Meyer, 1985), and maltodextrins or dextrins have also been used as coating agents for roasted nuts candy-coated with honey (Green and Hoover, 1979).

With regard to the freezer storage (in other words, 'cryo') stabilization of fabricated frozen foods (e.g. desserts such as ice cream, with smooth and creamy texture) against ice crystal growth over time, inclusion of low DE maltodextrins would elevate the composite $T'_g$ of the mix of soluble solids, which is typically dominated by low MW sugars. In practice, a retarded rate of ice recrystallization ('grain growth') at the characteristic freezer temperature ($T_f$) would result, along with an increase in the observed $T_r$. Such behaviour has been documented in the soft-serve ice cream patents of Cole *et al.* (1983, 1984) and Holbrook and Hanover (1983). In such products, ice recrystallization is known to involve a diffusion-controlled maturation process with a mechanism analogous to 'Ostwald ripening', whereby larger crystals grow with time at the expense of smaller ones which eventually disappear (Bevilacqua and Zaritzky, 1982; Harper and Shoemaker, 1983; Maltini, 1977). The rate of such a process, at $T_f$, would be reduced, as would also be $\Delta T$ ($= T_f - T'_g$), by formulating with low DE maltodextrins of high $T'_g$. In practice, the technological utility of the $T'_g$ and $W'_g$ results for sugars, polyols and SHPs (in *Tables 9.1, 9.2* and *9.3*) has been demonstrated (in combination with corresponding relative sweetness data) by the successful formulation of fabricated products (e.g. Cole *et al.* 1983, 1984) with an optimum combination of stability and softness at 0 °F freezer storage. Low DE maltodextrins are also used to stabilize frozen dairy products against lactose crystallization (another example of a diffusion-controlled collapse phenomenon) during storage (Kahn and Lynch, 1985).

Low DE maltodextrins and other high MW polymeric solutes (e.g. *see Table 9.3*) are well known as drying aids for processes such as freeze, spray, and drum drying (Flink, 1983; Karel and Flink, 1983; MacKenzie, 1981; Nagashima and Suzuki, 1985; Szejtli and Tardy, 1985). Through their simultaneous effects of increasing the composite $T'_g$ and reducing the unfrozen water fraction ($W'_g$) for freeze drying (or on other relevant $T_g/C_g$ for spray or drum drying), maltodextrins raise the observed $T_c$ (at any particular percentage moisture) relative to the drying temperature, thus facilitating drying without collapse or 'melt-back'. These attributes are illustrated by the findings of Nagashima and Suzuki (1985) on the freeze-drying behaviour of beef extract with added dextrin.

Thermomechanical stabilization refers to the stabilization of, for example, candy glasses against such collapse phenomena as recrystallization of sugars ('graining'), mechanical deformation and stickiness. Here, too, it has been shown (Cakebread, 1969; Lees, 1982; Vink and Deptula, 1982; White and Cakebread, 1966) that incorporation of low DE maltodextrins in low MW sugar glasses (to increase average $\bar{M}w$ of the solutes) increases $T_g$, thus increasing storage stability at $T < T_g$. Even when such a candy 'melt' is in the rubbery state at $T_g < T$ storage, maltodextrins are known to function as inhibitors of the diffusion-controlled propagation step in the sugar recrystallization process (Cakebread, 1969; White and Cakebread, 1966). Low DE maltodextrins are also used frequently to stabilize amorphous solids such as food powders against various collapse phenomena, as exemplified by two other recent patents. In Ogawa and Imamura (1985), 'dextrins' (SHPs $\geqslant$ tetrasaccharides of DE $\leqslant$25) were used as anti-caking/anti-browning agents in low moisture powders. In Miller and Mutka (1985), 10-DE maltodextrin was used as an additive in the production of stable amorphous juice solids compositions.

For the lower $T'_g$SHPs in *Figure 9.7* (as for many of the low $T'_g$ reducing sugars in *Figure 9.3*), sweetness and browning reactions are salient functional properties. A less familiar one involves the potential for cryoprotection of biological materials (Franks, 1982), for which the utility of various other low MW sugars and polyols is well known. The map of *Figure 9.7* predicts that such SHPs and other low MW carbohydrates, in sufficiently concentrated solution, can be quench-cooled to a completely vitrified state, wherein all the water is captured in the solute/unfrozen water glass. The essence of this cryoprotective activity, avoidance of ice formation in concentrated solutions of low MW solutes which have high $W'_g$ values, also has a readily apparent relationship to food applications involving soft, spoonable, or pourable-from-the-freezer products. One recent example is Rich's patented 'Freeze-Flo' beverage concentrate formulated with high fructose corn syrup (Kahn and Eapen, 1982).

The literature on SHPs as anti-staling ingredients for starch-based foods [reviewed by Slade (1984)], including the recent work of Krusi and Neukom (1984), reports that (non-entangling) SHP oligomers of $\overline{DP}n$ 3–8 are effective in inhibiting, *and* not participating in, starch recrystallization.

One could also postulate from the map of *Figure 9.7* that addition of a low MW sugar to a gelling maltodextrin should produce a sweet and softer gel. Addition of a glass-forming sugar to an encapsulating maltodextrin should enhance the collapse of the entangled network around the absorbed species (if collapse were desirable), but decrease the ease of spray drying. Furthermore, the map leads to two other intriguing postulates. The freeze-concentrated glass at $T'_g$ of an SHP cryostabilizer (of DE $\leqslant 6$) would contain entangled solute molecules, while in the glass at $T'_g$ of an SHP cryoprotectant, the solute molecules could not be entangled. By analogy, various high MW polysaccharide gums are claimed to be capable of improving freezer storage stability of ice-containing foods in some poorly understood way. The effect has been attributed to increased viscosity (Harper and Shoemaker, 1983; Keeney and Kroeger, 1974). Such gums may owe their limited success not only to their viscosity-increasing ability, which would be common to all glass-formers, but to their possible ability to undergo entanglement in the freeze-concentrated, non-ice matrix of the frozen food. Entanglement might enhance their limited ability to inhibit diffusion-controlled processes. In a related vein, the effects of entanglement coupling on the viscoelastic and rheological properties of random-coil polysaccharide concentrated solutions (Morris *et al.*, 1981) and gels (Braudo *et al.*, 1984; Mitchell, 1980), at $T > 0°C$, have been reported recently.

### *The role of SHPs in collapse phenomena and their mechanism of action*

In the last part of this discussion, we explore the critical role of SHPs in preventing structural collapse, within a context of the various collapse phenomena listed in *Table 9.4*. These phenomena include ones pertaining to processing and/or storage at $T > 0$ °C as well as ones involving the frozen state, all of which are governed by the particular $T_g$ relevant to the system and its content of plasticizing water. While for frozen systems, $T'_g$ of the freeze-concentrated glass is the relevant $T_g$ for describing the $T_g$/MW relationship (as illustrated by *Figure 9.7*), for amorphous dried powders and candy glasses the relevant $T_g$ pertains to a higher temperature/lower moisture state. It has been tacitly assumed that a plot of $T_g$ versus $\bar{M}n$ for dry SHPs would reflect the same fundamental behaviour as that shown in *Figure 9.7*. In fact, since $T_c$ for low moisture samples represents a good quantitative approximation of 'dry' $T_g$, To and Flink's (1978) results substantiate this assumption.

All the collapse phenomena mentioned in *Table 9.4*, as well as the glass transition itself (Ferry, 1980), are translational diffusion-controlled (many are also nucleation-limited) processes, with a mechanism involving viscous flow (Flink, 1983), under conditions of $T > T_g$ and $\eta < \eta_g = 10^{11} - 10^{14}$ Pa s (Downton, Flores-Luna and King, 1982). These kinetic processes are controlled by the variables of time, temperature, and moisture content (Tsourouflis, Flink and Karel, 1976). At the relaxation temperature, percentage moisture is the critical determinant of collapse and its concomitant changes (Karel and Flink, 1983) through the effect of water on $T_g$. This plasticizing effect of increasing moisture content at constant temperature (which is identical to the effect of increasing temperature at constant percentage moisture) leads to increased segmental mobility of polymer chains in the amorphous regions of both glassy and partially crystalline polymers. This in turn leads to the occurrence of the glass transition at decreased temperature (Cakebread, 1969).

The above concepts are well illustrated by the state diagram for poly(vinyl pyrrolidone)–water, shown in *Figure 9.8*. PVP is a much studied, water-miscible, amorphous polymer whose behaviour represents a good model of the analogous behaviour of polymeric SHPs. The state diagram for water–PVP ($\bar{M}n = 10\,000$, 44 000 and 700 000) in *Figure 9.8*, compiled from several sources (Franks, 1982; Franks *et al.*, 1977; MacKenzie, 1977; MacKenzie and Rasmussen, 1972; Olson and Webb, 1978) and augmented with our measurements of $T'_g$ and 'dry' $T_g$ (Levine and

**Table 9.4** 'COLLAPSE'-RELATED PHENOMENA WHICH ARE GOVERNED BY $T_g$ AND INVOLVE PLASTICIZATION BY WATER

| *I. Processing and/or storage at* T > *0 °C* | *References* |
|---|---|
| 1. Cohesiveness, sticking, agglomeration, sintering, lumping, caking, and flow of amorphous powders ≥ $T_c$ | Downton, Flores-Luna and King (1982)<br>Flink (1983)<br>Fukuoka *et al.* (1983)<br>Karel (1985)<br>Karel and Flink (1983)<br>Miller and Mutka (1985)<br>Moreyra and Peleg (1981)<br>Ogawa and Imamura (1985)<br>Passy and Mannheim (1982)<br>Peleg and Mannheim (1977)<br>Rosenzweig and Narkis (1981)<br>Tardos *et al.* (1984)<br>To and Flink (1978)<br>Tsourouflis *et al.* (1976)<br>White and Cakebread (1966) |
| 2. Plating of e.g. colouring agents or other fine particles on the amorphous surfaces of granular particles ≥ $T_g$ | Barbosa-Canovas *et al.* (1985)<br>Wuhrmann *et al.* (1975) |
| 3. (Re)crystallization in amorphous powders ≥ $T_c$ | Flink (1983)<br>Karel (1985)<br>Karel and Flink (1983)<br>Moreyra and Peleg (1981)<br>To and Flink (1978)<br>White and Cakebread (1966) |
| 4. Structural collapse in freeze-dried products (after sublimation stage) ≥ $T_c$ | Flink (1983)<br>Karel (1985)<br>Karel and Flink (1983)<br>To and Flink (1978)<br>Tsourouflis *et al.* (1976) |
| 5. Loss of encapsulated volatiles in freeze-dried products (after sublimation stage) ≥ $T_c$ | Downton, Flores-Luna and King (1982)<br>Flink (1983)<br>Karel (1985)<br>Karel and Flink (1983)<br>Szejtli and Tardy (1985)<br>To and Flink (1978) |
| 6. Oxidation of encapsulated lipids in freeze-dried products (after sublimation stage) ≥ $T_c$ | Flink (1983)<br>Karel (1985)<br>To and Flink (1978) |
| 7. Enzymatic activity in amorphous solids ≥ $T_g$ | Bone and Pethig (1982)<br>Morozov and Gevorkian (1985)<br>Poole and Finney (1983) |
| 8. Maillard browning reactions in amorphous powders ≥ $T_g$ | Ogawa and Imamura (1985) |
| 9. Stickiness in spray drying and drum drying ≥ $T_{sp}$ | Downton, Flores-Luna and King (1982)<br>Flink (1983)<br>Karel (1985)<br>Karel and Flink (1983)<br>To and Flink (1978)<br>Tsourouflis *et al.* (1976) |

**Table 9.4** 'COLLAPSE'-RELATED PHENOMENA WHICH ARE GOVERNED BY $T_g$ AND INVOLVE PLASTICIZATION BY WATER—*continued*

| *I. Processing and/or storage at* T > 0 °C | *References* |
|---|---|
| 10. Graining in boiled sweets $\geqslant T_g$ | Cakebread (1969)<br>Flink (1983)<br>Gueriviere (1976)<br>Herrington and Branfield (1984)<br>Lees (1982)<br>McNulty and Flynn<br>Soesanto and Williams (1981)<br>Vink and Deptula (1982)<br>White and Cakebread (1966) |
| 11. Sugar bloom in chocolate $\geqslant T_g$ | Chevalley *et al.* (1970)<br>Niediek and Barbernics (1981) |
| *II. Processing and/or storage at* T < 0°C | |
| 1. Ice recrystallization ('grain growth') $\geqslant T_r$ | Franks (1985) ($\geqslant T_g'$)<br>Franks (1982)<br>Franks *et al.* (1977)<br>MacKenzie (1977)<br>Maltini (1977) |
| 2. Lactose crystallization ('sandiness') in dairy products $\geqslant T_r$ | Flink (1983)<br>Franks (1982)<br>Kahn and Lynch (1985)<br>White and Cakebread (1966) |
| 3. Enzymatic activity $\geqslant T_g'$ | Levine and Slade (1984)<br>Morozov and Gevorkian (1985) |
| 4. Structural collapse, shrinkage, or puffing (of amorphous matrix surrounding ice crystals) during freeze drying (sublimation stage) = 'melt-back' $\geqslant T_c$ | Flink (1983)<br>Franks (1982)<br>Karel and Flink (1983)<br>MacKenzie (1977)<br>Maltini (1974)<br>Nagashima and Suzuki (1985)<br>To and Flink (1978)<br>Tsourouflis *et al.* (1976)<br>White and Cakebread (1966) |
| 5. Loss of encapsulated volatiles during freeze drying (sublimation stage) $\geqslant T_c$ | Flink (1983)<br>Karel and Flink (1983)<br>Szejtli and Tardy (1985)<br>To and Flink (1978) |
| 6. Reduced survival of cryopreserved embryos, due to cellular damage caused by diffusion of ionic components $\geqslant T_g'$ | Reid (1985) |

Slade, 1984), is the most complete one presently available for this polymer. It illustrates the dramatic effect of water on $T_g$ (especially at low moisture) and the smoothness of the $T_g$ curve, which ranges from 100 °C for dry PVP-44 to $\simeq -135$ °C for glassy water (Franks, 1982).

Whenever the glass transition and the resultant structural collapse occur on the same time scale (Franks, 1982), $T_g$ equals the minimum onset temperature for the

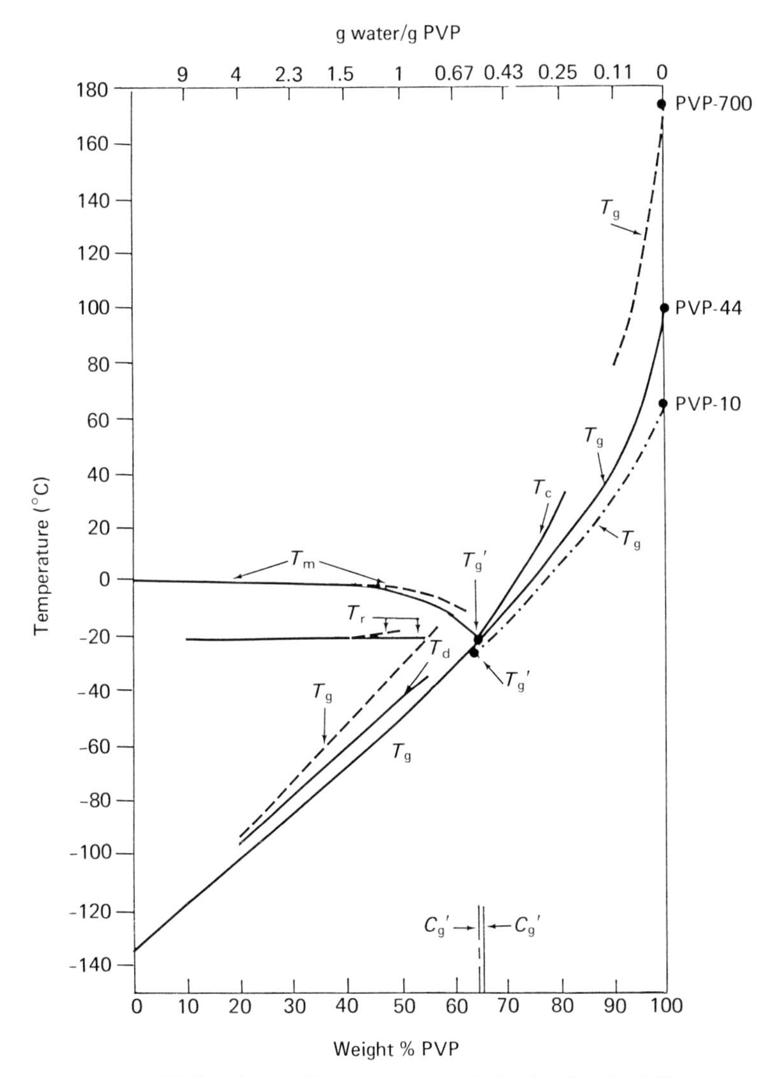


**Figure 9.8** Solid–liquid state diagram for water–PVP, showing the following transitions: $T_m$, $T_g$, $T'_g$, $T_d$, $T_r$, $T_c$. Key: ———. PVP-44; – – – – –, PVP-700; –·–·–·–·–, PVP-10; ●– – – – –●, data from Olson and Webb (1978); other ● refer to data from Levine and Slade (1984).

collapse phenomena in *Table 9.4*. Thus, a system is stable against collapse, within the period of the experimental measurements of $T_g$ and $T_c$, at $T < T_g$. Increasing percentage moisture leads to decreased stability and shelf-life at a particular storage temperature (Karel and Flink, 1983). The various phenomenological threshold temperatures (e.g. $T_c \cong T_r \cong T_{sp}$) are all equal to the particular $T_g$ (or $T'_g$) which corresponds to the solute(s) concentration for the situation in question. Thus, in *Figure 9.8*, for PVP-44, $T'_g = T_r = T_c \simeq -21.5$ °C and $C'_g \simeq 65$ w% PVP ($W'_g \simeq 0.54$ g unfrozen water/g PVP) (Franks, 1982; Levine and Slade, 1984; MacKenzie, 1977; MacKenzie and Rasmussen, 1972), while for PVP-700, $T_g \cong T_c \cong T_{sp} \simeq 120$ °C at

$\simeq$5% residual moisture (Levine and Slade, 1984; Olson and Webb, 1978). The equivalence of $T_r$ for ice or solute recrystallization, $T_c$ for collapse, and the concentration invariant $T'_g$ for an ice-containing system explains why $T_r$ and $T_c$ have always been observed in the past to be concentration-independent for all initial solute concentrations lower than $C'_g$ (Franks, 1982), as illustrated in *Figure 9.8*.

Our conclusion regarding the *fundamental* equivalence of $T_g$, $T_c$ and $T_r$ represents a departure from the previous literature. For example, while To and Flink (1978) acknowledged that 'the relationship between $T_c$ and MW is identical to the equation for $T_g$ of mixed polymers' and that 'collapse and glass transition are (clearly) phenomenologically similar events', they differentiated between $T_g$ and $T_c$ by pointing out that 'while glass transitions in polymeric materials are generally reversible, the collapse of freeze-dried matrices is irreversible'. While the latter facts may be true, the argument is misleading. At the molecular level, the glass-to-rubber transition for an amorphous thermoplastic material is reversible. That is, the glass at $T'_g/C'_g$ can be repeatedly warmed and recooled (slowly) over a completely reversible $T/C$ path between its solid and liquid states. The same is true for a completely amorphous (and non-crystallizable) freeze-dried material. The reason collapse is said to be irreversible for a porous matrix has nothing to do with reversibility between molecular states. Irreversible loss of porosity is simply a macroscopic, morphological consequence of viscous flow in the rubbery state at $T > T_g$, whereby the porous glass relaxes to a fluid (incapable of supporting its own weight against flow), which then becomes non-porous and more dense. Subsequent recooling to $T < T_g$ yields a non-porous glass of the original composition, which can thereafter be temperature-cycled reversibly. The only irreversible aspect of $T_g$-governed collapse is loss of porosity.

Recently, our conclusion about the fundamental identity of $T'_g$ with $T_r$ and $T_c$ was corroborated by Reid (1985). He reported a study in which $T'_g$, measured by DSC, corresponded well with the temperature at which a frozen aqueous solution, viewed under a cryomicroscope, became physically mobile. Reid remarked that '$T'_g$, the temperature at which a system would be expected to become mobile due to the appearance of the solution phase, has also been related to the collapse temperature in freeze drying, again relating to the onset of system mobility, which presumably allows for the diffusion of solution components'. Reid's study revealed another collapse-related phenomenon, governed by $T'_g$ of a frozen system, that has been added to *Table 9.4*: slow warming of cryopreserved embryos to $T > T'_g$ facilitates the detrimental diffusion of ionic components (salts), resulting in cellular damage due to high ionic strength, and in much reduced embryo survival.

*Table 9.3* shows a comparison of DSC-measured $T'_g$ values for a variety of food materials from our data bank of water-soluble monomers and polymers, and literature values (Franks, 1982; Luyet, 1960) for other 'collapse' transition temperatures. These results for the observed $T_c$ and $T_r$, which are usually measured (on an experimental time scale similar to that of our DSC method) by cryomicroscopy of frozen or vitrified aqueous solutions, are generally very close to, but almost always at a slightly higher temperature than, our values for $T'_g$. We take this as further support of our contention that $T'_g$ represents the minimum onset temperature for these subzero collapse phenomena.

## *A physicochemical mechanism for collapse based on WLF theory*

A universally applicable, quantitative mechanism for collapse is provided by the

WLF theory for the temperature dependence of the viscoelastic properties of amorphous polymers and glass-forming liquids within the temperature range for the rubbery (supercooled liquid) state from $T_g$ to $T_g + 100\,°K$ (Ferry, 1980). The WLF equation describes the kinetic nature of the glass transition, and defines the exponential temperature dependence of the rate of any diffusion-controlled relaxation, e.g. $\eta$, occurring at $T$, versus the rate at a reference temperature $T_g$ below $T$, in terms of $\log \eta T/\eta_g \; \alpha \; \Delta T$ (Ferry, 1980). The WLF equation depends critically on the appropriate reference $T_g$ (for a particular glass-forming system, be it $T'_g$ for a frozen system, or $T_g$ for a low moisture one), which is defined as an iso-free volume state and approximately as an iso-viscosity state of $\eta_g \simeq 10^{11} - 10^{14}$ Pa s (Soesanto and Williams, 1981).

The impact of WLF behaviour on the rates of relaxation processes in amorphous polymer–water systems (e.g. structural collapse, ice recrystallization) can be illustrated by the following relative rates (vs. rate = 1 at $T = T_g$, or $\Delta T = 0$) calculated from the WLF equation: for $\Delta T = 3$, 7, 11 and 21 °C, rate = 10, 100, 1000 and $10^5$, respectively. These calculations would also apply, for example, to retardation of the rate of the propagation step in the mechanism of recrystallization of an amorphous but crystallizable polymer (or monomer), where the rate would be zero at $T < T_g$ (i.e. immobility leads to inhibition of migratory diffusion of large main-chain segments), but would increase exponentially with increasing $\Delta T$ above $T_g$, up to the limit of $T_m$. In some collapse phenomena, such a recrystallization transition (from unstable amorphous liquid to crystalline solid) may occur after the glass transition (White and Cakebread, 1966), and its rate would be likewise defined by the WLF equation. Referring to the state diagram in *Figure 9.8*, one sees that a $T_g$ curve corresponds to a boundary between physical states in which the collapse phenomena in *Table 9.4* either can (at $T > T_g$) or cannot (at $T < T_g$) occur over realistic times, and that the WLF equation defines the rates of those molecular relaxations that will occur above $T_g$, in terms of an exponential function of $\Delta T$ above this boundary condition.

The controlled agglomeration of amorphous powders represents a specific example of a WLF-governed process related to caking. As reported by Downton, Flores-Luna and King (1982), and verified by Tardos, Mazzone and Pfeffer (1984), *spontaneous* agglomeration of solid powder particles occurs when $\eta$ of the liquid phase at the surface of the particle drops to $\simeq 10^7$ Pa s. This $\eta$ is $\simeq 10^5$ lower than $\eta_g$. From the WLF equation, this $\Delta\eta$ of $10^5$ Pa s corresponds to a $\Delta T$ of $\simeq 21$ °C between $T_g$ and the $T_{sp}$ for spontaneous agglomeration. Thus, on a state diagram of $T$ vs. percentage moisture, the $T_g$ and $T_{sp}$ curves would represent parallel iso-viscosity lines. The $T_{sp}$ curve for *fast* agglomeration during processing (reported by Downton, Flores-Luna and King, 1982) would lie above the $T_g$ curve for *slow* caking during storage, and the $\Delta T$ of 21 °C would reflect the different time scales for the two phenomena.

In practice, collapse (and all its different manifestations) can be prevented, and product quality and stability maintained, by the following three fundamental counter measures:

1. Storage at $T < T_g$ (White and Cakebread, 1966).
2. Formulation to increase $T_g$ to a temperature above the processing or storage temperature, by increasing the overall $\bar{M}w$ of the water-soluble solids in a product mixture. As described above, the latter is often accomplished by adding polymeric stabilizers such as lower DE SHPs (or other polymeric carbohydrate, protein, or cellulose and polysaccharide gum stabilizers, some of which are included in *Table 9.3*) to a formulation dominated by low MW sugars and/or polyols (Cakebread,

1969; Downton, Flores-Luna and King, 1982; Kahn and Lynch, 1985; Karel and Flink, 1983; Lees, 1982; Maltini, 1974; Miller and Mutka, 1985; Ogawa and Imamura, 1985; To and Flink, 1978; Tsourouflis, Flink and Karel, 1976; Vink and Deptula, 1982; White and Cakebread, 1966). The effect of increased MW on the $T_g$ of PVPs is also illustrated in *Figure 9.8*.
3. In low moisture amorphous food powders and other hygroscopic glassy solids, the latter including 'candy' glasses such as boiled sweets (Cakebread, 1969; Chevalley, Rostagno and Egli, 1970; Vink and Deptula, 1982; White and Cakebread, 1966), extruded melts (Gueriviere, 1976), candy coatings (Lees, 1982), sugar in chocolate (Niediek and Barbernics, 1981), and supersaturated sugar syrups (Downton, Flores-Luna and King, 1982; McNulty and Flynn, 1977; Soesanto and Williams, 1981):
   a) reduction of residual moisture content to $\leqslant 3\%$ during processing;
   b) packaging in superior moisture-barrier film or foil to prevent moisture pickup;* and
   c) avoidance of high $T$/high percentage RH($\gtrsim 20\%$ RH) conditions during storage (Cakebread, 1969; Flink, 1983; Gueriviere, 1976; Lees, 1982; White and Cakebread, 1966).

On the other hand, Karel (1985) pointed out that water plasticization (to depress $T_g$ below $T$ of the phenomenon) is not always detrimental to product quality. Examples of applications involving deliberate moisturization to produce desirable consequences include:

1. Controlled agglomeration or sintering (by limited heat/moisture/time treatment) of amorphous powders [as described above and in Flink (1983), Rosenzweig and Narkis (1981), Tsourouflis, Flink and Karel (1976)], and
2. Compression (without brittle fracture) of freeze-dried products after limited replasticization (Karel and Flink, 1983).

### *Prevention of enzymic activity and other chemical reactions at* $T < T_g$

One collapse-related phenomenon listed in *Table 9.4* but not previously discussed in this report involves enzymic activity in amorphous substrate-containing media which occurs only at $T > T_g$. Enzymic activity represents a pleasing case study with which to close this discussion, because it is potentially important in many food applications which cover the entire spectrum of processing/storage temperatures and moisture contents, and because examples exist (Karel, 1985) which elegantly illustrate the fact that activity is inhibited in low moisture amorphous solids at $T < T_g$, and in frozen systems at $T < T'_g$. Bone and Pethig (1982) studied the hydration of dry lysozyme powder at 20 °C, and found that, at 20 w% water, lysozyme becomes sufficiently plasticized so that measurable enzymic activity commences. We interpreted their results to indicate the following: a diffusion-controlled enzyme/substrate interaction is essentially prohibited in a glassy solid at $T < T_g$, but sufficient water plasticization depresses $T_g$ of lysozyme to < 20 °C, allowing the onset of enzymic activity in the

*Marsh and Wagner (1985) described a 'state of the art' computer model which can be used to predict the shelf life of particular moisture-sensitive products, based on the moisture-barrier properties of a packaging material and the temperature/humidity conditions of a specific storage environment.

lysozyme solution at $T > T_g$, the threshold temperature for activity (Levine and Slade, 1984). Our interpretation was supported by the results of a related study of 'solid glassy lysozyme samples' by Poole and Finney (1983), who noted conformational changes in the protein as a consequence of hydration to the same 20 w% level. They were 'tempted to suggest that this solvent-related effect is required before (enzymic) activity is possible'. More recently, Morozov and Gevorkian (1985) also noted the criticality of low temperature, water-plasticized glass transitions to the physiological activity of lysozyme and other globular proteins.

Within the context of cryostabilization, and the potentially critical role of low DE SHPs as cryostabilizers, we have verified the above conclusion in maximally frozen biological systems. By analogy, in such cases, the threshold temperature for onset of enzymic activity would be $T'_g$. Cryostabilization as a technology is a means of protecting freezer-stored and freeze-dried foods from the deleterious changes in texture (e.g. grain growth of ice, solute crystallization), structure (e.g. shrinkage, collapse), and chemical composition (e.g. flavour degradation, fat rancidity, as well as enzymic reactions) typically encountered. The key to this protection lies in controlling the physicochemical properties of the freeze-concentrated matrix surrounding the ice crystals. If this matrix is maintained as a mechanical solid (at $T_f < T'_g$), then the diffusion-controlled changes that typically result in reduced storage stability can be prevented or at least greatly retarded. If, on the other hand, a natural food is improperly stored at too high a $T_f$, or a fabricated product is improperly formulated, and thus the matrix is allowed to exist in the freezer as a rubbery fluid (at $T_f > T'_g$), then freezer storage stability would be reduced. Furthermore, the rates of the various deleterious changes would increase exponentially with the $\Delta T$ between $T_f$ and $T'_g$, as dictated by WLF theory.

The prevention of enzymic activity at $T < T'_g$ was demonstrated experimentally *in vitro* in a model system consisting of glucose oxidase, glucose, methyl red, and bulk solutions of sucrose, Morrex 1910 (10-DE maltodextrin), and their mixtures, which provided a range of samples with known values of $T'_g$ (Levine and Slade, 1986). The enzymic oxidation of glucose produces an acid which turns the reaction mixture from yellow to pink. Samples with a range of $T'_g$ values from − 9.5 to − 32 °C were stored at various temperatures: 25, 3, − 15 and − 23 °C. All the samples were fluid at the two higher temperatures, while all looked like coloured blocks of ice at − 15 and − 23 °C. However, only the samples for which the temperature of storage was above $T'_g$ turned pink! Even after two months' storage at − 23 °C, the samples containing the maltodextrin, with $T'_g > -23$ °C, were still yellow. The frozen samples which turned pink, even at − 23 °C, contained a concentrated enzyme-rich fluid surrounding the ice crystals, while in those which remained yellow, the non-ice matrix was a glassy solid. Significantly, enzymic activity was prevented by storage below $T'_g$, but the enzyme itself was not inactivated. When the yellow samples were thawed, they quickly turned pink. Thus, cryostabilization with a low DE SHP preserved the enzyme during storage, but prevented its activity below $T'_g$.

## Acknowledgements

We thank our colleagues at General Foods, Timothy Schenz, Allen Bradbury and Terry Maurice, for their contributions to the research programme from which the present work derives, and our consultant, Professor Felix Franks of the University of Cambridge for invaluable suggestions, discussions and encouragement.

## References

BARBOSA-CANOVAS, G.V., RUFNER, R. and PELEG, M. (1985). Microstructure of selected binary food powder mixtures. *Journal of Food Science*, **50,** 473–481

BEVILACQUA, A.E. and ZARITZKY, N.E. (1982). Ice recrystallization in frozen beef. *Journal of Food Science*, **47,** 1410–1414

BILIADERIS, C.G., PAGE, C.M., SLADE, L. and SIRETT, R.R. (1985). Thermal behavior of amylose–lipid complexes. *Carbohydrate Polymers*, **5,** 367–389

BILLMEYER, F.W. (1984). *Textbook of Polymer Science*, 3rd Edition. New York, Wiley-Interscience

BONE, S. and PETHIG, R. (1982). Dielectric studies of the binding of water to lysozyme. *Journal of Molecular Biology*, **157,** 571–575

BRAUDO, E.E., PLASHCHINA, I.G. and TOLSTOGUZOV, V.B. (1984). Structural characterisation of thermoreversible anionic polysaccharide gels by their elastoviscous properties. *Carbohydrate Polymers*, **4,** 23–48

BRAUDO, E.E., BELAVTSEVA, E.M., TITOVA, E.F., PLASHCHINA, I.G., KRYLOV, V.L., TOLSTOGUZOV, V.B., SCHIERBAUM, F.R. and RICHTER, M. (1979). Struktur und Eigenschaften von Maltodextrin-Hydrogelen. *Staerke*, **31,** 188–194

BRENNAN, W.P. (1973). *Thermal Analysis Application Study No. 8*. Norwalk, Perkin Elmer Instrument Division

BULPIN, P.V., CUTLER, A.N. and DEA, I.C.M. (1984). Thermally-reversible gels from low DE maltodextrins. In *Gums and Stabilizers for the Food Industry*, Volume 2, (Phillips, G.O., Wedlock, D.J. and Williams, P., Eds) pp. 475–484. Oxford, Pergamon Press

CAKEBREAD, S.H. (1969). Factors affecting the shelf life of high boilings. *Manufacturing Confectioner*, **49,** 41–44

CHEVALLEY, J., ROSTAGNO, W. and EGLI, R.H. (1970). A study of the physical properties of chocolate. V. *Revue Internationale du Chocolat*, **25,** 3–6

COLE, B.A., LEVINE, H.I., MCGUIRE, M.T., NELSON, K.J. and SLADE, L. (1983). Soft, Frozen Dessert Formulation. US Patent 4 374 154, Feb. 15

COLE, B.A., LEVINE, H.I., MCGUIRE, M.T., NELSON, K.J. and SLADE, L. (1984). Soft, Frozen Dessert Formulation. US Patent 4 452 824, June 5

DOWNTON, G.E., FLORES-LUNA, J.L. and KING, C.J. (1982). Mechanism of stickiness in hygroscopic, amorphous powders. *Industrial Engineering Chemistry Fundamentals*, **21,** 447–451

ELLIS, H.S. and RING, S.G. (1985). A study of some factors influencing amylose gelation. *Carbohydrate Polymers*, **5,** 201–213

FERRY, J.D. (1980). *Viscoelastic Properties of Polymers*, 3rd Edition. New York, Wiley and Sons

FLINK, J.M. (1983). Structure and structure transitions in dried carbohydrate materials. In *Physical Properties of Foods*, (Peleg, M. and Bagley, E.B., Eds), pp. 473–521. Westport, AVI Publishing

FLORY, P.J. (1953). *Principles of Polymer Chemistry*. Ithaca, Cornell University Press

FLORY, P.J. (1974). Introductory lecture—gels and gelling processes. *Faraday Discussions of the Chemical Society*, **57,** 7–18

FRANKS, F. (1982). The properties of aqueous solutions at subzero temperatures. In *Water: A Comprehensive Treatise*, Volume 7, (Franks, F., Ed.), pp. 215–338. New York, Plenum Press

FRANKS, F. (1985). Complex aqueous systems at subzero temperatures. In *Properties of Water in Foods* (Simatos, D. and Multon, J.L., Eds), pp. 497–509. Dordrecht, Martinus Nijhoff

FRANKS, F., ASQUITH, M.H., HAMMOND, C.C., SKAER, H.B. and ECHLIN, P. (1977). Polymeric cryoprotectants in the preservation of biological ultrastructure. I. *Journal of Microscopy*, **110,** 223–238

FUKUOKA, E., KIMURA, S., YAMAZAKI, M. and TANAKA T. (1983). Cohesion of particulate solids. VI. *Chemical and Pharmaceutical Bulletin*, **31,** 221–229

GRAESSLEY, W.W. (1984). Viscoelasticity and flow in polymer melts and concentrated solutions. In *Physical Properties of Polymers*, (Mark, J.E. *et al.*, Eds), pp. 97–153. Washington, ACS.

GREEN, W.M. and HOOVER, M.W. (1979). Honey Coated Roasted Nut Product and Method for Making Same. US Patent 4 161 545, July 17

GUERIVIERE, J.F. (1976). Recent developments in extrusion cooking of foods. *Industries Alimentaires et Agricoles*, **93,** 587–595

HARPER, E.K. and SHOEMAKER, C.F. (1983). Effect of locust bean gum and selected sweetening agents on ice recrystallization rates. *Journal of Food Science*, **48,** 1801–1806

HERRINGTON, T.M. and BRANFIELD, A.C. (1984). Physicochemical studies on sugar glasses. I. and II. *Journal of Food Technology*, **19,** 409–435

HOLBROOK, J.L. and HANOVER, L.M. (1983). Fructose-containing Frozen Dessert Products. US Patent 4 376 791, March 15

KAHN, M.L. and EAPEN, K.E. (1982). Intermediate-Moisture Frozen Foods. US Patent 4 332 824, June 1

KAHN, M.L. and LYNCH, R.J. (1985). Freezer Stable Whipped Ice Cream and Milk Shake Food Products. US Patent 4 552 773, November 12

KAREL, M. (1985). Effects of water activity and water content on mobility of food components, and their effects on phase transitions in food systems. In *Properties of Water in Foods*, (Simatos, D. and Multon, J.L., Eds), pp. 153–169. Dordrecht, Martinus Nijhoff

KAREL, M. and FLINK, J.M. (1983). Some recent developments in food dehydration research. In *Advances in Drying*, Volume 2, (Mujumdar, A.S., Ed.), pp. 103–153. Washington, Hemisphere Publishing

KEENEY, P.G. and KROEGER, M. (1974). Frozen dairy products. In *Fundamentals of Dairy Chemistry*, 2nd Edition, (Webb, B.H. *et al.*, Eds), p. 890. Westport, AVI Publishers

KRUSI, H. and NEUKOM, H. (1984). Untersuchungen uber die Retrogradation der Staerke in konzentrierten Weizenstarfkegelen. Staerke, **36,** 300–305

LEES, R. (1982). Quality control in the production of hard and soft sugar confectionery. *Confectionery Production*, Feb., 50–51

LENCHIN, J.M., TRUBIANO, P.C. and HOFFMAN, S. (1985). Converted Starches for Use as a Fat- or Oil-Replacement in Foodstuffs. US Patent 4 510 166, April 9.

LEVINE, H. and SLADE, L. (1984). Water as plasticizer—low moisture technology of polymers. Presented at Royal Society of Chemistry—Water Soluble Polymers: Chemistry and Application Technology Course, July 16–20, Girton College, Cambridge

LEVINE, H. and SLADE, L. (1986). A polymer physico-chemical approach to the study of commercial starch hydrolysis products. *Carbohydrate Polymers*, **6,** 213–244

LUYET, B. (1960). On various phase transitions occurring in aqueous solutions at low temperatures. *Annals of the New York Academy of Sciences*, **85,** 549–569

MACKENZIE, A.P. and RASMUSSEN, D.H. (1972). Interactions in the water–polyvinylpyrrolidone system at low temperatures. In *Water Structure at the Water–Polymer Interface*, (Jellinek, H.H.G., Ed.), pp. 146–171. New York, Plenum Press

MACKENZIE, A.P. (1977). Non-equilibrium freezing behaviour of aqueous systems. *Philosophical Transactions of the Royal Society of London*, **B278,** 167–189

MACKENZIE, A.P. (1981). Modelling the ultra-rapid freezing of cells and tissues. In *Microprobe Analysis of Biological Systems*, pp. 397–421. New York, Academic Press

MALTINI, E. (1974). Thermophysical properties of frozen juice related to freeze drying problems. *Annali Dell'Istituto Sperimentale Valorizzazione Technologica Prodotti Agricoli*, **5,** 65–72

MALTINI, E. (1977). Studies on the physical changes in frozen aqueous solutions by DSC and microscopic observations. *I.I.F.–I.I.R.–Karlsruhe*, 1977-1, 1–9

MARSH, K.S. and WAGNER, J. (1985). Predict shelf life. *Food Engineering*, August, 58.

MAURICE, T.J., SLADE, L., SIRETT, R.R. and PAGE, C.M. (1985). Polysaccharide–water interactions—thermal behavior of rice starch. In *Properties of Water in Foods*, (Simatos, D. and Multon, J.L., Eds), pp. 211–227. Dordrecht, Martinus Nijhoff

MCNULTY, P.B. and FLYNN, D.G. (1977). Force-deformation and texture profile behavior of aqueous sugar glasses. *Journal of Texture Studies*, **8,** 417–431

MEDCALF, D.G. (1985). Food functionality of cereal carbohydrates. In *New Approaches to Research on Cereal Carbohydrates*, (Hill, R.D. and Munck, L., Eds), pp. 355–362. Amsterdam, Elsevier

MEYER, D.R. (1985). Salt Substitute Containing Potassium Chloride Coated with Maltodextrin and Method of Preparation. US Patents 4 556 567, 4 556 568, December 3

MILES, M.J., MORRIS, V.J. and RING, S.G. (1985). Gelation of amylose. *Carbohydrate Research*, **135,** 257–269

MILLER, D.H. and MUTKA, J.R. (1985). Process for Forming Solid Juice Composition and Product of the Process. US Patent 4 499 112, February 12

MITCHELL, J.R. (1980). The rheology of gels. *Journal of Texture Studies*, **11,** 315–337

MOREYRA, R. and PELEG, M. (1981). Effect of equilibrium water activity on the bulk properties of selected food powders. *Journal of Food Science*, **46,** 1918–1922

MOROZOV, V.N. and GEVORKIAN, S.G. (1985). Low-temperature glass transition in proteins. *Biopolymers*, **24,** 1785–1799

MORRIS, E.R., CUTLER, A.N., ROSS-MURPHY, S.B. and REES, D.A. (1981). Concentration and shear rate dependence of viscosity in random coil polysaccharide solutions. *Carbohydrate Polymers*, **1,** 5–21

MURRAY, D.G. and LUFT, L.R. (1973). Low D.E. corn starch hydrolysates. *Food Technology*, **27,** 32–40

NAGASHIMA, N. and SUZUKI, E. (1985). The behavior of water in foods during freezing and thawing. In *Properties of Water in Foods*, (Simatos, D. and Multon, J.L., Eds), pp. 555–571. Dordrecht, Martinus Nijhoff

NIEDIEK, E.A. and BARBERNICS, L. (1981). Amorphisierung von Zucker durch das Feinwalzen von Schokoladenmassen. *Gordian*, **80,** 267–269

OGAWA, H. and IMAMURA, Y. (1985). Stabilized Solid Compositions. US Patent 4 547 377, October 15

OLSON, D.R. and WEBB, K.K. (1978). The effect of humidity on the glass transition temperature. *Organic Coatings and Plastics Chemistry*, **39,** 518–523

PASSY, N. and MANNHEIM, C.H. (1982). Flow properties and water sorption of food powders. II. *Lebensm.-Wiss. u.-Technol.*, **15,** 222–225

PELEG, M. and MANNHEIM, C.H. (1977). The mechanism of caking of powdered onion. *Journal of Food Processing and Preservation*, **1,** 3–11

POOLE, P.L. and FINNEY, J.L. (1983). Sequential hydration of a dry globular protein. *Biopolymers*, **22,** 255–260; *International Journal of Biological Macromolecules*, **5,** 308–310

RASMUSSEN, D. and LUYET, B. (1969). Complementary study of some non-equilibrium phase transitions in frozen solutions of glycerol, ethylene glycol, glucose and sucrose. *Biodynamica*, **10,** 319–331

REID, D.S. (1985). Correlation of the phase behavior of DMSO/NaCl/water and glycerol/NaCl/water as determined by DSC with their observed behavior on a cryomicroscope. *Cryo-Letters*, **6,** 181–188

RICHTER, M., SCHIERBAUM, F., AUGUSTAT, S. and KNOCH, K.D. (1976a). Method of Producing Starch Hydrolysis Products for Use as Food Additives. US Patent 3 962 465, June 8

RICHTER, M., SCHIERBAUM, F., AUGUSTAT, S. and KNOCH, K. D. (1976b). Method of Producing Starch Hydrolysis Products for Use as Food Additives. US Patent 3 986 890, October 19

ROSENZWEIG, N. and NARKIS, M. (1981). Sintering rheology of amorphous polymers. *Polymer Engineering and Science*, **21,** 1167–1170

ROWLAND, S.P. (Ed.) (1980). *Water in Polymers*, ACS Symposium Series 127. Washington, American Chemical Society

SCHENZ, T.W., ROSOLEN, M.A., LEVINE, H. and SLADE, L. (1984). DMA of frozen aqueous solutions. In *Proceedings of the 13th NATAS Conference, Sept. 23–26, Philadelphia, PA*, (McGhie, A.R., Ed.), pp. 57–62. Phiadelphia, NATAS

SEARS, J.K. and DARBY, J.R. (1982). *The Technology of Plasticizers*. New York, Wiley-Interscience

SLADE, L. (1984). Staling of starch based products. *American Association of Cereal Chemists 69th Annual Meeting, Minneapolis, MN*, Abstract No. 112

SLADE,L. and LEVINE, H. (1984). Thermal analysis of starch and gelatin. In *Proceedings of the 13th NATAS Conference, Sept. 23–26, Philadelphia, PA*, (McGhie, A.R., Ed.), p. 64. Philadelphia, NATAS

SLADE, L. and LEVINE, H. (1985). Intermediate moisture systems. Presented at Faraday Division, Royal Society of Chemistry, Industrial Physical Chemistry Group, Discussion Conference on Concept of Water Activity, July 1–3, Girton College, Cambridge

SLADE, L. and LEVINE, H. (1987). Polymer-chemical properties of gelatin in foods. In *Advances in Meat Research*, Volume 4, (Pearson, A.M., Dutson, T.R. and Bailey, A.J., Eds), pp. 251–266. Westport, AVI Publishers

SOESANTO, T. and WILLIAMS, M.C. (1981). Volumetric interpretation of viscosity for concentrated and dilute sugar solutions. *Journal of Physical Chemistry*, **85,** 3338–3341

SZEJTLI, J. and TARDY, M. (1985). Honey Powder Preserving its Natural Aroma Components. US Patent 4 529 608, July 16

TARDOS, G., MAZZONE, D. and PFEFFER, R. (1984). Measurement of surface viscosities using a dilatometer. *Canadian Journal of Chemical Engineering*, **62,** 884–887

TO, E.C. and FLINK, J.M. (1978). 'Collapse', a structural transition in freeze dried carbohydrates. I.–III. *Journal of Food Technology*, **13,** 551–594

TSOUROUFLIS, S., FLINK, J.M. and KAREL, M. (1976). Loss of structure in freeze-dried carbohydrate solutions. *Journal of Science of Food and Agriculture*, **27,** 509–519

VINK, W. and DEPTULA, R.W. (1982). High Fructose Hard Candy. US Patent 4 311 722, January 19

WHITE, G.W. and CAKEBREAD, S.H. (1966). The glassy state in certain sugar-containing food products. *Journal of Food Technology*, **1,** 73–82

WUHRMANN, J.J., VENRIES, B. and BURI, R. (1975). Process for Preparing a Colored Powdered Edible Composition. US Patent 3 920 854, November 18

10

# CREATION OF FIBROUS STRUCTURES BY SPINNERETLESS SPINNING

V.B. TOLSTOGUZOV
*Institute of Organoelement Compounds, USSR Academy of Science, Moscow, USSR*

## Introduction

The scientific basis for developing the spinneretless spinning process is the well known phenomenon of the deformation of emulsion droplets in flow (Kuleznov, 1980; Sherman, 1968; Tolstoguzov, Mzhel'sky and Gulov, 1974; Tolstoguzov *et al.*, 1985a). This phenomenon arises as follows. When an emulsion is flowing, the stresses developed in a dispersion medium can deform and orient the liquid-dispersed particles. The deformation of spherical droplets is counteracted by interfacial tension. Under certain specific conditions, the emulsion droplets in flow deform so much that they take the form of liquid filaments or cylinders. The latter are, however, not stable and tend to break down into small spherical drops. The competing process is drop coalescence, resulting in larger-sized dispersed particles. Hence, in a flowing emulsion a dynamic equilibrium may establish itself between the drop deformation and breakdown, on the one hand, and their coalescence, on the other hand.

The basic idea underlying the spinneretless spinning process is illustrated in *Figure 10.1*. When flowing, the initial emulsion becomes anisotropic due to the deformation

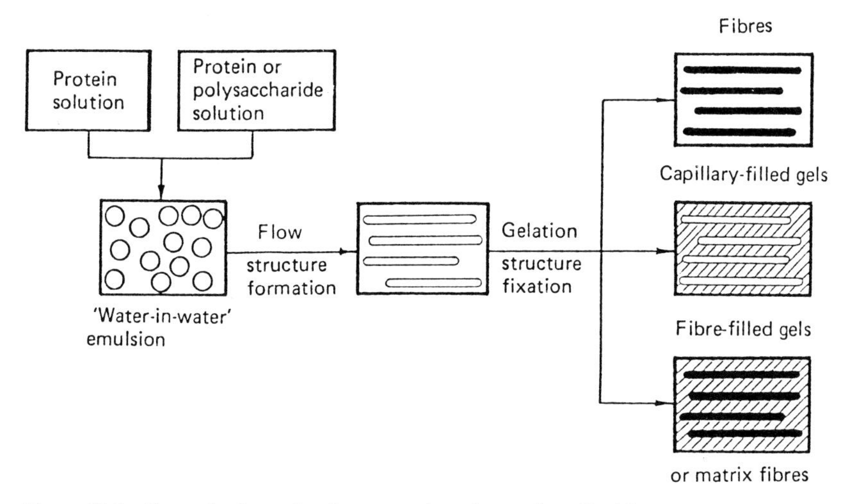


**Figure 10.1** General scheme for the processing of two-phase liquid systems by spinneretless spinning technology.

of dispersed particles. The problem is to avoid breakdown of the liquid cylinders and fix their form by the gelation of one or both of the two phases of the system (Tolstoguzov, 1974, 1978).

If the dispersed phase is gelled in flow, fibres are produced, whereas if the continuous phase is gelled the result is orientated liquid cylinders dispersed in a gel. The latter are called *capillary gels*. Gelation of the two phases of a two-phase liquid system in flow gives rise to gels filled with oriented fibres. All three cases of spinneretless spinning have been used. In this process each hole permits the production of many fibres, i.e. of fibre bundles or various fibrous anisotropic materials, simultaneously and in a single stage. This feature of spinneretless spinning makes it fundamentally different from the spinneret technology where each opening produces a single fibre and voluminous fibrous products are obtained in subsequent stages of the process (Tolstoguzov, 1974, 1978; Tolstoguzov, Grinberg and Gurov, 1985; Tolstoguzov *et al.*, 1985).

The dopes for spinneretless spinning of edible protein fibres should be two-phase liquid systems, namely water-in-water (W/W) emulsions. They contain protein or protein–polysaccharide mixtures. These biopolymers are needed to gel one or both phases of the system, to impart the required biological value to the product and to regulate the rheological and other physicochemical properties. The proteins and polysaccharides employed should concentrate primarily (through their physicochemical properties) in different phases of a W/W emulsion, i.e. they should be thermodynamically incompatible. Thus to understand and develop the process it is essential to study thermodynamic compatibility in mixtures of various proteins as well as of proteins and polysaccharides. Prior to our investigations we knew about the thermodynamic incompatibility of gelatin with soluble starch and several other water-soluble polymers. Thermodynamic incompatibility of globular proteins with other proteins or polysaccharides had remained virtually unstudied and it was not clear at all whether these systems could undergo phase separation. In addition the deformation of dispersed particles in W/W emulsions was also open to question. Deformation and breakdown of liquid-dispersed drops had been largely studied using individual droplets or with fairly dilute oil-in-water (O/W) or water-in-oil (W/O) emulsions. These emulsions contain surfactants and, hence, the viscosity of the interfacial layer will be high. This factor will interfere with the deformation of emulsion droplets.

Thus, to develop the spinneretless spinning process it was necessary to:

1. Establish the conditions for obtaining liquid two-phase systems containing globular proteins (Antonov, Grinberg and Tolstoguzov, 1979; Grinberg and Tolstoguzov, 1972; Polyakov, Grinberg and Tolstoguzov, 1980; Polyakov *et al.*, 1979, 1985a,b, 1986);
2. Examine the deformation conditions of dispersed droplets in W/W emulsions and find out how to fix the form of liquid filaments by gelation of one or two of the two phases (Antonov *et al.*, 1980; Tolstoguzov, 1974, 1978, 1986; Tolstoguzov, Mzhel'sky and Gulov, 1973, 1974; Tolstoguzov *et al.*, 1985);
3. Optimize the composition of the two-phase dopes and coagulating baths as well as studying the structure and properties of protein fibres and other anisotropic materials;
4. Develop the appropriate spinneretless spinning equipment, and find applications for the fibrous materials in various food systems (Antonov *et al.*, 1982b, 1985; Asafov *et al.*, 1985; Borisova *et al.*, 1985a,b; Suchkov *et al.*, 1987a,b).

The findings of these studies have been discussed in a number of review papers (Tolstoguzov, 1986; Tolstoguzov, Grinberg and Gulov, 1985; Tolstoguzov *et al.*, 1985) so we will confine ourselves here to a brief account.

## Preparation of two-phase dopes

We have shown it is feasible to form two-phase dopes by investigating the thermodynamic compatibility of mixtures of different protein types and mixtures of proteins and polysaccharides in aqueous media. These investigations were carried out with the main classes of proteins (Pr) as defined by Osborne according to their solubility, i.e. albumins, globulins, glutelins and prolamines, as well as with the main types of polysaccharides (P): neutral (nP) and anionic (aP) including carboxyl-containing, sulphated, and both linear and branched.

The most significant result is the discovery of thermodynamic incompatibility in (a) mixed solutions of all classes of proteins and polysaccharides, (b) mixed solutions of proteins belonging to different classes within the Osborne classification, and (c) mixed solutions of the same class of proteins having different conformations, for example mixtures of the native and denatured forms of the same protein. For instance, mixtures of rather concentrated solutions of soya bean globulins and casein, casein and ovalbumin, soya bean globulins and ovalbumin all separate into two liquid phases. The ovalbumin–bovine serum albumin–water system remains, on the contrary, single-phase up to a 50% total protein content over a wide range of pH values and salt concentrations. It is noteworthy that thermal denaturation of the protein may be accompanied by aggregation of polypeptide chains and produce soluble aggregates which are thermodynamically incompatible with the native protein. Thus, the ovalbumin aggregates formed when a 5% solution of the native ovalbumin is heated at 100 °C for 60 minutes are thermodynamically incompatible not only with bovine serum albumin but also with native ovalbumin (Polyakov *et al.*, 1985a,b). Hence heating provides a means for preparing two-phase dopes for spinneretless spinning from a single protein where each phase contains predominantly either native or denatured forms.

All the protein–polysaccharide mixtures studied showed phase separation under certain specific conditions. The general nature of the thermodynamic incompatibility phenomenon has been so far demonstrated with over 80 Pr–P–$H_2O$ systems and around 20 Pr–Pr–$H_2O$ systems. For 25 systems phase diagrams have been constructed. The most comprehensive results have been obtained with mixtures of casein and soya bean globulins and with mixtures of either of these proteins with a variety of polysaccharides. The effect of pH, temperature and salt concentration as well as the molecular weights of the components, has been studied (Antonov, Grinberg and Tolstoguzov, 1979; Polyakov *et al.*, 1980, 1986; Tolstoguzov *et al.*, 1985).

When these systems undergo a liquid phase separation, the two macromolecular components concentrate mainly in different phases. Phase separation occurs at pH values and salt concentrations which encourage the association between similar macromolecules, over interactions between the two different molecular species.

At pH values higher than the isoelectric point thermodynamic incompatibility (or, more correctly, limited compatibility) is observed in solutions of protein–anionic polysaccharide mixtures even in the absence of salts. In contrast, mixtures of protein and neutral polysaccharide solutions require ionic strengths of at least 0.1 M and protein–sulphated polysaccharide mixtures require ionic strengths of at least 0.5 M for phase separation. For all classes of polysaccharides their compatibility with proteins decreases in the series: albumins > globulins > glutelins > prolamines, i.e. albumins are most compatible, prolamines least compatible.

The phase diagrams for Pr–P–$H_2O$ show lower values of the critical point coordinates as well as lower bulk component concentrations at which phase separation occurs compared with Pr–Pr–$H_2O$ systems. This means that protein and polysaccharide mixtures are less compatible than mixtures of proteins. In Pr–P–$H_2O$ systems the total threshold concentration at which separation occurs is normally around 4%, while for Pr–Pr–$H_2O$ systems it is 12% or higher. The threshold concentration for phase separation reduces on transition from globular proteins to those with an unfolded conformation.

As a rule, the phase diagrams of the above systems are markedly asymmetric, particularly in the Pr–P–$H_2O$ case. The asymmetry shows up in the fact that mixtures of proteins and polysaccharide solutions separate into phases strongly differing in the concentrations of the macromolecular components. A highly concentrated predominantly protein phase is normally in equilibrium with a far more dilute predominantly polysaccharide phase. The difference between the macromolecular concentrations in the coexisting phases may be more than ten-fold. This phenomenon is basic to a new method for concentrating protein solutions called *membraneless osmosis* (Antonov *et al.*, 1982b; Tolstoguzov, 1986; Tolstoguzov, Grinberg and Gurov, 1985; Tolstoguzov *et al.*, 1985; Zhuravskaya *et al.*, 1986). The concentration of the protein phase may be so high that it becomes possible to form protein gels. The process of gel formation by concentrating a macromolecular solution has been given the name of *lyotropic gelation* (Tolstoguzov and Braudo, 1983). The phenomenon of membraneless osmosis is of great value for spinneretless spinning.

These findings on proteins and polysaccharides permit us to predict the preparation conditions for a two-phase dope system containing a particular protein, the water distribution between the system phases, the phase composition and which phase will be dispersed and which will be the continuous phase. Now we will consider the processing of this class of dope system.

## Structure and properties of two-phase dopes

In spinneretless spinning we are concerned with two aspects of spinnability, (spinnability in this context is the ability of a dope to experience substantial irreversible deformations without jet interruption):

1. Spinnability of the dispersed droplets, i.e. their ability to form into stable liquid filaments.
2. Spinnability of a two-phase liquid system as a whole.

The former aspect of spinnability is determined by the ability of the system to form an anisotropic structure of a two-phase liquid system in flow. The latter aspect is needed to realize this flow in spinneretless spinning. The two types of spinnability can be arbitrarily called *microspinnability* and *macrospinnability* respectively. We will now consider the major factors defining spinnability of two-phase dopes.

***Microspinnability of two-phase liquid systems***

Dispersed particles of two-phase liquid systems were made to deform in rotational flow and their shape was studied by small angle light scattering (Tolstoguzov, Mzhel'sky and Gulov, 1973, 1974; Tolstoguzov., 1974). The axial ratio of the equivalent ellipsoids in flow was used as a measure of the microspinnability of the two-phase system.

*Figure 10.2* shows the droplet asymmetry versus, respectively, the shear rate, the diameter of the initial spherical drops, the phase viscosity ratio, and the pH. The latter was studied for a two-phase gelatin–water–dextran system where the interfacial tension will vary with pH. The degree of asymmetry of the dispersed particles rises with both the shear rate and the initial diameter at a constant shear rate. In addition, the degree of asymmetry increases as the viscosities of the dispersed phase and of the dispersion medium approach each other and as the interfacial tension decreases. The gelatin–water–dextran system to which *Figure 10.2(d)* refers is two-phase between pH 4.5 and 5.1 and single-phase at pH values outside this range. As the critical pH values are approached the interfacial tension goes to zero and the asymmetry of the dispersed particles rapidly increases. When the interfacial tension decreases and the shear increases simultaneously, there is a decrease in turbidity of flow. This may be attributed to the fact that the diameter of liquid-dispersed filaments becomes comparable with the wavelength of light. The original optical properties are restored after the shear forces have been removed.

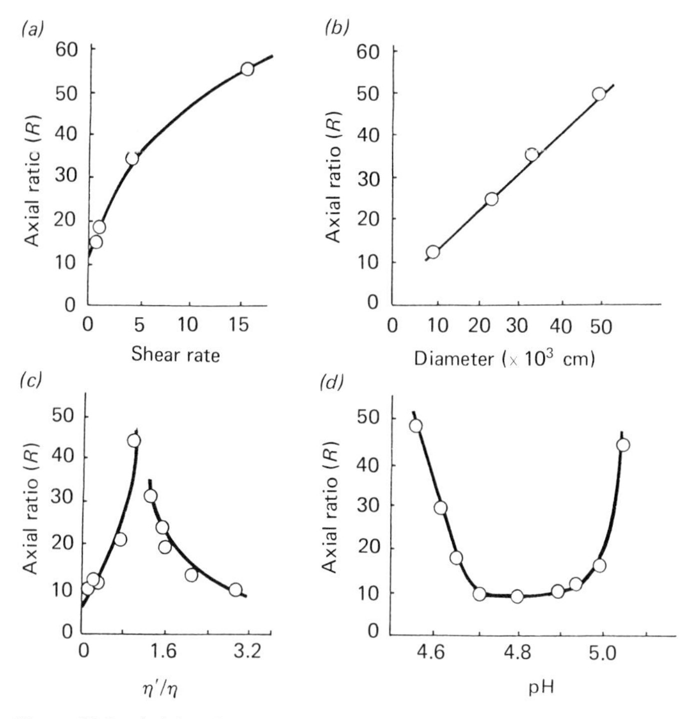


**Figure 10.2** Axial ratio ($R$) of dispersed particles versus (a) shear rate, (b) initial spherical particle diameter, (c) phase viscosity ratio where $\eta$ is the viscosity of the dispersed phase and $\eta'$ the viscosity of the dispersion medium, and (d) pH for the gelatin–dextran–water system.

Microspinnability of two-phase liquid systems depends heavily on the stability (lifetime) of liquid filaments formed upon deformation of dispersed droplets. Our results have shown that liquid filaments are not stable. They tend to break down into chains of similar size droplets. The lifetime of the liquid filaments increases with both the viscosity of the dispersed phase and the dispersion medium.

The form of a liquid cylinder can be fixed if the gelation time of one of the system phases does not exceed the cylinder lifetime. This condition is satisfied, for example, when a two-phase liquid system of gelatin (10%)–polyvinyl alcohol (5%) or soluble starch–water at a temperature of around 30 °C is poured into water at a temperature of 2–5 °C. Concurrently with the gelatin gelation, the continuous polyvinyl alcohol phase of the system is diluted with water. The result is an assembly of thin (3–10 μm diameter) gel-like fibres of gelatin (*see Figure 10.3*).

Thus, of critical importance in regard to the microspinnability of two-phase liquid systems are the phase viscosity ratio, interfacial tension, and rate of gelation of one of the phases.

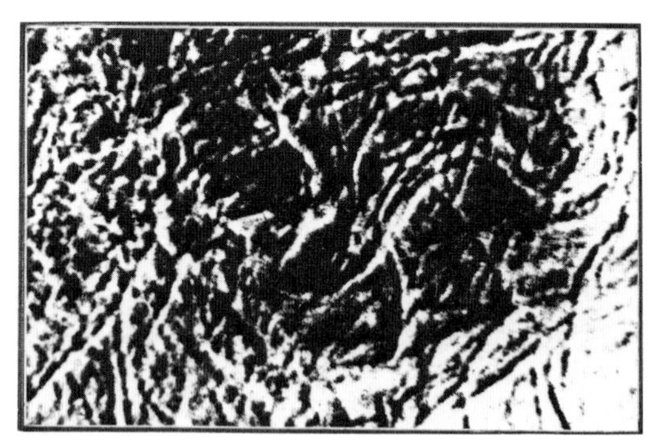

**Figure 10.3** Micrograph of gelatin microfibres spun from a two-phase liquid gelatin–polyvinyl alcohol–water system.

### *Macrospinnability of two-phase liquid systems*

As a measure of the macrospinnability we have examined the take-up velocity of the fibres at which jet interruption starts to occur (Dmitrienko *et al.*, 1978a,b; Suchkov, Grinberg and Tolstoguzov, 1980). *Figure 10.4* plots the spinnability of the two-phase casein–sodium alginate–water system versus the volume fraction of the protein phase. The spinnability of the two-phase system (*Figure 10.4*) is almost independent of phase composition. As the phase volume ratio varies, macrospinnability changes sharply from one virtually constant level to another. In other words, the macrospinnability of a two-phase system is determined solely by the spinnability of the continuous phase and is independent of the phase volume ratio until phase inversion. This has been confirmed by results on silicone oil emulsions in a sodium alginate solution. In this case only one emulsion phase, namely the sodium alginate solution, can form fibres. The spinnability is independent of the volume fractions of the oil up to the phase inversion point.

The variation in the volume fraction of the dispersed phase is equivalent to the variation of the effective cross-section of the continuous phase in the flowing emulsion. Hence, the established independence of spinnability on phase volume ratio

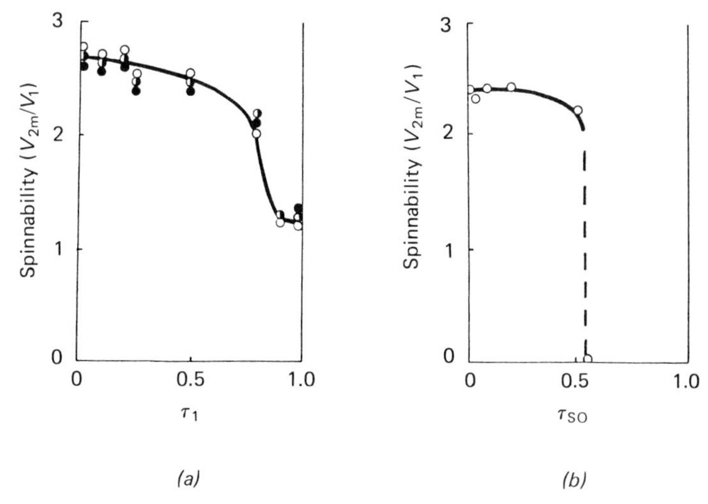


**Figure 10.4** (a) Relationship between the spinnability ($V_{2m}/V_1$) and the volume fraction of the protein phase for emulsions based on casein–sodium alginate–water.
(b) Relationship between the spinnability and volume fraction of the oil phase for an emulsion of silicone oil in a 2.2% sodium alginate solution. $V_{2m}$ is the maximum fibre take-up velocity; $V_1$ is the constant linear extrusion rate.

is equivalent to the independence of spinnability on the diameter of the spinneret hole.

Like microspinnability, the macrospinnability of two-phase systems is sensitive to the rate of gelation of one or of both phases in the system.

## Mechanical and other physicochemical properties of fibrous materials

In considering the properties of fibrous materials produced by spinneretless spinning we will confine ourselves to a brief description of the properties of gels filled with liquid and gel-like fibres, since they are the two main structurally different types of spinneretless spinning products.

The mechanical properties of gels filled with liquid cylinders oriented along the shear direction were studied using the anisometric indenter penetration technique. These gels, termed *capillary structure gels*, can be formed in a parallel plate cell of a rotational rheometer. Since the shear rate in the 'plate–plate' type rotational cell increases linearly with the distance from the axis of rotation, the relationship of droplet asymmetry and gel mechanical strength to shear rate can be examined using a single sample. The degree of drop asymmetry is found to rise gradually with the shear rate (*Figure 10.5*). In this case the transverse (with respect to the capillary orientation) strength of the capillary gels is higher than their longitudinal strength. Hence, the transverse-to-longitudinal strength ratio increases with the degree of asymmetry of the liquid drops (*Figure 10.5*). This ratio defines the strength anisotropy of capillary gels (Tolstoguzov, Mzhel'sky and Gulov, 1974).

The characteristic mechanical properties of gels filled with fibres are formed on conversion of both phases of a two-phase spin system in flow into the gelled state.

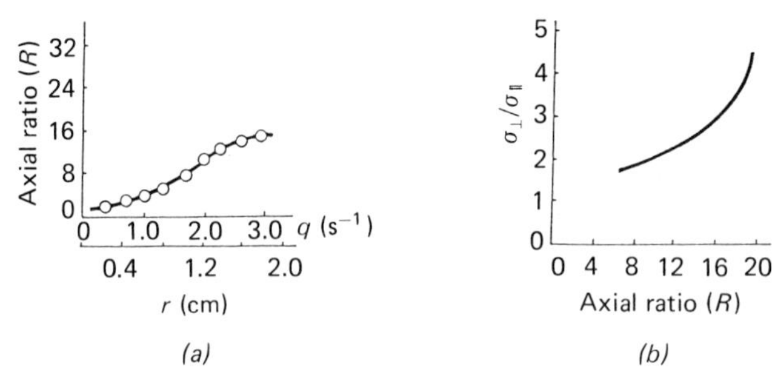


**Figure 10.5** (a) Axial ratio ($R$) of dispersed particles versus shear rate ($q$) in the formation of capillary gels. $q$ is proportional to distance from the centre of the rheometer ($r$) in parallel plate geometry; (b) mechanical anisotropy of gels (transverse/longitudinal modulus) versus degree of dispersed particle asymmetry for a dextran–gelatin–water system.

Using coagulating baths of the appropriate composition, *Figure 10.6* describes the mechanical properties of fibres produced by spinning a two-phase casein–sodium alginate–water system into coagulating baths of various compositions. The dope systems contained equilibrium phases of the same composition but in different proportions. In all cases, the elastic modulus and the tensile strength of fibres under simple elongation (*Figure 10.6*) are linear functions of the volume fraction of the dispersed phase (Suchkov *et al.*, 1987a).

A characteristic feature of fibre-filled and capillary structure gels is the anisotropy of structure and, hence, anisotropy of strength. Because of this, and in view of the relatively weak adhesion between the phases as well as the marked difference in the elasticity moduli of the two phases, anisotropic gels fibrillate readily, i.e. they split into fibres when the gel is deformed. The high anisotropy of strength and ready fibrillation of fibrous textured products generated by spinneretless spinning determine their specific functional properties. Splitting of anisotropic textured products (capillary structure or filled gels) on chewing in the mouth simulates the inhomogeneous

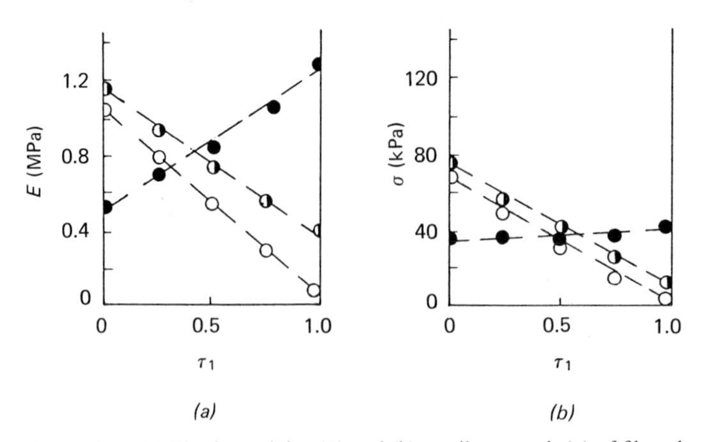


**Figure 10.6** (a) Elastic modulus ($E$) and (b) tensile strength ($\sigma$) of fibres based on a two-phase casein–sodium alginate–water system in different coagulating conditions versus protein volume fraction. The coagulating bath compositions: ●,pH 2.0, 1.0 M $CaCl_2$; ○,pH 4.5, 0.2 M $CaCl_2$; ○, pH 7.0, 0.1 M $CaCl_2$.

fibrous structure typical of meat products. Their visual appearance and consistency are also contributory factors to the simulation of traditional meat products.

An examination of the functional properties of fibres produced by spinneretless spinning and their mixtures with minced meat and fish suggests that they have considerable potential in the manufacture of combined meat products. Some examples are cited below. Further details are given in the following references: Antonov, Zhuravskaya and Tolstoguzov (1985), Antonov *et al.* (1982b), Asafov *et al.* (1985), Borisova *et al.* (1985b), Suchkov *et al.* (1987a).

A spinneretless spinning device with an output capacity of 10 kg $h^{-1}$ wet fibres was used to process a two-phase dope system containing 20% casein and 2% pectin at pH 6.7 and a temperature of 45 °C. The coagulating bath contained 16% $CaCl_2$ and 0.8% acetic acid. After water washing the fibres contained 26–31% protein, 0.3–0.5% calcium and the pH value was 6.0. The 0.2–1 mm diameter fibres were cut and dried. Their solubility in boiling water was under 3.5%. The water-holding capacity was 350%. The fibres were successfully used as replacement of 30% raw meat in the manufacture of beef steak-type products (Borisova *et al.*, 1985b). *Figure 10.7* shows the micrographs of the fibres taken by a scanning electron microscope with a magnification of × 3000. The fibre sections show many pores with a mean diameter of 0.1–0.3 μm. The structure of the fibres is stable to boiling water. The microrelief of the fibres displays many oriented microfibres with a diameter similar to the pore diameter seen in the fibre cross-section (Borisova *et al.*, 1985a).

(*a*)

(*b*)

**Figure 10.7** SEM micrographs of fibres based on a two-phase casein–pectin–water system.

Similar results have been obtained on the structure of fibres spun from two-phase systems (pH 7) containing a soya bean protein isolate and casein as well as a field bean protein isolate and casein.

Textured products generated by spinning protein mixtures have been added at up to the 40% level to a wide variety of products prepared from chopped meat and fish. They contribute to the water- and fat-holding capacity of the product, reduce the cooking loss, improve the texture and the general attractiveness of the finished products. Textured products based on protein mixtures are superior to those generated from the individual proteins studied and the level at which raw meat can be replaced is higher (Suchkov *et al.*, 1986a,b).

## Versions of the spinneretless spinning process

The general scheme for spinneretless spinning (*Figure 10.1*) presented earlier gave three possible versions of the process: formation of fibres, of capillary gels, and fibre-filled gels. Within each of these three types a wide range of products can be produced depending upon dope composition, the method for generating flow and the gelation procedure (Antonov *et al.*, 1980; Asafov *et al.*, 1985; Tolstoguzov, 1978). The gelation mechanisms available for food biopolymers fall into three groups: thermotropic, ionotropic and lyotropic gelation. Thermotropic gels form on heating or cooling of a liquid. Ionotropic gels arise from the interaction of metal ions with gelling agents. Lyotropic gels form, as indicated previously, in the concentration of liquid solutions or dispersed systems containing gelling agents. Actual processes generally rely on a combination of several gelation procedures. The choice of a particular gel-forming component of a system predetermines not only the gelation procedures to be employed but also the techniques for shaping a two-phase system and the design of the process equipment. In particular, the spinneret diameter is determined by the method and the conditions of gelation. For instance, in the case of thermotropic protein gelation spinneret holes can be rather large. On the contrary, in the diffusion-controlled processes of ionotropic or lyotropic gelation it is advantageous to have small openings in order to have fine jets of a spinning dope. The thermotropic (heat set) gelation of globular proteins denatured by heating is employed in both wet and dry spinning.

The dry spinning method does not involve a coagulating bath. Suspensions of protein–polysaccharide mixtures are plasticized by exposure to a high temperature in the extruder. They are deformed in the shear field and particles with different compositions do not coalesce due to their incompatibility. In this way textured fibres are produced, for example by extruding mixtures of soya bean and polysaccharide solutions.

Ionotropic gelation is useful in the processing of mixtures of proteins and polysaccharides, which form gels in the presence of calcium ions (e.g. alginates). We have investigated techniques for establishing the conditions for ionotropic gelation of solutions of polysaccharides and their mixtures with proteins (Tolstoguzov and Braudo, 1983).

As previously mentioned, lyotropic gelation involves the phenomenon of membraneless osmosis to concentrate the protein phase rapidly until it is gelled, while the polysaccharide phase is diluted. Spinneretless spinning is achieved by mixing the protein and polysaccharide solutions. The dispersed (protein) phase of the resultant W/W emulsion is concentrated and functions as a dope. Its liquid particles are deformed and gelled in a flowing dispersion medium. The latter functions, therefore, as a spinneret and a coagulating bath simultaneously. The processes of lyotropic protein gelation may also be accompanied by ionotropic gelation (Antonov, Zhuravskaya and Tolstoguzov, 1985).

The processes involving two-phase dopes can be subdivided into two groups according to the spinneret size. In spinneretless spinning the spinneret hole diameter is in excess of $10^{-4}$ m; if the hole dimensions are less than $10^{-4}$ m, the same process goes under the name of *spinneret matrix spinning* (Tolstoguzov, Grinberg, and Gurov, 1985). The distinction between these two groups is arbitrary because neither the spinnability of two-phase systems nor the microstructure of the product are dependent on the spinneret hole diameter. Spinneret matrix spinning is wet spinneret spinning of two-phase protein–protein and protein–polysaccharide mixtures and is a particular case of spinneretless spinning (Tolstoguzov, Grinberg and Gurov, 1985). Matrix spinning provides fibres of a fine fibrillar structure and, what is more, can utilize non-spinnable proteins to this end. To impart some spinnable properties to a protein solution, low level of anionic polysaccharides are incorporated. The low consumption of anionic polysaccharides in protein processing is due to the poor thermodynamic compatibility of the two biopolymer classes and to the fact that the phase equilibrium established in Pr–aP–$H_2O$ systems involves concentration of the protein phase and dilution of the polysaccharide phase. We should like to stress here that anionic polysaccharides give particularly easily spinnable solutions, with low critical concentrations for ionotropic gelation, high mechanical strength and good thermal stability (Dmitrienko *et al.*, 1978a; Tolstoguzov and Braudo, 1983).

The equipment (*Figure 10.8*) applicable to the spinneretless spinning of two-phase liquid systems is, as a rule, very simple in design. It includes a unit for preparing a mixture of macromolecular component solutions and a unit for shaping the prepared two-phase dope through nozzles with round or square openings. Fibres are generally spun vertically downwards into the air. The dope jet is taken up by a roller (vertical or horizontal) immersed completely or partially into a coagulating bath (Antonov *et al.*, 1980; Asafov *et al.*, 1985). Gel-like fibres are superimposed upon each other and closely packed on the roller. Having been in contact with the coagulating bath for a sufficient time, this layer of fibres is cut along the roller axis, washed and dried. The take-up roller together with the collected fibres can be transferred to another coagulating bath or baths for subsequent heat treatment and washing of fibres. The coagulating baths are acid–salt and salt solutions, and can be heated to temperatures exceeding protein denaturation temperatures. The fibres produced may be either capillary or filled gels, depending on the dope composition and on the gelation conditions.

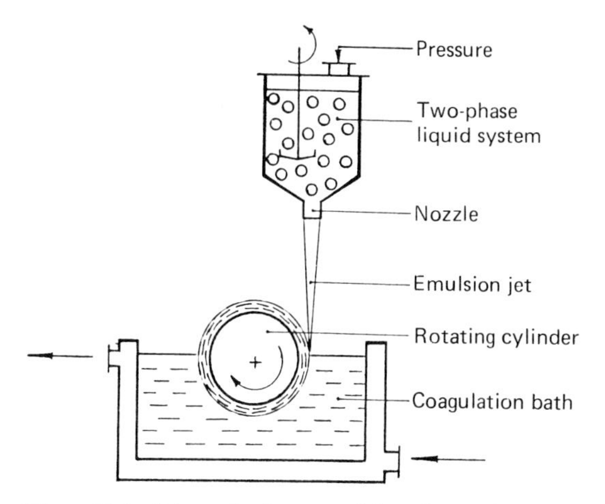


**Figure 10.8** Schematic diagram of the spinneretless spinning equipment for processing two-phase liquid systems.

In some instances, e.g. in spinneretless spinning of very concentrated protein mixtures, the equipment is similar to that employed in thermoplastic extrusion. The thermoplastic extrusion process is a particular case of spinneretless spinning (*see* below). Finally, matrix fibres are produced in the equipment designed for wet spinneretless spinning (Antonov *et al.*, 1982b; Suchkov, Grinberg and Tolstoguzov, 1980; Suchkov *et al.*, 1986a).

It is pertinent to mention the great variety of fibrous materials created by spinneretless spinning. Earlier (*Figure 10.3*) we used an example of short gelatin fibres. Infinite fibres are feasible, too. They are also produced as a result of deformation, orientation and coalescence of dispersed droplets flow, provided the volume fraction of the droplets is sufficiently large. *Figures 10.6* and *10.7* relate to matrix fibres of this type spun from the casein–sodium alginate–water system. Another fibre type is obtained by fibrillating anisotropic three-dimensional capillary gels or films. If the volume fraction of the dispersed phase is large, i.e. if the gel is filled with liquid cylinders to a high degree, the continuous phase of the gel between the cylinders is made of fibres although they are not cylindrical in shape.

So spinneretless spinning provides short, infinite and non-cylindrical fibres as well as matrix fibres filled with oriented liquid or gel-like filaments. Furthermore, anisotropic textured products can be produced from these fibres. Structurally these textured products are related to filled or capillary gels. The process can be applied to such a wide range of raw materials because it is based on two general fundamental phenomena:

1. Thermodynamic incompatibility in mixed solutions of proteins and of proteins and polysaccharides; and
2. Deformation of dispersed particles of two-phase aqueous systems in flow.

The advantages of spinneretless spinning technology compared to the classical wet spinning of protein fibres are the following. First of all, the process is flexible and fibres can be made from virtually any soluble protein. To process poorly spinnable proteins we can utilize dopes containing anionic linear polysaccharides. In this case proteins may concentrate in the dispersed phase and serve only as fillers in the continuous phase formed by the anionic polysaccharide solution. In fact, the general nature of a limited thermodynamic compatibility permits the incorporation of mixtures of various proteins which complement each other nutritionally into the spinning dopes. Dope components can be selected with the objective of reducing the final product cost. The high biological value and the relatively low costs of textured products created by spinneretless spinning are also attributable to the simple design and hence the low cost of the equipment used and to the mild operating conditions (neutral and weakly acid media, moderate temperatures). The spinneretless or matrix spinning process makes it possible to eliminate certain stages and simplify others involved in protein fibre formation by conventional wet spinning methods. These include preparation of a dope, its filtration and binding by means of the appropriate food binding system. The latter stage is no longer needed on account of the distinctly pronounced fibrous macrostructure and anisotropy of the strength of the three-dimensional capillary and fibre-filled gels.

Another advantage of spinneretless spinning is the stability of the two-phase dopes which is due to the relatively small difference in the phase densities and viscosities, the rapid establishment of phase equilibrium (thanks to the large interfacial area), and the dynamic equilibrium between the breakdown and coalescence of dispersed particles attained in the mixed system.

In conclusion the following features of spinneretless-spun fibrous textured products are particularly important: the difference in the rheological properties of the two phases, anisotropy of strength, easy fibrillation, and high porosity. These properties explain why these materials are fairly good meat analogues and can be added, in more substantial quantities than other textured proteins, to comminuted meat products.

## Spinneretless spinning as an element of food production technology

In the development of the spinneretless spinning process we have established and studied several general physicochemical phenomena. These include thermodynamic incompatibility of proteins with other proteins and polysaccharides in solution, concentration of protein solutions by the process of establishment of phase equilibrium with polysaccharide solutions, independence of the spinnability of the two-phase system on the volume fraction of the dispersed phase, the relationship of the anisotropy of these systems structure to the phase viscosity ratio and interfacial tension. As a rule, food manufacture involves preparation and shaping of liquid heterophase systems that are subsequently gelled (or converted to the glass state). That is why the physicochemical phenomena which are basic to the spinneretless spinning process are also of importance in the formation of structure in many traditional food products.

As an example let us consider textured proteins. Their manufacture involves processing a complex mixture of proteins falling under different classes defined by Osborne and exhibiting different degrees of denaturation, aggregation and dissociation. The processed systems may also contain some polysaccharide. When the protein content becomes high the system may undergo phase separation particularly in the presence of a polysaccharide because of thermodynamic incompatibility. This suggests that the liquid protein systems, which are the precursors of various textured products, are frequently two-phase. This two-phase structure may well be responsible for the fibrous structure of textured proteins produced by thermoplastic extrusion, dry spinning, stream texturization, shredding and other processes.

In thermoplastic extrusion two-phase liquid systems (W/W type emulsions) leaving the die are likely to form a three-phase structure as a result of the flashing off of water as a stream. Before the die the material contains cells of superheated water under pressure which are deformed in flow. The flash-off of superheated water as steam results in an anisotropic porous lamellar microstructure. The phase separation in the processed system is, in turn, responsible for fibrillation of the gel melt within the extruder barrel and for the fine fibrous microstructure of textured product lamellae.

Another type of textured protein with a structure similar to that of capillary gels is based on foams or oil-in-water emulsions when the continuous phase contains a heat-set protein. Textured products are fabricated on heating these systems in mixers or extruders. Anisotropic gels formed in flow are destroyed mechanically.

A further application of the suggested approach to the investigation of structure formation in heterophase food systems is the processing of dough. For instance, macaroni-type products are produced by extrusion of a concentrated suspension of starch granules in a high-viscosity protein dispersion medium. The structure and properties of the products produced seem to be governed by the amount of soluble starch, the rheological and gelation behaviour of the continuous protein phase as well as by the extrusion conditions. Liquid phase separation of highly concentrated protein systems occurs at relatively small quantities (under 1%) of added soluble starch. This separation is facilitated by the rather high surface hydrophobicity of the main wheat protein. It is particularly pronounced with hard wheat gliadins where the gluten complex has a higher lipid content.

Liquid phase separation of the extruded paste is expected to involve a competition for water between the protein and the polysaccharide phases. Concentration of the protein phase upon separation may result in increased viscosity and lyotropic gelation. If this process takes place in the extruder barrel, rather than in the die, the premature concentration of the protein phase may cause a mechanochemical destruction of the gluten, increased expansion of the jet, giving an extrudate with a rough surface and inferior mechanical properties. Presumably for this reason in the manufacture of high-quality macaroni products preference is shown for coarse-ground wheat, to keep the level of starch damage as low as possible.

Finally, the significance of these multicomponent systems for modelling various food processes and products should be mentioned. Most actual food systems are extremely complicated. Two-phase systems of a relatively simple composition containing proteins and polysaccharides are useful models of more complex foods. These systems exhibit a relatively low viscosity, reproducible composition and are often transparent. This makes it possibe to obtain reproducible results which are easy to interpret.

Thus, the applied significance of the above results is not restricted solely to the development of processes for spinneretless protein spinning. It lies also with possible modelling of some key physicochemical phenomena involved in protein texturization and in the production of traditional foodstuffs. It is no exaggeration to claim that spinneretless spinning, or better the physicochemical phenomena lying behind it, will play an important role in the processing of both traditional and novel food systems.

## References

ANTONOV, YU., GRINBERG, V. and TOLSTOGUZOV, V. (1979). Thermodynamische Aspekte der Verträglichkeit von Eiweisen und Polysacchariden in wässrigen Medien. *Die Nahrung*, **23**, 207–214, 597–610, 847–862

ANTONOV, YU., ZHURAVSKAYA, N. and TOLSTOGUZOV, V. (1985). Solubility of protein fibers obtained from casein solution and liquid two-phase water–casein–sodium alginate systems. *Die Nahrung*, **29**, 39–47

ANTONOV, YU., GRINBERG, V., ZHURAVSKAYA, N. and TOLSTOGUZOV, V. (1980). Liquid two-phase water–protein–polysaccharide systems and their processing into textured protein products. *Journal of Texture Studies*, **11**, 199–215

ANTONOV, YU., GRINBERG, V., ZHURAVSKAYA, N. and TOLSTOGUZOV, V. (1982a). Concentration of protein skimmed milk by the method of membraneless isobaric osmosis. *Carbohydrate Polymers*, **2**, 81–90

ANTONOV, YU., GRINBERG, V., ZHURAVSKAYA, N., SCHMANDKE, H. and TOLSTOGUZOV, V. (1982b). Mechanical properties and solubility of fibre obtained from liquid two-phase systems water–casein–sodium alginate. *Die Nahrung*, **26**, 9–13

ASAFOV, V., BORISOVA, L., MEL'NIK, N. and TOLSTOGUZOV, V. (1985). In *Enhancement of the Efficiency of Production and Utilisation of Milk–Protein Concentrates*, (Kostin, A. and Gronostaiskaya, N., Eds), pp. 37–45. Moscow, Agropromizdat

BORISOVA, L., ALEKSEEVA, N., PISMENSKAYA, V., TINYAKOV, V. and TOLSTOGUZOV, V. (1985a). Structure of casein-based fibrous textured products. *Molochnaya promyshlennost (Milk Industry)*, **11**, 30–32

BORISOVA, L., ASAFOV, V., GRINBERG, V., CHIMIROV, YU., BIKBOV, T., TOLSTOGUZOV, V., MUSHIOLIK G. and SCHMANDKE, H. (1985b). Utilisation of casein-based fibrous textured products in meat industry. *Myasnaya industriya SSSR (Soviet Meat Industry)*, **3**, 40–43

BORISOVA, L., SOKOLOVA, N., GRONOSTAISKAYA, N. and BIKBOV, T. (1985c). In *Enhancement of the Efficiency of Production and Utilisation of Milk–Protein Concentrates*, (Kostin, A. and Gronostaiskaya, N., Eds), pp. 45–49. Moscow, Agropromizdat

CHIMIROV, YU., SOLOGUB, L., BRAUDO, E., KOZMINA, E. and TOLSTOGUZOV, V. (1981). Utilisation of oilseed proteins in minced meat products. *Die Nahrung*, **25**, 255–259

DMITRIENKO, A., VARFOLOMEEVA, E., GRINBERG, V. and TOLSTOGUZOV, V. (1978a). The maximum take-up velocity as a spinnability criterion and its dependence on the composition of coagulation bath, illustrated by an example of sodium alginate. *Die Nahrung*, **22**, 391–400

DMITRIENKO, A., VARFOLOMEEVA, E., GRINBERG, V. and TOLSTOGUZOV, V. (1978b). Effect of the coagulation bath composition on the spinnability of casein solutions. *Die Nahrung*, **22**, 609–618

GRINBERG, V. and TOLSTOGUZOV, V. (1972). Thermodynamic compatibility of gelatin with some D-glucans in aqueous media. *Carbohydrate Research*, **25**, 313–321

KINSELLA, J. (1978). Texturized proteins: fabrication, flavoring, and nutrition. *CRC Critical Reviews in Food Science and Nutrition*, **10**, 147–207

KULEZNEV, V. (1980). *Polymer Blends*, p.303. Moscow, Chimia Press (in Russian)

POLYAKOV, V., GRINBERG, V. and TOLSTOGUZOV, V. (1980). Application of phase-volume-ratio method for determining the phase diagram of water–casein–soybean globulins system. *Polymer Bulletin*, **2**, 757–760

POLYAKOV, V., GRINBERG, V., ANTONOV, YU. and TOLSTOGUZOV, V. (1979). Limited thermodynamic compatibility of proteins in aqueous solutions. *Polymer Bulletin*, **1**, 593–597

POLYAKOV, V., KIREYEVA, O., GRINBERG V. and TOLSTOGUZOV, V. (1985a). Phase diagrams of some water–protein A–protein B systems. *Die Nahrung*, **29**, 153–160

POLYAKOV, V., POPELLO, I., GRINBERG, V. and TOLSTOGUZOV, V. (1985b). The effect of some physicochemical factors on thermodynamic compatibility of casein and soybean globulin fraction. *Die Nahrung*, **29**, 323–333

POLYAKOV, V., POPELLO, I., GRINBERG, V. and TOLSTOGUZOV, V. (1986). Studies on the role of intermolecular interaction in thermodynamics of the compatibility of proteins according to the data of solution enthalpics. *Die Nahrung*, **30**, 81–88.

SHERMAN, PH. (1968). In *Emulsion Science*, (Sherman, Ph, Ed.), pp. 197–312. London and New York, Academic Press

SUCHKOV, V., GRINBERG, V. and TOLSTOGUZOV, V. (1980). Study of the spinnability of emulsions based on a two-phase water–casein–sodium alginate system. *Die Nahrung*, **24**, 893–897

SUCHKOV, V., GRINBERG, V. and TOLSTOGUZOV, V. (1981). Steady-state viscosity of the liquid two-phase disperse system water–casein–sodium alginate. *Carbohydrate Polymers*, **1**, 39–53

SUCHKOV, V., GRINBERG, V., BIKBOV T. and TOLSTOGUZOV, V. (1987a). Non-spinneret formation and functional properties of fibrous texturates based on a liquid two-phase system water–casein–soya protein isolate. *Die Nahrung*, (in press).

SUCHKOV, V., POPELLO, I., BIKBOV, T., GRINBERG, V., DIANOVA, V., POLYAKOV, V., MUSCHIOLIK, G., SCHMANDKE, H., SCHUBRING, R. and TOLSTOGUZOV, V. (1987b). Non-spinneret formation and functional properties of fibrous texturates based on a liquid two-phase system water–casein–horse bean protein isolate. *Die Nahrung*, (in press).

TOLSTOGUZOV, V.B. (1974). Physikochemische Aspekte der Herstellung künstlicher Nahrungsmittel. *Die Nahrung*, **18**, 523–531

TOLSTOGUZOV, V.B. (1978). *Artificial Foostuffs*, p. 231. Moscow, Nauka (in Russian)

TOLSTOGUZOV, V.B. (1986). Functional properties of protein–polysaccharide mixtures. In *Functional Properties of Food Macromolecules*, (Mitchell, J. and Ledward, D., Eds), pp. 385–415. London and New York, Elsevier Applied Science

TOLSTOGUZOV, V.B. and BRAUDO, E. (1983). Fabricated foodstuffs as multicomponent gels. *Journal of Texture Studies*, **14**, 183–212

TOLSTOGUZOV, V.B., GRINBERG, V. and GUROV, A. (1985). Some physicochemical approaches to the problem of protein texturization. *Journal of Agricultural and Food Chemistry*, **33**, 151–159

TOLSTOGUZOV, V.B., MZHEL'SKY, A. and GULOV, V. (1973). On the preparation of anisotropic gels of capillary structure. *Vysokomolekulyarnye soedineniya*, **B15**, 824–827

TOLSTOGUZOV, V.B., MZHEL'SKY, A. and GULOV, V. (1974). Deformation of emulsion droplets in flow. *Colloid and Polymer Science*, **252**, 124–132

TOLSTOGUZOV, V.B., BELKINA, V., GULOV, V., TITOVA, E., GRINBERG, V. and BELAVTSEVA, E. (1974). Phasenzustand, Struktur und mechanische Eigenschaften des gelartigen Systems Wasser–Gelatine–Dextran. *Die Stärke*, **26**, 130–138

TOLSTOGUZOV, V.B., DIANOVA, V., MZHEL'SKY, A. and ZHVANKO, YU. (1978). The use of casein in food production. *Myasnaya industriya SSR (Soviet Meat Industry)*, **6**, 22–24

TOLSTOGUZOV, V.B., BRAUDO, E., GRINBERG,V. and GUROV, A. (1985). Physicochemical aspects of protein processing into foodstuffs. *Uspekhi khimii*, **54**,1738–1759

ZHURAVSKAYA, N., KIKNADZE, E., ANTONOV, YU. and TOLSTOGUZOV, V. (1986). Concentration of proteins as a result of the phase separation of water–protein–polysaccharide systems. *Die Nahrung*, **30**, 601–613

11

# DRY SPINNING OF MILK PROTEINS

J. VISSER
*Unilever Research Laboratory, Vlaardingen, The Netherlands*

## Introduction

The potential surplus of proteins, in particular those extracted from milk and soya beans, have led people to develop processes to transform these materials into a more palatable structure than the powdered form in which they are generally available. The two best known options in this respect are extrusion and spinning, both resulting in an ingredient which, after hydration and in combination with other ingredients such as a binder and flavour, can give a product resembling meat or fish. Since the basic structure of flesh foods is a bundle of fibres, protein spinning despite being more expensive than extrusion, is a more appropriate process for making meat analogues and similar products.

The classical process for spinning proteins was developed by Boyer in 1952 in which an alkaline protein solution is spun in an acid precipitation bath, followed by neutralization and rinsing the fibres formed. This process has been in operation on a commercial scale in the US and the UK but due to high costs, effluent problems and the restriction to the use of (soya) protein isolates of high purity, all such production has ceased to our knowledge.

To overcome the problems inherent in the Boyer process, the author has developed a new type of process combining some of the features of extrusion and spinning, i.e. a dry spinning operation, whereby a highly concentrated milk protein solution in water at a moderately elevated temperature (80°C) is forced through narrow openings into a drying chamber where the fibres harden by evaporation of the solvent (Visser, 1983). Use is made of the unique fibre-forming properties of casein, the major protein fraction in milk.

In this chapter, the basic principles of dry spinning will be described in detail and ways for scaling up to industrial production of fibres indicated. The potential of dry-spun fibres and other textured proteins as regards incorporation in products, costs and market potential will be given to demonstrate the criteria to be met by any potential new ingredient in this field.

In an introductory section, a survey of existing texturization processes is presented, including the typical protein functionalities which can be exploited for texturization, in order to place the work in the context of food texturization in general.

## Proteins, their functionality and traditional texturization processes

### *Protein sources*

The traditional major protein sources in the human diet in industrialized countries are milk, meat (including fish), wheat and, to a certain extent, eggs. In non-European countries such as Japan, the soya bean is an important additional source. In Europe in particular, there is a surplus of milk and consequently also of milk proteins, due to agricultural policies. Soya bean proteins and wheat gluten are available in large quantities as by-products from soya bean oil and starch production respectively. Other potential protein sources are meat offal and fish offal, blood from slaughter-houses and proteinaceous seeds other than soya such as rape, sunflower, etc. In addition, proteins may be obtained by newly developed processes, e.g. single cell and fungal proteins. In most cases, these proteins are available as dry powders with variable protein contents (*Table 11.1*).

**Table 11.1** SOURCES OF INDUSTRIAL FOOD PROTEINS

| *Source* | *Form* | *Protein level (%)* |
|---|---|---|
| Milk | Milk powder | 40 |
| | Caseinate | 90 |
| | Whey protein concentrates | 50–80 |
| Soya beans | Flour | 50 |
| | Concentrates | 70 |
| | Isolate | 90 |
| Wheat | Gluten | 90 |
| 'Man-made' | SCP/fungal proteins | — |
| Others | Blood/meat/fish | Variable |
| | Seed | Variable |
| | Egg | — |

Various extraction processes exist for the isolation of protein-enriched materials. *Figure 11.1* is a scheme for the isolation of the three major types of soya bean protein preparations, i.e. flour, concentrate and isolate, as a by-product of the oil extraction process; each has its own functionality depending on the level of protein present. For milk, the situation is different. Here, the proteins, except cheese whey and the corresponding whey proteins, are not a by-product but are specially isolated to give a particular functionality. *Figure 11.2* shows various ways of extracting the major milk proteins, each preparation having its typical application. Additional processing may be required depending on specific uses.

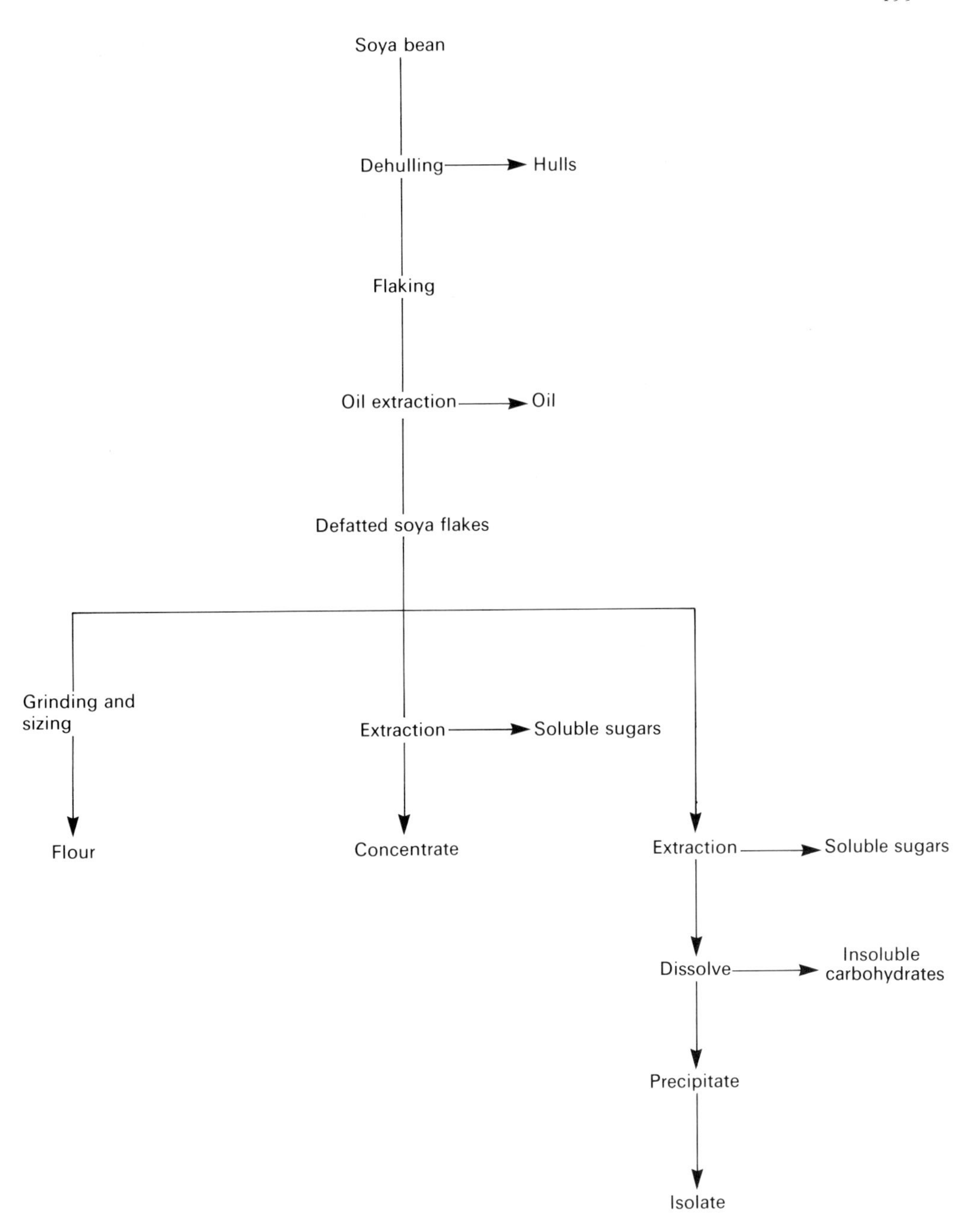


**Figure 11.1** Protein isolation: soya.

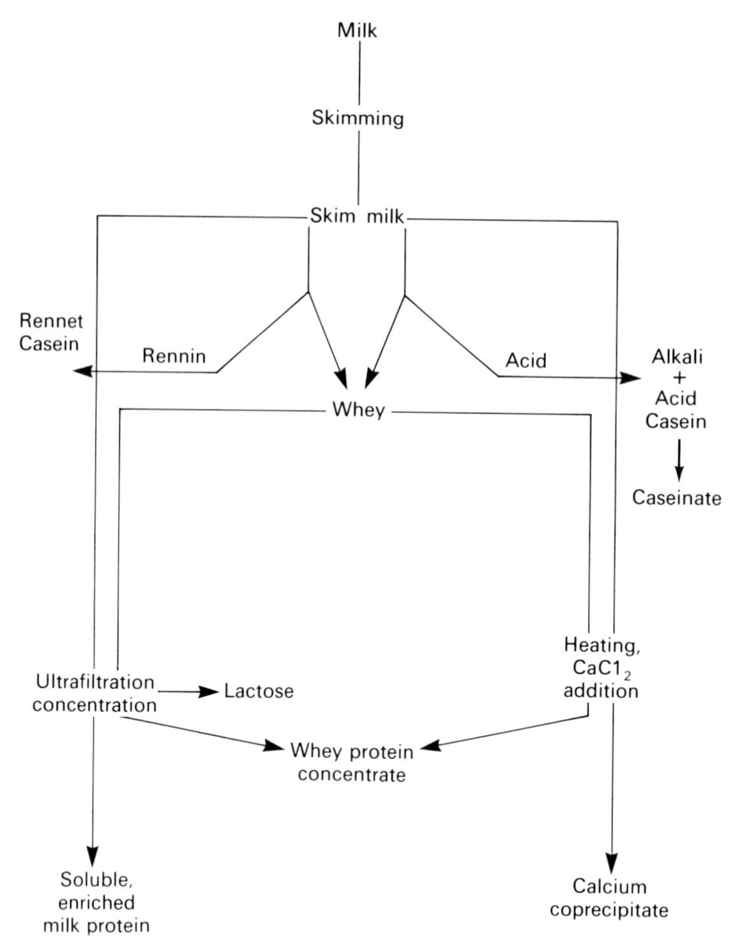


**Figure 11.2** Protein isolation: milk.

### *Traditional protein texturization processes*

Apart from milk, which is consumed in liquid form after mild pasteurization without texturization, all traditional protein sources in the diet are texturized before eating. This can be accomplished by processes as simple as cooking, boiling or frying in the case of eggs and meat. In general, more complicated processes are utilized such as those for cheese and bread making (*Table 11.2*). *Table 11.3* compares the properties of some of the textured proteinaceous food products. Of these, meat is by far the most attractive from a textural point of view, thus any *new* proteinaceous food should preferably be anisotropic and heterogeneous. This can only be achieved when the basic material is texturized in one form or another. The high price of meat (*Table 11.4*) and the low degree of protein conversion from seed to food (*Figure 11.3*) are major incentives for the development of imitation meat products. Of the proteinaceous sources that can be used, soya is particularly attractive because of its low cost and high yield/acre.

**Table 11.2** TRADITIONAL TEXTURIZATION PROCESSES AND PRODUCTS

| *Source* | *Method* | *Product* |
|---|---|---|
| Milk | Enzyme | Cheese |
| | Culture | Yoghurt |
| | Thickeners | Pudding |
| Soya | Enzyme | Natto |
| | Enzyme | Tempeh |
| | (Calcium chloride) | Tofu |
| Egg | Heat | Boiled egg |
| Flour | Heat/yeast | Bread |
| | Extrusion | Macaroni |
| Meat/fish/blood | Binder/heat | Sausage |
| | | Meat balls |

**Table 11.3** SOME TEXTURAL AND STRUCTURE PROPERTIES OF TRADITIONAL FOODS

| *Texture/structure* | *Product* |
|---|---|
| Pourable | Yoghurt |
| Chewable | Cheese |
| Isotropic | Cheese |
| Anistropic | Meat |
| Homogeneous | Yoghurt |
| Heterogeneous | Meat |

**Table 11.4** COST AND NUTRITION VALUE (NET PROTEIN UTILIZATION) OF VARIOUS PROTEINACEOUS FOODS

| | *Cost* (£ kg$^{-1}$) | *Net protein utilization(NPU)* |
|---|---|---|
| Egg | 3 | 95 |
| Meat | 5 | 76 |
| Milk protein | [1.5] | 90 |
| Soya bean protein | 1.5 | 70 |
| Whey | 2.5 | High |

## *Protein functionality*

In order to texturize proteins, full use should be made of their inherent functionality. In principle, one can distinguish between two broad classes:

1. Globular/temperature-sensitive proteins such as whey, soya bean and egg proteins; and
2. Random coil/temperature-insensitive protein such as casein, but globular proteins

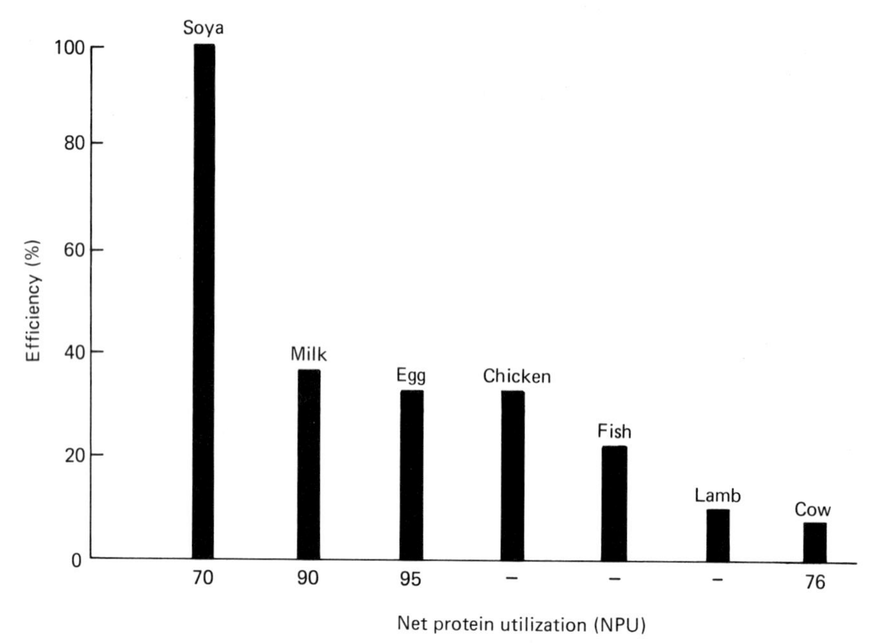


**Figure 11.3** Protein conversion efficiency and net protein untilization (NPU) values of typical meat products compared with soya and milk. Percentage efficiency = (protein as human food/protein fed to animals) × 100.

which have been treated by alkali at elevated temperatures also fall under this category, because they have then been changed into a random coil configuration.

The first class of proteins can be texturized simply by heating, whereupon a protein gel is formed. On the other hand, random coil proteins are well suited for texturization by spinning as use can also be made of their acid precipitation behaviour. To illustrate both methods, texturization of the two major proteins—soya bean protein and casein—will be described in more detail.

The first thing to notice is that both proteins are a mixture of proteins of different molecular weights and completely different in character.

1. The major components of soya bean proteins are conglycinin or 7S and glycinin 11S; the latter notation, derived from the sedimentation coefficient as measured under standard ultracentrifugation conditions, is a measure of the molecular weight of the protein. At neutral pH, the structure of glycinin, the most abundant and also the most functional of the soya bean proteins, can be described by a doughnut-like configuration composed of two hexamers on top of each other, based on proteins A and B (*Figure 11.4*). The high molecular weight is explained by the large number of participating subunits. Because of its compact structure it does not aggregate easily to form a network except by drastic means such as heating whereby exposure of SH groups allows the formation of intermolecular crosslinks. A similar behaviour also applies to the major whey proteins in milk, α-lactalbumin, a single globular molecule at neutral pH and β-lactoglobulin, present in dimeric form. In contrast to soya proteins, these proteins—when undenatured—are soluble at their isoelectric point so that denaturation is the only possibility left for their texturization.

| Soya proteins | MW |
|---|---|
| 7S (β-conglycinin) | 181 000 |
| 11S (glycinin) | 326 000 |
| 2S | 30 000 |
| 15S | >500 000 |

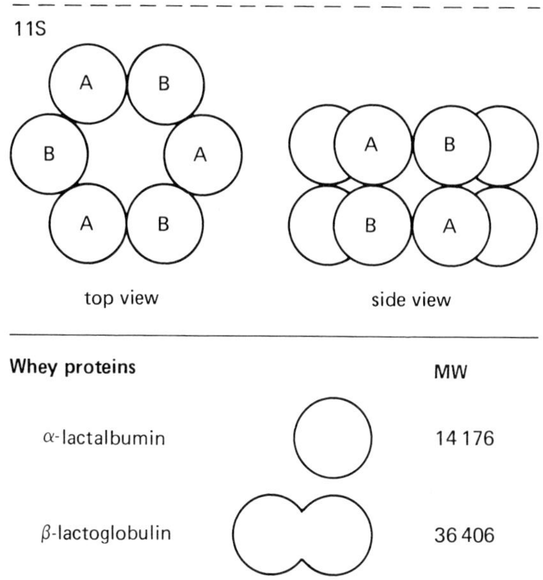

11S

A B B A A B

top view

A B B A

side view

| Whey proteins | MW |
|---|---|
| α-lactalbumin | 14 176 |
| β-lactoglobulin | 36 406 |

**Figure 11.4** Molecular weight and configuration of two globular proteins. A and B are glycinin subunits.

2. In milk, casein (the major protein fraction) is present in micellar form: aggregates of $\alpha$-$s_1$ and β-casein held together by amorphous calcium phosphate are stabilized against flocculation by a coat of κ-casein (*Figure 11.5*). The unique properties of κ-casein in combination with the enzyme chymosin allows a direct texturization into cheese curd after proteolysis of this casein. The casein micelles can be totally disrupted by dissolving the precipitate formed upon acidification of skim milk to pH 4.5 in sodium hydroxide, whereby sodium caseinate is formed. At pH 7, the caseins are totally solubilized and the system is ready for texturization either by acid precipitation at their isoelectric point or by network formation via calcium bridging. Due to the absence of sulphur-containing amino acids, however, texturization by heating is not possible. Texturization by acid precipitation of soya proteins is also possible provided these proteins are 'solubilized' first. This is only possible at extremely high pH values (above 11).

It is clear that milk casein and soya proteins are completely different in character, and different routes have to be applied for texturization. The typical fibre-forming properties of casein, related to its random coil structure, are particularly noticeable when cheese is melted. This has not been exploited so far and it is this particular functionality which formed the basis for developing a new protein texturization process: dry spinning.

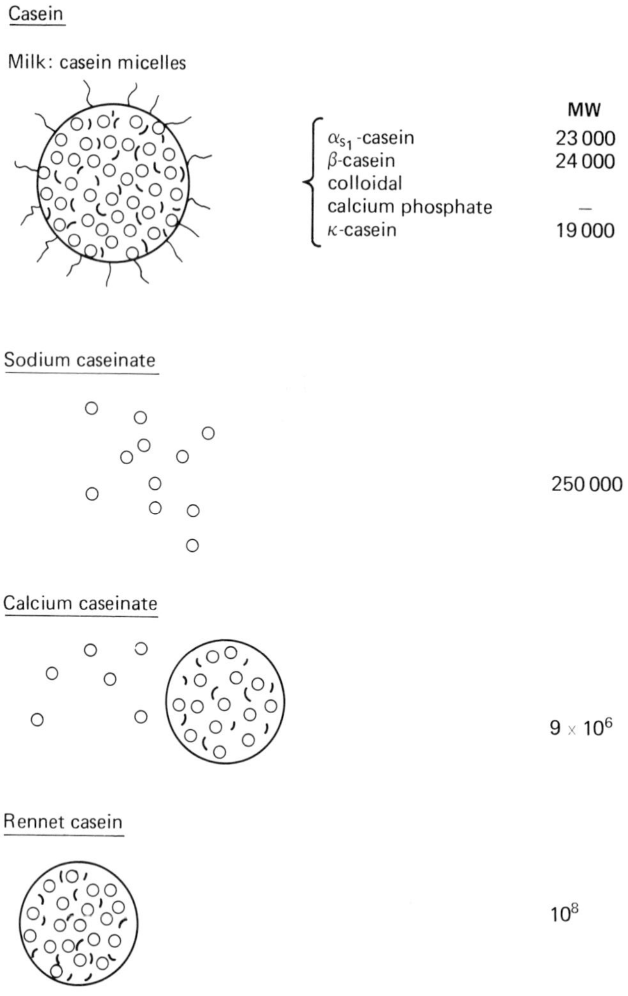

Casein

Milk: casein micelles

| | MW |
|---|---|
| $\alpha_{s1}$-casein | 23 000 |
| $\beta$-casein | 24 000 |
| colloidal calcium phosphate | – |
| $\kappa$-casein | 19 000 |

Sodium caseinate

250 000

Calcium caseinate

$9 \times 10^6$

Rennet casein

$10^8$

**Figure 11.5** Molecular weight, composition and configuration of casein.

# Protein texturization by wet spinning and extrusion cooking

## *Boundary conditions*

The markets for traditional texturized protein products (*Table 11.2*) are generally saturated, except for some specific market segments. In addition, the Western diet is not deficient in protein and people generally eat more protein than required. In view of this and considering the high market volume, the interest in high quality products and the cost sensitivity aspects, the meat industry is probably the first to be susceptible to using a new ingredient. In this context, texturization of surplus protein becomes attractive when costs are low and a product of good quality is obtained. Typical areas for utilization may, therefore, be found in upgrading low quality meat or in meat products with an inadequate fibrous texture. It is clear that only a fibrous ingredient is able to fulfil these criteria.

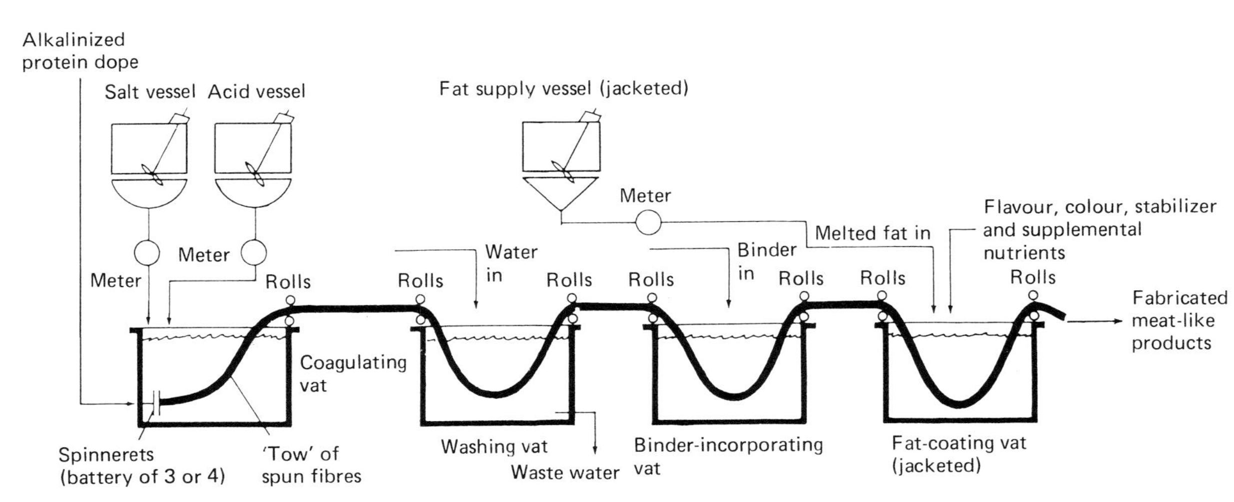


**Figure 11.6** Protein fibre spinning according to Boyer (1954). From Horan (1974).

### *The Boyer wet spinning process*

Boyer (1952) was the first researcher who adapted the existing wet spinning process for making textile fibres to obtain fibres for human consumption using soya proteins. To this end, soya protein isolate was dissolved in alkali to a pH of 11 and the liquid so obtained coagulated in an acid bath of pH 2 using a spinneret with up to 10 000 holes, 0.1 μm in diameter. The fibre bundle was then transported via rollers and passed through a series of neutralizing and rinsing baths. After cutting to the required length, the wet fibres were frozen for storage and after thawing they were ready for incorporation in a food product (*Figure 11.6*).

In the late 1970s, the Boyer process and a number of similar wet spinning operations were put into operation in the US (General Mills, Miles Laboratories), the UK (Courtaulds) and in France (Rhône-Poulenc) on a commercial basis. The major reasons for these operations having been stopped are: lack of demand as a result of lower meat prices, dryness in the mouth after chewing of the fibres (cotton wool effect), effluent costs and problems with the maintenance of the composition of the spinning baths. The fibrosity of the product was excellent. Too perfect an alignment of the fibres, however, gave an artificial impression whereas the beige colour restricted the utilization to non-white meats.

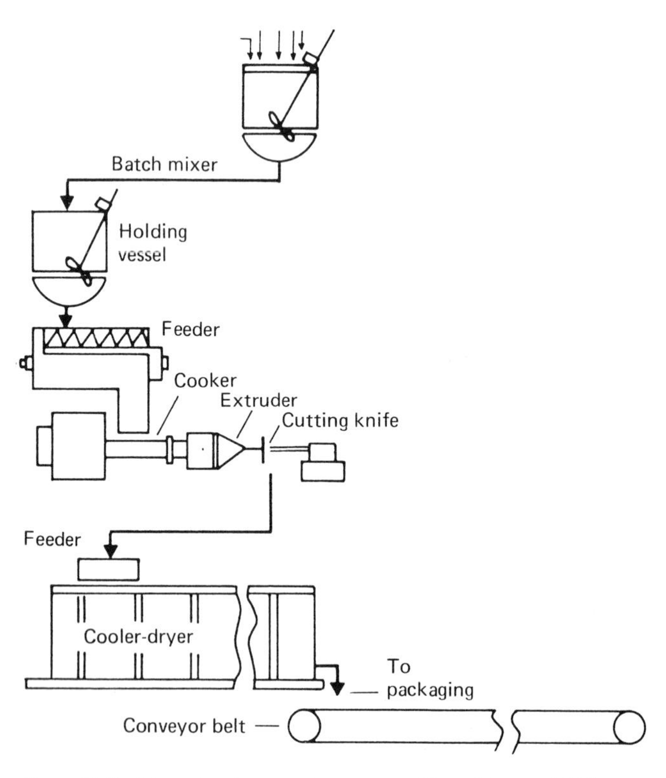


**Figure 11.7** The extrusion cooking process. From Horan (1974).

### *Extrusion*

A much simpler method of texturizing proteinaceous materials is extrusion cooking (*Figure 11.7*). Less refined heat-settable proteins such as soya protein concentrates can be used but the product obtained generally lacks the degree of fibrosity achieved in the spinning process. The much lower price, however, provides some compensation.

### *Other options*

To overcome the problems inherent in wet spinning and to improve the textures obtained by extrusion cooking, a large number of routes have been followed as becomes clear from a patent survey of the area. Examples are:

1. Shredding of extruded materials into thin fibres;
2. Texturization by roller milling or film formation;
3. Unidirectional freezing of a protein gel;
4. Enzymatic treatment of a protein solution in a shear field.

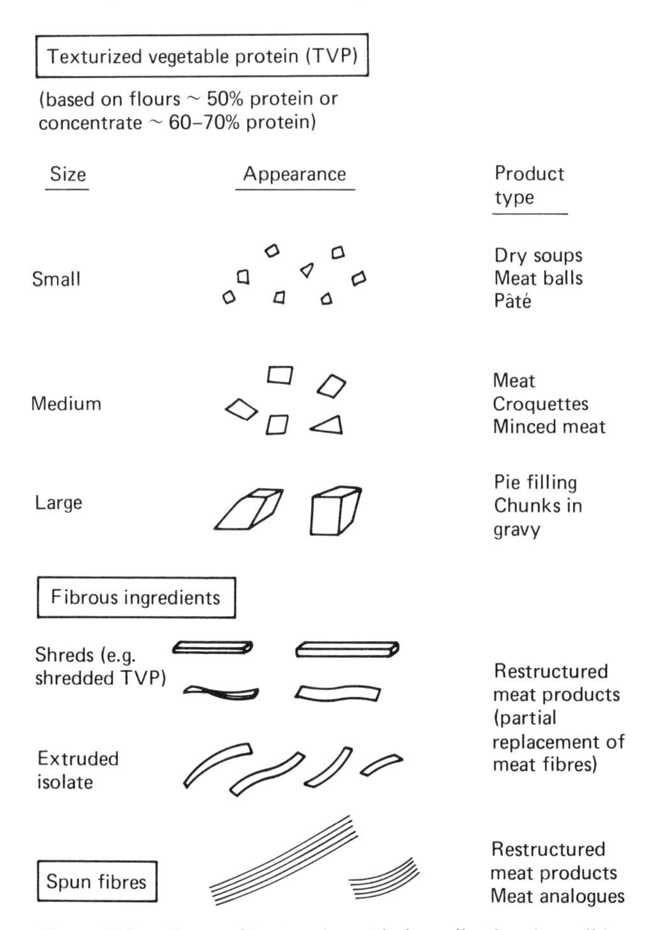


**Figure 11.8** Types of 'textured protein ingredient' and possible areas of application.

In most cases, soya proteins are considered as proteinaceous starting materials. Of these processes, only shredding has reached the stage of industrial production.

### *Texturized protein ingredients*

It is clear that the processing routes described above offer a variety of structure elements for incorporation in a food product. A summary of the various types of textured proteins and their possible areas of application are given in *Figure 11.8*. While it is obvious that every type has its typical application, only material with the highest degree of fibrosity can be used to imitate meat completely. Apart from the 'appearance' of the texturized material, other attributes are also essential for successful incorporation into a food product. Water uptake, colour and succulence, for example, should fulfil certain criteria, whereas the presence of off-flavour can make a product a complete failure. Any new process for making textured proteins should, therefore, be assessed on its capability to produce ingredients which can be incorporated in a food product; a positive contribution to the overall appearance and its sensory properties are essential. As yet, soya proteins have not been able to fulfil all these criteria whilst still giving a commercially viable product.

## Dry spinning of milk proteins

### *Background*

From the above discussion of the wet spinning process to obtain protein fibres and the inherent deficiencies of the texturized material produced by extrusion, it is clear that for any future commercial success of a protein texturization process a cheaper alternative route to obtaining protein fibres is essential, if possible using cheaper raw materials and at the same time imparting better eating characteristics to the products.

For this reason a completely new approach to obtain protein fibres was adopted by the Unilever Research Laboratory in Vlaardingen some 11 years ago. In view of the large surpluses and, hence, the low cost of skim milk powder in the EEC countries, it was decided to concentrate research initially on milk proteins, exploiting the observation that melted cheese as in cheese fondue is a very effective fibre former, a property which was also found for concentrated calcium caseinate solutions. By avoiding the use of a coagulating bath, it was anticipated that effluent problems—one of the major drawbacks of wet spinning—could be avoided.

### *The process*

On the basis of successful laboratory-scale experiments—in which it was found that casein when properly handled could be used to form fibres by dry spinning—a semi-pilot plant scale spinning unit was developed for the production of 2 kg dry fibres per hour (*Fibre 11.9*). It consists of a simple extruder, a spinneret holder, a spinneret, and a drying shaft (Visser, 1983). The extruder is either fed manually or by a powder-dosing unit containing pre-wetted powder. The gear pump shown between the extruder and the spinneret holder was found to be superfluous. The spinneret had 244 holes with a diameter of 0.25 mm. The process involves the preparation of a

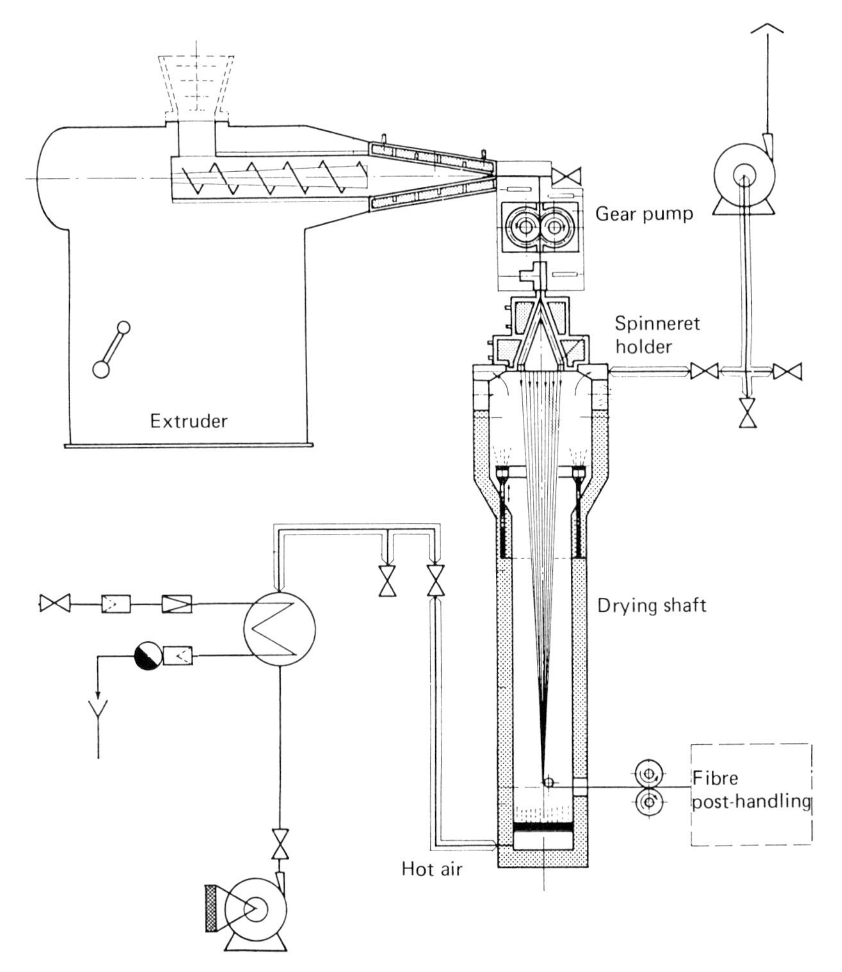


**Figure 11.9** Flowsheet of pilot plant for dry spinning of casein.

pumpable spin dope from dry ingredients (principally milk proteins; other ingredients such as starch, gluten, and soya proteins can be included up to a maximum level of approximately 40%) and water, followed by heating to about 80°C. The spin dope thus formed in the screw extruder is forced through the narrow orifices of the spinneret into a current of hot air emitted from the bottom of a vertical drying shaft. The water present in the fibres partly evaporates, giving them a surface hardness. The partly-dried fibres are then stretched by rollers and cut to a certain length, if desired. Drying of the fibres is completed by allowing them to stand overnight at ambient temperature and humidity. When dry, the fibres can be stored at room temperature and are ready for use.

### *Compositions*

It was found that all caseinates tested (sodium, calcium and rennet casein) could be spun by the above process. However, only certain specific compositions could be used to obtain fibres suitable for incorporation in meat-like products because the fibres needed to be resistant to boiling water, salt solutions, sterilization, etc.

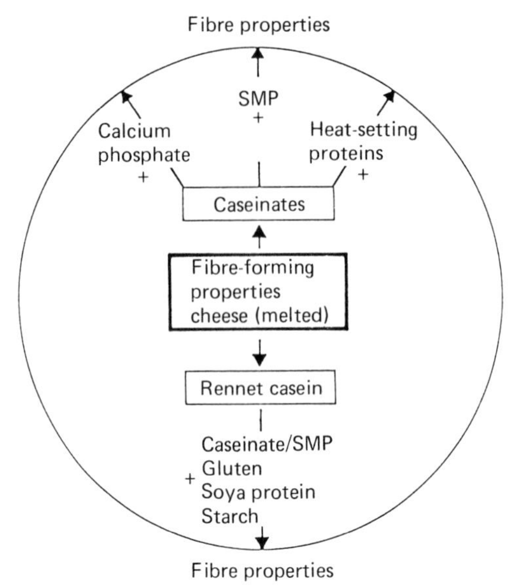


**Figure 11.10** Basic principles of the dry spinning process.

These specific compositions can be grouped into four patented categories (*see Figure 11.10*):

1. Caseinate + calcium phosphate
   By adding an appropriate amount of orthophosphate to calcium caseinate in the pH region 5.5–6.5, dry-spun fibres can be produced that are resistant to boiling water (Visser and Bey, 1977).
2. Skim milk powder (SMP) + sodium caseinate
   By adding an appropriate amount of sodium caseinate or any other ingredient able to reduce the calcium/casein ratio to that of category (1), dry-spun fibres can be produced with characteristics similar to those above, provided the correct level of phosphate is retained within the pH region indicated above (Visser *et al.*, 1978).
3. Calcium caseinate + heat-setting proteins
   By replacing part of the calcium caseinate by a heat-settable protein, dry-spun fibres can be obtained with a range of breaking strengths depending on the level incorporated, provided processing takes place below the heat-setting temperature of the protein added. Actual heat setting takes place when the final product is processed (Visser *et al.*, 1979).
4. Rennet casein + additive
   By replacing part of the rennet casein by any other material to reduce the extreme toughness of rennet casein alone, dry-spun fibres can be produced with any characteristic required for a given product application, as long as the spinning characteristics are retained. Materials tested include caseinate, gluten, skim milk powder, potato protein, soya protein concentrate, starch and their combinations. These were found to be the most versatile category and have, therefore, the highest potential (Visser and Bams, 1980). Typical combinations and their properties are given in *Tables 11.5* and *11.6*.

**Table 11.5** PHYSICAL PROPERTIES OF RENNET CASEIN-BASED FIBRES

| *Protein composition* | *Tensile properties* | | *Swelling properties*[a] | | | *Percentage rehydration* (Wt. wet fibres/Wt. dry fibres × 100) |
|---|---|---|---|---|---|---|
| | *Elastic modulus* (kn $m^{-2}$) | *Breaking strength* (kN $m^{-2}$) | *Dry cross-section* (sq mm) | *Wet cross-section* (sq mm) | *Percentage increase* | |
| 100% Rennet casein | 182 | 93 | 0.096 | 0.159 | 65 | 176 |
| 70% Rennet casein<br>30% Acid casein | 125 | 82 | 0.103 | 0.158 | 53 | 183 |
| 70% Rennet casein<br>30% Soya isolate | 195 | 167 | 0.064 | 0.163 | 153 | Not measured |
| 70% Rennet casein<br>30% Soya concentrate | 180 | 116 | 0.049 | 0.155 | 213 | 221 |
| 80% Rennet casein<br>20% Skim milk powder | 97 | 48 | 0.082 | 0.206 | 149 | Not measured |
| 70% Rennet casein<br>30% Skim milk powder | 49 | 42 | 0.063 | 0.367 | 480 | 238 |

[a] Swelling was performed by immersion in boiling tap water for 5 minutes.

### Product evaluation

Dry-spun fibres based on casein/soya protein isolate mixtures were tested by Unilever Research, UK in two different product concepts: (1) 'chicken fingers' and (2) restructured meat products. The results were compared against the same product type made from pure meat (control) and a product based on wet-spun fibres. The composition of the dry-spun fibre was optimized for the product application concerned. Results obtained by a trained panel (*Table 11.7*) show that for a number of product attributes, dry-spun fibres scored as high as the all-meat control. Overall, the dry-spun fibres scored higher than wet-spun fibres. The characteristic off-flavour of the latter fibres was found to be absent and the whiter colour gave a better, chicken meat-like appearance than the brown-grey colour of the wet-spun soya protein fibres.

**Table 11.6** EATING CHARACTERISTICS OF RENNET CASEIN-BASED FIBRES

| *Fibre* | *Chewing time prior to swelling* | *Appreciation* |
|---|---|---|
| (1) 100% Rennet casein | >15 | Tough, not very succulent |
| (2) 70% Rennet casein<br>30% Acid casein | 15 | Similar to (1) |
| (3) 70% Rennet casein<br>30% Soya isolate | 12–15 | Less tough and more succulent than (1) |
| (4) 70% Rennet casein<br>30% Soya concentrate | 12–15 | Less tough and more succulent than (1) |
| (5) 80% Rennet casein<br>20% Skim milk powder | 10–12 | Significantly less tough and more succulent than (1) and (2) |
| (6) 70% Rennet casein<br>30% Skim milk powder | 10–12 | Significantly less tough and more succulent than (1) and (2) |

**Table 11.7** PANEL ASSESSMENT OF PRODUCTS BASED ON DRY-SPUN AND WET-SPUN FIBRES COMPARED WITH AN ALL-MEAT CONTROL

| | |
|---|---|
| 1. *'Chicken fingers'* (Fibre composition 1: calcium caseinate (80%)–soya protein isolate (20%)–oil) | |
| Flavour: | Dry-spun fibre 1 ≡all-meat control >wet-spun fibre |
| Fibrousness: | Dry-spun fibre 1 ≡all-meat control >wet-spun fibre |
| Toughness: | Dry-spun fibre 1 <wet-spun fibre ≡all-meat control |
| 2. *'Restructured meat product'* (Fibre composition 2: calcium caseinate (60%)–soya protein isolate (40%)) | |
| Flavour: | All-meat control >dry-spun fibre 2 >wet-spun fibre |
| Succulence: | Dry-spun fibre 2 >all-meat control |
| Toughness: | Dry-spun fibre 2 ≡all-meat control |

### *Other product concepts*

In order to demonstrate the versatility of rennet casein-based fibres, a series of product concepts ranging from meat balls in gravy to a crab cocktail was developed by Unilever Research, UK. Some of the product characteristics are presented in *Table 11.8*. Visually, the products could not be distinguished from the real (all-meat) product. The meat balls in gravy were found to be superior in taste and texture to the commercial equivalent.

**Table 11.8** PRODUCT CONCEPTS INCORPORATING DRY-SPUN FIBRES

| *Product concept* | *Principles illustrated* |
|---|---|
| Crab cocktail style | Small pieces, delicately flavoured<br>White flesh<br>Cold eating |
| Fish pieces (scampi style) | Chunks of assembled fibres<br>Compatibility with battering |
| 'Chicken' pie | Firmer texture than fish application<br>Different flavouring<br>Good visual appearance of 'chicken piece' |
| 'Chicken breast' analogue | Large portion<br>Assembled fibre<br>Structure<br>Crumbed product |
| Reformed beef chunks | Structuring of comminuted meat<br>Stability to canning/sterilization |
| Meat balls in gravy | Extended meat<br>Stability to canning/sterilization |
| Beefburger style | Meat replacement<br>Compatibility with meat processing |
| Snack-style product | Ribbon extrusion<br>Expanded structure |

### *Scale-up*

To increase the production from 2 kg $h^{-1}$ to 20 kg $h^{-1}$, test runs were performed on three different extruders normally used for pasta and/or snacks extrusion. The idea behind these tests was that dry spinning can be seen as a kind of spaghetti-making process, except that thinner filaments are to be produced from a completely different raw material. The smallest size opening of a normal spaghetti die is 1.0 mm whereas for the typical product application of dry-spun fibres 0.25 mm is found to be optimal. Due to this small diameter, ambient drying is sufficient for caseinate fibres, in contrast to spaghetti production where special drying ovens and long drying times are required.

Spin dopes consisting of a mixture of 60% rennet casein, 30% soya protein concentrate and 10% gluten to which water was added to a final moisture content of 25–45% (depending on machine used and test conditions) were extruded on three different machines:

1. The Lalesse extruder (LAE)
   This single-screw extruder with automatic feeding has four heating elements for temperature control. A die with 8 × 0.5 mm diameter holes was specially made for one of the trials. In another trial, the spinneret of the semi-pilot plant scale machine (SPP) was coupled to the extruder exit. In both cases the fibres obtained were cut to a length of 50 cm and oven-dried.
2. The Braibanti Zambra pasta extruder (BRZ)
   This simple single-screw extruder was provided with a bronze die containing 560 holes with a diameter of 0.65 mm, the smallest hole size available at that time. For fibre drying, a bakery oven fitted with a rotating belt was positioned directly behind the spinneret head. The experiments were performed at TNO, Wageningen.
3. Gelatinizer–Former FG20 ex Mapimpianti
   This machine consists of a powder mixing unit, an extruder cooker (gelatinizer), including automatic water dosing and electrical heating where the actual spinning dope is formed, and a shaping press (the former) where the spin dope is forced through a die containing 1000 holes of 0.60 mm diameter. This unit was especially made for the test.

The conditions of the runs are given in *Table 11.9* and the conditions for the production of fibres of the same composition on the semi-pilot plant scale machine (SPP) are included for comparison. All machines were capable of producing a continuous stream of fibres. From the Braibanti experiments (*Figure 11.11*) it can be concluded that large scale production of dry-spun fibres is possible, provided some further optimization regarding fibre handling and positioning of the die is carried out.

**Figure 11.11** Extrusion of fibres from the Braibanti pasta extruder at TNO, Wageningen.

**Table 11.9** EXPERIMENTAL CONDITIONS FOR LARGE-SCALE FIBRE PRODUCTION TRIALS COMPARED WITH THOSE FOR THE UNILEVER RESEARCH (VLAARDINGEN) SEMI-PILOT PLANT SCALE SPINNING MACHINE (SPP).

| *Run* | *SPP* | *BRZ*[a] | | *LAE*[a] | | *FGM*[a] | | |
|---|---|---|---|---|---|---|---|---|
| | | *1* | *2* | *1* | *2* | *1* | *2* | *3* |
| Spin dope composition, | | | | | | | | |
| Moisture level (%) | 43 | 34–37 | 34–37 | 30 | 30 | 43 | 43 | 43 |
| Die, number of holes | 224 | 560 | 560 | 7 | 224 | 1000 | 1000 | 1000 |
| Die, diameter (mm) | 0.25 | 0.65 | 0.65 | 0.5 | 0.25 | 0.25 | 0.25 | |
| Temperature, barrel (°C) | 75 | 72 | 72 | various | 80 | 60 | 70 | 70 |
| Temperature, spinneret (°C) | 85 | 72 | 72 | — | — | — | — | — |
| Temperature, drying shaft oven (°C) | 140 | 100–120 | 100–120 | n.d. | n.d. | ambient | ambient | ambient |
| Fibre stretching | yes | no | yes | no | no | no | no | no |
| Fibre cutting, length (cm) | 50 | 50 | 50 | 30 | 30 | 150 | 150 | 150 |
| Fibre post-drying | ambient | oven | oven | oven | ambient | oven | oven | oven |
| Rotational speed barrel | | | | | | | | |
| (in % of maximum) | 25 | 50 | 35 | 40–80 | 40–60 | 100 | 100 | 100 |
| Pressure before die (bar) | 70 | n.d. | 50–70 | n.d. | ⩾250 | 185 | 185 | 280 |
| Capacity ($kg\,h^{-1}$) | 2 | 19 | 13 | 10 | 12 | 11 | 11 | 11 |
| Fibre diameter, dry (mm) | 0.3 | 0.8 | 0.5 | 0.8 | 0.3 | 0.3 | 0.3 | 0.3 |
| Fibre properties | reference | | | similar | similar | | visually | similar |
| Fibre production, | | | | | | | | |
| total/run (kg) | 3.5 | 1 | 5.5 | 3 | 2[b] | 3 | 3 | 0[b] |
| Date (1983) | — | | 10 February | April | 2 May | | 13 July | |

[a] BRZ = Braibanti Zambra pasta extruder; LAE = Lalesse extruder; FGM = Former–Gelatinizer ex Mapimpianti.
[b] Pressure exerted too high; experiment stopped.

### *Costs*

Earlier calculations showed that, on a protein basis, wet-spun fibres are approximately half the price of chicken meat and lean beef, whereas textured vegetable proteins (TVP) would on the same basis cost only one-quarter of the price of wet-spun fibres., demonstrating the higher costs of wet spinning over extrusion. Since dry spinning, particularly on a large scale, is akin to extrusion, costs will be low and in view of the relatively high costs of proteins, processing costs will amount to approximately 10% of the total costs. This implies that for dry-spun fibres containing 40% soya protein concentrate, the total price will at most be twice as high as that of TVP, i.e. half that of wet-spun fibres.

### *Advantages over wet spinning*

Product and process evaluations of the dry spinning process based on milk proteins as described above demonstrate that the process has the following advantages over wet spinning based on the Boyer process:

1. No effluent problems;
2. No pH adjustments;
3. Flexibility in ingredients;
4. Dry storage of the fibres;
5. Simplicity of processing (pasta extrusion);
6. Low production and ingredient costs.

The use of milk proteins at a minimum level of about 50% and the need to rehydrate the fibres prior to use are the only limitations.

## Evaluation of results

### *Market situation*

In 1973 it was expected that by 1980, 10% of the world meat market would be based on textured vegetable proteins. This has not proved to be the case and the main reasons for the failure of soya proteins to constitute a significant part of the Western diet are that meat prices have remained low in real terms and that inherent product deficiencies of TVP and wet-spun fibre could not be overcome: in particular, residual off-flavour, wrong colour, lack of succulence and fibrosity in the case of TVP and high processing and ingredient costs in the case of wet-spun fibres. The longer term potential of a new ingredient of this kind is, however, large.

### *Dry-spun fibres*

The technical feasibility of dry spinning fibres based on milk protein-derived materials and its exclusive technology have been established. A range of fibre types well suited for different product applications has been produced and the incorporation into a wide range of food product concepts has been and can be demonstrated. The disadvantages experienced when using other protein-based fibres, such as off-flavours and colour, were found to be absent. Moreover the economics of the process suggest that the selling price would be acceptable.

### *Future prospects*

Recently, the quality of soya bean protein preparations has been improved, but it is still impossible to produce an acceptable textured product at a reasonable cost. Dry-spun fibres allow the use of milk proteins outside their traditional markets. The high quality of the products made so far warrants further consumer-oriented evaluation in order to obtain some indication of their possible market acceptability and potential. The simplicity of processing and the low production costs (adapted pasta extrusion easy to scale up) makes most of the other recent developments in protein texturization less attractive, either due to high energy costs or to complicated processing.

## Acknowledgements

The work carried out to demonstrate the versatility of rennet casein-based fibres in a series of product concepts and the scale-up experiments were done in close collaboration with Unilever Research Vlaardingen, Unilever Research Colworth House, Dairy Crest Foods, TNO—Wageningen (Netherlands) and Mapimpianti, Galliera Veneto, Italy. It was only the changing economic environment within the dairy industry in Europe which made Dairy Crest Foods decide not to proceed to a further commercialization of the process.

I would like to thank all those who contributed so much to the overall progress of the work. I am grateful for the cooperation, in particular with Mr R. A. Dicker and Dr K. J. Burgess of Dairy Crest Foods (part of the Milk Marketing Board), Thames Ditton, UK, Mr G. Coton (formerly of MMB), Professor R. B. Leslie and Mr B. Giles of Unilever Research, Colworth House, Sharnbrook, UK, Drs Papotto and Pavan of Mapimpianti, Galliera Veneto, Italy and Mr P. Sluimer of the Instituut voor Graan, Meel en Brood, TNO, Wageningen, Holland. Without their help and enthusiasm the work would never have reached the present stage of development.

I would also like to thank the managements of Unilever and Dairy Crest Foods for giving me permission to publish the work in this form. In particular, I would like to thank Mr Francke and Dr Crossley of Unilever and Mr Dicker of Dairy Crest Foods for their encouragement and stimulating interest in the work.

## References

BOYER, R.A. (1954). High Protein Food Product and Process for its Preparation. US Patent 2 682 466

HORAN, F.E. (1974). Soy protein products and their production. *Journal of the American Oil Chemists Society,* **51,** 67A

VISSER, J. (1983). Spinning of food proteins. In *Progress in Food Engineering* (Cantarelli, C. and Peri, C., Eds), p. 689. Küssnacht, Switzerland, Forster Verlag AG

VISSER, J. and BAMS, G.W.P. (1980). Rennet Casein Containing Fibres. EP 0 051 423

VISSER, J. and BEY, E.P. (1977). Casein Filaments. UK Patent 1 474 179
VISSER, J. *et al.* (1978). Fibres from Skim Milk Powder. UK Patent 1 574 448
VISSER, J. *et al.* (1979). Mixed Protein Fibres. US Patent 4 118 520

12

# PROTEIN EXTRUSION—MORE QUESTIONS THAN ANSWERS?

D.A. LEDWARD and J.R. MITCHELL
*Faculty of Agricultural Science, University of Nottingham, UK*

Although extrusion cooking has been applied to the production of shaped pasta products and ready-to-eat breakfast cereals for over 50 years, its application to the texturization of vegetable proteins is a more recent innovation (Atkinson, 1970). However, the last 15 years or so have seen a great deal of activity in the development and production of textured products from protein-rich sources, especially soya flour (e.g. Smith, 1976; Lillford, 1986).

This activity has led to some commercial success but in spite of this the process must still be considered a developing art rather than a technology, and physicochemical studies relevant to this complex process are few and far between (Lillford, 1986). As Lillford (1986) has rightly stated, 'the art is developing purely by empirical rules relating the contribution of each of the process variables and ingredients to the properties of the final product'. This is obviously not satisfactory and an understanding of the changes taking place at the molecular level must be a major objective if the art/technology is to be optimized for efficiency of the process and quality of the final product.

In simple terms extrusion processing involves conditioning the proteinaceous material (usually defatted soya flour) to 15–40% moisture which is then fed through a feeder/hopper into the hollow barrel of the extruder where a rotating screw (or screws) forces the material towards the exit. The barrel is usually heated to a predetermined temperature. Generally the diameter of the screw rod increases towards the exit and thus material is compressed. This pressure increase is accompanied by a temperature rise as a result of both transfer of heat from the heated barrel jacket and conversion of mechanical into heat energy. The usual residence time in the extruder is 30–60 s after which the 'molten' mass passes into the die section before being squirted out to the atmosphere where spontaneous evaporation of water takes place. Smith (1976) has reviewed the process, product range and potential of extrusion processing.

The engineering aspects of the extrusion processing of proteins, i.e. the production of a continuous stream of expanded, textured material (Harper, 1981), the rheological properties of the melts (Otun, Crawshaw and Frazier, 1986) and the effect of such processing on the nutritional quality of the feed (Cheftel, 1986) are reasonably well documented. However, the changes taking place at the molecular level have received little constructive attention. It is usually assumed that with proteinaceous material the proteins denature and melt within the extruder barrel and are subsequently aligned by the action of the screw. Further alignment occurs within the die of the extruder prior to the mix being squirted out into the atmosphere where the instantaneous evaporation of the superheated water leads to an expanded, porous product.

Although this superficial description seems appropriate we have to appreciate that we are dealing with complex, reactive molecules under conditions of high temperature, pressure and shear and low moisture. Our knowledge of the chemical and physical changes that the proteins and the other components of the feed can undergo under these conditions is, to say the least, limited.

To try and gain some insight into this complex system it seemed appropriate to pose several questions regarding its chemistry. The questions chosen may not be the most pertinent and the answers will, by necessity, be rather speculative, but such a food science approach, as opposed to an engineering, or physical or nutritional one, will lead to new insights into this poorly understood process.

The questions we have chosen to ask are:

1. Is denaturation necessary?
2. Do proteins actually melt in the same way as plastics during extrusion to yield a relatively homogeneous phase with the water present or do they exist in a concentrated solution?
3. What conformations do the proteins adopt and do they, at the molecular level, actually align themselves parallel to the field?
4. What rǒle does the charge and its distribution on the protein play in extrusion?
5. What is the nature of the bonds stabilizing the aggregate and when do they form?
6. How does the changing chemical composition of the dough/melt affect the extrusion process?

In the following sections each of these questions, and the possible answers, will be considered and their relevance to protein extrusion discussed.

## Is denaturation necessary?

Although the word denaturation is widely used in the literature relating to protein chemistry it is not without ambiguity. *The Dictionary of Scientific and Technical Terms* (1984a) defines the verb 'to denature' as 'to change a protein by heating it or treating it with alkali or acid so that the original properties such as solubility are changed as a result of the protein molecular structure being changed in some way'. In dilute solution this is usually inferred to be a process leading to total loss of the secondary or tertiary structure of the native protein. This process is invariably associated with the rupture of the weak bonds (hydrogen bonds, hydrophobic interactions, etc.) maintaining the native structure. That such bonds will also rupture at the temperatures usually associated with the extrusion processing of proteinaceous materials is without doubt. Thus the proteins will be molecularly rearranged following extrusion and, if initially present in the native form, this conformation will be lost (Sheard, Mitchell and Ledward, 1986). Although denaturation (loss of native structure) was widely held to be a key step in the extrusion of soya (Harper, 1981) it is now known that native materials are not a necessary prerequisite for successful extrusion processing as products can be made from concentrates and isolates in which all 'native' structure has been destroyed by heat or solvents in the preceding purification steps (Lillford, 1986; Sheard, Mitchell and Ledward, 1986). Sheard, Mitchell and Ledward (1986) studied the extrusion behaviour of a whole range of commercially available soya isolates and found that those containing native material

**Table 12.1** DIAMETER AND FORCE NECESSARY TO SHEAR RETORTED EXTRUDATES OF SOYA PROTEINS PROCESSED AT 180 °C IN BRABENDER MODEL DN EXTRUDER USING A 4:1 COMPRESSION SCREW. RESULTS ARE RELATIVE TO NATIVE SOYA FLOUR

| *Sample* | *Conformation* | *Diameter* | *Peak force* |
|---|---|---|---|
| Flour | Native | 1.00 | 1.00 |
| Ardex D | Native | 1.49 | 0.22 |
| Pp860 | N + D[a] | 1.25 | 5.58 |
| Pp830 | Denatured | 1.40 | 0.71 |
| Pp710 | Denatured | 0.81 | 1.73 |
| Pp660 | Denatured | 1.40 | 0.70 |
| Pp610 | Denatured | 1.09 | 1.39 |
| Pp500E | Denatured | 1.09 | 0.33 |
| Pp220 | Denatured | 0.96 | 0.21 |
| Ardex DHV | Denatured | 1.59 | 0.09 |
| Ardex F | Denatured | 1.43 | 0.36 |

[a] 11S Native, 7S Denatured.
Full details described in Sheard, Mitchell and Ledward (1986).

**Table 12.2** GEL STRENGTH OF 18% SOYA PROTEIN GELS PREPARED BY HEATING AT 100 °C FOR 1 h. PRIOR TO USE 6% SUSPENSIONS OF THE NATIVE, LABORATORY PREPARED, ISOLATE WERE PREHEATED AT THE TEMPERATURE INDICATED FOR 1 h TO CAUSE PARTIAL OR TOTAL DENATURATION OF THE PROTEINS. SUBSEQUENT CONCENTRATION WAS BY FREEZE DEHYDRATION. FROM FELLOWS (1985)

| *Preheat treatment* | *Conformation* | *Gel strength*[a] (N) |
|---|---|---|
| None | Native | 16.2 ± 0.50 |
| 60 °C | Native | 7.1 ± 0.60 |
| 80 °C | 7S denatured, 11S native | 4.9 ± 1.00 |
| 100 °C | 7S denatured, 11S native | 3.9 ± 0.57 |
| 120 °C | Denatured | 4.2 ± 0.58 |

[a] Mean ± standard error of five gel preparations.

did not appear to behave differently to those in which the native structure had been totally destroyed during isolation (*Table 12.1*). Other properties of the feed materials were more important. Thus denaturation, as defined by loss of native structure, is not essential but as the proteins are reformed during processing, molecular rearrangements involving the rupture of some bonds and the formation of others undoubtedly takes place. In this respect it is of interest that, under appropriate conditions, several of the isolates which possessed, as judged by differential scanning calorimetry, no native structure formed gels on heating aqueous suspensions (Catsimpoolas and Meyer, 1970). Fellows (1985) also reported that gels could be formed from partially or totally denatured proteins (*Table 12.2*). It is tempting to suggest that the molecular rearrangements involved in gel formation and extrusion processing are not dissimilar.

## What is the nature of the 'melt' phase?

The process of 'melting' is defined in the *Dictionary of Scientific and Technical Terms*

(1984b) as 'to change a solid to liquid by the application of heat'. Such a process is usually endothermic and thus on scanning, with respect to temperature, pure metals or organic compounds a large endothermic peak is invariably seen at a temperature corresponding to the melting point of the material. When a mixture of soya flour (Sheard, Mitchell and Ledward, 1986) or isolate (Oates *et al.*, 1987) is heated at various moisture contents (from 10 to 95% water), sharp endothermic transitions corresponding to denaturation of the 7S and 11S globulins are observed but no peak corresponding to a 'melting' transition is seen. During extrusion processing the apparent change from a particulate feed to a homogeneous phase usually occurs at a position in the extruder barrel corresponding to a temperature between that at which the 7S and 11S globulins denature (Sheard, Mitchell and Ledward, 1986). Obviously in a scanning calorimeter it is not possible to duplicate the pressure and shear generated in an extruder but one might still expect to see some evidence of 'melting' at the lower pressure in the cell of the calorimeter, if such a phenomenon does occur in the extruder. In addition, if the feed to the extruder does melt during processing, it will 'freeze' on emergence from the extruder. Thus an extrudate should exhibit a melting transition when heated. No such transition has been observed in several soya extrudates (Sheard, 1985).

It must be remembered that, for example, a soya feed normally consists of about 30–40% water, about 30% of a mixture of proteins and a similar amount of a heterogeneous mixture of simple and complex polysaccharides. Little is known about the solubility of these compounds under the conditions prevailing during extrusion but certainly several of the forces favouring the formation of insoluble aggregates will be disrupted or weakened. Thus an alternative hypothesis may be that the high temperatures and pressures in the extruder plus the shearing action leads to the formation of a concentrated 'solution', i.e. the solids do not melt but dissolve in the superheated aqueous phase. If this does occur then heat will be absorbed or evolved as the molecules go into solution. This, though, will not occur at a fixed temperature but rather, as heating progresses, more of the components will go into solution. The relatively steady absorption (or evolution) of heat will not be detected in most scanning calorimeters since it could not be easily differentiated from the temperature dependence of the specific heat of the system.

Further work is obviously needed to examine the effect of heat and pressure on these multicomponent systems before the nature of the 'melt' phase can be unequivocally defined.

## Do the proteins align in the shear field?

Ferry (1948) stated that globular protein gel structures were based on side-by-side associations of highly unfolded peptide chains. Although this may be true for cold-set gelatin gels (Ledward, 1986), studies using a variety of techniques have shown that proteins in heat-set gels contain substantial tertiary and secondary structure (Clark and Lee-Tuffnell, 1986). This could be due to only partial unfolding on heating or partial renaturation of fully unfolded molecules (Bikbov *et al.*, 1981). The possible mechanisms are shown in *Figure 12.1*. Thus in most circumstances the heat-induced gelation of protein molecules gives rise to opaque gels stabilized by the aggregation of almost spherical molecules in a 'string of beads' type of organization (Tombs, 1970).

Even if the proteins are fully denatured the most probable conformations for random coils are as compact globules although, of course, an infinite number of

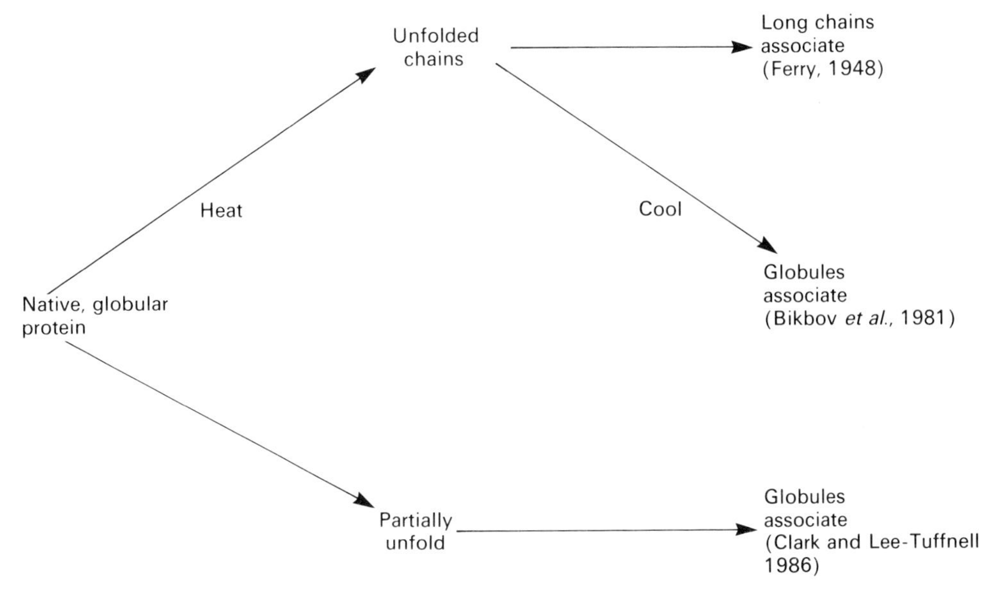


**Figure 12.1** Possible mechanisms for heat-induced gelation of globular proteins.

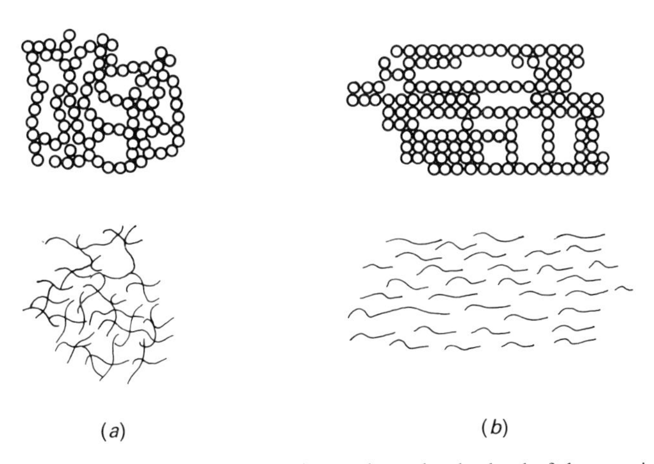


**Figure 12.2** Possible organization at the molecular level of the proteins in an extrudate; (a) no shear; (b) shear.

conformations are possible. Although the formation of a protein extrudate and a heat-set protein gel may well have features in common, great care must be exercised in drawing too close an analogy between the processes since the conditions for formation are so very different.

Thus the conditions in the extruder, with respect to protein stability, are extreme and may lead to complete unfolding (if native) or disaggregation (if denatured) of the protein complexes and the high shear may well force these to favour relatively extended chain conformations, which may then become orientated in the shear field. Even if the molecules associate as spherical particles, of either partially or fully denatured material, the shear might be expected to cause some orientation of the aggregate. Idealized schemes for the organization at the molecular level of the proteins in an extrudate are shown in *Figure 12.2*.

It seems likely that the molecules in the actual extrudates have no unique conformation but are intermediate between those shown in *Figure 12.2*. Unfortunately estimates of the likely shapes of molecules in dilute solution are not really applicable to these concentrated systems.

## How important are charge effects?

Unlike most synthetic polymers proteins, depending on their environment, may carry either a *net* positive or negative charge, and at their isoelectric point will carry a *net* charge of zero. In dilute solutions where the molecules are fully hydrated, repulsive and attractive electrostatic forces between these molecules, although of significance, are only one of many possible interactions. At the high temperatures and low moisture contents found in extrusion processes many of the forces usually deemed to be of importance in protein–protein interactions will not operate (e.g. hydrogen bonds and hydrophobic interactions). Electrostatic forces, though, will still be very strong and it is possible that, within the barrel and die of the extruder, these become the dominant forces in controlling the mode of interaction between the macromolecules. Unfortunately the pK values of the various groups on a protein at the temperatures and pressures attained in extrusion processing are not known but it is not unreasonable to assume they are not too dissimilar from those observed in dilute solution at normal temperatures and pressures (Edsall and Wyman, 1958). A further problem is that the proteins present in soya flour or isolate are heterogeneous with respect to both size and amino acid composition and during extrusion the various subunits may dissociate or aggregate leading to increased heterogeneity. Thus at any given pH the total and net charge on the different components will differ, although obviously with increasing pH all will become more negatively charged. However, the major protein components present in soya (the 7S and 11S globulins) have isoelectric points around 4.7. At this pH, therefore, it would be expected that during extrusion the overall average *net* charge on the proteins would be minimal. If this is so attractive forces between the proteins should be maximized permitting the molecules to approach one another very closely. Limited evidence suggests that this is so as Ardex R, a soya isolate prepared by isoelectric precipitation and which, when suspended in distilled water has a pH of 4.7, yields a far tougher textured extrudate than those prepared from isolates with solution pH values in the range 6.3–7.0 (Sheard, Mitchell and Ledward, 1986) when extruded under similar conditions. Thus the extrudates described in *Table 12.1*, which were all prepared from isolates yielding solution pH values of 6.3–7.1, had a peak shear force value 15–180 times less than the value found for an extrudate prepared from Ardex R (*Table 12.3*). Not unexpectedly the diameter of the extrudate from Ardex R was less than that of those made from the other isolates. The diameter was 4.1 mm compared with values of from 4.3 to 8.4 mm for the other extrudates (*Table 12.3*).

Rhee, Kuo and Lusas (1981) in an extensive study of the factors affecting the extrusion behaviour of soya protein showed that the pH of the feed (adjusted with M HCl or M NaOH) markedly affected the properties of the extrudate. Thus in the range 5.3–9.0, extrudates prepared from the feeds of lower pH had stronger texture in terms of Instron stress and resilience values and also smaller diameters. The two texture parameters decreased by a factor of about 1.8 on increasing the pH from 5.3

**Table 12.3** PRODUCT DIAMETER AND OTTOWA TEXTURE MEASURING SYSTEM (OTMS) TEXTURE FOR SOYA PROTEINS EXTRUDED AT 120 °C AND 180 °C. FROM SHEARD, MITCHELL AND LEDWARD (1986)

| | | *120 °C* | | *180 °C* | |
|---|---|---|---|---|---|
| *Sample* | *Solution pH* | *Diameter* (mm) | *Peak force* (N) | *Diameter* (mm) | *Peak force* (N) |
| Ardex R | 4.7 | 4.1 | 4500 | 4.1 | 4500 |
| Mean of seven isolates | 6.3–7.0 | 4.6–6.6 | 26–1290 | 5.1–8.4 | 22–1300 |

to 8.0 whilst the diameter increased by a similar margin. These results would support the contention that increasing net charge on the proteins gives rise to repulsive forces during extrusion which do not permit the molecularly rearranged proteins in the heated extruder to interact too strongly.

Similarly, chemical modification of the charged groups on the protein markedly affects the extrusion behaviour of soya protein. Thus treatment of soya flour with ninhydrin or acetic anhydride results in an increase in the net negative charge on the protein as the free amino groups acquire a neutral charge. Both treatments have been shown to significantly inhibit texture formation as measured by Warner–Bratzler shear (Simonsky and Stanley, 1982), although in contrast Sheard (1985) actually found that ninhydrin treatment of feeds appeared to reduce the product diameter and increase the textural strength of hydrated extrudates.

More realistic estimates of the effect of charge may be achieved if the *net* charge is varied whilst keeping the *total* charge constant and both Rhee, Kuo and Lusas (1981) and Simonsky and Stanley (1982) attempted to do this by treatment of soya flour with succinic anhydride, which converts positively charged amino groups to negatively charged carboxylate ones and thus, at pH values above the isoelectric point, markedly increases the net negative charge on the protein whilst keeping the total constant. Both authors found this treatment led to a marked decrease in textural strength (by a factor of two).

If charge effects are of such importance in extrusion processing it would be expected that the addition of neutral salts to the feed should markedly affect extrusion behaviour. Rhee, Kuo and Lusas (1981) examined the effects of both sodium and calcium chloride on the extrusion of processing of soya and at low concentrations ($<0.02\%$) observed little effect. However, at higher concentrations (1.0–2.0%) significant effects were noted and NaCl was found to inhibit texture formation with the result that the extrudates could not survive retorting whilst $CaCl_2$ increased the textural strength of the extrudates, which were retort-stable. Obviously salt-specific effects operate and complicate the issue. It must also be remembered that most workers have determined the textural properties of the rehydrated extrudate and thus the effect of protein charge and ionic strength on the swelling of the extrudate, as well as on the interactions between the proteins within the extruder, must be considered when discussing their effects on texture. No such considerations affect the discussion with regard to product diameter and in this respect it is interesting that Smith (1984) found that the addition of 1% NaCl to soya flour led to a small but significant reduction in product diameter following extrusion. This would be expected if the salt shielded the charges on the protein.

However, in view of the difficulty in modifying a feed in such a way that only the charge on the protein is changed, the importance of this property during extrusion must still be rather speculative.

## What bonds stabilize the aggregate and when do they form?

Of all the questions posed in this chapter this is possibly the one that has received most attention and consequently to which the least speculation needs to be attached.

Rhee, Kuo and Lusas (1981), Jeunink and Cheftel (1979) and Hager (1984) all concluded, on the basis of solubility work using sodium dodecyl sulphate (SDS) or urea—compounds capable of destroying hydrogen bonds and hydrophobic interactions—and β-mercaptoethanol (ME) or dithiothreitol—compounds which disrupt disulphide linkages—that disulphide and non-covalent linkages were responsible for the insolubility of soya extrudates in water. Thus it was concluded that these types of linkages were responsible for the 'texture' of the product. Above extrusion temperatures of 95 °C, Rhee, Kuo and Lusas (1981) found that about 20%, 40% and 100% of the protein could be solubilized in buffered solutions of 1% SDS, 1% ME and a solution containing 1% SDS and 1% ME and we have found very similar results (*Table 12.4*).

**Table 12.4** PROTEIN SOLUBILITIES (%) IN 1% SDS, 1% ME AND 1% SDS + 1% ME FOR A SOYA FLOUR AND SOYA ISOLATE AND THE EXTRUDATES PREPARED FROM THEM. FROM SHEARD (1985)

| | *SDS* | | | *ME* | | | *SDS + ME* | | |
|---|---|---|---|---|---|---|---|---|---|
| | *Solvent protein concentration (%)* | | | *Solvent protein concentration (%)* | | | *Solvent protein concentration (%)* | | |
| *Sample* | *1* | *2* | *5* | *1* | *2* | *5* | *1* | *2* | *5* |
| Native soya flour | 84 | 78 | 83 | 64 | 61 | 61 | 104 | 103 | 105 |
| Extruded soya flour | 41 | 44 | 38 | 21 | 23 | 22 | 91 | 92 | 85 |
| Native isolate | 97 | 89 | 80 | 83 | 83 | 86 | 101 | 97 | 99 |
| Extruded isolate | 45 | 50 | 45 | 23 | 20 | 16 | 89 | 89 | 80 |

Although disulphide linkages and hydrophobic and other weak interactions are primarily responsible for the textural quality of the cooled, rehydrated extrudates, other covalent linkages may form (Oates *et al.*, 1987a; Simonsky and Stanley, 1982) but these are not numerous enough to lead to insolubility in SDS and ME. However, as will be discussed in the following section, the reactions giving rise to these types of bonds may markedly affect the rheological properties of the 'melt' and be of importance in allowing a protein to be successfully extruded.

Although disulphide and hydrophobic bonds are major determinants of the texture of the extrudates, it does not seem reasonable that they are formed in the extruder itself as most such interactions are thermally labile. The majority presumably form on cooling and thus it is tempting to speculate that electrostatic and to some extent steric factors and more specific chemical reactions (discussed in the next section) govern the alignment and organization of the proteins in the extruder and the disulphide and non-covalent bonds are subsequently formed on cooling.

## Is the changing chemical composition of the 'melt' of importance?

It has recently been demonstrated that during heat treatment at temperatures similar to those attained in extrusion processing volatiles (assumed to be water) are produced (Oates *et al.*, 1987a). *Figure 12.3* summarizes the apparent increase in water content of a soya isolate following heating in a differential scanning calorimeter. The results were obtained as follows:

1. Samples of soya isolate with and without added high mannuronic acid alginate were equilibrated above saturated salt solutions of different water activity.
2. The moisture content of an aliquot was determined by heating for 48 h in a vacuum oven at 70 °C for 48 h.
3. The equilibrated material was sealed in aluminium pans and heated in a differential scanning calorimeter to a temperature of about 150 °C.
4. After cooling the pans were punctured and the moisture content redetermined.

It is seen from *Figure 12.3* that addition to the isolate of only 2% of an alginate rich in mannuronic acid residues leads to the formation of increased amounts of water and this may well explain the marked decrease in the viscosity observed during the extrusion of soya samples containing this polysaccharide (Berrington *et al.*, 1984; Oates *et al.*, 1987a). No such effects are given by alginates rich in guluronic acid residues or with a whole range of other polysaccharides. Recent work in our laboratory has also suggested that the formation of water, or other volatiles, is also dependent on the nature of the protein, both gluten and soya exhibiting this phenomenon (Oates *et al.*, 1987b). Since both these proteins are known to extrude relatively successfully it may well be that the chemical reactions giving rise to the increased 'water' are of importance in extrusion. We are currently investigating the possible nature of these reactions but as the water must arise from degradative and/or condensation-type reactions it may be that a few stable non-sulphide covalent linkages are established in the extruder which, although insufficient in number to render the protein insoluble in SDS + ME (Jeunink and Cheftel, 1979; Rhee, Kuo and Lusas, 1981), initiate in conjunction with electrostatic forces the whole texturiza-

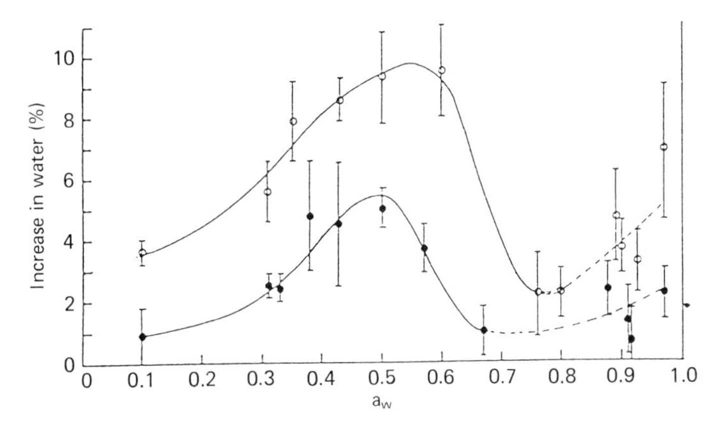


**Figure 12.3** Increase in apparent water content (determined by drying *in vacuo* at 70 °C) as a function of initial water activity ($a_w$) for soya isolate + 2% high mannuronate alginate (○) and soya isolate alone (●). All values are means and the error bars are the standard errors of the means. Reprinted with permission from Oates *et al.* (1987a).

tion process. If such linkages are formed during soya extrusion, and the brown colour and decrease in nutritional quality of the protein following extrusion (Cheftel, 1986) certainly suggests that some 'Maillard'-type pigments are formed, then their identification could be of key importance in understanding the factors responsible for the successful extrusion of proteinaceous feeds.

## Discussion

Thus for a protein to extrude it is possible that:

1. The component polypeptides must have the correct charge, size and shape to allow the molecules to come into the correct juxtapositions in the extruder and, perhaps, form a few key non-disulphide covalent linkages.
2. To yield a stable extrudate the proteins must, in the molecularly rearranged 'concentrated solution', have appropriate groups in the correct positions to form the requisite number of stabilizing interactions on cooling.

Such considerations may to some extent explain the fact that some proteins will yield stable, well-expanded extrudates (e.g. soya) while others will not.

Whatever the faults and imperfections of the above questions and the naïvety of some of the suggested answers we do believe that the further development of our understanding of protein extrusion depends on looking at the chemistry of these complex systems rather than further time and effort being devoted to the physical and engineering aspects. Obviously all three disciplines have contributions to make but the chemistry of these systems has, until now, been sadly neglected.

## References

ATKINSON, W.T. (1970). Meat-like protein food product. US Patent No. 348 870

BERRINGTON, D., IMESON, A.P., LEDWARD, D.A., MITCHELL, J.R. and SMITH, J. (1984). The effect of added alginate on the extrusion behaviour of soya grits. *Carbohydrate Polymers*, **4**, 443–460

BIKBOV, T.M., GRINBERG, V. YA., SCHMANDKE, H., CHAIKA, T.S. VAINTRAUB, I.A. and TOLSTOGUZOV, V.B. (1981). A study on gelation of soy bean globulin solutions: viscoelastic properties and the structure of thermotropic gels of soy bean globulins. *Colloid Polymer Science*, **259**, 536–547

CATSIMPOOLAS, N. and MEYER, E.W. (1970). Gelation phenomena of soybean globulins. 1. Protein–protein interactions. *Cereal Chemistry*, **47**, 559–570

CHEFTEL, J.C. (1986). Nutritional effects of extrusion cooking. *Food Chemistry*, **20**, 263–283

CLARK, A.H. and LEE-TUFFNELL, C.D. (1986). Gelation of globular proteins. In *Functional Properties of Food Macromolecules* (Mitchell, J.R. and Ledward, D.A., Eds), pp. 203–272. London, Elsevier Applied Science

*Dictionary of Scientific and Technical Terms*, 3rd Edition (1984a) (Parker, S.P., Ed.), p. 428. New York, McGraw-Hill

*Dictionary of Scientific and Technical Terms*, 3rd Edition (1984b) (Parker, S.P., Ed.), p. 993. New York, McGraw-Hill

EDSALL, J.T. and WYMAN, J.J. (1958). *Biophysical Chemistry*, Volume 1. New York, Academic Press

FELLOWS, A. (1985). Effect of heat treatment on the functional properties of soy protein isolates. BSc Hons. Thesis. University of Nottingham

FERRY, J.D. (1948). Protein gels. *Advances in Protein Chemistry*, **4**, 1–78

HAGER, D.F. (1984). Effects of extrusion upon soy concentrate solubility. *Journal of Agricultural and Food Chemistry*, **32**, 293–296

HARPER, J.M. (1981). *Food Extrusion*, Volumes I and II. Florida, CRC Press

JEUNINK, J. and CHEFTEL, J.C. (1979). Chemical and physicochemical changes in field bean soybean protein texturisation. *Journal of Food Science*, **44**, 1322–1325, 1328

LEDWARD, D.A. (1986). Gelation of gelatin. In *Functional Properties of Food Macromolecules* (Mitchell, J.R. and Ledward, D.A., Eds), pp. 171–201. London, Elsevier Applied Science

LILLFORD, P.J. (1986). Texturisation of proteins. In *Functional Properties of Food Macromolecules* (Mitchell, J.R. and Ledward, D.A. Eds), pp. 355–384. London, Elsevier Applied Science

OATES, C.G., LEDWARD, D.A., MITCHELL, J.R. and HODGSON, I. (1987a). Physical and chemical changes resulting from heat treatment of soya and soya alginate mixtures. *Carbohydrate Polymers*, **7**, 17–33

OATES, C.G., LEDWARD, D.A., MITCHELL, J.R. and HODGSON, I. (1987b). Glutamic acid reactivity in heat protein and protein alginate mixtures. *International Journal of Food Science and Technology*, **22** (in press)

OTUN, E.L., CRAWSHAW, A. and FRAZIER, P.J. (1986). Flow behaviour and structure of proteins and starches during extrusion cooking. In *Fundamentals of Dough Rheology* (Faridi, H. and Faubion, J.M., Eds), pp. 37–53. St. Paul, Minnesota, American Association of Cereal Chemists

RHEE, K.C., KUO, K. and LUSAS, E.W. (1981). Texturisation. In *Protein Functionality in Foods* (Cherry, J.P., Ed.), pp. 51–88. American Chemical Society Symposium Series, No. 47

SHEARD, P.R. (1985). Role of protein and carbohydrate in soya extrusion. PhD Thesis. University of Nottingham

SHEARD, P.R., MITCHELL, J.R. and LEDWARD, D.A. (1986). The extrusion behavior of different soya isolates and the effect of particle size. *Journal of Food Technology*, **21**, 627–641

SIMONSKY, R.W. and STANLEY, D.W. (1982). Texture–structure relationships in textured soy protein. V. Influence of pH and product acetylation on extrusion texturisation. *Canadian Institute of Food Science and Technology Journal*, **15**, 294–301

SMITH, J. (1984). Protein–polysaccharide interactions in textured foods. PhD Thesis. University of Nottingham

SMITH, O.B. (1976). Extrusion cooking. In *New Protein Foods*, Volume 2 (Altschus, A.M., Ed.), pp. 86–121. London, Academic Press

TOMBS, M.P. (1970). Alterations to proteins during processing and the formation of structures. In *Proteins as Human Foods* (Lawrie, R.A., Ed.), pp. 126–138. London, Butterworths

13

# REFORMED MEAT PRODUCTS—FUNDAMENTAL CONCEPTS AND NEW DEVELOPMENTS

P.D. JOLLEY and P.P. PURSLOW
*AFRC Institute of Food Research, Bristol, UK*

## Introduction

Reformed meat products represent a rapidly increasing sector of the total meat market. In the UK the sale of reformed grillsteaks alone, worth approximately £70 million per annum, grew by 29% in 1985.

These products offer many advantages. For the producer there is the benefit of better utilization of carcass meat. Small offcuts may be incorporated into a product and, because meat reforming aims to upgrade textural and visual quality of lower grade cuts, the economic value of the carcass can be maximized. To the consumer these products offer convenience, portion control and reasonably consistent eating quality.

Historically, the largest area of meat reforming was the production of sectioned and formed pork products from trimmed whole muscles (the modern ham) or large muscle chunks (e.g. pressed shoulder 'ham'). At the other end of the meat particle size scale, the manufacture of hamburgers and beefburgers from finely ground or comminuted muscle is familiar and well established. Recently there has been a new series of products appearing; chopped and shaped products commonly called reformed steaks or, in the UK, grillsteaks. The aim is to arrive at a product whose eating quality closely emulates that of whole meat such as steak. They are thus seen to be more appealing than beefburgers and are of intermediate value between UK-style beefburgers and whole muscle steaks.

The manufacture of all these products has largely arisen empirically. Most process development and product formulation seems to have been by extending existing methods on almost a trial and error basis. This applies even to the more recent flaked products; the commonest flaking machine used in their manufacture was originally marketed as a device to chop vegetables. However, some of the newer products seem to have been 'designed' for improved textural quality by the application of some well-founded concepts.

In this review we wish to discuss some of the key aspects of processes common to the manufacture of reformed meat where basic principles are becoming more clearly understood. We also intend to point out areas where the mechanisms of action of some processes are not well characterized or are confused and, hopefully, shed some light on possible reasons for this. The aim of basic research on reformed meats is to understand how each unit operation within the whole process interacts with raw material characteristics to determine end-product quality on a fundamental cause and effect basis. Only by gaining this fundamental understanding will we be able to control and modify these products to the best advantage, as well as facilitating successful new product development.

## Muscle structure

A common aim of meat reforming is to make a product in which the objectionable textural attributes of poor quality meat have been eliminated, whilst retaining desirable attributes. This provides a fairly precise target; to alter and remake the structure of the material so that only the undesirable mechanical properties are changed. In order to achieve this aim, it is first necessary to understand the underlying relationships between mechanical properties and meat structure.

A schematic diagram of the general structure of a skeletal muscle is shown in *Figure 13.1*. The whole muscle is ensheathed in a connective tissue layer, the epimysium. This collagenous network is continuous with the tendons and, in some muscles, is quite thick (e.g. *M. obliquus internus*, *M. infraspinatus*, *M. rectus abdominis*). Internally, the muscle is divided into bundles of muscle fibres, typically in the order of 1–10 mm in diameter. Another collagenous structure, the perimysium, surrounds each muscle fibre bundle. Perimysia merge with each other and form a continuous network of connective tissue across the muscle, which connects into the epimysium at the muscle periphery. The perimysium consists of laminae or piles of parallel, crimped collagen

**Table 13.1** TYPICAL CHEMICAL COMPOSITION OF SKELETAL MUSCLE

| *Components* | *Wet Weight (%)* |
|---|---|
| Water | 75.0 |
| Lipid | 2.5 |
| Carbohydrate | 1.2 |
| Vitamins | trace |
| Miscellaneous non-protein solutes (nitrogenous and inorganic ions) | 2.3 |
| Protein | 19.0 |
| Connective tissue and organelle (stromal) | 2.0 |
| Myofibrillar | |
| Myosin | 4.9 |
| Actin | 2.5 |
| Titin (connectin) | 1.2 |
| Tropomyosin | 0.6 |
| Troponin | 0.6 |
| Nebulin | 0.6 |
| C-protein | 0.2 |
| M-protein | 0.2 |
| α-actinin | 0.2 |
| Others | 0.5 |
| | 11.5 |
| Sarcoplasmic (mainly glycolytic enzymes and myoglobin) | 4.9 |
| Haemoglobin and other extracellular proteins | 0.6 |

Data taken from Lawrie (1985) and Ohtsuki, Maruyama and Ebashi (1986).

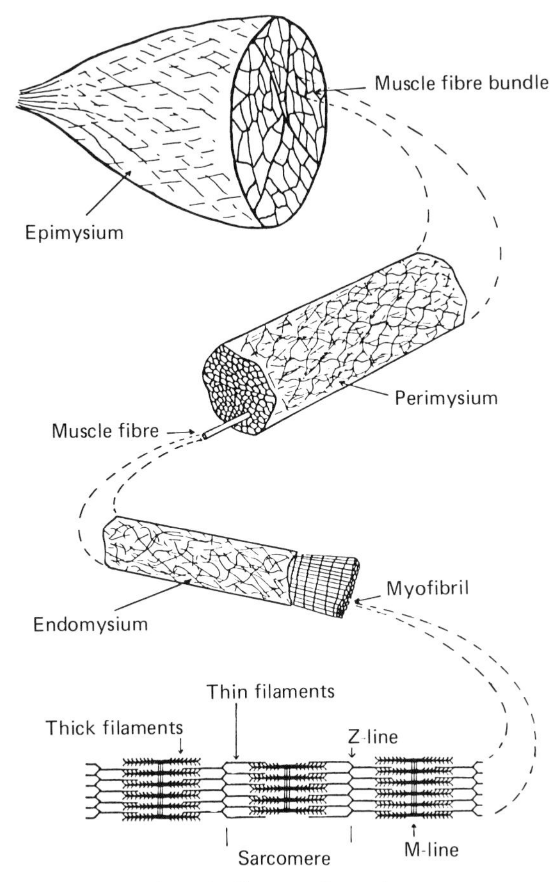


**Figure 13.1** Schematic diagram of muscle structure.

fibres arranged in some form of crossed-helical arrangement with respect to the long axis of the muscle fibres (Rowe, 1974). Perimysial collagen fibre diameters vary from muscle to muscle within the same animal (Light *et al.*, 1985).

Within each bundle there are many individual muscle fibres, each one a giant multinucleated cell, often running the entire length of the muscle. Muscle fibres are typically 10–100 μm in diameter. Surrounding each muscle cell outside the cell membrane (sarcolemma) is a connective tissue sheath, the endomysium.

Within each muscle fibre there are in the region of 1000 myofibrils, the contractile organelles, occupying approximately 80% of the cell volume (Bagshaw, 1982). The myofibrils are in turn made up of an aligned and regularly organized assembly of myofilaments, as shown in the bottom element of *Figure 13.1*. *Table 13.1* shows the major protein components of the myofibril (myofibrillar proteins), which comprise about 70% of the protein in a typical muscle. Most of the remaining 30% of total protein is composed of proteins soluble in low ionic strength buffers, and loosely termed the sarcoplasmic proteins. Many sarcoplasmic proteins are enzymes (Lawrie, 1985; Scopes, 1970).

## Mechanical properties of meat

The sensory perception of meat texture mainly involves the physiological transduction or sensing of a variety of its mechanical properties during chewing. Since the act of chewing breaks down the meat structure, it is the fracture properties of meat that are most important in determining perceived texture.

As described above, muscle tissue is a complex, hierarchical organization of mainly fibrous proteins (both myofibrillar and collagenous) in a well-ordered and aligned structure. It is therefore not surprising that the mechanical properties of meat are highly directional (anisotropic). The tensile strength (breaking stress in tension) along the muscle fibre direction is much greater than at right angles to it, both in raw meat (Munro, 1983) and cooked meat (Bouton, Harris and Shorthose, 1975; Penfield, Barker and Meyer, 1976; Purslow, 1985). Cooked meat toughness, as measured by the force required to drive a blunt blade through the meat (a Warner–Bratzler 'shear' test) is greater when cutting across the muscle fibres than along them (Bouton, Harris and Shorthose, 1975). Using a similar 'bite' test (in which the meat is cut between two opposing blunt wedges), a reasonable correlation between connective tissue content and toughness of different beef muscles has been established (Dransfield, 1977). Despite the relatively small connective tissue content by weight of most muscles (typically 1–10% of dry fat-free matter; Bendall, 1967), the very high stiffness and strength of collagen fibres means that collagen is likely to have a disproportionately great contribution to the mechanical properties of the whole tissue. Purslow (1985) has recently shown some of the structural events associated with the fracture of meat. This initially involves separation of muscle fibre bundles from each other. It seems likely that this results from a debonding at the junction between perimysia and adjacent endomysia, rather than splitting of the perimysium, as shown by *Figure 13.2*. The continuous nature of the network of perimysial sheaths, joined together at 'nodes' between muscle fibre bundles in samples containing a sufficient number of fibre bundles in a cross-section means that, subsequent to endomysial/perimysial debonding, the perimysial network strands then have to be broken (Purslow, 1987). The muscle fibre bundles may then have to be transversely fractured, depending on the orientation of the sample and the way it is deformed.

Although we do not as yet fully understand the sequence of deformations and the resultant structural degradations of cooked meat in the mouth during successive chews (and still less the relationship between these and sensorily perceived texture), the following would seem to be a reasonable working hypothesis. In the initial stages of chewing (i.e. the first few bites), the principal fracture event is the breakdown of long-range connective tissue structure holding muscle fibre bundles together. The integrity of the perimysial network becomes progressively reduced and, at later stages of mastication, some sensory appreciation of the anisotropic nature of the muscle fibres and fibre bundles is involved. From this hypothesis, the rationale behind reforming tough meat can be more clearly formulated; namely, to reduce or remove the naturally occurring thick, coarse connective tissue structures that ordinarily make the poor-quality muscle objectionably tough by their very high resistance to fracture in the initial stages of chewing, whilst still retaining as much as possible the anisotropic nature of muscle fibres/bundles that is apparently a positive textural attribute. This is practically achieved by cutting out gross connective tissue inclusions and/or cutting the tissue into sufficiently small pieces to destroy long-range intramuscular connective tissue structure, and then sticking the cut pieces back together. The adhesive junctions between the meat pieces must provide less resistance to

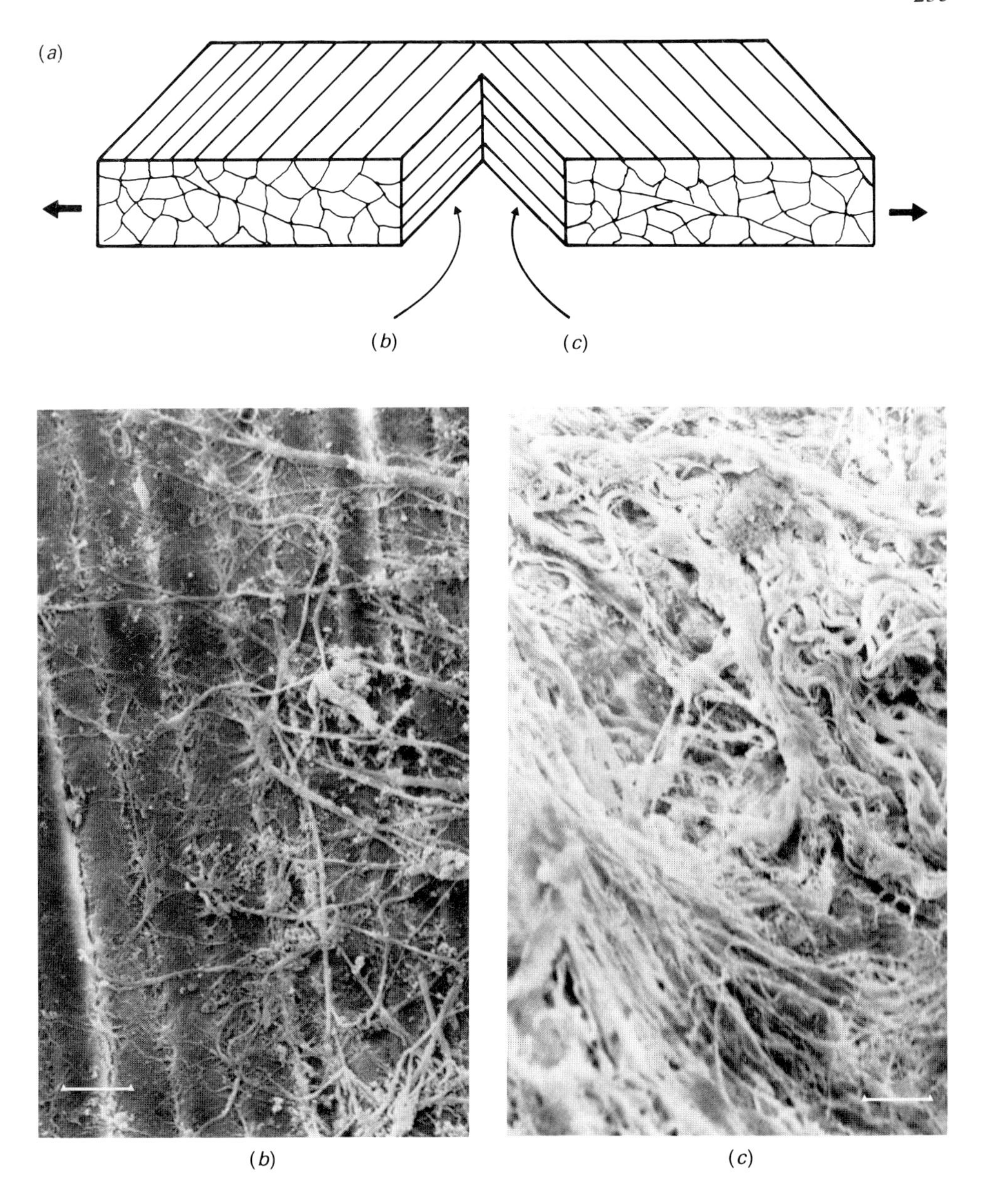


**Figure 13.2** Scanning electron micrographs of complementary fracture surfaces from a longitudinal slice of cooked *M. semitendinosus* muscle pulled perpendicular to the fibre direction. (a) Schematic diagram to show origin of surfaces shown in (b) and (c). (b) The first surface shows individual muscle fibres with intact endomysia overlain by a sparse covering of light connective tissue. (c) The complementary surface is covered in thick, coarse perimysial connective tissue. Bars denote (b) 62.5 μm, (c) 130 μm. Adapted from Purslow (1987) with permission.

fracture than the natural (perimysial) initial fracture sites for the textural quality of the meat to be upgraded.

## Adhesion between meat pieces: principles

Together with optimal raw material breakdown, good adhesion between meat pieces ('bind') is a key, quality-determining property of the final product. It is therefore important to understand it as fully as possible, in order to be able to create and manipulate it more successfully. The adhesive is usually considered to be solubilized muscle protein (Asghar, Samejima and Yasui, 1985; Schmidt and Trout, 1984) that, on heating, sets the assembly of meat pieces together into a solid and provides sufficient cohesive and adhesive strength for this solid cooked product to resist fracture and remain whole, up to the point of being cut up and eaten.

### *Mechanisms of adhesion*

A great deal of what is known about the way in which adhesive junctions are formed, and the mechanical consequences of this, can be traced through work in the engineering adhesives and polymer science fields. Kinloch (1980, 1982) gives excellent reviews of these areas.

The first aspect of the formation of a good adhesive joint is the promotion of intimate contact between adhesive and the surfaces of the pieces to be stuck together (substrate). An effective joint will be promoted by displacing from the substrate surface any substances inhibiting adhesion; in meat reforming this principally means the displacement of any trapped air bubbles that would spoil good adhesive–meat substrate contact. In the great majority of reformed meat products, the adhesive is a heat-setting 'exudate' of muscle fibre fragments and soluble proteins (principally myosin) formed from the meat particles themselves (Maas, 1963). This method of adhesive production naturally provides little problems in the areas of adhesive–substrate compatibility ('wettability'), even coating, and air displacement; intuitively it seems that, because the adhesive consists of some of the same molecules as does the meat substrate, adhesive–substrate compatability is likely to be very good. Solubilization and extraction of protein all over the surface of the meat piece also promotes an even adhesive coating and, because the solubilized protein is extracted from the interior onto the exterior of the meat piece, it naturally displaces air from the meat surface in the process. The use of vacuum during processing probably has a similar effect.

Adhesive–substrate interfacial characteristics therefore naturally promote strong adhesive bonding, the only potentially poor sites being where fat and connective tissue occur on the meat surface. Fat may be not coated well by the aqueous solution, and connective tissue does not seem to interact well with the solubilized myofibrillar proteins to form a good adhesive bond (McGowan, 1970).

In general, when an adhesive has been set or cured by, for example, heat-induced gelation or chemically-induced polymerization reactions, the formation of a mechanically strong bond between the adhesive and substrate is thought to be due to one or more of the following mechanisms (Kinloch, 1980):

1. Mechanical interlocking (i.e. physical entanglement) between adhesive and substrate polymer molecules.

2. Diffusion of the adhesive molecules into the surface, to intermingle with substrate molecules at considerable depths from the nominal adhesive–substrate interface.
3. Electrostatic forces due to junction potentials at the adhesive–substrate interface.
4. Chemical bonding, including adsorption (Van der Waal's forces) and chemisorption (formation of ionic, covalent and hydrogen bonds).

Again in general, it seems that, for load-bearing adhesives, mechanism (4) is by far the most mechanically important (Kinloch, 1980). However, the sticking together of meat pieces by meat proteins solubilized from them may additionally have some contribution from a special case of mechanism (2), that is termed 'autohesion'. This is a process whereby the addition of a solvent solubilizes substrate polymer molecules which then migrate through the interface and intermingle with their identical molecular species in the substrates on either side of the adhesive junction, before being repolymerized or brought out of solution by chemical or thermal treatments. This is clearly analogous to the usual process of reformed meat binding by extracted meat proteins.

### *Mechanical aspects of adhesive fracture*

*Figure 13.3* is a schematic simplified representation of a reformed meat product. Separation of this composite material into two or more separate pieces (i.e. fracture) will involve either:

1. Cohesive failure completely within the gel, and/or
2. Adhesive failure of the meat particle–gel interface, and/or
3. Cohesive failure within the meat particles.

Adhesive fracture can either occur as tensile fracture (the gel and meat can be pulled apart) or as shear fracture (they can slide past each other), just as (cohesive) fracture of the gel or of the substrate can occur in tension or shear. The exact site of fracture of protein gel–meat particle junctions, or their preferred mode of failure (i.e. whether

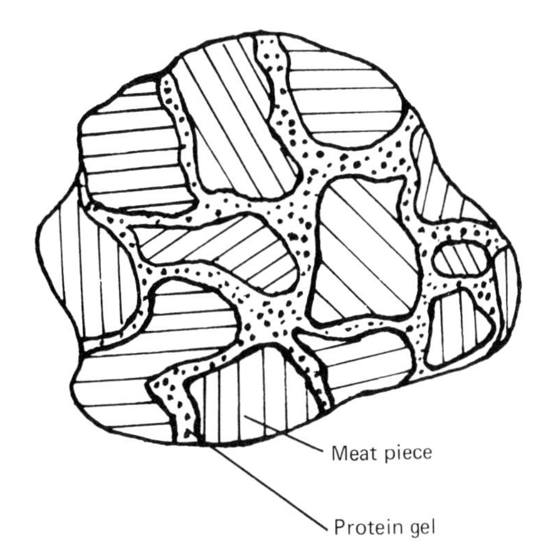


**Figure 13.3** Schematic diagram of a reformed meat product.

they are easier to break in tension or shear) has not been previously studied at all to our knowledge.

### *Methods of assessing binding*

All published methods of assessing bind have measured some sort of strength, i.e. the maximum stress achieved when the sample is deformed to breaking point. Another approach which has proved very useful in generally understanding the fundamental mechanisms of adhesive failure is to measure adhesive fracture resistance in terms of the energy required to propagate fracture. No meat product studies have yet taken advantage of this approach, although the application of this technique to other adhesives is already established; for example, the peel test is a British Standards technique for measuring glue effectiveness (BS5350).

In studies on the ability of given protein solutions to bind two meat pieces together (i.e. 'model' adhesive junctions), the method of test has generally been to apply tensile loads perpendicular to the plane of the adhesive junction (MacFarlane, Schmidt and Turner, 1977; Siegel and Schmidt, 1979a,b; Siegel, Church and Schmidt, 1979; Turner, Jones and MacFarlane, 1979). The testing of binding ability in meat products, as opposed to these 'model' adhesive junctions, has been subject to a wider range of techniques, some quite empirical in nature. Again, tensile testing (of slices or strips) has been a popular technique (Booren *et al.*, 1981a,b; Gillett *et al.*, 1978; Huffman *et al.*, 1984; MacFarlane, McKenzie and Turner, 1984; MacFarlane, Turner and Jones, 1986). Gripping of the specimens in these tensile tests is largely either by pneumatic grips (e.g. Booren *et al.*, 1981b) or by spiked clamp plates (e.g. Gillett *et al.*, 1978).

Other methods used in estimating binding ability have been the Kramer shear press, which basically consists of a series of regularly spaced blades that are pushed through a slotted cage holding the specimen (Booren *et al.*, 1981a,c), plunger (punch) penetration (Fukazawa, Hashimoto and Yasui, 1961), and a shearing blade (Berry *et al.*, 1986). These sorts of test yield peak 'shear' force values that are to some extent dependent on the geometry of the apparatus used (e.g. blade thickness, rounded or flat punches). It is also less likely that failure in some of these tests will be purely at the adhesive junction; meat pieces may also be sheared if, for example, they are big enough to bridge the slots in the Kramer press cage. Some workers have measured binding ability in terms of the force required to break rolls or patties in three-point bending (Ford *et al.*, 1978; Pepper and Schmidt, 1975; Siegel *et al.*, 1978). In this technique, the specimen is placed so as to bridge two end supports and is then broken by pushing a horizontal bar down through the specimen at a point mid-way between the supports. Specimen fracture is most probably due to tensile stresses in bending created on the lower, convex, side of the specimen exceeding the tensile adhesive strength. In this bending test, the maximum tensile stress generated in bending for a given applied load depends very much on the length between the two supports, as well as specimen thickness.

In some recent work in our laboratory (Purslow, Donnelly and Savage, 1987), we have used meat–myosin gel 'model' adhesive junctions to show the effect of various test parameters on binding strength. The tensile adhesive strength (TAS) of these junctions has been shown not to be a constant, characteristic of a given meat–myosin gel adhesive junction, but to vary with cross-sectional size and shape. This is explained on the basis of previously described differences in stress distributions across the width of an adhesive joint (Kinloch, 1982). The TAS of a given meat–myosin gel junction was found to be an increasing function of applied deformation rate, and varies with the orientation of muscle fibres in the surface of the meat piece substrates, as summarized in *Figure 13.4*. The strongest joint was found to occur when muscle fibres in both meat pieces ran perpendicular to the adhesive junction plane under the binding conditions used. In reformed meat products where meat pieces are randomly orientated with respect to each other and the junction between them, many junctions will be of the weaker kind formed when muscle fibres in one or both meat pieces run parallel to the junction. It may be these weaker junctions that, by fracturing at low stresses, determine the breaking strength of the product.

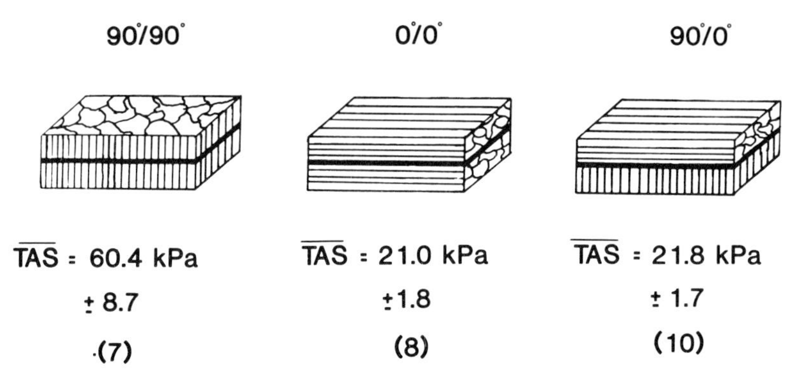


**Figure 13.4** Diagram to show the three orientations of muscle fibres used in model binding junctions. Mean tensile adhesive strength ($\overline{TAS}$) ± one standard error for each orientation is shown. Figures in brackets refer to the number of samples contributing to each mean. Adapted from Purslow, Donnelly and Savage (1987) with permission.

## Factors affecting the extraction of protein from meat

Considerable research has been directed at establishing the main processing factors and raw material characteristics which influence the amount and composition of protein extracted from meat in order to predict the composition of the adhesive exudate, and thereby predict its effectiveness. Such studies have led to the inclusion of 'bind constants' in programmes for determining least-cost formulations (Acton, 1979). Since these programmes are still in use the practical value of such work may be said to have stood the test of time.

The early work in the field of muscle biology on extraction, characterization and quantification of muscle proteins is obviously germane. In particular, such studies led to confirming that myosin was confined to the thick filaments of skeletal muscle (Hanson and Huxley, 1953), and highlighting the changes that occur to the properties of myosin solutions when actin is also present (Banga and Szent-Györgyi, 1941; Straub, 1942: both as cited by Szent-Györgyi, 1951). Szent-Györgyi (1951), Weber and Portzehl (1952) and Helander (1957) all provide good introductory reviews of this research. Much of this information was obtained from rabbit muscle treated pre-rigor, and its total applicability to post-rigor beef, pork, lamb and poultry is questionable. We intend, therefore, to concentrate on data more obviously relevant to present-day meat reforming.

It is essential to be aware of the problems of relating results between papers and extrapolating findings from such work to conditions pertaining to meat reforming.

Firstly, we need to consider how the extracted protein has been characterized. There are three aspects of the composition of extracted protein which are likely to influence the effectiveness of adhesion; the total protein concentration of the extract, the concentration of myosin, and the ratio of myosin to actin—an increase in any of these factors will increase adhesion, as will be discussed later—but we are not aware of any paper which reports all three characteristics. A common method of categorization is to determine the total amount of protein extracted by (and soluble in) water containing high concentrations (typically 0.6 M–1.2 M) of NaCl and subtracting from this value either the concentration of proteins precipitated by lowering ionic strength following extractions (such as the dialysis fractionation of Trautman, 1964, or, more commonly, dilution procedures such as that of Awad, Powrie and Fennema, 1968) or the quantity of protein extracted at low ionic strength from a second sample of the meat (e.g. Helander, 1957). These methods are thus an attempt to distinguish the salt-soluble myofibrillar proteins (principally myosin and actin) from the sarcoplasmic proteins which will also be present in the initial extract. Since the sarcoplasmic proteins comprise almost 30% of the protein of a typical muscle (*Table 13.1*) and are poor adhesives (MacFarlane, Schmidt and Turner, 1977; Ford *et al.*, 1978) the importance of this distinction will be appreciated.

Secondly, we need to be aware that all extraction procedures use a higher solvent:meat ratio than occurs during any meat processing. As one might expect, the amount of protein extracted increases with solvent:meat ratio, eventually approaching a plateau at a solvent:meat ratio of 10:1 (Helander, 1957) or 20:1 (Regenstein and Rank Stamm, 1979). Saffle (1968) made the important observation that the percent of total protein which was salt-soluble for any meat was a constant for solvent:meat ratios of 4:1 and above, and that the percentage difference in amount of extracted protein among different types of meat remained the same above this value.

Finally, the majority of data has been obtained from very finely comminuted material, and may or may not be equally applicable to products containing large meat pieces.

With these caveats in mind, the following practical considerations concerning protein extraction from meat will now be discussed: the properties of the raw meat ingredients, processing conditions, and the amount of added salt and/or linear phosphates. The emphasis will be on those factors over which the manufacturer may realistically expect to exercise some degree of control.

### *Properties of raw meat ingredients*

Protein solubilization in salt solutions depends on the physiological state of the muscle and, as a consequence, generally declines with time post-mortem. The myosin:actin ratio will decrease with time post-mortem because of increasing extraction of actin. Unfortunately, this information is of little practical value to the manufacturer of reformed meat products because the opportunity to use pre-rigor meat is not generally available. The possibility of using pre-rigor muscle extracts to hold together pieces of post-rigor meat has been suggested (Atteck, 1983; Ford *et al.*, 1978), but has received little or no commercial success.

The choice of types of meat used as raw material is determined by the desired characteristics of the product, and economics. Large differences in protein solubility exist between a variety of meat sources [*see*, for example, Saffle and Galbreath (1964) or Gillett *et al.* (1977)] which are not attributable simply to the protein content of these cuts; to give some idea of the extremes in this variation, about 45% of the protein in pork cheek meat can be extracted in 3% saline whilst only about 4% of the protein in pork snouts is similarly extracted (Saffle and Galbreath, 1964). The possibilities of taking advantage of such differences (for example, using cheaper meats to achieve the same bind) are currently limited in meat reforming except in products where the morphological identity of the meat used is lost to the eye. There is undoubtedly scope for more effective selection from the better quality processing cuts (most obviously cow forequarter), for example on a muscle basis. There is little information comparing the same muscle from either different species or animal age within species. Major differences exist between different muscles of different phyla (Regenstein and Rank Stamm, 1979).

In practice, the choice between using fresh chilled, frozen, or frozen and thawed meat is often more restricted than it may seem due to market structure and economics. Generally, freezing and thawing reduces the amount of salt-soluble protein (Saffle and Galbreath, 1964). There is further reduction during frozen storage (Awad, Powrie and Fennema, 1968; Helander, 1957; Miller, Ackerman and Palumbo, 1980) which presumably represents increasing amounts of denaturation, probably of myosin. However, Acton and Saffle (1969) found no effect of freezing on protein extractability.

### Processing conditions

Reducing meat particle size (i.e. increasing comminution) increases protein extraction (Acton, 1972; Helander, 1957). Proportionally more myofibrillar proteins are extracted from smaller particles (Acton, 1972). Although Huxley and Hanson (1957) state that myosin extraction is less complete from fibres than myofibrils, myosin is extracted in its purest form from muscle mince; further size reduction (for example, to myofibrils) leads to increasing amounts of actin in the preparation (Weber and Portzehl, 1952), as does grinding with blunt blades (R.L. Starr and G.W. Offer, personal communication). If size reduction is progressive and accompanied by heat input (as may occur in bowl chopping, for example), protein solubility may start to decline through denaturation (Hamm and Grabowska, 1979).

The longer the period allowed for extraction, the more protein is extracted (e.g. Bard, 1965; Gillett *et al.*, 1977; Helander, 1957). However, longer extraction periods also increase the amount of actin extracted from muscle (Banga and Szent-Györgyi, 1941, as cited by Szent-Györgyi, 1951) and myofibrils (Cheng and Parrish, 1978). On the other hand, Solomon and Schmidt (1980) found a linear increase in 'crude myosin' (i.e. solubilized protein which became insoluble when the salt concentration was lowered) with mixing time, although the total protein concentration did not change.

*Figure 13.5* illustrates the important disagreement concerning the effect of temperature on protein solubilization. Gillett *et al.* (1977) found a maximum amount of protein was extracted in 1.28 M NaCl at 7.2 °C, about one-third more than that at –3.9 °C. In direct contrast, Bard (1965) reported a small but gradual increase in the amount of salt-soluble protein as temperature was decreased from 30 °C to about 0 °C, with increases of three- to four-fold with further reduction of temperature to –5 °C. The ranges studied in both papers include the initial freezing point of lean meat but unfortunately neither paper reports if the meat used below this point was already frozen before the extraction began.

Many items of meat machinery are now supplied with the option of drawing a

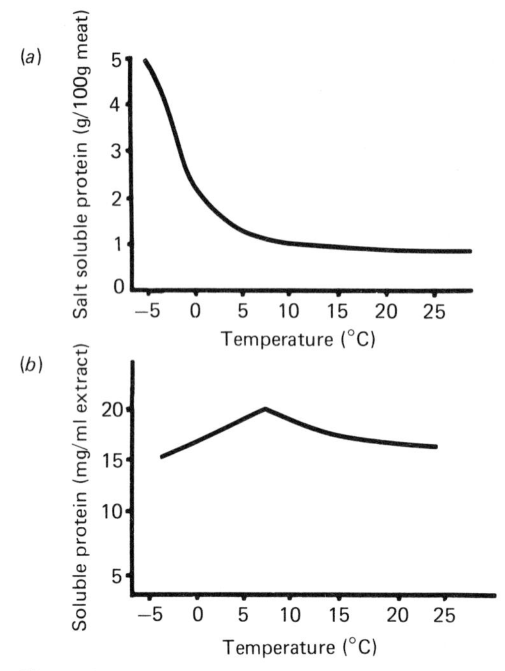


**Figure 13.5** Conflicting results on the relationship between temperature and solubilization of protein. (a) From Bard (1965) by courtesy of the American Meat Institute; (b) From Gillett *et al.*, (1977) with permission of the publishers

vacuum during processing. Solomon and Schmidt (1980), using extraction conditions not typical of meat reforming, found that although vacuum had little effect on the amount of protein in the supernatant, it did alter the composition, resulting in a higher yield of crude myosin.

The application of high pressure, for example of the order of 150 MN $m^{-2}$, has been shown to extract and solubilize muscle protein (MacFarlane, 1974). This effect may have commercial application, particularly since lower pressures have been found to improve product cohesion (MacFarlane, McKenzie and Turner, 1984).

### *Level of added salt and/or linear phosphates*

Increasing salt concentration increases the amount of extracted protein in a generally linear fashion to a maximum before falling again (e.g. Bard, 1965; Callow, 1931; Grabowska and Hamm, 1979). Flavour constraints limit the amount of salt that can be added, and increasing public concern over levels of sodium chloride in the diet is likely to lead to further reduction. However, in the manufacture of most meat products the salt is added either as a concentrated solution or as crystals of sodium chloride. The meat initially in contact with the salt is therefore exposed to far higher ionic strength than is usually considered in this context, and it seems unreasonable to expect that this has no effect on the nature of protein extracted. The amounts of water-soluble protein extracted are generally independent of salt concentration (Grabowska and Hamm, 1979).

Linear phosphates are frequently used in meat reforming to enhance cooked yield and ensure good adhesion, and are of particular interest in the context of protein solubilization. Many studies have reported that di- and triphosphates, in the presence of sodium chloride, increase protein extraction (e.g. Bendall, 1954; Kotter, 1960; Grabowska and Hamm, 1979; Turner, Jones and MacFarlane, 1979; van den Oord and Wesdorp, 1978a,b; Yasui, Sakanishi and Hashimoto, 1964). Turner, Jones and MacFarlane (1979) examined various combinations of salt and sodium triphosphate (STP) and found a maximum yield of crude myosin with 1 M NaCl in combination with 0.25% sodium triphosphate. The yield of crude myosin decreased with higher levels of sodium triphosphate, even though these higher levels of STP increased the pH of the extracting solutions. The proportion of actin present in the crude myosin increased markedly in the absence of STP, or in solutions containing 2% STP.

There is some disagreement—not to say controversy—concerning the underlying mechanisms which lead to increased protein extraction through the use of linear phosphates. In normal practice, linear phosphates will increase pH (typically by 0.10–0.25 pH units), which, as a general rule, will lead to increased protein extraction and solubilization (e.g. Grabowska and Hamm, 1979; Helander, 1957; Trautman, 1966; van den Oord and Wesdorp, 1978a; Yasui, Sakanishi and Hashimoto, 1964), although the myosin:actin ratio might be expected to be reduced (Hasselbach and Schneider, 1951, as cited by Weber and Portzehl, 1952). Protein extraction and solubilization will also be enhanced by the resultant increase in ionic strength, although arriving at a true value for this latter effect is complicated (van den Oord and Wesdorp, 1978b). These effects (raising pH and buffering it at the increased value, and increasing ionic strength) are thought by some authors to account totally for observed increases in extracted protein (Swift and Ellis, 1956), related functional properties such as bind (Trout and Schmidt, 1984) or structural changes (Lewis and Jewell, 1975; Lewis, Groves and Holgate, 1986). In contrast, Offer and Trinick (1983) compared the effect on removal of A-band material (mainly myosin) from isolated rabbit myofibrils in the presence or absence of 10 mM diphosphate at pH 5.5; diphosphate was more effective in extracting the A-band than can be accounted for by the increase in ionic strength. This is a particularly striking example of the synergism between salt and diphosphate in extracting protein which was also shown by van den Oord and Wesdorp (1978b), and has similarly been found for related functional properties such as swelling (Bendall, 1954) or water holding capacity (Hellendorn, 1962).

Diphosphate has long been known to dissociate the actomyosin complex (e.g. Hanson and Huxley, 1953; Weber and Portzehl, 1952) and it has generally been assumed that any synergistic increase in protein solubilization is due to a similar effect taking place in meat, with the dissociation of myosin heads from the thin filaments facilitating solubilization of myosin; attention was drawn to this ATP-like action in this context by Bendall (1954), Kotter (1960); Kotter and Prändl (1956); Kotter and Fischer (1975) and Hellendorn (1962), amongst others. Removal of A-band material has been demonstrated in the presence of diphosphate in many muscle systems (Hanson and Huxley, 1953; Huxley and Hanson, 1957; Offer and Trinick, 1983; Voyle, Jolley and Offer, 1984). Although triphosphates are also known to dissociate actomyosin (Szent–Györgyi, 1951; Yasui *et al.*, 1964), the work of Yasui *et al.* (1964) shows that this effect probably occurs after diphosphate has been produced by enzymatic hydrolysis of the triphosphate; triphosphate is commonly used in meat reforming in preference to diphosphate because of its higher solubility, particularly in the presence of salt. Linear phosphates of chain length greater than four do not show synergism with salt (Bendall, 1954; Hellendorn, 1962).

## The adhesive properties of muscle protein gels

We shall now consider the compositional and environmental factors known to affect the properties of gels formed from solubilized meat proteins on heating. As described above, a number of these factors also affect the level of solubilization of proteins from meat pieces, and so can be said to promote binding by extracting more adhesive, but we are concerned here only with the way in which these factors affect the properties of gels formed from the already extracted proteins. There is a clear distinction to be made between the effects of some of these factors on properties of the bulk gel (setting point, shear stiffness (rigidity) and (cohesive) gel fracture properties) and their effects on the adhesive, binding properties of the same gel. This distinction is not artificial, nor merely semantic; these properties are independent characteristics which, from a theoretical mechanics point of view, need not be correlated. This point is worth stressing because, in the literature, some confusion about factors affecting bind seems to occur, perhaps because of a lack of emphasis on this distinction.

The following are the main factors known to affect both the bulk properties of meat protein gels and also their adhesive strength in binding meat pieces:

1. Protein species.
2. Protein concentration.
3. Level of added salts.
4. pH.
5. Temperature of heating.

### *Protein species*

MacFarlane, Schmidt and Turner (1977) showed that the tensile adhesive strength of myosin in binding meat pieces together was higher than that of the actin–myosin complex (actomyosin). The addition of sarcoplasmic protein depressed the adhesive strength of the myosin–meat junctions. Siegel and Schmidt (1979a) similarly showed that the tensile adhesive strength of myosin–meat junctions was higher than for junctions between meat and a whole muscle homogenate, with or without fat and sarcoplasmic proteins in the homogenate. In any one medium used to extract myofibrillar proteins, they also found that a higher mole ratio of myosin to actin resulted in a higher tensile adhesive strength. Schmidt and Trout (1984) suggest that the only contribution of sarcoplasmic proteins to adhesive strength results from their effects in the ionic environment of other, myofibrillar, proteins in the gel.

The superior binding ability of myosin over actomysin contrasts with the rigidity and cohesive strength of the gels alone. Nakayama and Sato (1971a,b) reported that the tensile strength of isolated gels of myosin A is lower than that of myosin B (myosin B contains far more actin than myosin A, which is comparatively pure myosin). Yasui, Ishioroshi and Samejima (1980) also showed that the rigidity of gels of myosin alone was lower than that of actomyosin gels. The interaction between actin and myosin molecules, one of myosin heads attaching to binding sites on the actin (crossbridge formation), was shown to be responsible for the increased rigidity of an actomysin gel compared to a myosin gel by blocking this interaction (Ishioroshi *et al.*, 1980).

***Protein concentration***

The adhesive strength of myosin gels in binding meat pieces increases with protein concentration (Siegel and Schmidt, 1979b). Cohesive tensile gel strength of both myosin A and B also increases with protein concentration (Nakayama and Sato, 1971a). The shear stiffness of myosin gels increases with protein concentration to the power 1.8 (Ishioroshi, Samejima and Yasui, 1979).

***Level of added salts***

Here again, the adhesive properties of a myosin gel–meat interface reacts differently to changes in the ionic environment than do the bulk properties of the gel alone. The adhesive strength of myosin gel–meat junctions increases with increasing sodium chloride concentration from 0–6% sodium chloride; the addition of 0.5% STP also increases adhesive strength (Siegel and Schmidt, 1979b). MacFarlane, Schmidt and Turner (1977) show a similar effect of sodium chloride concentration in a 5% myosin gel, and also in actomyosin gels. Turner, Jones and MacFarlane (1979) report increasing myosin gel–meat adhesive strength with increasing sodium chloride concentration from 0–1 M sodium chloride. However, the rigidity of myosin gels reacts differently to increasing ionic strength; rigidity is high in gels formed from myosin solutions containing 0.1–0.2 M potassium chloride, but decreases rapidly in the range 0.2–0.4 M potassium chloride and remains low at higher chloride concentrations (Ishioroshi, Samejima and Yasui, 1979).

***pH***

Siegel and Schmidt (1979b) could demonstrate no effect of pH in the myosin solutions on the adhesive strength of meat–myosin gel junctions. They ascribe this to the buffering capacity of the meat pieces, which would tend to bring the pH of all of the applied solutions to a common value in their experiments.

In terms of the bulk properties of isolated gels, Yasui *et al.* (1979) and Ishioroshi, Samejima and Yasui (1979) show that pH does affect the shear stiffness of myosin gels, with maximum rigidity at pH 6. These stiff gels were found to have a fibrous structure, whereas the less stiff gels formed at pH 7 have an aggregated, spongy appearance (Yasui *et al.*, 1979). Hermansson, Harbitz and Langton (1986) also report fine stranded and coarse, aggregated myosin gel structures, but at different ionic strengths rather than pH values (low ionic strengths producing the fibrous gel, which again is stiffer than the aggregated gel). Samejima, Ishioroshi and Yasui (1981) suggest that it is interaction of the heads of myosin molecules that is responsible for the formation of spongy aggregates, and that the helical tails of the molecules under different conditions are responsible for the fibrous network formation.

***Temperature effects***

Because the gelation of solubilized myofibrillar proteins is a heat-initiated reaction, this factor has obvious importance, and again there are differences in the effect of heating temperature on the adhesive strength of meat–myosin junctions as opposed to gel rigidity. Ishioroshi, Samejima and Yasui (1979) showed that maximum gel rigidity is generated by heating myosin solutions to between 60 °C and 70 °C. Yasui, Ishioroshi and Samejima (1980) found an increase in myosin gel stiffness between the onset of gelation, at 40 °C, and 60 °C, above which stiffness decreased. This is in contrast to the adhesive strength of myosin gel–meat junctions, which was found to increase steadily from 40 °C right up to 80 °C, remaining constant thereafter up to 95 °C (Siegel and Schmidt, 1979b).

So, of the five main factors listed above, only increasing protein concentration has a similar effect on both cohesive and adhesive gel properties. The other four factors all affect cohesion and adhesion very differently. This suggests (a) that the mechanism of binding junction failure is unlikely to be simple cohesive failure within the gel, and (b) that cohesive strength or rigidity measurements on isolated gels should not be taken as indicators of true binding strength.

## Adhesion in the absence of sodium chloride, and adhesive properties of non-meat gels

Some adhesion can be effected in the absence of any additives and early studies on meat reforming (e.g. Chesney, Mandigo and Campbell, 1978; Popenhagen and Mandigo, 1978) do not comment on poor product cohesion in the absence of salt; indeed, US-style burgers do not usually contain added salt or binder. Sarcoplasmic proteins presumably contribute to adhesion in greater part here than when sodium chloride is present (*cf.* MacFarlane, Schmidt and Turner, 1977). However, it seems to us unlikely that this is the only factor contributing to additive-free adhesion, and other mechanisms may be involved such as physical entanglement of pieces. It is difficult to envisage an adhesive action of extracted, non-solubilized (acto)myosin.

*Table 13.2*, based on data taken from Siegel, Church and Schmidt (1979) shows the adhesive strength of gel–meat junctions formed with a variety of non-meat proteins.

**Table 13.2** COMPARATIVE BINDING ABILITIES OF VARIOUS NON-MEAT PROTEINS, IN RELATION TO CRUDE MYOSIN

| *Protein solution* | *Tensile adhesive strength* (g cm$^{-2}$) |
|---|---|
| 3% Crude myosin | 842.5[a] |
| 13% Wheat gluten | 350.8[b] |
| 10% Egg white | 240.6[b] |
| 13% Calcium reduced dried skim milk | 149.0[b] |
| 7% Bovine blood plasma | 143.8[b] |
| 13% Isolated soy protein | 133.4[b] |
| 13% Sodium caesinate | 0[b] |
| Control (salt and phosphate solution only) | 214.0[b] |

[a]In 6% NaCl, 2% Sodium triphosphate. Data from Siegel and Schmidt (1979a) with permission of the publishers
[b]In 8% NaCl, 2% Sodium triphosphate. Data from Siegel, Church and Schmidt (1979) with permission of the publishers

In general, non-meat proteins result in much lower adhesive strengths than do myofibrillar protein gels.

The use of salt to effect adhesion has its drawbacks. Colour stability is reduced because metmyoglobin production is accelerated, and texture changes are induced to an extent determined by the degree of blending. Although some adhesion does occur without heating, binding is generally a heat-induced phenomenon and, as a consequence, most reformed meats that rely on the use of salt are usually offered for sale either frozen or cooked. Recently, two very different 'adhesives' have been used in apparently successful attempts to overcome these problems, either of which could lead to more reformed meats being sold chilled. The first (reported by Paardekooper, 1986) is an enzymatically-produced gel resulting from the interaction between thrombin, fibrinogen and a transglutaminase. Plasma enriched with fibrinogen is mixed with meat pieces and other ingredients and the mixture is allowed to stand in a mould overnight. A cohesive product results. This ingenious application of the blood clotting mechanism might engender adverse aesthetic response in some countries, and studies on consumer acceptance of such products are anticipated. The second is a chemical gelation resulting from the well-known interaction between guluronic acid moieties of alginate (a hydrocolloid extracted from seaweed) and calcium (Means and Schmidt, 1986; Schmidt and Means, 1986). The novelty of this approach is that, because of the formation of a thermally stable gel at low temperatures, the algin–meat junction has appreciable adhesive strength in the raw, uncooked product, as well as in the cooked state. However, no mechanical measurements of algin binding strength have yet been published.

## Current manufacturing practice: key aspects

Outline schemes of the manufacture of two broad categories of product are shown in *Figures 13.6* and *13.7*. Most stages in both types of process have been subject to numerous investigations, developments and refinements of either a conceptual or pragmatic nature; however, the contribution of each unit operation can be highly interactive. We shall concentrate on developments in our understanding of the stage in each process which could be considered the most important in determining end product quality.

The first process (*Figure 13.6*) applies to products from large meat pieces ('sectioned and formed' meats), the prime example being the modern or reformed ham (BMMA, 1985). The raw material used here is usually of relatively good quality, so that comminution of the muscle to break up connective tissue is often unnecessary although mechanical treatment (including tenderization) is a common pre-treatment (Huffman, 1980; McGowan, 1970; Schlamb, 1970; Seiffhart, 1983). The main purpose of reforming these sorts of product is to offer convenience, variety and portion control. The key step in this process is the formation of the adhesive through the action of added salts and mechanical treatment.

The second scheme (*Figure 13.7*) applies to those products reformed from muscle tissue that has been mechanically comminuted (e.g. reformed steaks) and is more clearly aimed at upgrading poor quality meat by breaking up long range connective tissue structures. Although the adhesion between meat particles provided by solubilized meat proteins obviously has a quality-determining role in these products too, we shall in this manufacturing scheme concentrate on the essential breakdown of the raw material by pre-breaking and primary comminution.

## *The formation of an adhesive exudate in 'sectioned and formed' product processing*

### *Manufacturing practice*

In the process described by *Figure 13.6*, solubilization and extraction of meat proteins is achieved by the action of added salt and mechanical agitation. The most common method of introducing salt in cured products is by multi-needle or single stitch injection of brine. The brine will also typically contain linear phosphates, nitrite (with cured meats, frequently in association with ascorbate to accelerate the formation of nitrosylmyoglobin), monosodium glutamate (or other flavour enhancers), and sugars. Various proteins, for example from soya or milk, may also be added to improve yield and adhesion.

It is quite common for the salt to be introduced alternatively during the period of mechanical agitation. In addition to effecting good adhesion, mechanical agitation also increases the speed of equilibration of the added ingredients. Salt and other additives may be mixed with the meat as dry ingredients, or as a brine; with some products, especially those associated with a high yield, brine incorporation may be via both an initial injection and subsequent inclusion during mechanical agitation. It is perhaps inevitable that at least one machine exists that can inject brine directly into meat pieces during agitation (Langen, 1977).

Machinery for effecting the mechanical agitation tends to be grouped together in two classes, termed tumblers and massagers. The simplest way to envisage a tumbler is to think of a cement mixer. Baffles are usually present inside a drum which revolves around its cylindrical axis. The baffles pick up meat and carry it to the top of the drum when it drops onto the other meat pieces at the bottom. This impact damages the muscle fibres on the surface of the meat pieces which leads to the production of the adhesive exudate. However, excessive tumbling leads to extensive tissue destruction

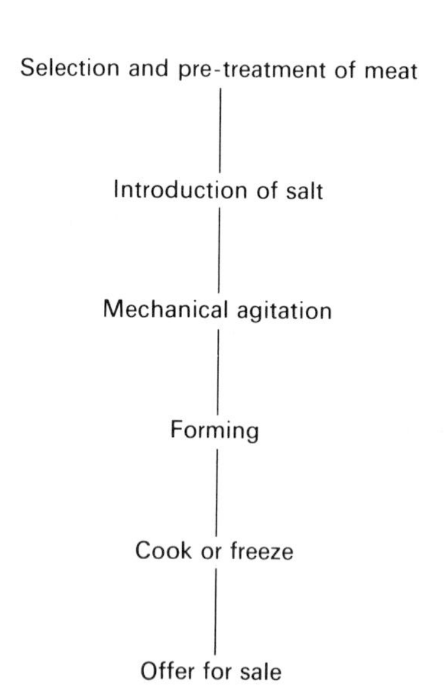


**Figure 13.6** Schematic diagram of a common method for the manufacture of reformed meat products made from muscle pieces ('sectioned and formed' products).

with loss of piece identity and reduced textural quality, and massagers were developed to overcome these problems.

Early massagers (e.g. Jespersen and Engman, 1977) may be thought of as scaled up, slowed down kitchen mixers. Slowly rotating paddles rub the meat pieces over each other and the sides of the massager, the resulting friction causing the damage to the muscle fibres. Over the last few years, possibly in response to the need for increased batch loads and ease of offering vacuum facilities, massagers have come to resemble tumblers externally, but lacking the internal baffles typically present in early tumblers. Consequently, as the drum revolves, the meat is 'jostled' rather than picked up and dropped.

There can be no doubting the considerable experience, expertise and ingenuity that now exists in this area of meat technology, but one is still struck by the empirical nature of this knowledge on the one hand, and the debt owed to the original observation of a mechanically produced sticky exudate by Maas (1963) on the other.

### *Structural changes produced by tumbling or massaging of meat pieces*

The production of reformed meats by tumbling or massaging leads to quite extensive structural damage, particularly if linear phosphates are included in the brine. Spaces develop between fibres (Rejt, Kubick and Pisula, 1978) which leads to longitudinal splitting (Theno, Siegel and Schmidt, 1977). There is extensive disruption to the cell membrane (Cassidy *et al.*, 1978; Theno, Siegel and Schmidt, 1978b) and myofibrils separate from each other longitudinally (Theno, Siegel and Schmidt, 1977, 1978b). As processing continues, breaks occur transversely across myofibrils (Rahelić, Pripis and Vicević, 1974), such that myofibrillar pieces are broken off (Theno, Siegel and Schmidt, 1978b). Quite extensive destruction within fibres occurs (Rahelić, Pripis and Vicević, 1974), although superficially they may appear intact (Siegel, Tuley and Schmidt, 1979). Theno, Siegel and Schmidt (1978b) reported that fibres took on a wave-like form after extensive massaging.

If the salt concentration is high enough, and especially in the presence of linear phosphates, the myofibrils appear to be coated (Theno, Siegel and Schmidt, 1978b), and there is predominantly amorphous material surrounding fibre and fibre bundles (Siegel, Tuley and Schmidt, 1979). The composition of this material at any time during processing is unknown. Cassidy *et al.* (1978) observed that tumbling and linear phosphates both reduced the clarity of striations of muscle. Although some loss of clarity could arise through myofibrils swelling to accommodate added water (Offer and Trinick, 1983) and/or an obscuring effect of precipitated protein, the most likely explanation for this reduction of clarity is loss of A-band material. It cannot be assumed, however, that this is through preferential extraction of myosin over more general myofibrillar destruction; indeed, both Lewis (1979) and Rahelić and Milin (1979) are of the opinion that the A-band is usually the last area of the myofibril to be extracted.

It is obvious that the structural changes will be manifested as textural changes and, in particular, that loss of perimysial and endomysial integrity should make the meat pieces more pliable (Maas, 1963; Theno, Siegel and Schmidt, 1978b). Structural breakdown through mechanical action is enhanced in higher concentrations of salt and the presence of linear phosphates, either through the composition of the brine (Theno, Siegel and Schmidt, 1978b) or by ensuring more even distributions of curing salts through intermittent tumbling (Cassidy *et al.*, 1978). Although it is not usually possible to separate the individual effects on texture of mechanical breakdown, protein solubilization and uptake and retention of bulk water, it is generally true that most processes attempt to optimize the beneficial effect of tumbling and massaging without inducing too pronounced structural damage and adverse texture. It is presumably for this reason that many processes are intermittent, rest period(s) allowing for the equilibration of added ingredients and the development of cured colour (if appropriate), without continual damage to structure.

It is clear from these studies that the composition of the adhesive exudate is determined by two mechanisms. Firstly, there is the generalized breakdown of the fibres on the surface of the muscle pieces. Fibre fragments are present in the exudate in the early stages of massaging (Theno, Siegel and Schmidt, 1978a) and throughout tumbling (Jolley and Savage, 1985). Secondly, proteins extracted from deeper tissue may migrate to the surfaces along the 'channels' formed by longitudinal splitting through a 'squeezing' action of the mechanical working. It is this second mechanism which would seem the more likely to accommodate any preferential extraction of proteins from the myofibril, rather than the more generalized extraction which would result from breakdown of fibres on the surface of the meat pieces. Thus, although there is no direct evidence in the literature to support the oft repeated phrase that tumbling/massaging 'brings salt-soluble proteins to the surface of the meat pieces', these structural studies offer indirect evidence that such a mechanism could exist.

A further point arises from a consideration of the structural changes that occur during tumbling or massaging. If we assume that the adhesive exudate and the extracted protein in deeper tissue are continuous, then this implies an enormous effective increase in surface area and quite considerable 'keying' into the depth of substrate either side of the apparent interface.

### *Composition of the adhesive exudate produced during the manufacture of reformed meats*

Investigations of exudate compositions have aimed to increase our basic understanding of factors affecting protein solubilization, not in a model system, but actually during product manufacture. Whilst much of the literature in this area is clear cut, some confusion also exists.

The following major aspects of exudate composition during massaging are clear cut. The percentage of sodium and water in the exudate decreases and protein concentration increases as massaging progresses (Siegel *et al.*, 1978; Theno, Siegel and Schmidt, 1977), presumably as the brine penetrates into the meat. The total amount of exudate increases with massaging time (Evans, Lewis and Ranken, 1979; Theno, Siegel and Schmidt, 1978a). The level of added salt seems not to influence greatly the relative proportions of myosin, actin, tropomyosin, α-actinin and C-protein present in the exudate of massaged hams (Siegel, Theno and Schmidt, 1978); duration of massaging also has little influence on the relative proportions of these proteins (Siegel, Theno and Schmidt, 1978) as has also been found for other meat-mechanical agitation regimes (Booren *et al.*, 1982; Evans, Lewis and Ranken, 1979; Jolley and Savage, 1985; Reichert, Färber and Flachman, 1984).

There is less agreement about the influence of linear phosphates on the composition of exudate. Reichert (1986) reported that the percentage of solubilized myosin and actin present in the exudate of tumbled pork increased in the presence of diphosphate, and Jolley and Savage (1985) presented evidence that the presence of linear phosphates leads to preferential extraction of myosin in the exudate of tumbled hams. Siegel, Theno and Schmidt (1978) found that the most marked effect of the presence of linear phosphates was to decrease the relative percentage of tropomyosin. They argued that this decrease could reflect a concomitant increase in extraction of actin and myosin, therefore offering only indirect evidence that linear phosphates lead to increased extraction of actin and myosin in a practical environment. There was also no suggestion of preferential extraction of myosin.

One reason for the apparent lack of preferential extraction of myosin in the work of Siegel, Theno and Schmidt (1978) was suggested by Voyle, Jolley and Offer (1984). They pointed out that the material examined by Siegel, Theno and Schmidt (1978a) was material that had not been solubilized plus any myofibrillar protein solubilized in the exudate but precipitated by reduction in ionic strength by the washing procedure. Preferential extraction of myosin in the presence of linear phosphates may have been demonstrated by Siegel, Theno and Schmidt (1978) if the discarded supernatant had been checked for myosin.

Re-examination of the data presented by Siegel, Theno and Schmidt (1978) however, leads to more direct evidence for the action of linear phosphates. The percentage of total protein accounted for by the five proteins reported varies from 54 to 94%; direct comparison of relative percentages is therefore potentially misleading. If the data are recalculated such that the sum of the five proteins examined at each sampling point is equated to 100%, and the relative percentages of each protein calculated accordingly, preferential extraction of myosin is demonstrated in 20 of the 24 combinations of massaging time and final salt level examined (*Table 13.3*).

**Table 13.3** RELATIVE PERCENTAGE OF MYOSIN IN EXUDATE FOR VARIOUS LEVELS OF SALT, PHOSPHATE AND MASSAGING (CONCENTRATIONS OF PROTEINS EXAMINED EQUATED TO 100%)

| *Hours of massaging* | *0% salt* | | *1% salt* | | *2% salt* | | *3% salt* | |
|---|---|---|---|---|---|---|---|---|
| | *0% P* | *0.5% P* | *0% P* | *0.5% P* | *0% P* | *0.5% P* | *0% P* | *0.5% P* |
| 0 | 19 | 38 | 25 | 38 | 18 | 36 | 23 | 27 |
| 1 | 25 | 29 | 28 | 42 | 20 | 33 | 22 | 42 |
| 2 | 26 | 35 | 26 | 30 | 19 | 33 | 31 | 40 |
| 4 | 30 | 30 | 28 | 30 | 27 | 32 | 39 | 38 |
| 8 | 32 | 29 | 27 | 33 | 24 | 32 | 29 | 32 |
| 24 | 28 | 30 | 30 | 30 | 23 | 34 | 33 | 36 |

Calculated from data of Siegel, Theno and Schmidt (1978). P = commercial blend of linear phosphates.

### *The pre-breaking and comminution stages in reformed steak production*

Two typical manufacturing pathways for the production of reformed steaks (and, with some simple modifications particularly at the forming stages, reformed meat slices, dice, roasts, and rib products, amongst others) are shown in *Figure 13.7*. The scheme on the left of *Figure 13.7* is a simplified flow chart of a procedure followed by probably the majority of the literature on the subject. The potential economic importance of meat products reformed in this manner has no doubt contributed to the large number of recent reviews on the subject (Breidenstein, 1982; Franklin and Cross, 1982; Pearson and Tauber, 1984). However, few real principles arise from the

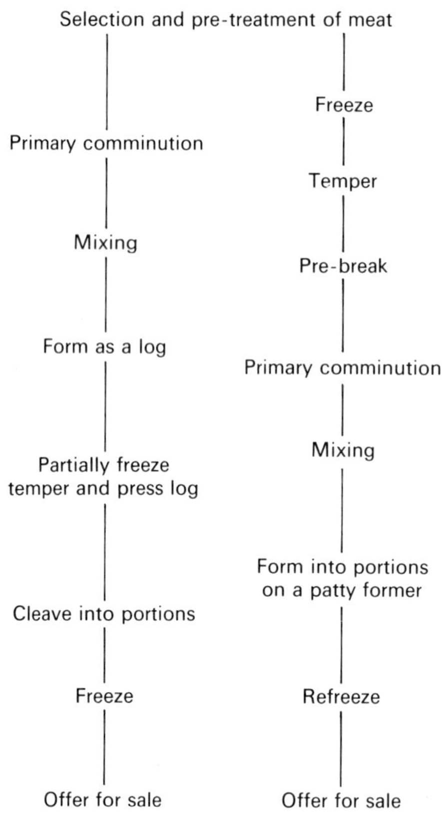


**Figure 13.7** Schematic diagram of two common methods of manufacture of reformed meat products from mechanically comminuted meat. 'Chunked and formed', 'flaked and formed', and 'chopped and shaped' products may all be produced following either pathway.

studies reviewed by these several authors. The scheme on the right of *Figure 13.7* illustrates a common process followed by many manufacturers in the UK, where such products enjoy an increasingly important share of the retail meat products market.

A major difference between the two schemes is the use of frozen meat as the starting material in the UK. The fracture behaviour of the meat used as raw material in both the pre-breaking and primary comminution stages is dependent on the intrinsic characteristics of the meat (e.g. more or less tough, fat content, etc.), temperature and rate of breaking or cutting. Different conditions will produce different sizes and shapes of particles and different types of fractured surface; these factors will be important in determining both the amount of adhesive exudate extracted and, by providing varying amounts of particle surface area and surface roughness, may affect adhesion between particles. Munro (1983) has demonstrated the importance of temperature and deformation rate in this respect, by measuring some tensile properties of frozen and thawed lean beef. Anisotropy of tensile strength according to fibre orientation was far less marked in frozen beef than in thawed. Two distinct fracture behaviours were observed in frozen beef. At high temperatures and low strain rates viscoelastic fracture occurred, whilst at lower temperatures and high strain rates fracture was more of a brittle nature.

The following specific examples of the growing understanding of each of the two unit operations (pre-breaking and comminution) involving meat breakdown serve to illustrate some of these basic considerations.

### *Pre-breaking*

Pre-breaking is the process of taking large meat pieces (especially in the form of frozen blocks, of up to 30 kg each) and producing smaller pieces which can be handled by the primary comminution procedure. Before the meat enters the process, it is usually 'tempered' i.e. brought to a temperature below the initial freezing point where the mechanical properties of the semi-frozen meat are such that the meat can be handled well by the machinery used.

Although the practical effects of using meat either too cold (undertempered) or too warm (overtempered) have been summarized by Bezanson (1975) and Koberna (1986), the influence of tempering and pre-breaking on product quality has received less attention than the later stages of manufacture. What literature there is has been restricted to the use of meat grinders for pre-breaking.

Because frozen meat becomes increasingly brittle with decreasing temperature (Munro, 1983), pre-breaking at low temperatures induces a degree of shattering which results in an increasing proportion of smaller particles. This can be simply but dramatically demonstrated by the number of particles present in a constant mass of sample, as shown in *Figure 13.8*. This effect on particle size is of direct relevance because perception of particle size is a key aspect of texture in this type of product (Dransfield, Jones and Robinson, 1984).

There is good agreement that product cohesiveness is adversely affected when meat is pre-broken by grinding at lower temperatures (Ellery, 1985; Gumpen, 1978; Harbitz and Egelandsdal, 1983; Jolley *et al.*, 1986). Increased fat loss during cooking from products pre-broken at low temperatures, which is of concern to the manufacturer, probably arise through damage to the cells of the adipose tissue (Evans and Ranken, 1975). It is less easy to explain in structural terms why products should be less cohesive since, in the presence of salt, myofibrillar proteins should be easier to extract from smaller pieces (Acton, 1972); one possibility is that during cooking the liquid fat interferes with adhesion between meat pieces.

Allowing meat to temper to a particular temperature prior to pre-breaking requires time and space. Microwave tempering, although not a new development (it was described by Bezanson in 1975), is becoming increasingly common in the UK, and this technology has evolved considerably over the last 10 years (Meredith, 1986). Alternatively, some pre-breaking machines powerful enough to handle deep-frozen 30 kg blocks in their entirety are claimed not to have any adverse effect on bind. Objective information on these claims is not available, although without a better understanding of the cause of reduced cohesion they cannot be dismissed.

### *Primary comminution*

Probably all of the recognized procedures for size reduction of meat have been used to

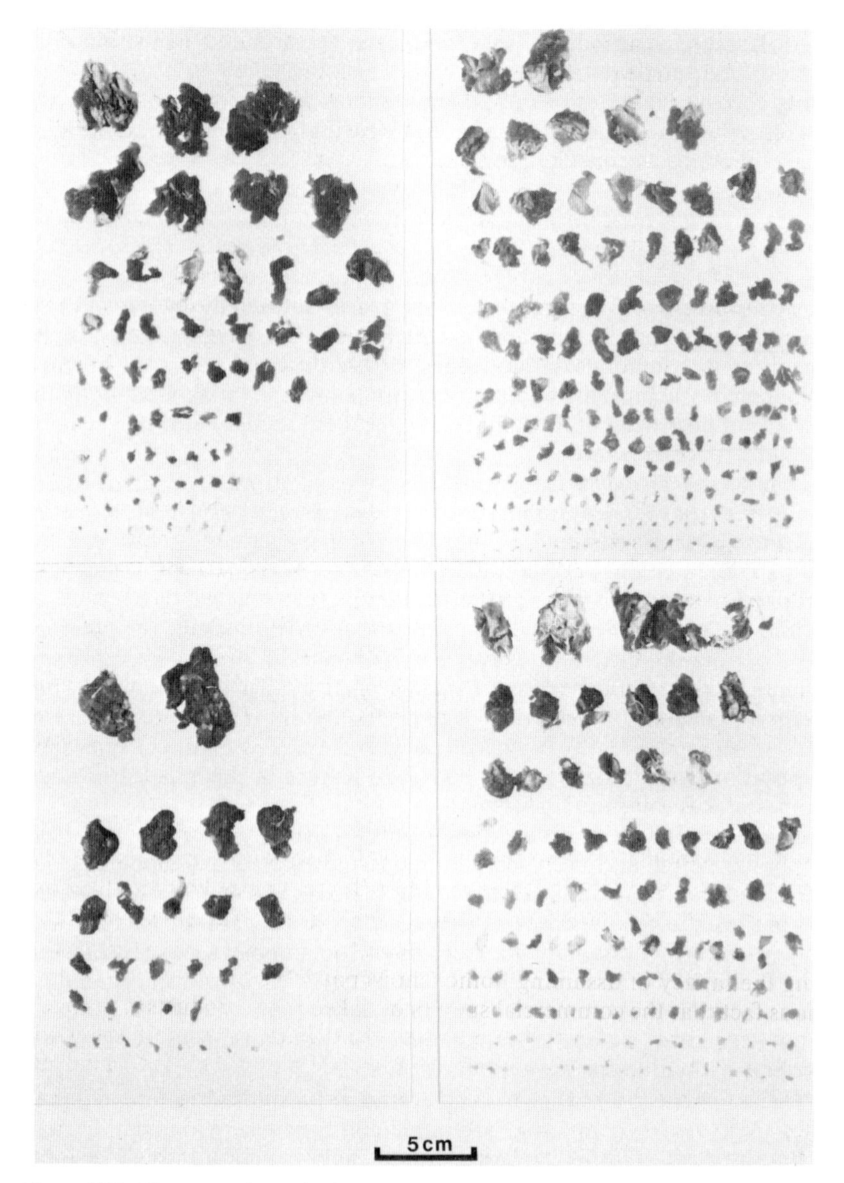


**Figure 13.8** Two examples each of beef forequarter pre-broken by grinding through a kidney plate at –3 °C (left) or –8 °C (right). There are 25 g meat in each photograph.

make reformed meat, even if many of these processes have not gone past the development stage. In the UK, the term 'chopped and shaped' is generally applied to a group of products where the degree of comminution, however effected, has been such that the meat particles in the final product are not easily discernible by eye (UKAFFP, 1975, 1977). To the best of our knowledge, the majority of steak-, chop- and rib-type products include a high proportion of flaked meat. Less favoured methods of comminution include mincing (or grinding) and bowl chopping. We are not aware of either dicing or slicing being used for primary comminution other than in patents (e.g. Bradshaw and Hughes, 1986; Gagliardi, 1985; Huffman, 1980; Mart, 1978; O'Connell, 1985) and in development work, although of the existing methods these would appear to be the ones that lend themselves most easily to produce particles with preferred fibre orientation, and therefore might be expected to produce a more meat-like texture when reformed. Some reformed roasts have been manufactured using flaked beef, whilst reformed diced-type products tend to be made from predominantly flaked or bowl-chopped material; in the latter case, the meat is usually of low quality (such as cheek meat) and is very finely comminuted.

Much research has been published on the manufacture of steaks and roasts from flaked material ('flaked and formed' products), and to a lesser extent slicing. The phrase 'chunked and formed' is used to refer to products with very coarsely comminuted meat pieces, sometimes produced by hand butchery or dicing but more usually by coarse grinding.

*Flaking using the Urschel Comitrol*

The use of Urschel Comitrols to flake meat for subsequent reforming was first advocated in print by Fenters and Ziemba (1971), and the subsequent commercial acceptance of this method of primary comminution has been remarkable. Meat is cut by impelling it against a stationary circle of blades through the action of a high speed impeller. From the claims made (Anon, 1980), it is possible to infer that the principle of this process is to cut connective tissue cleanly (thereby reducing long range toughness) without destroying the underlying anisotropy resulting from myofibrillar structure. Other existing methods of comminution are seemingly less successful at achieving both objectives simultaneously.

If meat is flaked below its initial freezing point (ifp), discrete flakes are produced altogether different in appearance to meat ground at similar temperatures. The difference in cutting action between flaking and grinding is less obvious with chilled or thawed material, particularly with small apertures (under 6 mm) where meat 'strings' are usually produced by either method. *Figure 13.9* illustrates the difference in effect in flaking meat through a variety of apertures above or below the ifp, as well as demonstrating the fallacy of assuming homogenous particle size.

It is a curious fact that the commercial success of flaked meat products is in spite of their generally poor performance in objective studies which compare flaking with grinding (Chesney, Mandigo and Campbell, 1978; Costello *et al.*, 1981; Randall and Larmond, 1977). Perhaps the problem with these sorts of comparison is that like is not truly compared with like. If we assume that the particles produced by flaking are different from those produced by grinding, then it seems illogical to expect them to behave similarly; there is therefore little justification for standardizing on mixing time, for example.

For the same exit aperture and temperatures below ifp, flaking meat produces smaller particles than grinding it (Sheard, personal communication 1987). For a given combination of type of meat, impeller speed, and impeller design, flake size will depend on pre-history (e.g. the method and temperature of pre-breaking), the flake head selected (i.e. the size of the individual exit apertures), and the temperature of flaking. Little attention has so far been paid to the effects of pre-history or flaking temperature, other than comparing meat flaked above or below the ifp (Chesney, Mandigo and Campbell, 1978; Costello *et al.*, 1981; Popenhagen and Mandigo, 1978) when meat flaked at about +2 °C tended to produce reformed products of better

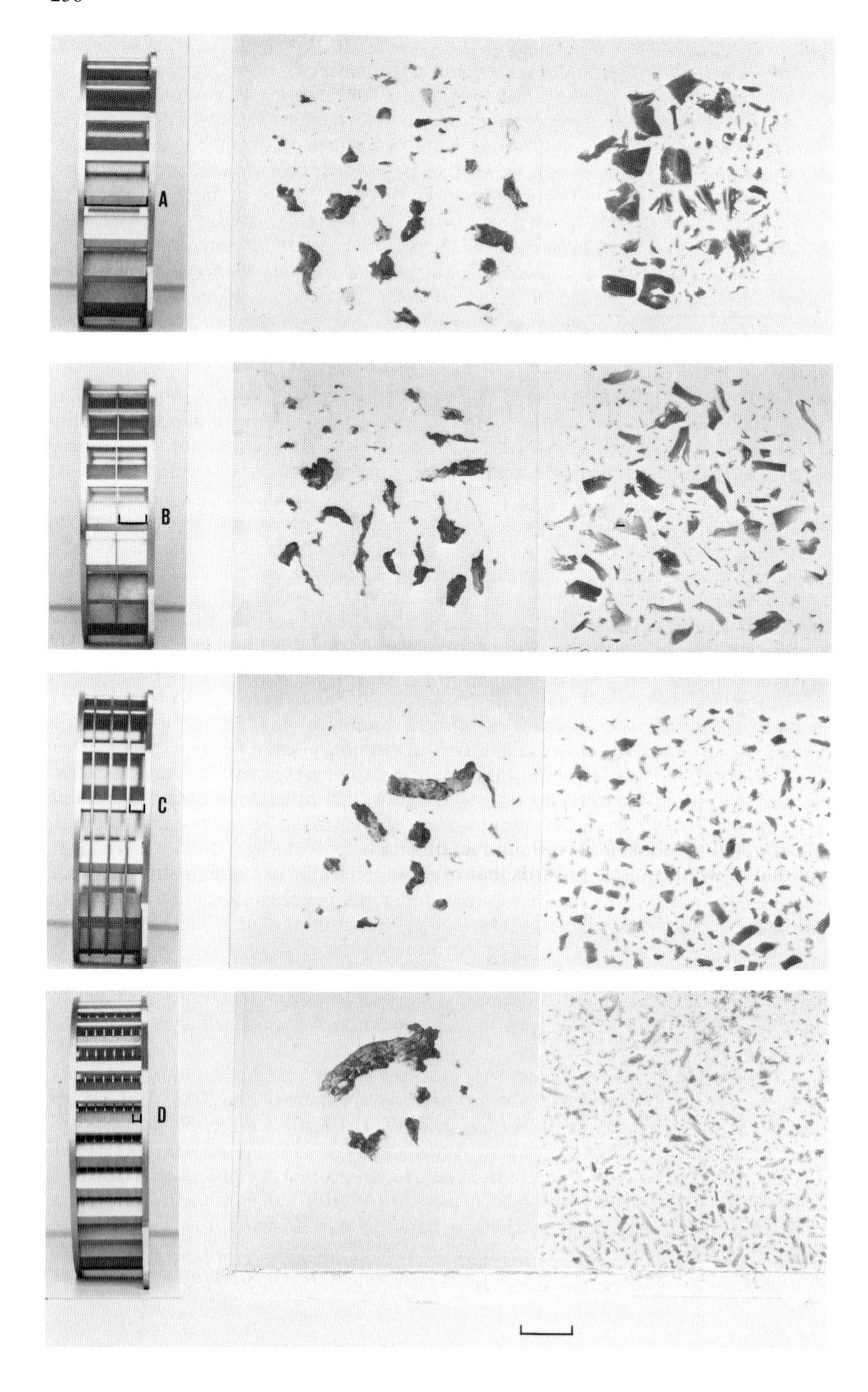
A
B
C
D

quality than products reformed from meat flaked at about –5 °C. Several authors (Cardello *et al.*, 1983; Chesney, Mandigo and Campbell, 1978; Durland *et al.*, 1982; Popenhagen and Mandigo, 1978) have examined the influence of comminuting with different flake heads on eating quality of reformed products. The variety of head sizes, raw materials, processing conditions, nature of sensory test, and sometimes experimental design makes comparison between these studies all but impossible. However there is some evidence that smaller flake head sizes lead to increased disruption of connective tissue in the work of Cardello *et al.* (1983; experiments 2 and 3) and Durland *et al.*, (1982). Fat becomes less detectable visibly in the raw (Durland *et al.*, 1982) or cooked (Cardello *et al.*, 1983; experiment 3) product as aperture size is reduced.

## The success of meat reforming operations

The techniques of reforming meat, however empirically derived or poorly understood, have led to a greater variety of products and better utilization of poorer quality meat or trimmings. Realistically, the only true test of the success of a reforming process is whether it can compete economically in the market place; idealistically, one can ask:

1. Has meat with tough connective tissue been upgraded, and
2. How close to meat of good eating quality is the reformed product?

Very few papers address either question.

The reformed steaks produced by Costello *et al.* (1981) were generally rated better than whole muscle steak from the lean meat raw material (cow inside rounds), and therefore these processes were successful in upgrading this raw material, and the subcutaneous fat which was also included in formulation.

The answer to the second question is, perhaps, disappointingly predictable. The results of Cardello *et al.* (1983)—and, to a lesser extent, Durland *et al.*, (1982)—showed generally that the larger the meat pieces used in the reformed steaks, the closer the product was to whole muscle steak, although with noticeably more chewiness and perceived gristle. These undesirable attributes were reduced with size reduction, but in terms of overall product quality there was a concomitant shift away from that of whole muscle towards that of ground beef. Thus it seems still to be true that the meat used as raw material determines how close the reformed product is to meat of good quality.

## References

ACTON, J.C. (1972). The effect of meat particle size on extractable protein, cooking loss and binding strength in chicken loaves. *Journal of Food Science*, **37**, 240–243

**Figure 13.9** Close-up detail of four flake heads commonly used in our laboratory in the manufacture of flaked and formed reformed meats (left), and comparisons of the corresponding particles produced from beef chuck and blade flaked above the initial freezing point (centre; + 13 °C) and below the initial freezing point (right; –3 °C). The scale bar represents 5 cm for all photographs of meat particles, each of which depicts 25 g meat. Bars in the left column show exit aperture width in mm, A = 40.6, B = 19.1, C = 9.9, D = 4.6.

ACTON, J.C. (1979). Functional properties of sausage raw materials (related to comminuted meat product systems). *Proceedings of 21st Annual Meat Science Institute, University of Georgia, Athens, Georgia*, 14–22

ACTON, J.C. and SAFFLE, R.L. (1969). Preblended and prerigor meat in sausage emulsions. *Food Technology*, **23**, 367–371

ANONYMOUS (1980). *Facts, Flakes and Fabricated Meats.* Valparaiso, Indiana, Urschel Laboratories

ASGHAR, A., SAMEJIMA, K. and YASUI, T. (1985). Functionality of muscle proteins in gelation mechanisms of structured meat products. *CRC Critical Reviews in Food Science and Nutrition*, **22**, 27–106

ATTECK, L.A.G. (1983). Meat Protein Product and Process. US Patent 4 406 831

AWAD, A., POWRIE, W.D. and FENNEMA, O. (1968). Chemical deterioration of frozen bovine muscle at –4 °C. *Journal of Food Science*, **33**, 227–234

BAGSHAW, C.R. (1982). *Muscle Contraction.* London, Chapman and Hall

BARD, J.C. (1965). Some factors influencing extractability of salt-soluble proteins. In *Proceedings of Meat Industry Research Conference*, (Koliar, O.E.and Aunan, W.J. Eds), pp. 96–98. Chicago, American Meat Science Association and American Meat Institute Foundation

BENDALL, J.R. (1954). The swelling effect of polyphosphates on lean meat. *Journal of the Science of Food and Agriculture*, **5**, 468–475

BENDALL, J.R. (1967). The elastin content of various muscles of beef animals. *Journal of the Science of Food and Agriculture*, **18**, 553–558

BERRY, B.W., SMITH, J.J., SECRIST, J.L and ELGASIM, E.A. (1986). Effects of pre-rigor pressurisation, method of restructuring and salt level on characteristics of restructured beef steaks. *Journal of Food Science*, **51**, 781–785

BENZANSON, A. (1975). Thawing and tempering frozen meat. *Proceedings of the 11th Meat Industry Research Conference*, pp. 51–62. Chicago, AMIF

BMMA (Bacon and Meat Manufacturers' Association)(1985). *Code of Practice on the Labelling of 'Re-formed' Cured Meat Products.* London, BMMA

BOOREN, A.M., MANDIGO, R.W., OLSON, D.G. and JONES, K.W. (1981a). Vacuum mixing influence on characteristics of sectioned and formed beef steaks. *Journal of Food Science*, **46**, 1673–1677

BOOREN, A.M., JONES, K.W., MANDIGO, R.W. and OLSON, D.G. (1981b). Effects of blade tenderisation, vacuum mixing, salt addition and mixing time on binding of meat pieces into sectioned and formed beef steaks. *Journal of Food Science*, **46**, 1678–1680

BOOREN, A.M., MANDIGO, R.W., OLSON, D.G. and JONES, K.W. (1981c). Effect of muscle type and mixing time on sectioned and formed beef steaks. *Journal of Food Science*, **46**, 1665–1672

BOOREN, A.M., MANDIGO, R.W., OLSON, D.G. and JONES, K.W. (1982). Characterisation of exudate proteins involved in binding meat pieces into a sectioned and formed beef steak. *Journal of Food Science*, **47**, 1943–1947

BOUTON, P.E., HARRIS, P.V. and SHORTHOSE, W.R. (1975). Possible relationships between shear, tensile and adhesion properties of meat and meat structure. *Journal of Texture Studies*, **6**, 297–314

BRADSHAW, N.J. and HUGHES, D. (1986). Meat Product. European Patent Application 0 175 397

BREIDENSTEIN, B.C. (1982). *Intermediate Value Beef Products.* Chicago, National Livestock and Meat Board

CALLOW, E.H. (1931). The solubility of the proteins of pork-muscle in salt solutions at 0 °C. *Report of the Food Investigation Board*, 144–145

CARDELLO, A.V., SEGARS, R.A., SECRIST, J., SMITH, J., COHEN, S.H. and ROSENKRANS, R.C. (1983). Sensory and instrumental texture properties of flaked and formed beef. *Food Microstructure*, **2**, 119–133

CASSIDY, R.D., OCKERMAN, H.W., KROL, B., VAN ROON, P.S., PLIMPTON, R.F.JNR. and CAHILL, V.R. (1978). Effect of tumbling method, phosphate level and final cook temperature on histological characteristics of tumbled porcine muscle tissue. *Journal of Food Science*, **43**, 1514–1518

CHENG, C.S. and PARRISH, F.C. JNR. (1978). Molecular changes in the salt-soluble myofibrillar proteins of bovine muscle. *Journal of Food Science*, **43**, 461–463, 487

CHESNEY, M.J., MANDIGO, R.W. and CAMPBELL, J.F. (1978). Properties of restructured pork product as influenced by meat particle size, temperature and comminution method. *Journal of Food Science*, **43**, 1535–1537

COSTELLO, W.J., SEIDEMAN, S.C., MICHELS, J.D. and QUENZER, N.M. (1981). Effect of comminution method and pressure on restructured beef steaks. *Journal of Food Protection*, **44**, 425–429

DRANSFIELD, E. (1977). Intramuscular composition and texture of beef muscles. *Journal of the Science of Food and Agriculture*, **28**, 833–842

DRANSFIELD, E., JONES, R.C.D. and ROBINSON, J.M. (1984). Development and application of a texture profile for UK beefburgers. *Journal of Texture Studies*, **15**, 337–356

DURLAND, P.R., SEIDEMAN, S.C., COSTELLO, W.J. and QUENZER, N.M. (1982). Physical and sensory properties of restructured beef steaks formulated with various flake sizes and mixing times. *Journal of Food Protection*, **45**, 127–131

ELLERY, A. (1985). Changes in ice-content during the manufacture of grill-steaks and their influence on product quality. MSc Thesis, University of Bristol.

EVANS, G.G. and RANKEN, M.D. (1975). Fat cooking losses from non-emulsified meat products. *Journal of Food Technology*, **10**, 63–72

EVANS, G.G., LEWIS, D.F. and RANKEN, M.D. (1979). Cooking losses in meat products Part VII. Some effects of tumbling pork leg meat. *British Food Manufacturing Industries Research Association Research Report*, Number 320, Leatherhead, Surrey

FENTERS, W. and ZIEMBA, J.V. (1971). New way to fabricate meats. *Food Engineering*, January, 64–65

FORD, A.L., JONES, P.N., MACFARLANE, J.J., SCHMIDT, G.R. and TURNER, R. (1978). Binding of meat pieces: objective and subjective assessment of restructured steakettes containing added myosin and/or sarcoplasmic protein. *Journal of Food Science*, **43**, 815–818

FRANKLIN, F. and CROSS, H.R. (Eds) (1982). *Proceedings of the International Symposium, Meat Science and Technology*, pp. 222–309. Chicago, National Livestock and Meat Board

FUKAZAWA, T., HASHIMOTO, Y. & YASUI, T. (1961). Effect of some proteins on the binding quality of an experimental sausage. *Journal of Food Science*, **26**, 541–549

GAGLIARDI, E.D. (1985). Restructured Meat Products and Methods of Making the Same. UK Patent Application GB2 156 650A

GILLETT, T.A., BROWN, C.L., LEUTZINGER, R.L., CASSIDY, R.D. and SIMON, S. (1978). Tensile strength of processed meats determined by an objective Instron technique. *Journal of Food Science*, **43**, 1121–1124

GILLETT, T.A., MEIBURG, D.E., BROWN, C.L. and SIMON, S. (1977). Parameters affecting meat protein extraction and interpretation of model system data for meat emulsion formation. *Journal of Food Science*, **42**, 1606–1610

GRABOWSKA, J. and HAMM, R. (1979). Protein solubility and water binding under the conditions obtaining in Brühwurst mixtures IV. Effects of NaCl concentration, pH value and diphosphate. *Fleischwirtschaft*, **59**, 1166–1172

GUMPEN, S.A. (1978). Grinding of deep-frozen meat. Effect on binding properties. *Proceedings of the 24th European Meeting of Meat Research Workers*, Kulmbach, D7:1–D7:6

HAMM, R. and GRABOWSKA, J. (1979). Protein solubility and water binding under the conditions existing in Brühwurst mixtures V. Post mortem changes—Conclusions. *Fleischwirtschaft*, **59**, 1338–1344

HANSON, J. and HUXLEY, H.E. (1953). Structural basis of the cross-striations in muscle. *Nature*, **172**, 530–532

HARBITZ, O. and EGELANDSDAL, B. (1983). Technological properties of frozen meat ground on a frozen meat grinder. *Proceedings of the 29th European Meeting of Meat Research Workers*, 313–319

HELANDER, E. (1957). On quantitative muscle protein determination. *Acta Physiologica Scandinavica*, **41**(Supplement), 141

HELLENDORN, B.W. (1962). Water-binding capacity of meat as affected by phosphates. I. Influence of sodium chloride and phosphates on the water retention of comminuted meat at various pH values. *Food Technology*, **16**, 119–124

HERMANSSON, A-M., HARBITZ, O. and LANGTON, M. (1986). Formation of two types of gels from bovine myosin. *Journal of the Science of Food and Agriculture*, **37**, 69–84

HUFFMAN, D.L. (1980). Process for Production of a Restructured Fresh Meat Product. US Patent 4 210 677

HUFFMAN, D.L., MCCAFFERTY, D.M., CORDAY, J.C. and STANLEY, M.H. (1984). Restructured beef steaks from hot and cold boned carcasses. *Journal of Food Science*, **49**, 164–167

HUXLEY, H.E. and HANSON, J. (1957). Quantitative studies on the structure of cross-striated myofibrils I. Investigations by interference microscopy. *Biochimica et Biophysica Acta*, **23**, 229–249

ISHIOROSHI, M., SAMEJIMA, K. and YASUI, T. (1979). Heat induced gelation of myosin: factors of pH and salt concentrations. *Journal of Food Science*, **44**, 1280–1283

ISHIOROSHI, M., SAMEJIMA, K., ARIE, Y. and YASUI, T. (1980). Effect of blocking the myosin–actin interaction in heat-induced gelation of myosin in the presence of actin. *Agricultural and Biological Chemistry*, **44**, 2185–2194

JESPERSEN, K. and ENGMAN, T.E. (1977). Process for Pickling Meat Sections. US Patent 4 038 426

JOLLEY, P.D., ELLERY, A., HALL, L. and SHEARD, P.R. (1986). Keeping your temper—why it is important for the manufacturer. *Proceedings of the Subject Day: 'Meat Thawing/Tempering and Product Quality'*. Institute of Food Research, Bristol

JOLLEY, P.D. and SAVAGE, A.W.J. (1985). Protein solubilisation and adhesion in reformed meats. Proceedings of the Reiner Hamm Symposium 'Current Meat Research', *Mitteilungsblatt der Bundesanstalt fur Fleischforschung, Kulmbach*, **91**, 6781–6788

KINLOCH, A.J. (1980). The science of adhesion, Part 1. Surface and interfacial aspects. *Journal of Materials Science*, **15**, 2141–2166

KINLOCH, A.J. (1982). The science of adhesion, Part 2. Mechanics and mechanisms of failure. *Journal of Materials Science*, **17**, 617–651

KOBERNA, F. (1986). Tempering temperature requirements for cutting and processing equipment. *Proceedings of the Subject Day: 'Meat Thawing/Tempering and Product Quality'*. Institute of Food Research, Bristol

KOTTER, L. (1960). *Zur Wirkung kondensierter Phosphate und anderer Salze auf tierisches Eiwei.* Hannover, Verlag M u H Schaper

KOTTER, L. and FISCHER, A. (1975). The influence of phosphates or polyphosphates on the stability of foams and emulsions in meat technology. *Fleischwirtschaft*, **55**, 365–368

KOTTER, L. and PRÄNDL, O. (1956). Einfluss der Fleisch-zerkleinerung auf die Zustandsänderung der fibrillären Muskeleiweisskörper. *Fleischwirtschaft*, **8**, 688–689

LANGEN, C.P. (1977). Method of and Apparatus for Treating Meat, More Particularly Ham Meat. US Patent 4 029 824

LAWRIE, R.A. (1985). *Meat Science*, 4th Edition. Oxford, Pergamon Press

LEWIS, D.F. (1979). Meat products. In *Food Microscopy*, (Vaughan, J.G., Ed.), pp. 233–272. London, Academic Press

LEWIS, D.F. and JEWELL, G.G. (1975). Structural alterations produced on processing meat. Part II. The effect of heat and polyphosphate brines on meat structure. *British Food Manufacturing Industries Research Association Research Report* Number 212, Leatherhead, Surrey

LEWIS, D.F., GROVES, K.H.M. and HOLGATE, J.H. (1986). Action of polyphosphates in meat products. *Food Microstructure*, **5**, 53–62

LIGHT, N.D., CHAMPION, A.E., VOYLE, C. and BAILEY, A.J. (1985). The role of epimysial, perimysial and endomysial collagen in determining texture in six bovine muscles. *Meat Science*, **13**, 137–149

MAAS, R.H. (1963). Processing Meat. US Patent 3 076 713

MACFARLANE, J.J. (1974). Pressure-induced solubilization of meat proteins in saline solution. *Journal of Food Science*, **39**, 542–547

MACFARLANE, J.J., SCHMIDT, G. R. and TURNER, R.H. (1977). Binding of meat pieces: a comparison of myosin, actomyosin and sarcoplasmic proteins as binding agents. *Journal of Food Science*, **42**, 1603–1605

MACFARLANE, J.J., TURNER, R.H. and JONES, P.N. (1986). Binding of meat pieces: influence of some processing factors on binding strength and cooking losses. *Journal of Food Science*, **51**, 736–741

MACFARLANE, J.J., MCKENZIE, I.J. and TURNER, R.H. (1984). Binding of comminuted meat: Effect of high pressure. *Meat Science*, **10**, 307–320

MART, C. (1978). Meat Product and Method of Making Same. US Patent 4 072 763

MCGOWAN, R.G. (1970). Method of Preparing a Poultry Product. US Patent 3 503 755

MEANS, W.J. and SCHMIDT, G.R. (1986). Algin/calcium gel as a raw and cooked binder in structured beef steaks. *Journal of Food Science*, **51**, 60–65

MEREDITH, R.J. (1986). Tempering and thawing of meat, fish and poultry using microwave energy. *Proceedings of the Subject Day: 'Meat Thawing/Tempering and Product Quality'*. Institute of Food Research, Bristol

MILLER, A.J., ACKERMAN, S.A. and PALUMBO, S.A. (1980). Effects of frozen storage on functionality of meat for processing. *Journal of Food Science*, **45**, 1466–1471

MUNRO, P.A. (1983). Tensile properties of frozen and thawed lean beef. *Meat Science*, **9**, 43–61

NAKAYAMA, T. and SATO, Y. (1971a). Relationships between binding quality of meat and myofibrillar proteins III. Contribution of myosin A and actin to rheological properties of heated minced-meat gel. *Journal of Texture Studies*, **2**, 75–88

NAKAYAMA, T. and SATO, Y. (1971b). Relationships between binding quality of meat and myofibrillar proteins IV. Contribution of native tropomyosin and actin in myosin B to rheological properties of heat set minced-meat gel. *Journal of Texture Studies*, **2**, 475–488

O'CONNELL, P.M. (1985). Structured Meat Product. US Patent 4 544 560

OFFER, G. and TRINICK, J. (1983). On the mechanism of water holding in meat: the swelling and shrinking of myofibrils. *Meat Science*, **8**, 245–281

OHTSUKI, I., MARUYAMA, K. and EBASHI, S. (1986). Regulatory and cytoskeletal proteins of vertebrate skeletal muscle. *Advances in Protein Chemistry*, **38**, 1–67

PAARDEKOOPER, E.J.C. (1986). *Slaughtering Techniques in the Future*. Netherlands Centre for Meat Technology Department, Zeist

PEARSON, A.M. and TAUBER, F.W. (1984). *Processed Meats*, 2nd Edition. Westport, Connecticut, AVI Publishing Company

PENFIELD, M.P., BARKER, C.L. and MEYER, B.H. (1976). Tensile properties of beef semitendinosus muscle as affected by heating rate and end point temperature. *Journal of Texture Studies*, **7**, 77–85

PEPPER, F.H. and SCHMIDT, G.R. (1975). Effect of blending time, salt, phosphate and hot-boned beef on binding strength and cook yield of beef rolls. *Journal of Food Science*, **40**, 227–230

POPENHAGEN, G.R. and MANDIGO, R.W. (1978). Properties of restructured pork as affected by flake size, flake temperature and blend combinations. *Journal of Food Science*, **43**, 1641–1645

PURSLOW, P.P. (1985). The physical basis of meat texture: observations on the fracture behaviour of cooked bovine *M. semitendinosus*. *Meat Science*, **12**, 39–60

PURSLOW, P.P. (1987). The fracture properties and thermal analysis of collagenous tissue. In *Advances in Meat Research*, Volume 4, (Pearson, A.M., Dutson, T.R. and Bailey, A.J., Eds), Chapter 10. Westport, Connecticut, AVI Publishing Co.

PURSLOW, P.P., DONNELLY, S.M. and SAVAGE, A.W.J. (1987). Variations in the tensile adhesive strength of meat–myosin junctions due to test configurations. *Meat Science*, **19**, 227–242

RAHELIĆ, S. and MILIN, J. (1979). Electron microscopic findings of the changes in the muscle under the influence of brine. *Fleischwirtschaft*, **59**, 971–972

RAHELIĆ, S., PRIBIS, V. and VICEVIĆ, Z. (1974). Influence of mechanical treatment of cured muscles on some characteristics of pasteurized canned pork. *Proceedings of the 20th European Meeting of Meat Research Workers*, 133–135

RANDALL, C.J. and LARMOND, E. (1977). Effect of method of comminution (flake-cutting and grinding) on the acceptability and quality of hamburger patties. *Journal of Food Science*, **42**, 728–730

REGENSTEIN, J.W. and RANK STAMM, J. (1979). Factors affecting the sodium chloride extractability of muscle proteins from chicken breast, trout white and lobster tail muscles. *Journal of Food Biochemistry*, **3**, 191–204

REICHERT, J.E. (1986). Cooked products. *Proceedings of the 32nd European Meeting of Meat Research Workers*, 327–331

REICHERT, J.E., FARBER, D. and FLACHMAN, A. (1984). On the cohesion of boiled ham slices. *Fleischerei* (10) V—VII, (11) VII and VIII, (12) V and VI (1985) (1) V–VIII

REJT, J., KUBICKA, H. and PISULA, A. (1978). Changes of physical and chemical properties and of histological structure of meat subjected to massage under vacuum. *Meat Science*, **2**, 145–153

ROWE, R.W.D. (1974). Collagen fibre arrangement in intramuscular connective tissue. *Journal of Food Technology*, **9**, 501–508

SAFFLE, R.L. (1968). Meat emulsions. *Recent Advances in Food Science*, **16**, 105–160

SAFFLE, R.L. and GALBREATH, J.W. (1964). Quantitative determination of salt-soluble protein in various types of meat. *Food Technology*, **18**, 1943–1944

SAMEJIMA, K., ISHIOROSHI, M. and YASUI, T. (1981). Relative roles of the head and tail portions of the molecule in heat-induced gelation of myosin. *Journal of Food Science*, **46**, 1412–1418

SCHLAMB, K.F. (1970). Methods of Binding Large Pieces of Poultry. US Patent 3 499 767

SCHMIDT, G.R. and MEANS, W.J. (1986). Process for Preparing Algin/Calcium Gel Structured Meat Products. US Patent 4 603 054

SCHMIDT, G.R. and TROUT, G.R. (1984). The chemistry of meat binding. In *Recent Advances in the Chemistry of Meat*, Volume 12, (Bailey, A.J., Ed.), pp. 231–245. London, Royal Society of Chemistry

SCOPES, R.K. (1970). Characterization and study of sarcoplasmic proteins. In *The Physiology and Biochemistry of Muscle as a Food*, Volume 2, (Briskey, E.J., Cassens, R.G. and Marsh, B.B., Eds), pp. 471–492. Madison, University of Wisconsin Press

SEIFFHART, J.B. (1983). Process for Treating Meat. US Patent 4 409 704

SIEGEL, D.G. and SCHMIDT, G.R. (1979a). Crude myosin fractions as meat binders. *Journal of Food Science*, **44**, 1129–1131

SIEGEL, D.G. and SCHMIDT, G.R. (1979b). Ionic, pH and temperature effects on the binding ability of myosin. *Journal of Food Science*, **44**, 1686–1689

SIEGEL, D.G., CHURCH, K.E. and SCHMIDT, G.R. (1979). Gel structure of non-meat proteins as related to their ability to bind meat pieces. *Journal of Food Science*, **44**, 1276–1279

SIEGEL, D.G., THENO, D.M. and SCHMIDT, G.R. (1978). Meat massaging: The effects of salt, phosphate and massaging on the presence of specific skeletal muscle proteins in the exudate of a sectioned and formed ham. *Journal of Food Science*, **43**, 327–330

SIEGEL, D.G., TULEY, W.B. and SCHMIDT, G.R. (1979). Microstructure of isolated soy protein in combination ham. *Journal of Food Science*, **44**, 1272–1275

SIEGEL, D.G., THENO, D.M., SCHMIDT, G.R. and NORTON, H.W. (1978). Meat massaging: the effects of salt, phosphate and massaging on cooking loss, binding strength and exudate composition in sectioned and formed ham. *Journal of Food Science*, **43**, 331–333

SOLOMON, L.W. and SCHMIDT, G.R. (1980). Effect of vacuum and mixing time on the extractability and functionality of pre- and postrigor beef. *Journal of Food Science*, **45**, 283–287

SWIFT, C.E. and ELLIS, R. (1956). The action of phosphates in sausage products 1. Factors affecting the water retention of phosphate-treated ground meat. *Food Technology*, **10**, 546–552

SZENT-GYÖRGYI, A. (1951). *Chemistry of Muscular Contraction*, 2nd Edition. New York, Academic Press

THENO, D.M., SIEGEL, D.G. and SCHMIDT, G.R. (1977). Meat massaging techniques. *Proceedings of the Meat Industry Research Conference*, pp. 53–68. Chicago, AMIF

THENO, D.M., SIEGEL, D.G. and SCHMIDT, G.R. (1978a). Meat massaging: effect of salt and phosphate on the microstructural composition of the muscle exudate. *Journal of Food Science*, **43**, 483–487

THENO, D.M., SIEGEL, D.G. and SCHMIDT, G.R. (1978b). Meat massaging: effects of salt and phosphate on the ultrastructure of cured porcine muscle. *Journal of Food Science*, **43**, 488–492

TRAUTMAN, J.C. (1964). Fat-emulsifying properties of prerigor and postrigor pork proteins. *Food Technology*, **18**, 1065–1066

TRAUTMAN, J.C. (1966). Effect of temperature and pH on the soluble proteins of ham. *Journal of Food Science*, **31**, 409–418

TROUT, G.R. and SCHMIDT, G.R. (1984). Effect of phosphate type and concentration, salt level and method of preparation on binding in restructured beef rolls. *Journal of Food Science*, **49**, 687–694

TURNER, R.H., JONES, P.N. and MACFARLANE, J.J. (1979). Binding of meat pieces: an investigation of the use of myosin containing extracts from pre and post-rigor bovine muscle as meat binding agents. *Journal of Food Science*, **44**, 1443–1446

UKAFFP (United Kingdom Association of Frozen Food Producers) (1975). Code of Practice for the labelling of certain quick frozen meat products made from or containing substantial quantities of reformed meat. London, UKAFFP

UKAFFP (United Kingdom Association of Frozen Food Producers) (1977). Memorandum of interpretation, AM/13/77. London, UKAFFP

VAN DEN OORD, A.H.A. and WESDORP, J.J. (1978a). Solubility of meat proteins: interdependence of pH, sodium chloride, pyrophosphate and intrinsic properties of the muscle. *Proceedings of the 24th Meeting of European Meat Research Workers*, Kulmbach, D12:1–D12:6.

VAN DEN OORD, A.H.A. and WESDORP, J.J. (1978b). The specific effect of pyrophosphate on protein solubility of meat. *Proceedings of the 24th Meeting of European Meat Research Workers,* Kulmbach, D13:1–D13:5

VOYLE, C.A., JOLLEY, P.D. and OFFER, G.W. (1984). The effect of salt and pyrophosphate on the structure of meat. *Food Microstructure*, **3**, 113–126

WEBER, H.H. and PORTZEHL, H. (1952). Muscle contraction and fibrous muscle proteins. *Advances in Protein Chemistry*, **7**, 161–252

YASUI, T., ISHIOROSHI, M. and SAMEJIMA, T. (1980). Heat induced gelation of myosin in the presence of actin. *Journal of Food Biochemistry*, **4**, 61–78

YASUI, T., SAKANISHI, M. and HASHIMOTO, Y. (1964). Effect of inorganic polyphosphates on the solubility and extractability of myosin B. *Agricultural and Food Chemistry*, **12**, 392–398

YASUI, T., FUKAZAWA, T., TAKAHASHI, K., SAKANISHI, M. and HASHIMOTO, Y. (1964). Specific interaction of inorganic polyphosphates with myosin B. *Agricultural and Food Chemistry*, **12**, 399–404

YASUI, T., ISHIOROSHI, M., NAKANO, H. and SAMEJIMA, K. (1979). Changes in shear modulus, ultrastructure and spin-spin relaxation times of water associated with heat-induced gelation of myosin. *Journal of Food Science*, **44**, 1201–1204

14

# SURIMI-BASED FOODS—THE GENERAL STORY AND THE NORWEGIAN APPROACH

K. FRETHEIM*
*The Innovation Centre, Oslo*
B. EGELANDSDAL
*Norwegian Food Research Institute, Ås*
E. LANGMYHR
*Norwegian Herring Oil and Meal Industry Research Institute, Bergen*
O. EIDE and R. OFSTAD
*Institute of Fishery Technology Research, Tromsø, Norway*

## Introduction

This chapter can be divided into three parts:

1. An introduction to 'surimi';
2. Structure formation in surimi and surimi-based foods;
3. The objectives and scope of current Norwegian work with respect to surimi and products derived from it.

Surimi is a Japanese development and the majority of work carried out on surimi and surimi products has been done by the Japanese. However, since 1983 much work has been carried out in Norway—both scientific and commercially oriented—which should lead to further significant developments in surimi technology. This will be discussed at the end of this chapter.

## What *is* surimi?

The quick and easy definition of surimi is *washed fish mince*. This is perhaps a deceptively simple definition and it is helpful to expand on each of the three words.

### *Fish*

Although more than 60 species are used for *fresh* surimi, 95% of *frozen* surimi is made from Alaska pollock (*Theragra charcogramma*), of which 40–45% is shore-processed and 50–55% is ship-processed. Alaska pollock offers the right physicochemical properties, and it is there to be harvested in an annual amount of about two million tons.

### *Mince*

It is important to realize that the production of surimi on an industrial scale requires the mechanical removal of head, guts, skin and bones prior to mincing. This rather obvious statement carries an important implication since for a fish to be mechanically

*Present address: Dyno Industrier A.S., Oslo.

decapitated and eviscerated it has to be of reasonable size, and huge schools of fish are not used for surimi because the fish are too small. This will be discussed in more detail later.

### *Washed*

The fish flesh is washed 1–5 times with fresh, cold water. Thorough washing is a key step in the surimi production process. The mechanically separated fish flesh is repeatedly washed until it becomes odourless and colourless as a range of water-soluble substances are removed. The extent of the washing (water volume, number of cycles) is determined from consideration of fish species, condition of the fish, desired quality of the surimi, and efficiency of the washing unit on hand. The acceptable maximum temperature of the wash water depends on the fish species, specifically on the thermal stability of the myofibrillar fish proteins.

When processing at sea only one washing cycle is used, compared with 3–5 cycles in shore-based processing. There are good reasons for surimi produced on board ship to have undergone only one cycle of washing, one reason being that fresh water needs to be generated by desalinization. The fact that the raw material is fresh to the point of flopping when entering the process is also highly advantageous, and the removal of belly flaps and backbone in a separate operation allows for a mince that is easily purified by washing. Finally, modern processing lines include a refiner which removes residual black skin, bone and scale by sifting prior to dehydration by pressing.

The above provides a reasonable introduction to what surimi *is*. There is, however, one particular aspect of its commercial production which warrants special mention. In spite of the clean-up during processing, the fresh material normally becomes unacceptable after about three days. Frozen storage is therefore essential and it is important to realize that surimi production could never have attained its present-day proportions without the discovery of additives which allow frozen storage while retaining functional properties.

Initially as a result of Japanese work sucrose at the 8% level was used as the cryoprotectant but this was then changed to a combination of 4% sucrose and 4% sorbitol to alleviate the sweetness and colour change (on storage) caused by sucrose alone (Tamoto *et al.*, 1961). The details of the mechanisms by which sugars exert their protective effect will not be discussed here. Suffice it to say that protein stabilization has been attributed to the ability of the sugars to increase the amount of bound water on the protein molecule and, perhaps more importantly, to increase the surface tension of water (Arakawa and Timasheff, 1982).

It has been widely accepted that polyphosphates enhance the cryoprotective effect of sugars, and 0.2% of tri- or pyrophosphate is presently used commercially. However, Lanier and Hamann (1985) maintain that polyphosphates merely improve gel formation and water binding during the manufacture of surimi-based products, i.e. addition after frozen storage may have the same effect. *Figure 14.1.* outlines the various steps in the production of stabilized surimi.

Even when cryoprotective agents are added, for long-term frozen storage to have no detrimental effect on the gel-forming ability of surimi, certain requirements must be met. According to acknowledged production practice, the fish must be no more than 1–2 days old and in good condition when entering the processing line. However, recent findings (Holmes, 1986) suggest that if the catch is kept at 1.5 °C or lower and

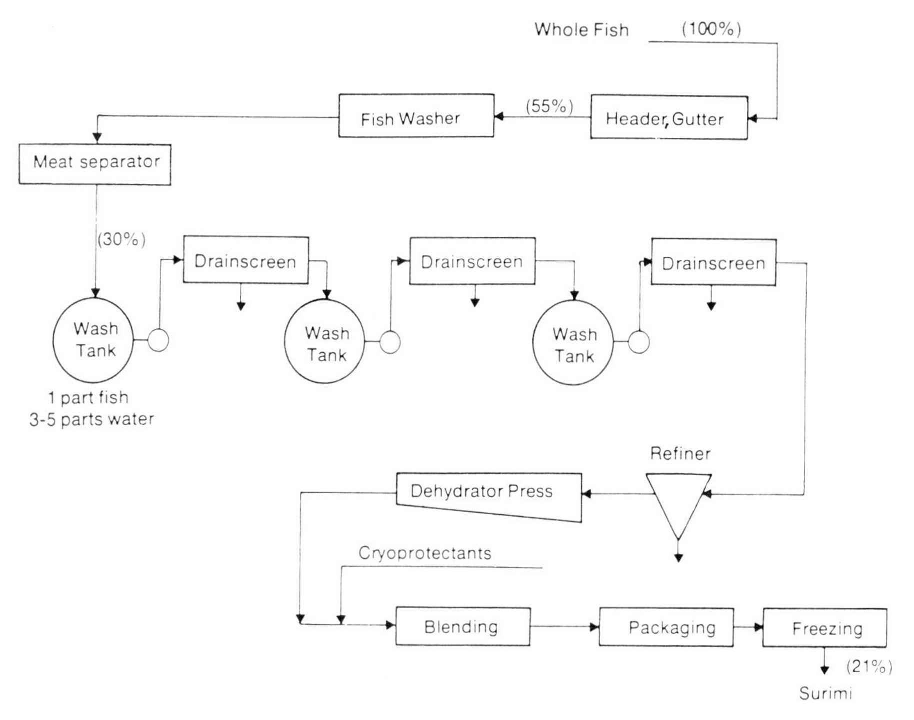


**Figure 14.1** Outline of the surimi production process, indicating yields.

an improved washing system is used, Alaska pollock as old as 115 h may yield high quality surimi. In any event, the storage temperature needs to be constant, below −20 °C (Iwata *et al.*, 1968, 1971). At −10 °C the surimi is useless after about three months whatever the quality of the original raw material (Iwata *et al.*, 1971). Addition of salt improves storage stability but may give rise to undesirable gelation during storage (Iwata and Okada, 1971).

## Structure formation in surimi and surimi-based foods

### *Quality evaluation of surimi: gel-forming properties*

The gel-forming properties of surimi are measured on a model product. The surimi is mixed with 2–3% salt and the mixture is heated, usually at 90 °C, to form a gel; other temperatures may be used, depending on the aim of the investigation. It is a fact that large deformations are the most relevant with respect to extrusion processing and consumption of final products, so most of the grading methods involve large deformation techniques of various kinds.

In the Japanese grading system the texture of a surimi gel is determined by three tests: a puncture test, a sensory test, and a folding test. For the puncture test, the sample (2.5 cm high, 3 cm in diameter) is penetrated axially with a plunger (5 mm in diameter) until the gel breaks; gel strength is given as the product of force times the distance at failure. The sensory test addresses both strength and elasticity of the gel and requires trained panelists. The folding test is a crude but simple test: a 3 mm thick slice is folded, first into halves, then into quarters, and the ability to withstand crack formation determines the grade.

A tensile test has also been developed (Suzuki, 1981): a ring is stamped out from a slice of the surimi gel and subsequently stretched between two hooks until breakage occurs. The gel strength, or breaking strength, calculated from the force and distance of stretch at failure, correlates well with sensory texture as determined by a panel.

The Japanese convention of expressing gel strength as the product of force and deformation obscures the relative contribution of these two variables; a soft and cohesive gel may give the same value for gel strength as a firm and brittle gel.

A torsion test is recommended by Lanier, Hamann and Wu (1985) for specifying the gel-forming properties of surimi: a cylindrical test specimen is given the shape of a dumbbell and subsequently twisted (*Figure 14.2*). The twisting moment and twist (angle) at failure are used to calculate shear stress and strain. Thus, this method gives two basic rheological parameters as opposed to the Japanese methods which are empirical. Results from the torsion test are clearly better correlated with sensory textural measurements than are results from the puncture test. Significant errors may arise from penetration measurements when evaluating samples of high gel-forming ability; gels of medium strength yield results from penetration and stress/strain measurements, respectively, that are well correlated.

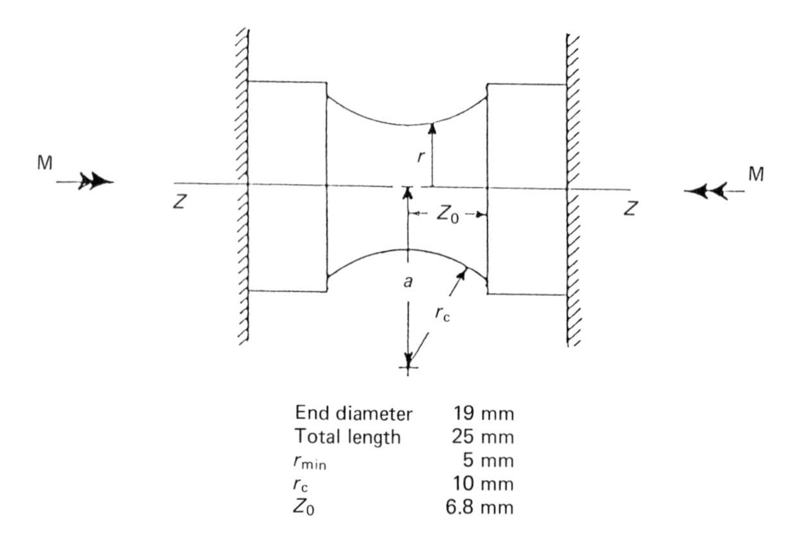


**Figure 14.2** Specification of sample shape for the torsion test of surimi gel quality.

The most promising criterion for classifying gels is the relationship between torsional rigidity (stress/strain) and strain, both determined at failure. Rigidity is the most sensitive indicator of concentration effects while strain is more sensitive to the gel-forming ability of the protein (Lanier, Hamann and Wu, 1985). Thus, the functional properties of surimi can be conveniently specified. Preparing test gels at three temperatures helps to describe surimi properties more fully: preincubation at 40 °C to detect the ability to form gel at low temperatures, heating at 60 °C as a check for the presence of proteolytic enzymes, and heating at 90 °C to assess the gel formation obtainable during rapid heating.

### *Gel formation of surimi: fundamental and practical aspects*

The definition of surimi discussed initially is rather academic. In commercial/practical terms surimi is an intermediate product which can easily be processed into various ready-to-eat products. When mixed with water, salt and other ingredients surimi gives a paste, and the quality of the surimi determines the ease with which the paste can be shaped, as well as the textural properties resulting from gel formation of the final product.

At the molecular level the actomyosin and myosin complexes/molecules in the aqueous phase of surimi (or other minced meat) are essential with regard to gel-forming properties and (water) binding. In post-rigor myofibrils part of the actin stays bound to myosin, necessitating several per cent of salt to obtain solubilization at the pH values in question. It should be kept in mind that there are wide variations in the properties of muscle proteins from different species of fish (Shimizu, Machida and Takenami, 1981), but fish meat offers the advantage of having a much lower content of insoluble stroma proteins than does mammalian meat (Suzuki, 1981). Hence surimi consists almost entirely of myofibrillar proteins.

The rheological properties and the texture of surimi and its products thus depend on:

1. The physical state and concentration of actomyosin/myosin.
2. The amount (and type) of natural 'contaminants'.
3. Parameters such as pH and ionic strength (salt concentration) during processing.
4. Other ingredients/additives.

We shall now discuss each of these items.

#### *Physical state and concentration of actomyosin/myosin*

Pure actomyosin is a complex system which denatures in several steps with concomitant rheological changes (Montejano, Hamann and Lanier, 1984b; Wu, Lanier, and Hamann, 1985). One implication of the observed complexity is that the rheological properties of the surimi paste after cooking depend on the heating rate employed (Lee, 1984; Okada, 1963). A paste which is heated slowly through the temperature region 40–50 °C will become firmer than a paste which is passed quickly through that temperature region. The effect of heating at 40–50 °C is much more pronounced in the case of fish myofibrillar proteins than with the mammalian proteins since the former are innately less thermally stable (due to lower body temperature).

The actomyosin system is subject to denaturation as well as to proteolysis upon storage, even at sub-zero temperatures. The slow denaturation which takes place at low temperatures is usually observed as having a negative effect on subsequent heat gelation properties.

As would be expected, increasing the surimi content of the paste (i.e. lower moisture content) generally leads to improved rheological properties of the gel (Hamada and Inamasu, 1983).

*Amount and type of natural 'contaminants'*

Contaminants which affect the textural properties of surimi include primarily water-soluble proteins, especially proteolytic enzymes (Okada, 1964). These enzymes originate from the gut and may contaminate the muscle either during pre-process storage or during deboning or, in some species, they are endogenous to the meat (Lanier, 1985). Washing only partially removes the proteases and since they are active in the range 50–75 °C (Lanier *et al.*, 1981; Suzuki, 1981), surimi product preparation should be designed to raise the temperature rapidly through this region.

Proteolysis results in decreased elasticity of the heat-treated fish paste. The extent of the effect depends both on the type and concentration of proteases, as well as on the temperature, pH, ionic strength, etc. during storage. Obviously, storage time is also important but the relation between proteolytic breakdown and gel formability is highly complex. It has been suggested that some of the observed decreases in rigidity may stem from protein transitions or a temperature effect after completion of protein aggregation (Montejano, Hamann and Lanier, 1984a).

*Effect of pH and ionic strength during processing*

As the solubility of myosin in particular is sensitive to changes in pH and ionic strength, these two parameters are of decisive importance for the rheological properties of raw and cooked surimi. In actual practice the two parameters are placed within a rather narrow range by the need for certain sensory properties of the final product, i.e. the desirable pH is usually between 6 and 7 and the salt concentration 2–3%. To maximize the concentration and hence the dissolving power of salt it is added to the surimi at the very beginning of final product preparation. This procedure also ensures homogeneous distribution of (acto)myosin in the continuous phase of the paste.

*Other ingredients/additives*

If textural or binding properties are of particular importance in the final product, or the surimi on hand is of unsatisfactory quality, hydrocolloids might be added. Addition of starch increases the storage modulus of the fish paste (Hamada and Inamasu, 1984). However, gels produced with unmodified wheat or potato starches have poor freeze–thaw stability; modified starches in combination with egg white perform better (Lee, 1984). Nonetheless, it should be realized that the proteins of high quality surimi yield gels of about the same rigidity as do egg white proteins at the same concentration (Burgarella, Lanier and Hamann, 1985). Also, it has been maintained, and disputed, that the rheological properties obtained with high quality surimi may be simulated, to a large extent, by increasing the concentration of lower quality surimi.

It should be pointed out that for monitoring (structural) changes in gelling samples as a function of time, temperature, etc., non-destructive evaluation techniques are quite useful. Thus, most of the rheological properties referred to in this section were studied using such methods. Clearly, the large deformation techniques employed for grading are quite different, and caution must be exercised when attempting to view technological properties in the light of observations pertaining to the molecular level.

Lee (1984) in his comprehensive review on surimi technology, pointed to the following as important processing considerations:

1. Highly elastic and resilient texture, as in fibrous products, necessitates the use of top quality surimi. Therefore, each batch of surimi should be checked for its gel-forming ability and ability to produce a glossy and tacky paste.
2. The better the surimi, the greater the amount of water that can be added (maximum 80% moisture). Addition of hydrocolloids allows for higher total amounts of water.
3. Chopping aids in solubilization of (acto)myosin from the myofibrils; 15–20 minute chopping time is recommended. However the temperature must be kept below 10 °C to avoid problems due to actomyosin lability.
4. Setting of the paste by heating should occur slowly, since the longer the setting takes, the better the product becomes.
5. Water-insoluble colourings are to be preferred. Such pigments can be dispersed uniformly and will not bleed during further processing and heating.

### *Products made from surimi*

In 1985 the world production of surimi-based foods amounted to about one million tons, worth about £2400 million. Shellfish analogues accounted for roughly 10% of the total, half of which was sold in the USA.

The traditional Japanese processing of surimi essentially included five steps: chopping of the surimi, stone grinding with salt and spices, straining, shaping of the individual product units and steaming (*Figure 14.3*). The product was called kamaboko. Today it is customary to name all products made from surimi kamaboko products. Up to a few years ago Japan was the only market for kamaboko. Outside Japan the interest in such products was raised by the development of seafood analogues: imitation crab legs, scallop and shrimp analogues.

#### *Fibreized products*

Simulated shellfish products probably command the greatest interest. The general recipe includes the ingredients shown in *Table 14.1*. The paste obtained by treating this mixture in a silent cutter is extruded as a thin sheet onto a conveyor belt and is heat-treated, using gas and steam, to achieve partial setting. Subsequent to cooling a strip cutter subdivides the sheet into strings which pass through a rope former. The 'rope' is coloured and suitably cut, i.e. straight in the case of imitation crab legs and obliquely to obtain flake and chunk type products. Steam cooking yields the finished product (*see Figure 14.4*).

#### *Moulded products*

These are analogous to the traditional kamaboko products. The surimi (plus other ingredients) paste is moulded into the desired shape and heated to set. Extrusion is the usual procedure; if a multi-opening nozzle is employed a meat-like texture can be

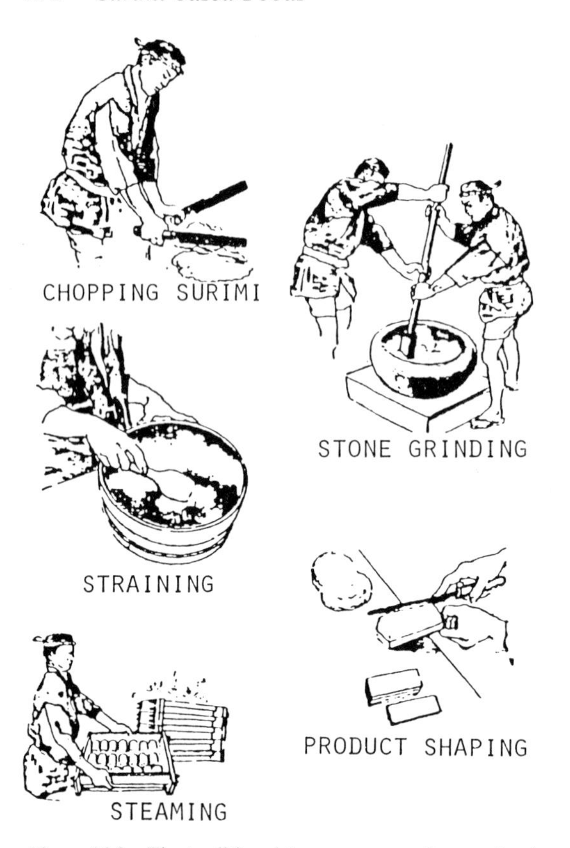


**Figure 14.3** The traditional Japanese way of preparing kamaboko products. Reproduced by permission of Suzuhiro Kamaboko Kogyo Co. Ltd, Japan.

**Table 14.1** INGREDIENTS OF SIMULATED SHELLFISH PRODUCTS PREPARED FROM SURIMI

| | |
|---|---|
| Essential ingredients: | Surimi<br>Water<br>Salt |
| Product-dependent ingredients: | Starch<br>Egg white<br>Shellfish meat<br>Shellfish flavour<br>Flavour enhancer |

achieved. Alternatively, composite-moulded products may be deemed the better choice. Suitable lengths of strings (intermediate in the production of fibrous products by 'rope' formation) are mixed with surimi paste and extruded, giving a better bite than the somewhat rubbery mouth-feel resulting from simple extrusion. By adding diced ham to the surimi paste prior to extrusion, a product called 'fish ham' has been made (Lee, 1984).

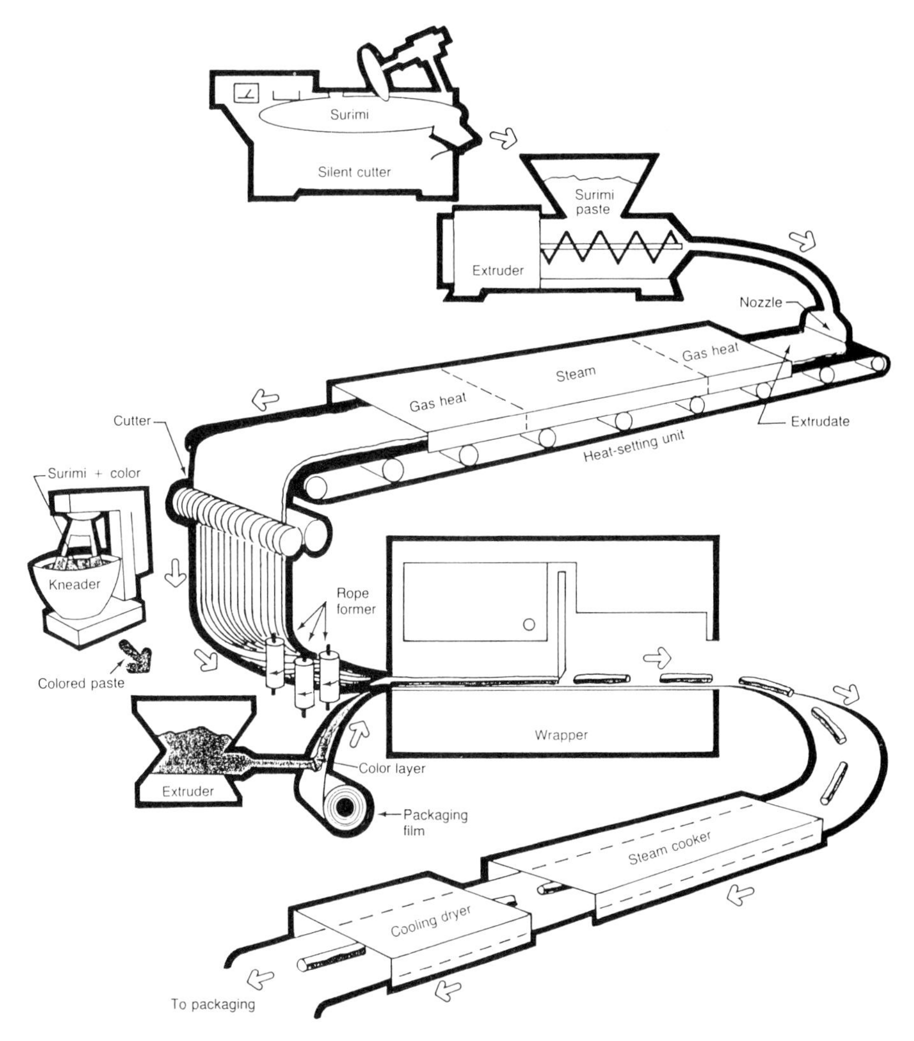


**Figure 14.4** Processing line for preparing simulated crab leg from surimi. Reprinted from Lee (1984) with permission of the publishers.

## *Emulsion-type products*

If surimi is mixed with fat (< 10%, plus other ingredients) and processed in analogy with meat, stable emulsions are obtained. By stuffing into casings and cooking a variety of products are made. The Japanese market has received emulsion-type products well, while in the West work has so far been limited to laboratory experiments (Lee, 1984).

## *Dried products*

Flaked, dry 'crab' leg is thinly cut and requires no soaking before eating as it is or as an ingredient in salads. Another snack-type product has cheese sandwiched between layers of dried kamaboko.

## Recent Norwegian work on surimi

The Norwegian Fishery Research Council decided in 1983 to explore the possibilities for surimi production, primarily from their own industrial fish resources. The research programme which was developed comprised two main lines of work: (1) establishing reliable methods for measuring gelling properties and (2) pilot plant production of surimi for screening purposes. As a consequence, a surimi pilot plant was installed at The Institute of Fishery Technology Research (FTFI) in Tromsø. The specific purpose of this installation was to adapt technology and to improve Norwegian know-how for the production of functional, storable fish mince both from conventional fish species (cod, saithe, haddock) and from industrial fish species like capelin, herring, mackerel and trash fish.

### *Initial screening*

In the course of 1985/86 surimi was produced from cod, saithe, blue whiting, redfish and herring. The results demonstrate that both cod and blue whiting, which are white, lean fish species, yield surimi with good gelling properties. However, the gel-forming quality depends very much on the freshness of the raw material. Surimi from saithe also proved to have good gelling properties, but the colour was slightly greyish. Both redfish and herring are fatty fish species, giving surimi with poorer gelling properties.

Further work will include studies on the effects of seasonal variations and storage and should, hopefully, result in standardized surimi processes for the different fish species. Also, considerable efforts will be devoted to improving the methods for analysis (composition, functional properties) of surimi. Special emphasis will be placed on rheological techniques.

### *General product development*

As has already been pointed out, surimi is an intermediate product. Studies pertaining to preparation of consumer-ready products have so far been centred around a recently installed 'Incruster' machine. This equipment works like a co-extruder, i.e. the products are made from two different constituent materials, one coating the other. For example, 'shrimps' are given red stripes by using red-coloured surimi paste as the coating material. Also, flakes (0.3–0.5 mm thick) of heat-set surimi paste have been mixed with non-heated paste and the mixture subsequently extruded into the desired shape.

Not surprisingly, it has been confirmed that first grade surimi is a prerequisite for the production of shellfish analogues. Fresh cod and saithe, being white and lean fish species, constitute good raw materials while small and fatty, or frozen, fish yield surimi with poor gel-forming properties when processed in the conventional manner.

***Fideco technology***

The technology developed and owned by Fideco (Fishery Development Company of Norway) may change this situation entirely. The salient facts are as follows.

Capelin and herring are the most important industrial fish species in Norway; the

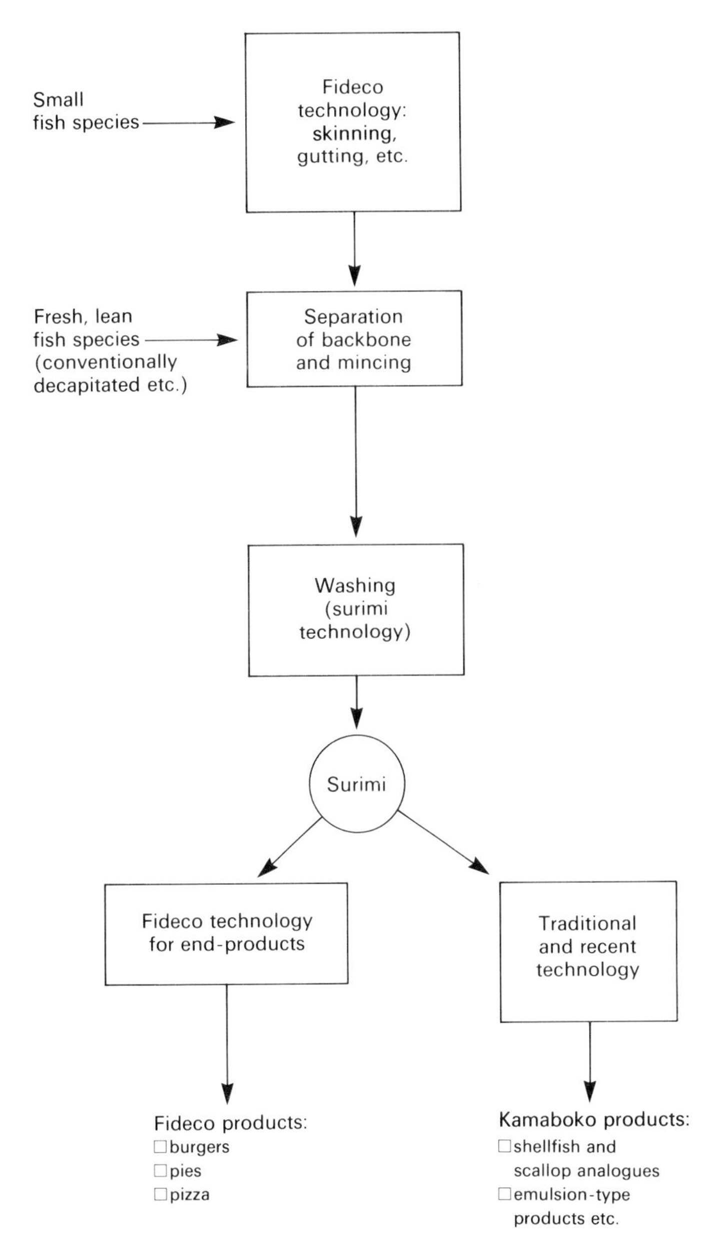


**Figure 14.5** Production of surimi and end products as planned by Fideco (Fishery Development Company of Norway).

total catch amounts to 1–2 million tons annually. Production of a functional mince from these small and fatty, pelagic fish species has for many years been the main goal in our fish processing research. The challenge has consisted in removing skin, viscera, black belly-lining and fat in a bulk process without destroying the functional properties of the fish proteins. Many attempts have been made in the course of the last ten years, but none have proved successful until now. Fideco technology will form the basis for new processing plants which will be built in the northern part of Norway in the coming years; a pilot plant has been in successful operation at FTFI for some time.

The process includes the following steps (*see Figure 14.5*):

1. Gutting, skinning etc.: Head, skin, viscera, black belly-lining and fat are removed by an automatic, bulk process (the details of which cannot be disclosed).
2. Separation: Mechanical removal of backbone followed by mincing.
3. Washing: Surimi technology (washing) removes fat, sarcoplasmic proteins, fish flavour, blood and other impurities.
4. Product process: The fish mince is mixed with additives, formed and heat-treated. A variety of new products have been developed such as burgers, combination products and patés. The products are then frozen and packed.

The merits of the Fideco technology are:

1. It can use frozen fish as a raw material.
2. It can process small fish species on a large industrial scale.
3. It can upgrade fish mince with poor functional properties to resemble fish mince with good gelling properties through end-product technology; in this respect a wide range of innovative fish mince products have been developed by Fideco.
4. It has a high degree of flexibility and can easily be combined with established fish mince technology.

The combination of Japanese surimi/kamaboko technology and Fideco technology is expected to make all the Norwegian fish resources available for production of new, high quality fish mince products.

## References

ARAKAWA, T. and TIMASHEFF, S.N. (1982). Stabilization of protein structure by sugars. *Biochemistry*, **21**, 6536

BURGARELLA, J.C., LANIER, T.C. and HAMANN, D.D. (1985). Effects of added egg white or whey protein concentrate on thermal transitions in rigidity of croaker surimi. *Journal of Food Science*, **50**, 1588

HAMADA, M. and INAMASU, Y. (1983). Influences of temperature and water content on the viscoelasticity of kamaboko. *Bulletin of the Japanese Society of Scientific Fisheries*, **49**, 1897

HAMADA, M. and INAMASU, Y. (1984). Influences of temperature and starch on the viscoelasticity of kamaboko. *Bulletin of the Japanese Society of Scientific Fisheries*, **50**, 537

HOLMES, K. (1986). Ship vs. shore processing: new information arises. *The Lodestar*, **IV**, 3

IWATA, K. and OKADA, M. (1971). Protein denaturation in stored frozen Alaska pollock muscle. I. Protein extractability and kamaboko forming ability of frozen surimi. *Bulletin of the Japanese Society of Scientific Fisheries*, **37**, 1191

IWATA, K., OKADA, M., FUJII, Y. and MIMOTO, K. (1968). Influences of storage temperatures on quality of frozen Alaska pollock surimi. *Reito* [*Refrigeration*], **43**, 1145

IWATA, K., KANNA, K., UMEMOTO, S. and OKADA, M. (1971). Study of the quality of frozen stored Alaska pollock surimi. I. The influence of freshness of the material and changes in storage temperature. *Bulletin of the Japanese Society of Scientific Fisheries*, **37**, 626

LANIER, T.C. (1985). Menhanden: Soybean of the sea. *University of North Carolina Sea Grant Publication*, pp. 85–102

LANIER, T.C. and HAMANN, D.D. (1985). Readers disagree on surimi standards. *Food Technology*, **39**, 28 (letter)

LANIER, T.C., HAMANN, D.D. and WU, M.C. (1985). Development of methods for quality and functionality assessment of surimi and minced fish to be used in gel-type food products. Alaska Fisheries Development, Anchorage, Alaska

LANIER, T.C., LIN, T.S., HAMANN, D.D. and THOMAS, F.B. (1981). Effects of alkaline protease in minced fish on texture of heat-processed gels. *Journal of Food Science*, **46**, 1643

LEE, C.M. (1984). Surimi process technology. *Food Technology*, **38**, 69

MONTEJANO, J.G., HAMANN, D.D. and LANIER, T.C. (1984a). Thermally induced gelation of selected comminuted muscle systems—rheological changes during processing, final strengths and microstructure. *Journal of Food Science*, **49**, 1496

MONTEJANO, J.G., HAMANN, D.D. and LANIER, T.C. (1984b). Final strengths and rheological changes during processing of thermally induced fish muscle gels. *Journal of Rheology*, **27**, 557

OKADA, M. (1963). Studies of elastic property of kamaboko. *Bulletin of the Tokai Regional Fisheries Research Laboratory*, **36**, 21

OKADA, M. (1964). Effect of washing on the jelly forming ability of fish meat. *Bulletin of the Japanese Society of Scientific Fisheries*, **30**, 255

SHIMIZU, Y., MACHIDA, R. and TAKENAMI, S. (1981). Species variations in the gel-forming characteristics of fish meat paste. *Bulletin of the Japanese Society of Scientific Fisheries*, **47**, 95

SUZUKI, T. (1981). *Fish and Krill Protein: Processing Technology*. London, Applied Science

TAMOTO, K., TANAKA, S., TAKEDA, F., FUKAMI, T. and NISHIYA, K. (1961). Studies on freezing of surimi and its application. IV. On the effect of sugar upon the keeping quality of frozen Alaska pollock meat. *Bulletin of the Hokkaido Regional Fisheries Research Laboratory*, **23**, 50

WU, M.C., LANIER, T.C. and HAMANN, D.D. (1985). Rheological and calorimetric investigations of starch-fish protein systems during thermal processing. *Journal of Texture Studies*, **16**, 53

15

# STRUCTURED FAT SYSTEMS

G.G. JEWELL* AND J.F. HEATHCOCK
*Cadbury Schweppes, Reading, Berkshire, UK*

## Introduction

Fats are present in foodstuffs in a variety of different forms and contribute to food structure in many different ways. In considering their physical state at room temperature (~ 20 °C) fats may be completely liquid, i.e. an oil such as a cooking oil, or they may be a hard solid e.g. cocoa butter, or spreadable plastic masses e.g. vegetable oil shortenings. This difference relates to the basic molecular structure—the triglyceride molecule. In simple terms the triglyceride consists of a glycerol backbone to which three fatty acid chains are attached. The chains may be different either in their length of in their degree of saturation. One or other or both of these chemical differences occur in fats from natural sources and lead to marked differences in their physical properties.

An important property of many fats is that they can exist in a plastic or a solid state and this relates to the ability of the triglycerides to form crystals which can then associate often forming a network. Plasticity in fats is best thought of by reference to the modern day margarines which will spread straight from the fridge. Butter, in comparison, is very hard at fridge temperatures and will not spread until it approximates to room temperature. This difference is related to the amount of solid fat compared with the liquid fat content of the system which, in turn, reflects the higher quantity of polyunsaturated fats in the margarine, and the fact that glycerides with high levels of polyunsaturated fatty acids remain liquid, since the shapes of their molecules preclude packing into crystal lattices.

Fats may also be derived from either animal or vegetable sources. In the natural tissues of both animals and plants, the fats are usually laid down as discrete cellular deposits and the main function of the fat is to act as an energy store. In considering foodstuffs and in animal tissues, in particular, the quantity and type of fat present will have significant impact on cooking processes such as roasting and frying both in terms of modification of texture and structure and also the creation of flavours. There are obviously a very wide range of systems containing fat, from natural plant and animal tissues to emulsions where only a small amount of fat has an important influence on the structure. However, since the structure and role of fat in meat products (Chapter 13) and in emulsions (Chapter 4) have been described elsewhere in this book they will not be included here. This contribution will concentrate,

*Present address: Quaker Oats Limited, Bridge Road, Southall, Middlesex.

principally, on the role of those fats that have already been extracted from either animal or plant tissues and it will describe the influence they have on the creation and stabilization of structure.

The word 'structure' is obviously the common theme of these chapters. It is probably worth, however, giving a definition in the context of 'structured fats'. Structure results from the increasingly complex association or building up of simple discrete units. The units may be identical in their nature or may differ. The structures themselves may be relatively simple, i.e. all of one component or they may be complex, e.g. multicomponent and often multifunctional. The structure is possibly seen by the naked eye or may only be evident when instruments of increasingly greater resolution are used, e.g. the light microscope (resolution = 0.2 μm), transmission electron microscope (resolution = 0.2 nm) or X-ray diffraction which is capable of resolving structures with dimensions of the order of 0.1 nm.

The structure of fat systems and foods containing fat have been discussed by Berger, Jewell and Pollitt (1979), De Man (1982) and Larsson (1982) and a recent review by Timms (1984) discusses the physical chemistry of fats.

## The structure and crystallization of triglycerides

It has frequently been noticed that fats have multiple melting points and, as early as 1852, Duffy showed that a certain triglyceride, tristearin, had three melting points. Much later, Malkin and his associates (1954) established that multiple melting was due to polymorphism, i.e. the occurrence for a given compound of different crystalline forms distinguishable by their X-ray diffraction pattern. Unfortunately, a great deal of confusion and controversy arose from the early work, since different authors used the same form of nomenclature ($\alpha$, $\beta'$, $\beta$), but assigned them to the various crystalline forms on the basis of different criteria. Thus, the British school, led primarily by Malkin (1954), assigned them in terms of increasing melting point, so that the form with the highest melting point was designated $\beta$ whilst the Americans, led principally by Lutton (1950) used X-ray diffraction data as their basis. The two are not always strictly comparable and hence, 2-stearodipalmitin, one of the triglycerides often found in fats was $\beta$ according to Malkin (1954) and $\beta'$ according to Lutton (1950). The confusion continued until Chapman (1962), in a definitive review of the polymorphism of glycerides, demonstrated that infrared absorption spectroscopy could be used to study polymorphic forms and this technique did allow the controversy to be resolved. Larsson (1966) then suggested a nomenclature based on both X-ray and infrared data, and it is this terminology which is the most generally accepted.

The nature of the triglycerides and thermal history are the key factors controlling the type of crystal formed. Thus, the first thing to consider is the structure of triglycerides and then their possible modes of crystallization. *Triglycerides* are formed when three fatty acids react with glycerol. The fatty acids are long chain compounds with 4–20 or more carbon atoms in the chain. The chain may be fully saturated as in palmitic ($C_{16}$) and stearic ($C_{18}$) acids, or contain one or more double bonds and be called unsaturated as in oleic acid ($C_{18}$ one double bond) or polyunsaturated as in linoleic acid ($C_{18}$ two double bonds). When double bonds are present, the configuration across the double bond gives rise to two possible *isomers*, e.g. both oleic and elaidic acids have the structure $C_{18}$ with one double bond, in the oleic acid it is in the *cis* form, whilst elaidic is in the *trans* form. In the *cis* form both ends of the chain are on the same side of the double bond, whilst in the *trans* form the chains project on opposite sides of the double bond. The *cis* form gives rise to a sharp kink and overall curve in the chain, whereas the *trans* produces a straight chain. Whether the fatty acids are *cis* or *trans* has a marked influence on the crystallization of the fats. The shape of the triglyceride molecule is also important and from X-ray diffraction studies on single crystals it was established that the molecule has a characteristic shape which is usually described as either chair- or tuning fork-shaped and that it is always the 2-position which projects away from the other 2. When fats crystallize, the individual triglyceride molecules become associated to form linear arrays of bilayers as depicted in *Figure 15.1*. The simplest form of packing is that in which the fatty acid chains pack parallel with each other and perpendicular with respect to the plane of the glyceryl groups. The chains are thus orientated laterally in a hexagonal array (as a stack of pipes when viewed end on). Because of the random rotational distribution of the hydrogen atoms, the centres of the chains are separated equidistantly.

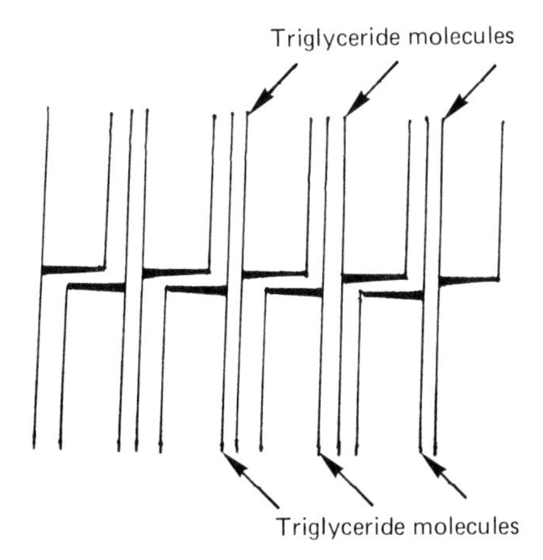


**Figure 15.1** Schematic diagram of the bilayer packing in triglycerides.

Using the generally accepted values for interatomic distances and angles, it can be calculated that the spacings of this molecular orientation will be 4.15 Å in all three lateral directions. Structures with this lattice type are designated α and the principal X-ray line is at 4.15 Å; this is a so-called *short spacing* as it provides information about the lateral arrangement of the triglyceride. Additionally, the overall length of the triglyceride will be dependent upon the length of the individual fatty acid chains, so that a triglyceride with only $C_{10}$ or $C_{12}$ length chains will obviously be shorter than a glyceride containing only $C_{18}$ fatty acids.

This effect influences the dimension of the bilayer discussed above, and gives rise to a further characteristic dimension which can be deduced from the X-ray data and called the *long spacing* (*see Figure 15.2*); typical values would be between 30 and 60 Å. The *α form* described above is the simplest crystal form and has the lowest density. The next crystal type has a more densely packed crystal structure, and this is achieved by the fatty acid chains being tilted relative to the glyceryl chain, at an angle of about 70 degrees. In this so-called *β′ type*, the axes of the molecules are spaced at 3.80 Å in two directions and at 4.20 Å in the third direction. The long spacings are similar to the α form but slightly shorter due to the tilt of the chain axis.

In the third and final crystal type, designated β, the angle of tilt relative to the

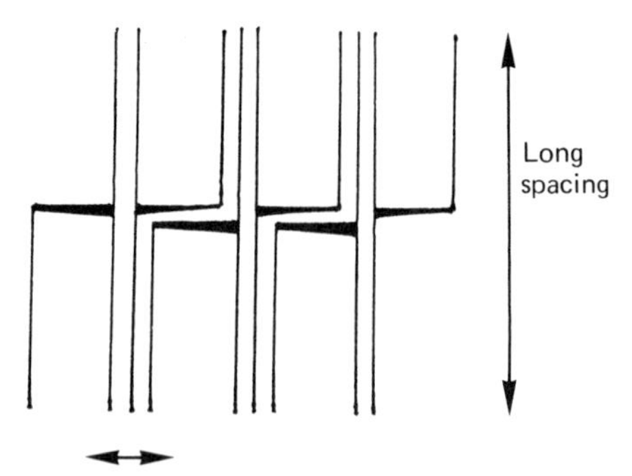


**Figure 15.2** Schematic diagram showing the relationship of short and long spacings in α triglycerides.

glyceryl chain is even greater, being of the order of 60 degrees. In this polymorph the molecules are spaced at 4.58, 3.85 and 3.65 Å and these are the observed principal short spacings. The alignment which gives rise to the long spacings can produce two types of structure. Firstly, a form where the long spacings are similar to both the α and β′ type, but slightly shorter due to the increased angle of tilt. This type of structure can be described as *double packing* as the long spacing is related to the sum of the lengths of the fatty acids on chains 1 and 2.

In the second type of packing, the length of the bilayer is increased by the fatty acids at the 2-position being aligned preferentially adjacent to each other (*see Figure 15.3*). This packing can be described as a *triple packing*, and results in an increase in the long spacing value of the order of 50% compared to the β′ and β double forms (double packing has typical values of 40 Å whilst β triple is 60 Å). The majority of the naturally occurring vegetable fats which contain an unsaturated fatty acid have the *cis* configuration as in oleic acid ($C_{18}$). Also in nearly all these fats the oleic acid is found in the 2-position. This enables a very precise form of molecular packing which for a 2-

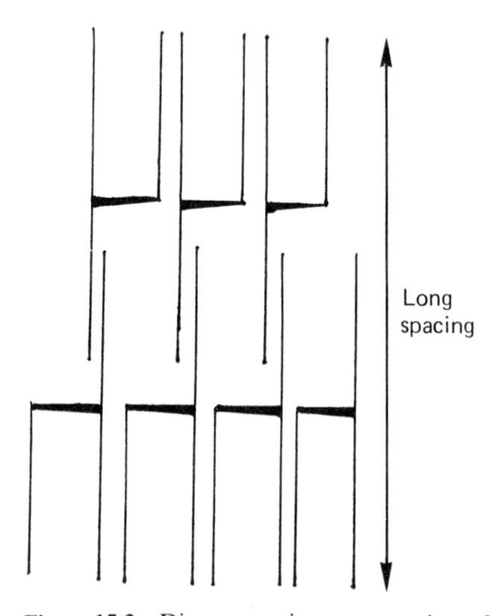


**Figure 15.3** Diagrammatic representation of the triple packing in triglycerides.

oleic triglyceride in the β form is as depicted in *Figure 15.4*. If some triglycerides are introduced in which the 2-position contains a $C_{18}$ acid of *trans* configuration (e.g. elaidic acid) it is clear that the orientation around the double bond will make uniform packing between *cis* and *trans* type virtually impossible, and extensive crystallization will be prohibited. This effect is most pronounced with the β crystal and is of less significance with the α type.

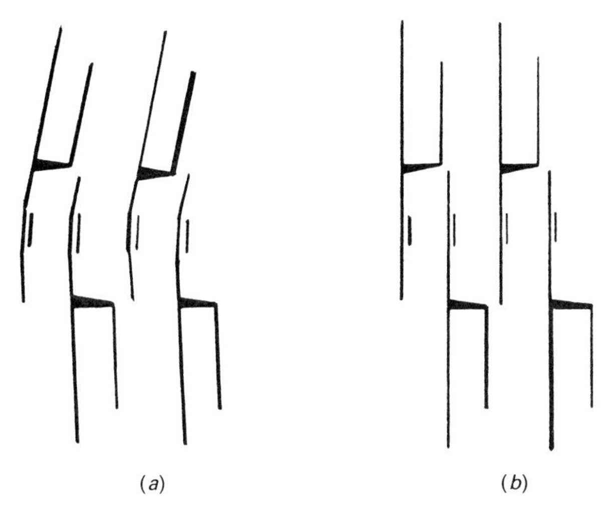


**Figure 15.4** Schematic diagram showing the structure of *cis* and *trans* β packing: (a) *cis* 2-oleo glyceride; (b) *trans* 2-elaido glyceride.

We may now consider how these bilayers of glycerides are built up into crystals. In simple terms, the bilayers simply stack upon themselves and join end-to-end to produce a layered crystal structure. In general, the α crystals, which are the simplest, grow most quickly. The β′ type which contains tilted chains needs to grow a little more slowly to achieve the precise packing, whilst in the β form, which requires even greater tilt and chain extension, the growth rates will be slowest of all. The different packing and growth rates will result in differently shaped crystals, and this has been demonstrated by Hoerr (1960) using polarized light microscopy, and Jewell (1974) using electron microscopy.

One major factor controls the ease with which the bilayers can stack, and that is the type of fatty acid present. It is a consequence of the arrangement of the fatty acids around the glyceryl group that the ends of the bilayers are not perfectly planar, but terraced. The height of the terrace steps is directly related to the length of the constituent fatty acid; triglycerides containing similar or identical fatty acids (i.e. $C_{16}$ and $C_{18}$) will have a less marked step height than those with markedly different fatty acids, i.e. $C_{18}$ and $C_{12}$ (*see Figure 15.5*). It has been shown by Larsson (1972) that bilayers with pronounced step heights (more correctly termed *methyl terraces*) do not build properly to produce crystals, since the contour surface of the terrace allows certain end groups (methyl groups) to become too closely associated and this is not a stable conformation.

Thus it can be seen that the fatty acid composition and configuration are the key factors which control the type of crystal formed:

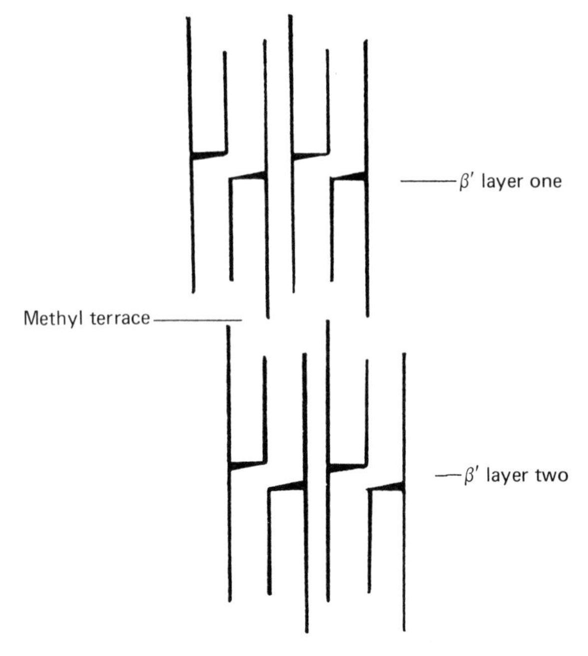


**Figure 15.5** Diagrammatic representation of the methyl terraces in adjacent layers in a β′ crystal.

1. The α crystal has the lowest molecular packing, has the fastest growth rate, and consequently is the least stable and has the lowest melting point.
2. The β′ type with a more densely packed tilted molecular structure than the α is more stable, and has a higher melting point.
3. The β crystal with the highest molecular packing and tilting is the most stable form, and has the highest melting point.

On the basis of the foregoing discussion only three different types of crystal should be found for a given fat. Unfortunately, certain fats have been found to have more types, and the most prominent exception is cocoa butter which has been shown to have six crystal types (Wille and Lutton, 1966). Faced once again with a dilemma over nomenclature, Wille and Lutton (1966) chose to number the crystal forms numerically from I–VI in order of increasing melting point. Form I is the least stable and form VI the most stable. The explanation for the apparent disparity between the α, β′, β concept and cocoa butter is that cocoa butter has certain minor variations in crystal packing which permit slightly different crystals to be formed within a given crystal type. A comparison of melting point and nomenclature for the different crystal forms of cocoa butter is given in *Table 15.1*.

One of the most difficult aspects of the results reported in *Table 15.1* is the fact that form I (the least stable form) is a β′ type of crystal. The explanation is that the short spacings clearly indicate β′ structure, whilst the long spacings indicate that the bilayer ordering is fairly short range. In simple terms, this means that although the lateral arrangements (short spacings) are typical of a true β′ form, the number of bilayers that have built up is very restricted, thus giving a poorly crystalline material of very low stability (melting point). Forms II, III, IV and V then adhere to the α, β′, β sequence with the proviso that form III is a mixture of α and β′. The second aspect of note is the occurrence of two β crystals, i.e. forms V and VI. In this case, a very slight

**Table 15.1** MELTING POINTS AND NOMENCLATURE FOR THE VARIOUS CRYSTALS OF COCOA BUTTER

| *Melting point* (°C) | *Wille and Lutton category* | *X-ray category* |
|---|---|---|
| 21.3 | I | β′ |
| 23.3 | II | α |
| 22.5 | III | α and β′ |
| 27.5 | IV | β′ |
| 33.8 | V | β |
| 36.4 | VI | β |

change in the packing at the methyl terrace occurs which results in a marginal shift in the short spacings, but an increase in the melting point of some 3 °C. In general terms, this transformation of form V to form VI occurs very slowly taking some 12 weeks at 26 °C. In terms of chocolate technology the transition from form V to VI results in bloom formation (*Figure 15.6*), the bloom being the more stable form VI polymorph (Jewell, 1974). It has been demonstrated (Cruickshank and Jewell, 1977) that form VI crystals can be grown quickly in about 15 h provided they are grown from the liquid state, rather than by transformation from the solid form V. Milk fat which has well known antibloom properties is traditionally thought to work by forming a mixed glyceride crystal with cocoa butter which, when solidified as form V, is stabilized and thereby prevented (possibly by steric hindrance) from transforming to form VI.

Whilst certain fats like cocoa butter exhibit more than three crystal types, others exhibit less—usually only two, the α and β′. The so-called 'self-tempering' fats fall into this category as, under all but the most rapid cooling conditions, the fat can only crystallize in the β′ form. The α form only occurs after quenching with coolants like liquid nitrogen.

Although the foregoing discussion has considered the various crystal types (*polymorphic forms*) which can be formed during a crystallization procedure, an important aspect is that forms of low stability will transform to those of high stability.

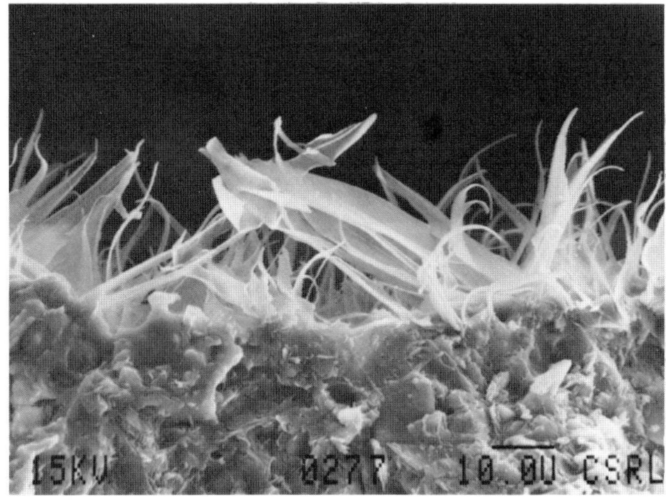


**Figure 15.6** Scanning micrograph showing long bloom crystals projecting out of the surface of dark chocolate. Bar ≡ 10 μm.

The transformation will occur during appropriate thermal conditioning whilst the materials are still in the solid phase; true melting does not occur. The transformations are not thermally reversible so that the changes from form II, III, IV and V which occur during holding at 26 °C, cannot be reversed by holding form V at low temperatures. However, a reversal could appear to take place, since a further quantity of material could crystallize from the liquid phase at low temperature and this would be of a lower polymorphic form than V. However, this is not reversal of polymorphism, but crystallization of a lower stability crystal.

A further important aspect to be considered is the interaction of triglycerides when fats of different origin are mixed together, e.g. cocoa butter and milk fat, or cocoa butter, cocoa butter substitute and milk fat. Conventionally, these types of interactions are studied by means of a so-called phase diagram in which the variation in thermal properties (such as melting point and setting point) of a mixture is plotted against its composition. The major difficulty is in trying to relate the behaviour of a binary (two-component) or even ternary (three-component) phase diagram to real fats which will inevitably contain more than two or three triglycerides. Cocoa butter, for example, is relatively simple but contains approximately 10 major triglycerides, while milk fat has been estimated to contain over 2 500 000 triglycerides. The phase diagrams of triglyceride systems have been reviewed by Rossell (1967). The majority of the information is related to binary systems, and hence can only provide very broad guides to possible interactions between simple natural fats rich in one or two components. Some information concerning certain ternary systems is also given, including a mixture of POP, POS and SOS, the three principal glycerides of cocoa butter (Andersson, 1963). However, it has been found to be difficult to relate directly the predictions from the phase diagram to the behaviour of true cocoa butter for two reasons. Firstly, the cocoa butter contains about 10 triglycerides, and the remaining seven can markedly influence the interaction of the three listed above. Secondly, insufficient attention was paid, in the early work, to the problem of polymorphism and the reported melting points may not have all been of the stable β form but more probably a mixture of β and β′.

A much better way of studying the interaction of fats would be to use the technique of isodilatation as suggested by Rossell (1973). Dilatation is a study of the increase in volume as a substance melts, and is a technique which has found particular application in fat technology. The increase in volume can be related to the change from the solid state to the liquid state as melting proceeds, and hence information can be obtained on the relative proportions of solid and liquid at any given temperature. In the isodilatation methods, a dilatation curve is measured and drawn for each of a series of mixtures representing, say, 10% composition intervals of fat A with fat B. A suitable dilatation value, e.g. D = 1000 is chosen, and the temperature at which each mixture has this D-value is read off and graphed against compositions. The procedure is repeated for other D-values, and a series of contour lines are obtained, showing the temperatures at which the various mixtures exhibit a similar volume, and hence a similar solid/liquid ratio.

If all the lines are fairly parallel with each other, this result indicates that fat A will form a compatible mixture with fat B. If, however, the lines diverge and produce large troughs, this indicates that at certain compositions, fat A is not compatible with fat B and the resulting mixture has a markedly lower melting point and the level of solid fat would undoubtedly be low as well. However, even this technique does not provide all the answers for, as Rossell (1973) has indicated, the polymorphic behaviour of the system has to be considered when interpreting the dilatation data.

A recent study (Hicklin, Jewell and Heathcock, 1985) made use of three complementary techniques (X-ray, differential scanning calorimetry (DSC) and electron microscopy) on the same set of samples of cocoa butter in order to examine the six polymorphs. The X-ray data are shown in *Table 15.2* and the transmission electron microscopy results in *Figure 15.7(a)–(f)*. Form I shows evidence of lamellar structure, but with only limited ordering. In form II the structure becomes more ordered with the development of layered sheets of lamellae. X-ray shows the characteristic α structure. Interesting developments in form III are the distinct tubular structures frequently clustered together against a featureless background. More distinct crystals are evident in the development of form IV, with needle-like structures ranging in size from 0.5–2.0 μm. X-ray diffraction now shows the typical β′ structure with the presence of a doublet and a shift in the long spacing to 46 Å. As has already been mentioned, in chocolate it is the form V polymorph of cocoa butter that is normally

**Table 15.2** X-RAY DIFFRACTION DATA

| | *Short spacing* | | *Long spacing* |
|---|---|---|---|
| Polymorph I | 3.70 (S) | | 34 (W) |
| | 4.19 (VS) | | |
| Polymorph II | 4.25 (S) | | 16.6 (M) |
| | | | 49 (VS) |
| Polymorph III | 3.87 (M) | | 15.24 (M) |
| | 4.25 (S) | | 16.6 (S) |
| | 4.63 (M) | | 49 (VS) |
| Polymorph IV | 4.17 (VS) | | 14.9 (S) |
| | 4.35 (VS) | | 46 (VS) |
| Polymorph V | 3.68 (W) | 4.6 (VS) | 8.08 (W) |
| | 3.76 (M) | 5.43 (M) | 13.15 (W) |
| | 3.88 (W) | | 16.20 (W) |
| | 3.99 (M) | | 34 (S) |
| | | | 66 (S) |
| Polymorph VI | 3.71 (S) | 4.6 (VS) | 8.18 (W) |
| | 3.88 (S) | 5.16 (W) | 13.2 (W) |
| | 4.04 (M) | 5.47 (M) | 35 (S) |
| | 4.28 (W) | | 66 (S) |

S = strong; VS = very strong; M = medium; W = weak

present in the finished product. Clearly defined crystals are seen by microscopy which are regular, plate-like and measure 1 μm in length. Such crystals have the ability to pack well together and it is this structure that gives the product its good gloss, texture and mouthfeel. Form VI is characterized by elongated crystals (2–20 μm) which commonly project from the surface. In chocolate, as already mentioned, the transition from form V to form VI results in bloom (*see Figure 15.6*).

## Lard and vegetable shortenings

The polymorphic forms of fats are important in the creation of structure, but equally important is the influence of crystal size. Lard and vegetable shortenings are known to have very different properties of aeration when used in cakes and biscuits. This difference can be explained partly in terms of their crystal size (Jewell and Meara,

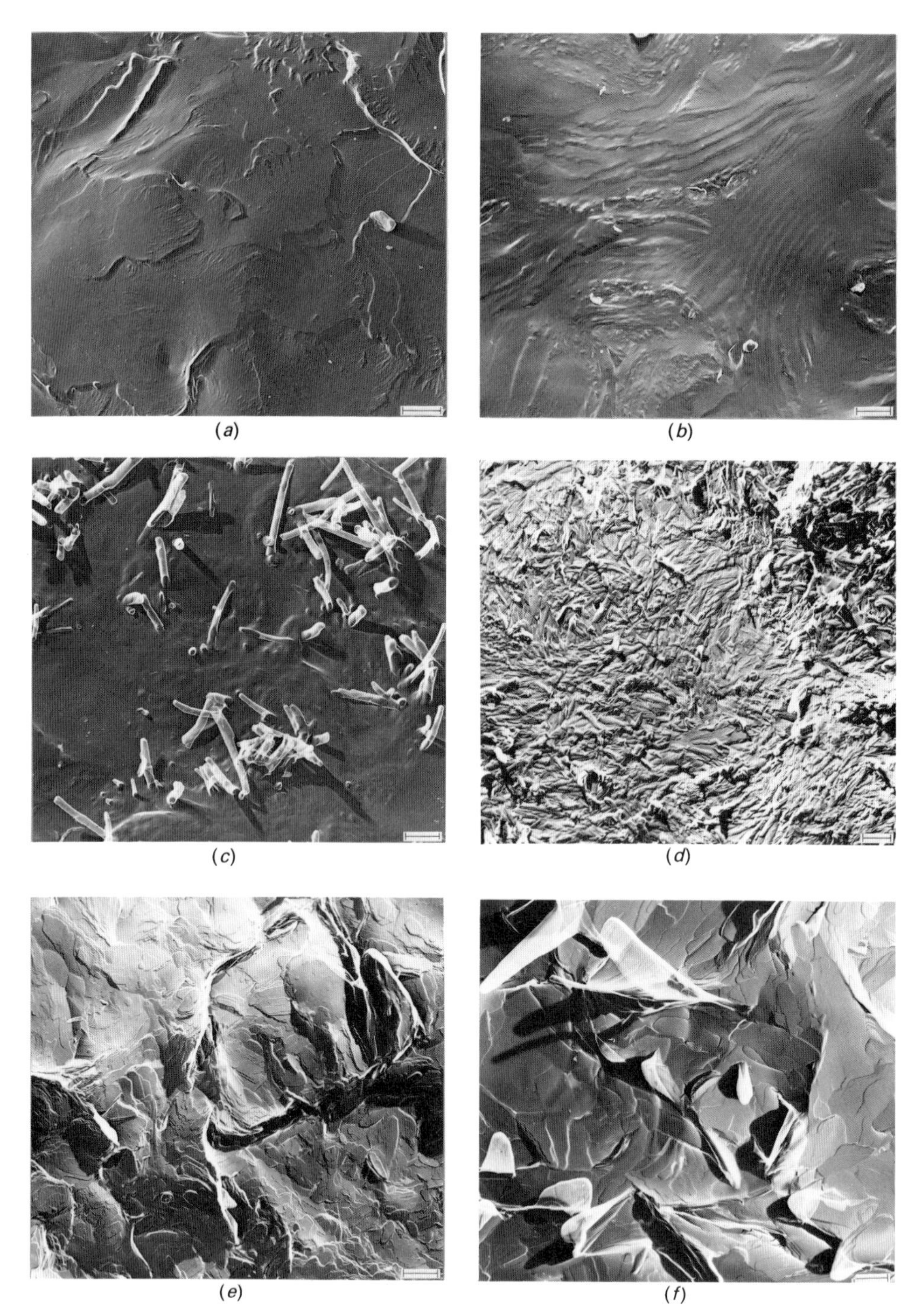


**Figure 15.7** (a)–(f) Freeze fracture preparations of the six polymorphs (I–VI) of cocoa butter. Bar ≡ 1 μm. Pictures reproduced by courtesy of SEM Inc. (Chicago).

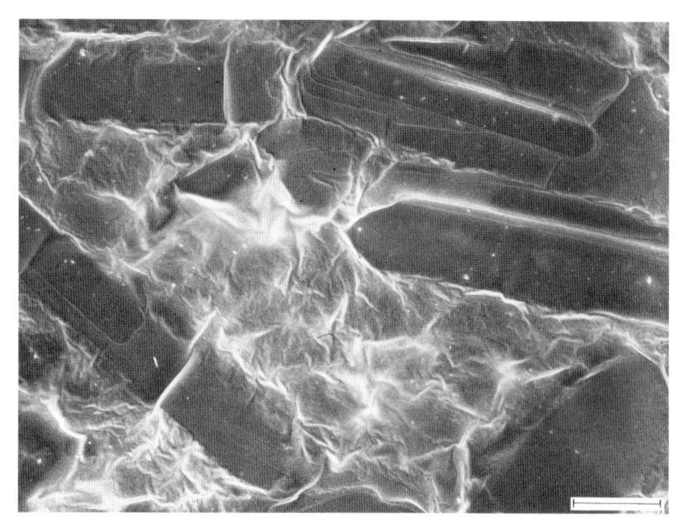

**Figure 15.8** Electron micrograph showing the large fat crystals in lard. Bar ≡ 1 μm.

1970). Lard has distinct, large crystals measuring 1 μm wide and often up to 3–5 μm in length (*Figure 15.8*). This micrograph represents a freeze fracture preparation and the liquid fat has been removed by mild detergent to reveal the crystal structures. This is a good illustration of how both solid and liquid components of the fat may be present within a single product.

In contrast to the large crystals of lard, vegetable fat shortenings show small (1 μm) fairly uniform crystals and have very good aerating properties. A cake batter prepared from such a fat shows air cells 2–3 μm in size, which are stabilized by a layer of solid fat (*Figure 15.9*). This layer consists of small crystals packed closely together around individual air cells. Such crystals are also present in clusters in the amorphous matrix. Lard cannot perform this aeration process effectively because the longer

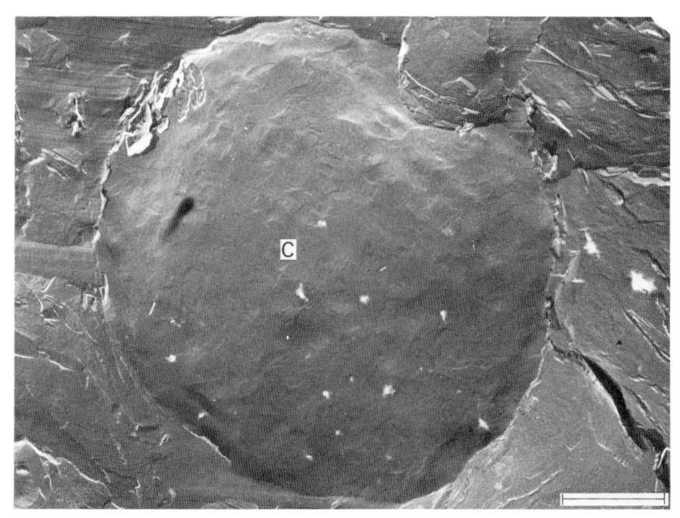


**Figure 15.9** Electron micrograph of a replica of cake batter showing a large air cell (C) stabilized by fat. Bar ≡ 1 μm.

crystals cannot align themselves to the curved surface of the air cells. The net result is a product with little air incorporated and the few bubbles present are large. Interesterified lard is an interesting product as, although it too has large crystals, it does incorporate air fairly effectively. The explanation, as described by Berger, Jewell and Pollitt (1979) appears to be related to the fact that the crystals themselves have surface defects (*Figure 15.10*). The beating process probably breaks up the long crystals at these points of weakness resulting in crystals sufficiently small to be accommodated around the air cells and thus leading to a stable structure.

**Figure 15.10** Electron micrograph of interesterified lard showing large fat crystals with numerous surface defects (arrows). Bar ≡ 1 μm.

## Chocolate

As already mentioned, one of the key ingredients of chocolate is the fat, constituting up to 30% of the product. It is understood that the fat acts as the continuous phase and the dispersed solid phase is made up of sugar and cocoa (providing flavour and colour) and for milk chocolate, additional milk fat and protein.

Many of the properties of the chocolate change if a different source of cocoa butter is used. This relates, in general, to the climate in which the cocoa is grown. In hot climates, for example, there is an increase in the level of saturated fatty acids and a hard fat is produced. The quantity of solid fat may be measured using wide line nuclear magnetic resonance (NMR). The variation of the level of solid fat in a Brazilian cocoa butter during a season is shown in *Figure 15.11*.

The physical properties of cocoa butter and of the chocolate made from it are also modified when other fats are added. One effect of milk fat addition has already been referred to. Milk fat is important in chocolate in order to give more desirable flow and textural properties compared to plain chocolate, but it is a much softer fat and, as shown by NMR, not only lowers the melting point, but markedly lowers the solid fat content (*Figure 15.12*). Other fats are frequently added to the product to replace a portion of the cocoa butter. In the UK it is permissible to replace up to 5% of the butter (15% of the fat) with what are termed CBEs (cocoa butter equivalents) or CBSs

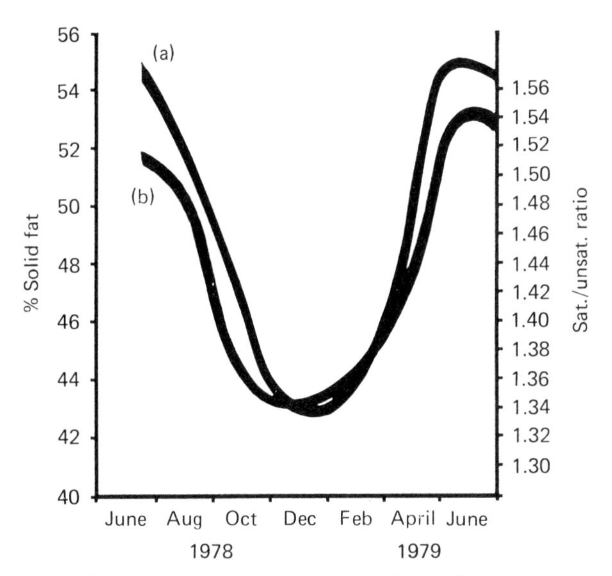


**Figure 15.11** The variation in the level of solid fat and the ratio of saturated to unsaturated fatty acids in a Brazilian cocoa butter during a season. (a) Percentage solid fat at 30 °C versus month; (b) sat./unsat. ratio versus month.

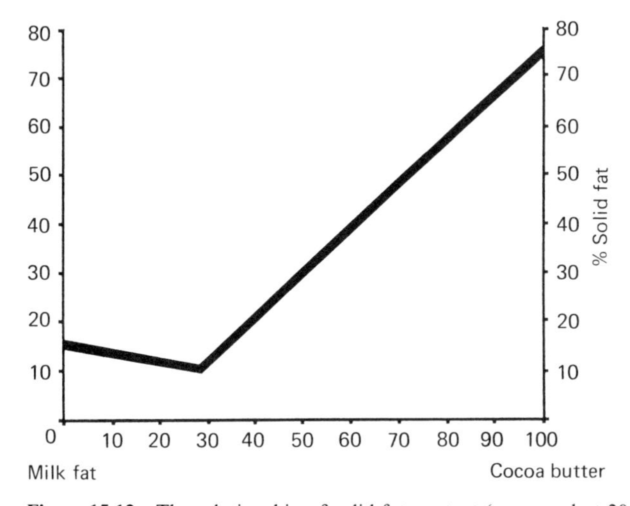


**Figure 15.12** The relationship of solid fat content (measured at 20 °C) with the level of milk fat in mixtures of cocoa butter/milk fat.

(cocoa butter substitutes), the distinction being whether the added fats have similar or different triglycerides to cocoa butter.

The properties of these types of blends have also been investigated using X-ray, DSC and microscopy (Hicklin, Jewell and Heathcock, 1985). It is interesting that microscopical examination of freshly prepared samples shows poorly developed and ill-defined crystals in the incompatible blends. Subjecting the blends to conditions that would promote bloom development leads to a phase separation and the appearance of two separate, distinct types of crystals corresponding to the two fat sources. The reason for the incompatability is the presence of a high proportion of $C_{18}$ *trans* acid rather than the *cis* form which has a different configuration. As shown schematically in *Figure 15.13*, the *cis* configuration of the cocoa butter chain packs poorly with the *trans* region of the vegetable fat molecule.

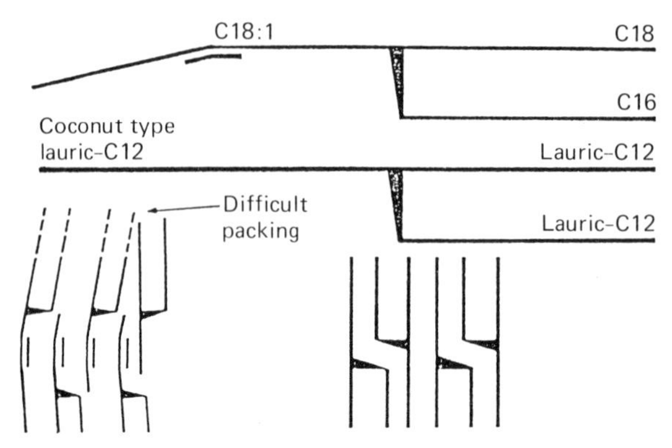


**Figure 15.13** A schematic representation of the poor packing of a coconut glyceride with a β crystal of cocoa butter.

## Caramels and fudges

The structures of caramels and fudges are quite distinct from chocolate and the role fat plays in the product is rather different. One of the important functions of the fat is considered to be as a lubricant in the mouth. In the traditional recipes, where butter is present in place of the more commonly used vegetable fats, the fat would also make a noticeable contribution to the flavour.

In these products, fat is no longer in the continuous phase but is present as a

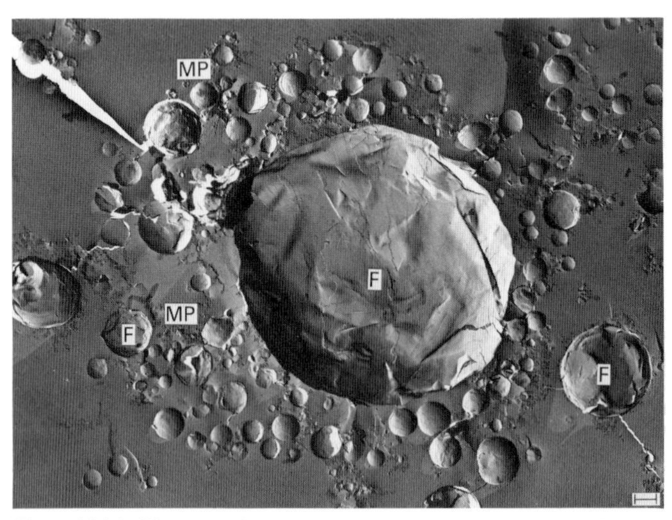


**Figure 15.14** Electron micrograph of a freeze etch preparation of caramel. Spherical fat globules (F) are associated with milk protein (MP) and grouped into flocs. The sugar matrix is relatively smooth and featureless. Bar ≡ 1 μm.

dispersion of globules within the sugar matrix. The globules may range in size from < 1 μm up to hundreds of microns, this size being dependent on the degree of emulsification and of subsequent coalescence of the fat. Recipe and process both influence these parameters. Milk proteins, present in caramels and fudges, also affect the role that the fats play in creating the structure (Dodson *et al.*, 1984a,b; Heathcock, 1985). In the initial stages of processing, the protein remains well dispersed in the sugar matrix but, depending on the type and level of protein, it may then become associated with the surface of the fat globules and promote flocculation (*Figure 15.14*). This is common in caramels containing sweetened condensed milk or milk powders. If whey protein alone is present, there is limited association with the fat and flocculation is significantly reduced. This leads to a reduced viscosity and is also thought to be related to the phenomenon of 'cold flow'. In caramels the sugar matrix, which is predominantly glucose, remains amorphous throughout the process to the final product. By comparison, fudge contains more sucrose and is encouraged to grain, i.e. to develop sugar crystals in the matrix. This frequently results in the distortion of the spherical globules into more irregular, fatty masses.

## Ice cream and dessert toppings

Further, more complex fat-containing systems are ice cream and many of the commercially available dessert toppings. Ice cream represents a frozen foam in which fat and water are suspended in an amorphous matrix. Fat is present at 6–12% by weight, whereas water, mainly as ice crystals, constitutes greater than 60% (Berger and White, 1979). The remainder is milk solids and sugar. In the first stages of ice cream production, the fat is homogenized which results in the formation of a large number of emulsified droplets. In the aerated and frozen product, these fat droplets are dispersed throughout the continuous matrix but also are seen projecting from the matrix into the cavity of the large air cells; some liquid fat is thought to line these cells (*Figure 15.15*). Mixing and aeration of the dessert mixtures also leads to the

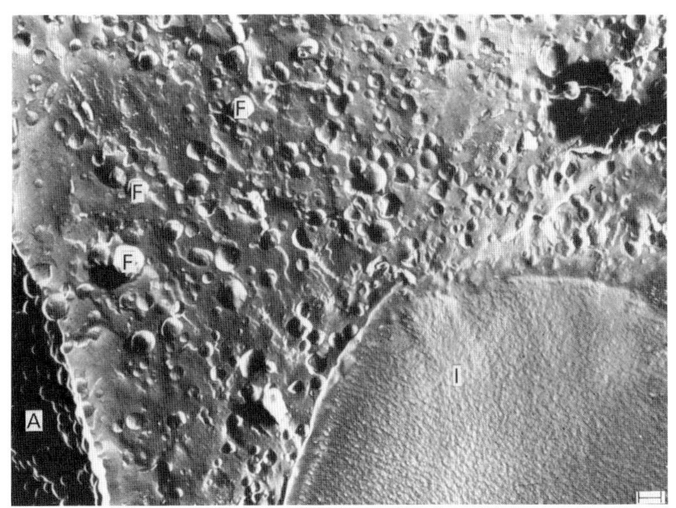


**Figure 15.15** Electron micrograph of a freeze etch preparation of ice cream. Fat globules (F) are dispersed in the continuous phase and also coat the air cells (A). Large ice crystals (I) are observed. Bar ≡ 1 μm.

development of fat droplets and again the appearance of the droplets projecting from the surface of the air cells.

## Acknowledgements

The authors wish to thank their colleagues at Cadbury Schweppes for useful discussions during the preparation of this chapter, Janet Wendy Bell for her secretarial skills and David Spybey for photographic assistance.

## References

ANDERSSON, W. (1963). Fat bloom and phase changes. *International Chocolate Review*, **18,** 92–98

BERGER, K. and WHITE, G.W. (1979). Ice cream. In *Food Microscopy*, (Vaughan, J.G., Ed.), pp. 499–528. London, Academic Press

BERGER, K., JEWELL, G.G. and POLLITT, R.J.M. (1979). Oils and fats. In *Food Microscopy*, (Vaughan, J.G., Ed.), pp. 445–495. London, Academic Press

CHAPMAN, D. (1962). The polymorphism of glycerides. *Chemistry Reviews*, **62,** 433–456

CRUICKSHANK, D.A. and JEWELL, G.G. (1977). Structural studies on tempered cocoa butter. *Leatherhead Food RA Research Report*, No. 256

DE MAN, J.M. (1982). Microscopy in the study of fats and emulsions. *Food Microstructure*, **1,** 209–222

DODSON, A.G., BEACHAM, J., WRIGHT, S.J.C. and LEWIS, D.F. (1984a). Role of milk proteins in toffee manufacture. Part I. Milk powders, condensed milk and wheys. *Leatherhead Food RA Research Report*, No. 491

DODSON, A.G., BEACHAM, J., WRIGHT, S.J.C. and LEWIS, D.F. (1984b). Role of milk proteins in toffee manufacture. Part II. Effect of mineral content and casein to whey ratios. *Leatherhead Food RA Research Report*, No. 492

DUFFY, P. (1852). On certain isomeric transformations of fats. *Journal of the Chemical Society*, **5,** 197

HEATHCOCK, J.F. (1985). Characterisation of milk proteins in confectionery products. *Food Microstructure*, **4,** 17–27

HICKLIN, J.D., JEWELL, G.G. and HEATHCOCK, J.F. (1985). Combining microscopy and physical techniques in the study of cocoa butter polymorphs and vegetable fat blends. *Food Microstructure*, **4,** 241–248

HOERR, C.W. (1960). Morphology of fats, oils and shortenings. *Journal of the American Oil Chemists Society*, **37,** 539–543

JEWELL, G.G. (1974). Electron microscopy of chocolate. Part 1. The structure of bloom on plain chocolate. *Leatherhead Food RA Research Report*, No. 202

JEWELL, G.G. and MEARA, M.L. (1970). New and rapid method for the electron microscopic examination of fats. *Journal of the American Oil Chemists Society*, **47,** 535–538

LARSSON, K. (1966). Classification of crystal forms. *Acta Chemica Scandinavica*, **20,** 2255–2258

LARSSON, K. (1972). Molecular arrangement in glycerides. *Fette Seifen Anstrichmittel*, **74,** 136–142

LARSSON, K. (1982). Some effects of lipids on the structure of foods. *Food Microstructure*, **1,** 55–62

LUTTON, E.S. (1950). Review of the polymorphism of saturated even glycerides. *Journal of the American Oil Chemists Society*, **27,** 276–281

MALKIN, T. (1954). *Progress in Chemistry of Fats and Other Lipids*, Volume 2. London, Pergamon Press

ROSSELL, J.B. (1967). *Advances in Lipid Research*, Volume 5. London, Academic Press

ROSSELL, J.B. (1973). Interactions of triglycerides and of fats containing them. *Chemistry and Industry*, **17,** 832–835

TIMMS, R.E. (1984). Phase behaviour of fats and their mixtures. *Progress in Lipid Research*, **23,** 1–38

WILLE, R.L. and LUTTON, E.S. (1966). Polymorphism of cocoa butter. *Journal of the American Oil Chemists Society*, **43,** 491–498

16

# STRUCTURED SUGAR SYSTEMS

M.G. LINDLEY
*Tate and Lyle, Reading, Berkshire, UK*

## Introduction

Sucrose is the most abundant free sugar in plants and its use has been referred to in documents spanning many centuries. Despite this it is only a relatively recent development for populations to have a ready access to it. This access was made possible by the discovery and transportation of sugar cane from India to many parts of the world, and also by the development of sugar beet as a crop during the Napoleonic wars. This latter advance came about in response to the blockade of continental ports which prevented sugar from being imported. Hence Napoleon encouraged development of an indigenous supply of sugar to sate the French palates.

The old adage that a spoonful of sugar helps the medicine go down epitomizes one of the main uses of sucrose, i.e. as an aid to palatability. Sweetness is a generally pleasurable sensation and sugar is used in many food and beverage products, simply for that purpose. However, in reality, sweetness is just one of the characteristics of sugar which are responsible for its widespread use within the food industry. Sugar is an extremely versatile food ingredient and its use has led to statements such as 'even if sucrose did not taste sweet, it would continue to be a major food ingredient in many food products' (MacKay, 1978).

Food technology has developed as a science to the state where it now has a major influence on the world's food supplies. A feature of the growth of food technology has been the transference of food preparation from the kitchen to the factory. Consequently, product development technologists have utilized fully the physical and the chemical properties of food ingredients such as sugar in order to provide better, cheaper and more attractive food. Hence the uses of sugar within the food industry have grown considerably in recent years, but despite this the development of two principal food market sectors has depended almost totally on the use of sugar as a major ingredient—the baked goods market sector and the sugar confectionery market sector.

These industries grew up using manufacturing techniques which had been transferred from the kitchen to the production line, and although there have been major modifications to production techniques over the years, the basic formulations used today are little changed. Thus the recipes were developed empirically but with the clear objective that they should be capable of producing satisfactory products, rather than as a consequence of a clear understanding of the precise technological rôles of the individual ingredients. Thus we are continuing to develop our understanding of the rôles of the individual ingredients and the influence of their interactions on the products in which they are incorporated. Hence our understanding of the precise and complete rôles of sugar in these products is as yet incomplete.

## Available sugars

Sugar (sucrose) is a disaccharide comprising one molecule of glucose and one molecule of fructose linked glycosidically. Although single crystals appear colourless and transparent, sugar in bulk is white owing to reflected light. Sugar is obtained almost entirely from sugar cane and sugar beet, and in use is virtually identical from whichever source it is obtained. The only differences in performance of beet and cane sugar become apparent when the impurity level reaches several per cent.

The food manufacturer is offered a very wide range of dry crystalline sucroses which differ in particle size and in colour. In addition, there is a limitless range of liquid sugar blends comprising sucrose, invert sugar and glucose syrups. Typical of these is a selection of sugars available to the UK food industry.

*Raw sugar* is the first product obtained by boiling the concentrated beet or sugar cane juices and centrifuging off the syrup. Raw sugar is only rarely sold as a consumer product because of its variable quality and unsatisfactory storage properties. Its principal rôle is to serve as the raw material for sugar refineries which operate throughout the year. Raw cane sugar has a pleasant flavour and may well be used to serve as a characteristic flavour component, providing it meets the necessary hygienic standards, whereas the taste of raw beet sugar excludes it from being used directly as a human food.

*Brown sugars* are probably one of the oldest sugars which are still of great interest to both baker and confectioner. They have characteristic flavours as well as variations in sweetness intensity and come in a wide range of colours and crystal sizes. Certain of the brown sugars are particularly suitable for the confectionery and baking industries because of their soft and fine crystal structure. Their high solubility eases manufacture and ensures effective mixing. Their colouring, slightly astringent sweet taste and flavour-enhancing properties ensure their wide application.

*Granulated sugar*, in contrast, is the standard high purity crystalline product which constitutes the greater part of bulk deliveries to the food industry. It is a general purpose sweetener and bulking agent used in all sectors.

*Caster sugar* is similar to granulated sugar in purity, but with a smaller mean crystal size. It is the smallest boiled crystal produced and the smaller size makes for easy dissolution. It is used particularly extensively in both confectionery and baked products applications.

*Icing sugar* is pulverized and sieved white sugar with an added anti-caking agent. It is very rapidly soluble because of its small particle size and hence finds most use within the industry as a dusting on a variety of products.

*Liquid sugars* are convenient to handle, transport, store and use. They eliminate the necessity of time-consuming dissolution and its control, and in addition there is the possibility of a cost advantage to the manufacturer. The saturation point of sucrose at the standard temperature of 20 °C is 67% w/w. This is the concentration limit at which sucrose syrups may be offered to the manufacturer, although the use of blends with invert and glucose syrup effectively can increase the solids concentration. For example, the concentration of solutions which contain 50% sucrose and 50% invert may be increased to a total nearing 80% w/w without any crystallization taking place. Sugar syrups are susceptible to microbiological spoilage, particularly in the lower purity range; therefore strict control with appropriate handling and spoilage precautions are essential.

## Physical characteristics

### *Crystallinity*

The crystallinity of sucrose is particularly important to the acceptability of certain foods. Sucrose crystallizes readily at a rate of at least an order of magnitude faster than dextrose and fructose (Nicol, 1977). Pure sucrose is classified as a member of the monoclinic system. This system is known alternatively by the names monosymmetric, clinorhombic or oblique. The three axial lengths are all unequal in the monoclinic system. Two of the axes are at 90° and one is different (*Figure 16.1*).

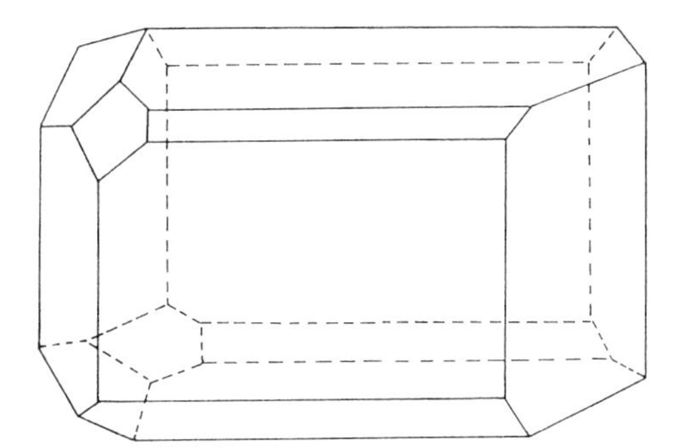

**Figure 16.1** Monoclinic sucrose crystal.

Many properties of the crystal suggest that sucrose, like dextrose, is very extensively hydrated and it is believed that a hexahydrate is formed under certain circumstances (Lees, 1965). Sucrose solutions supersaturate readily, and in this state when work is done, for example by beating, nucleation and crystal growth are immediate. On the other hand it is possible by spray-drying, or freeze-drying sucrose solutions, to have a non-crystalline amorphous dry sugar (Mathlouthi, 1975).

### *Crystallization*

The crystallization of sucrose proceeds in accordance with certain general chemical and physical principles. The development of a crystal structure, and indeed crystallization itself, cannot be simply explained and neither are the theories as yet fully developed. However, much has been learned through empirical observation.

Supersaturated solutions may or may not deposit crystals on cooling. In sugar manufacture the supersaturation of the solution is defined (Anon, 1958) by the following formula:

$$S = \frac{\text{g sucrose/100 g water at temperature } t}{\text{g sucrose/100 g water when saturated at temperature } t}$$

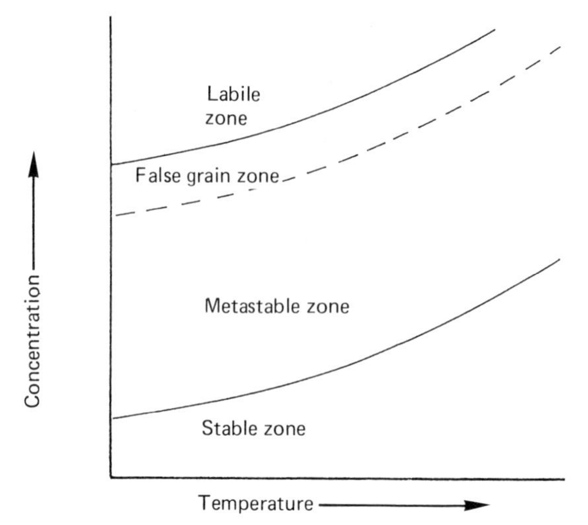


**Figure 16.2** Crystallization zones for supersaturated solutions.

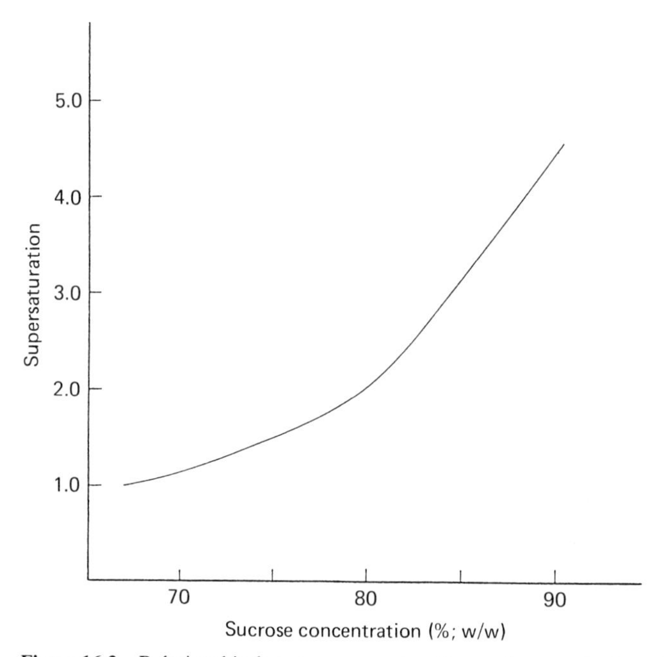


**Figure 16.3** Relationship between supersaturation and concentration of sucrose at 20 °C (Lees, 1965).

Ostwald (1897) suggested that during the cooling of a saturated solution the solute passes through a series of zones, and the ability of the solute to crystallize depends on its presence in a particular zone (*Figure 16.2*). In the stable zone crystallization will not occur, no matter how the solution is treated. In the metastable zone, which for sucrose is from $S = 1.0$–$1.3$, spontaneous crystallization does not normally occur, although a crystal which is present will grow. In the upper portion of this zone

($S = 1.2$–$1.3$) a false grain may be formed. A false grain occurs when there is a sharp rapid growth of crystal at supersaturations not normally expected to give rise to spontaneous crystallization. In the labile zone above $S = 1.3$, crystallization may occur spontaneously. It must be emphasized, however, that these zones are approximate and refer to pure sucrose. Crystallization will occur in the metastable zone as the solution is agitated (Lees, 1965).

The relationship between supersaturation and concentration of sucrose solutions is shown in *Figure 16.3*. The ability to control crystallization repeatedly forms the basis of the sugar refining process and it is also of critical relevance to the manufacturer of many forms of confectionery products. The degree to which crystallization can be retarded governs the shelf life and acceptability of the product.

## *Solubility*

Sucrose is one of the more soluble of the carbohydrates as is shown in *Figure 16.4*. There appears to be some debate as to the precise solubility of pure sucrose at 20 °C. The value of 66.6% is that found by Hinton (1925), and differs from the 67.1% which is often quoted (Herzfeld, 1892). Charles (1960) has recently confirmed the 66.6% value. The rate of solution of sucrose varies directly with temperature, degree of agitation, degree of undersaturation and inversely with crystal size. As has already been indicated, the total solids of a solution can be increased by mixing sugars. As an illustration, by combining sucrose with invert sugar in a 50:50 ratio, the overall solubility can be increased to almost 80% (w/w).

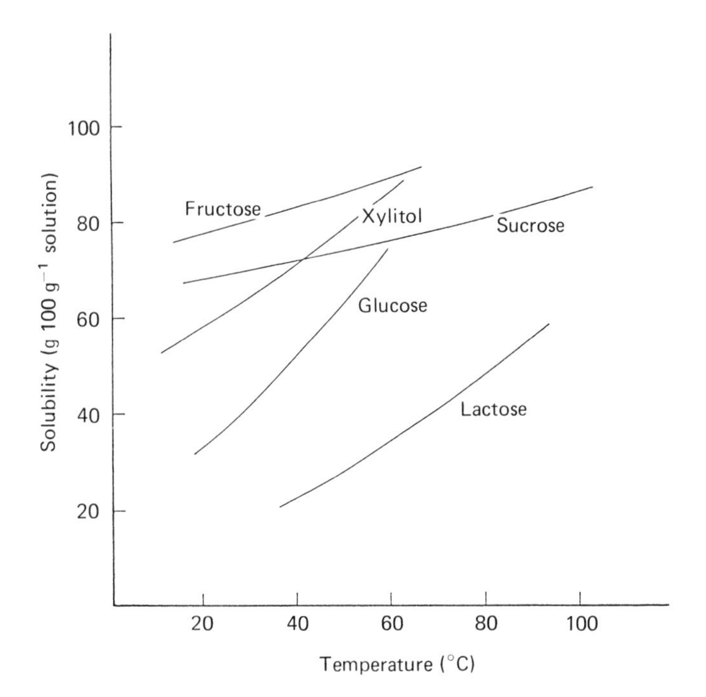


**Figure 16.4** Solubility of carbohydrate sweeteners.

### *Viscosity*

The viscosity of sucrose solutions in most circumstances is Newtonian. Viscosity varies directly with concentration and inversely with temperature. At all practical solution concentrations, viscosity plays a significant part in controlling the texture of foods. This is exemplified by the smooth mouthfeel in soft drinks and the immobile glass of boiled sweets.

### *Osmotic and Vapour Pressures*

Sugar solutions generate high osmotic pressure and this is the major factor in the use of sugars as preservatives (*Table 16.1*) (Sourirajan, 1970). The significance of this to the confectionery industry is that the practical manifestation of the vapour pressure effect is elevating boiling points. The impact of sucrose concentration on the elevation of boiling point is illustrated in *Figure 16.5* (Jones, 1959).

**Table 16.1** OSMOTIC PRESSURE OF SUCROSE AT 25 °C. ADAPTED FROM SOURIRAJAN (1970)

| *Sucrose concentration* (%, w/w) | *Osmotic pressure* (Pa × $10^3$) |
|---|---|
| 10 | 8.26 |
| 20 | 19.19 |
| 30 | 34.50 |
| 40 | 57.11 |
| 50 | 92.52 |
| 60 | 152.34 |

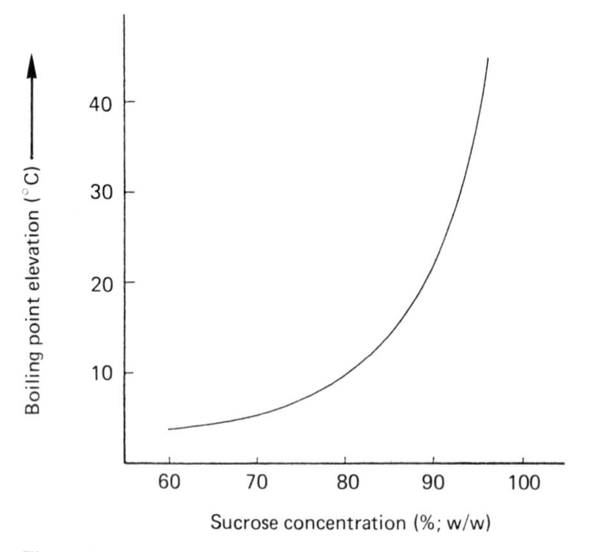


**Figure 16.5** Boiling point elevation of sucrose solutions at atmospheric pressure (Jones, 1959).

Freezing point depression is a related phenomenon which is important to the ice cream industry in particular.

### *Water activity*

Commercially available white sugar has a moisture content of approximately 0.02%. Below a relative humidity of approximately 85% the sucrose crystal will be stable. Above 85% it will begin to absorb water from the air. Closely allied with vapour pressure effects is the water activity or equilibrium relative humidity (ERH). The moisture in crystalline sugar depends on the relative humidity of the surrounding air. This explains why sucrose can cake if the humidity exceeds 86%. The crystals absorb water and the surface becomes sticky. When it falls below 86% the crystals dry out and are cemented together.

## Sugar confectionery

It is the bulk provided by sucrose and its ability to exist in different amorphous or crystalline states which, in addition to sweetness and ready solubility, provide the foundations for its largest manufacturing use in chocolate and sugar confectionery. Chocolate has already been addressed in Chapter 15 and will not be considered here. In reality the distinction between chocolate and sugar confectionery is that chocolate confectionery contains chocolate, whilst sugar confectionery does not. As regards sugar confectionery it is not easy to stipulate precise groupings within the classification. Possibly the best that can be done is to propose two classifications; those in which sugars are present in non-crystalline form and those in which some or all of the sugars are crystallized. Boiled sweets are an example of the first category, and fondants an example of the second. These products will form the basis of the future discussion within this section.

### *Boiled sweets*

Boiled sweets, sometimes referred to as 'high boilings', make use of the property that sugars in suitable combinations develop a 'glass'. They are, in fact, highly viscous, supersaturated solutions of sugar produced by boiling to yield a product of low residual moisture content. They are usually prepared from a mixture of sucrose and confectioner's glucose or invert sugar or both. By boiling under vacuum it is possible to extract the maximum amount of water compatible with ease of handling, shelf life, minimal chemical breakdown and low colour formation. A typical high boiled sweet will contain up to 98% total sugar solids and 2% residual moisture. The smaller the amount of water which is left in the product, the less will be the tendency for the boiling to undergo any physical or chemical changes. In essence, because of the very low moisture content, the sugar molecules in the sweet are effectively immobilized in the glass state and hence are unable to aggregate and come together in order to reorder themselves in a crystal structure.

The key physical measurement that can be applied to high boiled sugars is that of their viscosity. The viscosity of high boiled sweets is a particularly important factor in the design of pumping and other manufacturing equipment. Data are available

(Hudson, Roberts and Elson, 1980) to show that high sugar boilings all show Newtonian behaviour in viscosity (rate of shear proportional to stress), irrespective of the sucrose, glucose syrup, invert ratios.

The keeping qualities of boiled sweets are affected by the development of defects such as sweating and graining, and it has been shown that the absorption of moisture from the atmosphere is the principal cause of these defects. When boiled sweets are held in a humid atmosphere they will immediately begin to absorb water. In doing so the non-crystalline boil of mixed sugars is rapidly diluted by the absorbed water. The sugars in this syrup on the surface of the sweet will then reach a sufficiently low concentration and viscosity for crystallization to occur. The surface film then falls further in viscosity because of this crystallization and consequent lowering of concentration. As a result the film of syrup is able to dilute more of the boiled syrup mass, and crystallization again takes place progressively through the sweet until complete graining has occurred (*Figure 16.6*).

This description of the graining process (Campbell and Clothier, 1926) suggests that there is a single process at work in the development of grain within boiled sweets. However, Cramer (1950) considers that two types of crystallization can occur in boiled sweets. He describes the occurrence of a surface grain which develops slowly throughout the high boiling on storage and the occurrence of a complete grain which occurs during manufacture.

A dull, opaque coating with accompanying changes in colour is a consequence of surface grain. Flavour losses can also occur due to transfer to the surface by capillary

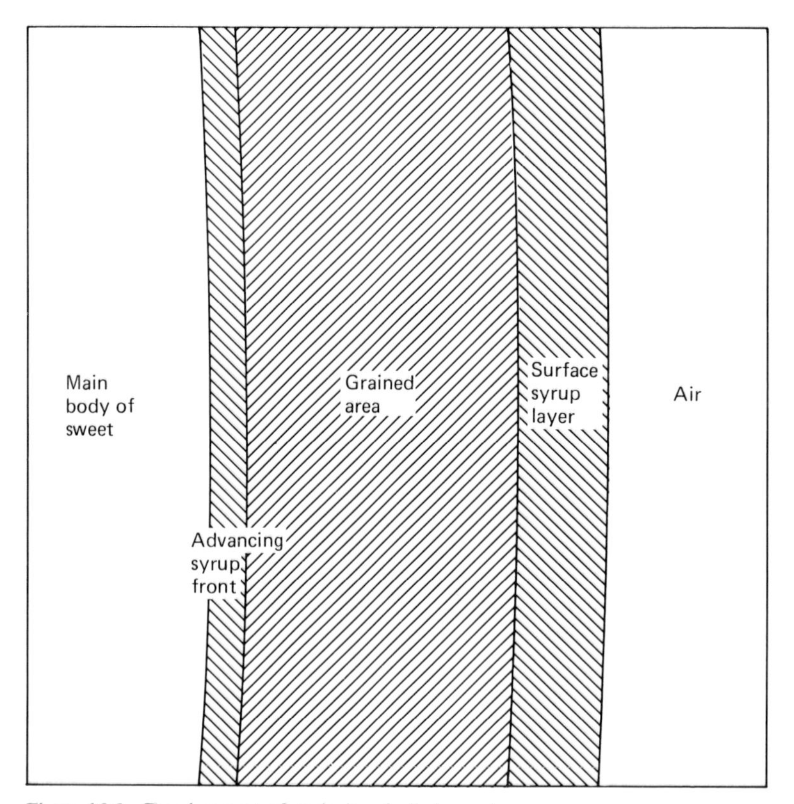


**Figure 16.6** Development of grain in a boiled sweet.

action. A boiled sweet with complete grain has a fine, closely-bound, sandy texture. In order to reduce the tendency of boiled sweets to undergo graining, a change in the choice of raw material and the ratio of the sugars in the recipe must be considered. It has been shown (Holmes, 1970) that glucose syrup is remarkably effective in reducing the rate of moisture absorption compared with other humectants. The consequence of this is a reduction in the tendency for products containing significant proportions of glucose syrups to undergo graining. The most likely explanation for this effect is the high viscosity which is associated with glucose syrup, and which has the effect of retarding the migration of moisture from the surface through the body of the sweet, although it is also possible that the higher saturated solids content may have some role to play in this.

## *Fondants*

In contrast to boiled sweets, where crystallization, should it occur, is an important defect, fondants actively require that a proportion of the mixture is crystalline and a portion remains in the syrup state. Fondant is normally prepared by heating sucrose and glucose syrup and/or invert sugar to a high temperature and agitating the syrup after it has been cooled. Prior to agitation the cooked syrup must be cooled to a highly supersaturated state so as to obtain the most favourable conditions for crystallization (Woodruff and Van Gilder, 1931). This is necessary to produce very fine crystals arising from the development of a great number of crystallization points at the same time. There are a number of factors which affect the extent of crystallization and the quality of the resultant fondant. These are:

1. The degree of supersaturation;
2. The temperature at which crystallization is induced;
3. The proportions of the syrup and crystal phases;
4. The particle size of the sugar crystals;
5. The viscosity of the syrup phase;
6. The presence of grain inhibitors (Lees, 1965).

Basic fondant is made either of mixed sucrose/glucose syrup or sucrose/invert sugar or sucrose/glucose syrup invert sugar. Haliday and Noble (1947) found that the addition of 6% invert sugar imparts the optimum desirable texture to a fondant and this figure is that given by Woodruff and Van Gilder (1931). Lees (1960) suggests that 10% invert sugar is an optimum and that 20% glucose syrup results in a fondant of the desired characteristics.

The temperature at which fondants are beaten influences greatly the crystal sizes which result (Griswold, 1962; Lowe, 1955; Woodruff and Van Gilder, 1931). There is a critical temperature for crystallization below which the viscosity inhibits that crystallization, and there is a general acceptance (Griswold, 1962) that a beating temperature of around 40 °C results in optimal crystal size, although it should be realized that slightly higher temperatures can give quicker and easier production.

The crystal size of fondants changes during maturation and this has an impact on the acceptability of the resulting product. Consequently the original formulation must be decided on the basis of the crystal size that will result at the end of the acceptable storage period. The glucose syrup and/or invert sugar component in a fondant influences the crystal size and the growth of those crystals during storage (*Table 16.2*).

**Table 16.2** GROWTH OF CRYSTALS IN A FONDANT WITH TIME. FROM DESMARAIS AND GANZ (1962)

| *Beating temperature* (°C) | *Crystal size* (% below 16 μm) | | |
|---|---|---|---|
| | *1 day* | *7 days* | *28 days* |
| 42 | 98 | 95 | 45 |
| 66 | 98 | 50 | 30 |
| 119 | 90 | 20 | 5 |

## Non-sucrose sweeteners in confectionery products

The recent availability of a range of new sweeteners presents the confectionery industry with the opportunity either to replace conventional sugars in existing products or to develop new products. Dodson and Pepper (1985) have assessed the applicability of new sugars as alternatives to sucrose and have considered sorbitol, mannitol, xylitol, hydrogenated glucose syrup and polydextrose. The results of their evaluations are summarized in *Table 16.3* with respect to their effectiveness in boiled sweet and fondant applications.

Clearly sucrose is the basic ingredient for traditional sugar confectionery, and the industry has evolved around its physical and chemical properties. These properties give sucrose-based confectionery their essential features and on the basis of this work

**Table 16.3** SUMMARY OF WORK ON ALTERNATIVES TO SUCROSE. FROM DODSON AND PEPPER (1985)

| *Product* | *Alternative* | | | |
|---|---|---|---|---|
| | *Sorbitol/ mannitol* | *Xylitol* | *Hydrogenated glucose syrup* | *Polydextrose* |
| Boiled sweets | Produces transparent, glassy products which are actually crystalline; best results obtained with a sorbitol–mannitol mixture; requires high boiling temperature | Does not form a glass | Forms a tough glass; poor flavour carrier | Forms a tough glass; poor flavour carrier; high boiling temperature necessary; very high viscosity |
| Fondants | Produces crystal phase; imparts a waxy texture | Good crystal phase; best results require other sugars as liquid phase | No crystal phase | Works well with glucose; texture not as plastic as sucrose-based fondants |

the new alternative carbohydrates are unable to match in full the properties conferred by sugar in these applications.

## Baked products

Whereas sugar confectionery and fondants can truly be described as structured sugar systems, it is not really possible to discuss baked products in the same light. Sugar is indeed the principal component in confectionery products, yet in baked goods complex carbohydrates, proteins and fats all have important roles to play in the creation of baked structures. It is probably because of this extra level of complexity in baked goods resulting from the greater range of ingredients and the interactions between them that there has been a greater level of scientific endeavour targetted at understanding which parameters control the quality of the finished products. This is in contrast to the somewhat empirical approach that is a feature of the experimental work recorded in the literature on sugar confectionery and fondants. Just as the confectionery section of this chapter focussed on boiled sweets and fondants, so the section on baked goods will address the role of sugar in biscuits and in cakes.

### *Biscuits*

The simplest form of biscuit is just a mixture of flour and water, but the additional use of fats and sugars permits the production of a substantial range of biscuit products of differing textures, appearances and tastes. The function of the fats in biscuits is one of shortening. This is a method of preventing the development of gluten in a dough by coating the flour particles with fat and thus preventing the gluten-forming proteins from becoming wetted. The consequence of this action is a general softening of the texture of the finished product. The addition of sugar to biscuits is done for a variety of reasons. The main function of sugar is to sweeten the biscuit and this simple fact must not be forgotten, but in addition to sweetening biscuits, sugar has a considerable effect on biscuit structure.

Firstly, sugar has an effect on the flour gluten, softening it and rendering it more extensible. In the absence of sugar the gluten network develops fully and a biscuit of this type shrinks in length after cutting and also during baking. The resulting product is a very hard type of biscuit. An increase in the sugar concentration results in the disruption of the gluten network and even mechanical interference in the network by sugar crystals. Secondly, sugar confers hardness to a biscuit unless sufficient shortening is added to counteract it. The hardness is caused by the sugar in solution during baking becoming saturated, and on cooling some of this sugar recrystallizes and some sets in a glass form, not dissimilar to that which occurs during the manufacture of boiled sweets. A third function of sugar, by virtue of its softening action on gluten, and the fact that it will melt during baking and form a syrup mass, is the effect that it has on biscuit flow. Sugar also has an impact on the gelatinization temperature of the starch (Abboud and Hoseney, 1984).

The type of sugar used in the manufacture of biscuits is also important. Particle size has been shown to have a significant impact, both on the spread of biscuits and on the degree to which checking occurs during baking. Caster sugar, because of its fine crystallinity and easy solubility, is used extensively, whereas granulated sugar, which is much coarser, is seldom used in biscuit doughs because of the difficulty of dissolving it in the low water content.

*Checking* is the term used to describe a phenomenon in which a hairline crack appears across the diameter of the biscuit. The crack may or may not extend across the full width of the biscuit and it usually passes through, or very close to, the centre. The structural strength of a cooled biscuit is highly dependent on its sugar content. Any stresses which are set up by the migration of moisture would be then opposed by the rigidity of the structure within the biscuit. Since the only way to increase the structural strength of a biscuit effectively is by increasing the sugar content, this might seem to be a ready solution to the checking problem. However, in the majority of cases this would seriously alter the character of the biscuit beyond the level of acceptability. Hence the practical solution is to keep the biscuits warm and therefore plastic for as long as possible after they have left the oven. In this way any moisture migration which is still taking place occurs when the biscuit is able to make compensatory movements to remove the stresses set up within it.

### *Cakes*

In cake making the ingredients fall into three principal groups:

1. Flour and eggs provide strength and structure;
2. Sugars, fats and baking powder open the texture and increase tenderness;
3. Milk and water close the structure and produce a rubbery texture.

The extent to which the various ingredients contribute to cake structure has been measured by many approaches, including recording physicochemical changes in model batter systems (Donovan, 1977) and measuring characteristics of the final product (Frazier, Brimblecombe and Daniels, 1974). These studies have concluded that the main function of sucrose in cakes is to raise the denaturation temperature ranges of the starch and the protein (*Figures 16.7* and *16.8*).

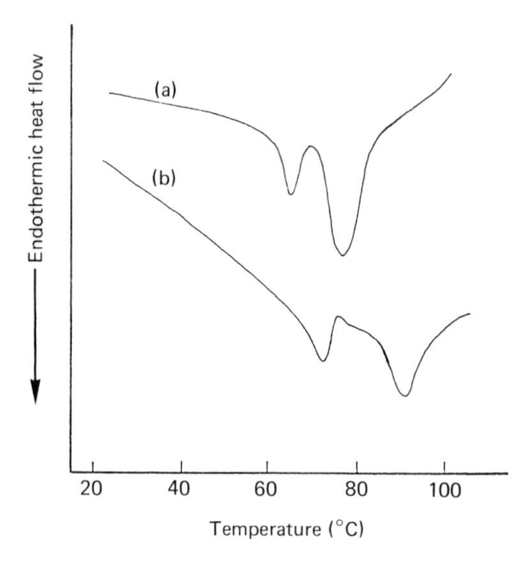


**Figure 16.7** Effect of sucrose on the denaturation of egg white proteins: (a) egg white; (b) egg white plus sucrose. Reproduced from Donovan (1977) with permission of the publisher.

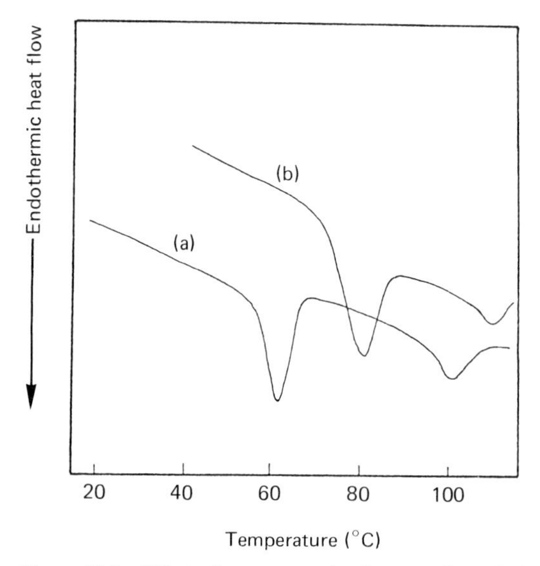


**Figure 16.8** Effect of sucrose on the denaturation of wheat starch: (a) wheat starch and water; (b) wheat starch and water, plus sucrose at 30% weight/total weight. Reproduced from Donovan (1977) with permission of the publishers.

The cohesive forces occurring during baking and the influence of sugar on them have been studied by Paton, Larocque and Holme (1981). They concluded that all levels of added sucrose resulted in good air incorporation and that the volume index was affected only at the highest sucrose level. In common with liquid shortening, sucrose was shown to exhibit an overall tenderizing effect. The maximum cohesive force was negatively correlated with increasing levels of sucrose.

The particle size of the sucrose crystals is also important. When the fat and sugar are creamed initially, air is trapped, which subsequently expands on cooking and aerates the cake. Caster sugar has been shown to be best for this purpose because granulated sugar is too coarse and has too low a surface area (Nicol, 1977). The level of sucrose has also been shown to be important (Mizukoshi, Maeda and Amano, 1980) to the starch gelatinization temperature and hence to the gas retention and cake volume.

### *Alternatives to sugar in baked products*

Recognizing that the gelatinization temperature of wheat starch was an important measurement which governs the acceptability of baked products, Bean, Yamazaki and Donelson (1978) studied the effects of different sugars on that measurement. They were able to identify the carbohydrate to flour ratio for each of the sugars which resulted in the optimal product. Thus there are some indications that glucose and fructose can substitute satisfactorily for sugar, particularly in cake products, although fructose-sweetened cakes have always been shown to be smaller in volume and have a tendency to undergo more rapid browning reactions than sucrose cakes (Hyvonen and Koivistoinen, 1982).

Glucose and fructose have also been shown (Varo *et al.*, 1979) to function in a similar way to sucrose in yeast-raised dough products. Sugar alcohols, however, were found to cause severe textural problems even at very low concentration, and their use as sweeteners in this type of product is not feasible when yeast is used as a leavening agent.

The use of lactose in bakery products has been examined by Holmes and Lopez (1977). They concluded that lactose was perfectly capable of replacing a part of the sugar in a wide range of baked products without adversely affecting the texture, appearance or flavour of the products. Lactose was, however, not able to substitute completely for sugar in the full range of baked products examined.

## Conclusions

Sucrose is an extremely versatile and valuable food ingredient to both the confectionery and baked goods industries. As well as contributing sweetness to these products, sugar has an important rôle to play in both the development and maintenance of the texture in these classes of products. Although alternatives can be used under certain circumstances, there is no single alternative low molecular weight carbohydrate which has the required versatility to fulfil all of the functional requirements that are placed on sucrose in these structured sugar systems.

## References

ABBOUD, A.M. and HOSENEY, R.C. (1984). Differential scanning calorimetry of sugar cookies and cookie doughs. *Cereal Chemistry*, **61,** 34–37

ANON (1958). Twelfth Session of the International Commission for Uniform Methods of Sugar Analysis (ICUMSA), Washington DC, p. 50

BEAN, N.M., YAMAZAKI, W.T. and DONELSON, D.H. (1978). Wheat starch gelatinisation in sugar solutions II. Fructose, glucose and sucrose: cake performance. *Cereal Chemistry*, **55,** 945–952

CAMPBELL, L.E. and CLOTHIER, G.L. (1926). Boiled goods and their keeping qualities. *BFMIRA Research Records*, No. 8

CHARLES, D.F. (1960). Solubility of pure sucrose in water. *International Sugar Journal*, **62,** 126–131

CRAMER, A.B. (1950). Crystallisation in boiled sugar confectionery. *Food Technology*, **4,** 400–403

DESMARAIS, A.J. and GANZ, A.J. (1962). Effect of cellulose gum on sugar crystallisation and its utility in confections. *Manufacturing Confectioner*, **42,** 33–36

DODSON, A.G. and PEPPER, T. (1985). Confectionery technology and the pros and cons of using non-sucrose sweeteners. *Food Chemistry*, **16,** 271–280

DONOVAN, J.W. (1977). A study of the baking process by differential scanning calorimetry. *Journal of the Science of Food and Agriculture*, **28,** 571–578

FRAZIER, P.J., BRIMBLECOME, F.A. and DANIELS, N.W.R. (1974). Rheological testing of high ratio cake flours. *Chemistry and Industry*, **24,** 1008

GRISWOLD, R.M. (1962). *Experimental Study of Foods*. Boston, Houghton Mifflin

HALIDAY, E.G. and NOBLE, I.T. (1947). *Hows and Whys of Cooking*. Chicago, University of Chicago Press

HERZFELD, A. (1892). Die Löslichkeit von Saccharose. *Zeitschrift der Deutschen Zuckerindustrie*, **42,** 232

HINTON, C.L. (1925). *BFMIRA Research Records*, No. 3.

HOLMES, A.W. (1970). The use of humectants in sugar confectionery. *BFMIRA Layman's Guide*, No. 8

HOLMES, D.G. and LOPEZ, J. (1977). Lactose in bakery products. *Bakers' Digest*, **51,** 21–26, 57

HUDSON, J.B., ROBERTS, R.T. and ELSON, C.R. (1980). The viscosity of high boiled sugars. *Food RA Research Report*, No. 338

HYVONEN, L. and KOIVISTOINEN, P. (1982). In *Nutritive Sweeteners*, (Birch, G.G. and Parker, K.J., Eds), pp. 133–144. London, Applied Science

JONES, N.R. (1959). The boiling point of sugar syrups. *BFMIRA Technical Circular*, No. 165

LEES, R. (1960). Fondant manufacture. *Confectionery Production*, **26,** 69–73

LEES, R. (1965). Factors affecting crystallization in boiled sweets, fondants and other confectionery. *BFMIRA Scientific and Technical Surveys*, No. 42

LOWE, B. (1955). *Experimental Cookery*. New York, Wiley

MACKAY, D.A.M. (1978). In *Health and Sugar Substitutes*, Proceedings of ERGOB Conference, (Guggenheim, B., Ed.) p. 321. Basel, Karger

MATHLOUTHI, M. (1975). Physical state of sucrose after lyophilisation. *Industries Agricoles et Alimentaires*, **92,** 1279–1285

MIZUKOSHI, M., MAEDA, H. and AMANO, H. (1980). Model studies of cake baking II. Expansion and heat set of cake batter during baking. *Cereal Chemistry*, **57,** 352–355

NICOL, W.M. (1977). Sucrose and food technology. In *Sugar, Science and Technology*, (Birch, G.G. and Parker, K.J., Eds), pp. 211–230. London, Applied Science

OSTWALD, W. (1897). *Zeitschrift für Physikalische Chemie*, **22,** 289

PATON, D., LAROCQUE, G.M. and HOLME, J. (1981). Development of cake structure: influence of ingredients on the measurement of cohesive force during baking. *Cereal Chemistry*, **58,** 527–529

SOURIRAJAN, S. (1970). *Reverse Osmosis*. Ongar, Logos

VARO, P., WESTERMARCK-ROSENDAHL, C., HYVONEN, L. and KOIVISTOINEN, P. (1979). The baking behaviour of different sugars and sugar alcohols as determined by high pressure liquid chromatography. *Lebensmittel Wissenschaft und Technologie*, **12,** 153–156

WOODRUFF, S. and VAN GILDER, H. (1931). Photomicrographic studies of sucrose crystals. *Journal of Physical Chemistry*, **35,** 1355–1367

17

# ELEMENTS OF CEREAL PRODUCT STRUCTURE

J. M. V. BLANSHARD
*University of Nottingham School of Agriculture, Sutton Bonington, UK*

## Introduction

Although many food grains are accurately described as cereals, we shall confine our attention almost exclusively in this chapter to wheat on the grounds that wheat gluten is particularly effective in the generation of three-dimensional elastic structures. Further, the wide variety of wheat-based cereal products produced and purveyed by the food industry reflects the subtle ways in which the elastic, rubbery properties of gluten are modified through the presence of starch, water, fat and sugar to produce the myriad products which are purchased by the consumer.

It is also our intention to consider cereal product structure from the physical and physicochemical viewpoint. Such an approach has been the subject of a comparatively limited number of studies, especially when we move away from the wheat flour/water (bread) system. Those that have been attempted have brought the investigators face to face with the enormous complexity of such systems. Firstly the molecular structures and properties of the major constituents of baked products that are significant in baked systems will be examined; thereafter some of the structures that develop and the factors that contribute to them will be looked at and finally some more recent studies of time-dependent changes occurring post-baking will be outlined.

Before embarking upon a full-scale discussion of the field, it is important to appreciate the physicochemical significance of a number of terms that will be used to describe different states and also the impact of temperature and solvent concentration on these regimes.

All amorphous polymers, whether synthetic or natural, in the absence of solvent and at sufficiently low temperatures are stiff and glassy. This is the *glassy state*, sometimes called the *vitreous state* when inorganic materials are under consideration and is characterized by molecular immobility, a high viscosity ($\eta \geqslant 10^{10}$ Pa s) and a correspondingly low molecular diffusivity. On increasing the temperature, the internal energies of the molecules increase and over a comparatively narrow range ($\sim$ 20 °C) there is a significant change in specific volume, specific heat and molecular mobility leading to rubber-like behaviour. The temperature at which this occurs is termed the *glass transition temperature*. Solvents frequently exert a plasticizing effect such that increasing the concentration of the solvent decreases the glass transition temperature. Where the rubber-like polymer/solvent system contains a substantial proportion of water and is viscoelastic in character it is termed a *gel*. This connotation is widely adopted with biopolymer/water systems.

## Molecular structure and properties of the major components of baked systems

### *Protein*

Though protein occurs typically within wheat flour as a minor constituent in the range 8–18%, nevertheless its properties and manipulation are of fundamental importance affecting water absorption, reduction/oxidation, rheology, gas retention and the ultimate product texture. Some 90% of this protein is termed gluten which is the functionally important element and represents a complex of two largely insoluble protein groups, the glutenins (with polypeptide subunits ranging in molecular weight from ~40 000 to 150 000) and the gliadins (average molecular weight ~40 000) in approximately equal quantities, which have been the subject of extensive research. These two proteins appear from electrophoretic studies to be multicomponent and polydisperse and, rather significantly, to have a similar amino acid composition but quite different physical properties when hydrated. Few of the carboxy groups are free to ionize while the low levels of lysine, histidine and arginine which are capable of acquiring positive charges in solution result in the low solubility of the proteins. In contrast the gliadin fraction behaves very much like a viscous liquid and is responsible for the extensible properties. Since the glutenin fraction behaves like a cohesive elastic solid it is not surprising that it is this component which accounts in large part for differences in gluten quality between wheat varieties (Miflin, Field and Shewry, 1983) and that the high molecular weight (HWM) elements are particularly effective. The amino acid composition (which is notable for a high level of proline, glutamate and amide nitrogen) and sequences are significant in that computer predictions of the conformational structures point to the existence of regions rich in α-helices at the N- and C-terminal ends of the polypeptides but with a central domain of numerous, repetitive β-turn structures. The existence of the latter has been demonstrated by circular dichroism spectroscopy (Tatham, Shewry and Miflin, 1984). The demonstration of such structures has led to the formulation of a model of dough behaviour where the glutenin contains both α-helical and repetitive β-turn units, the latter being responsible for the elasticity (Ewart, 1977, 1979) as shown in *Figure 17.1*.

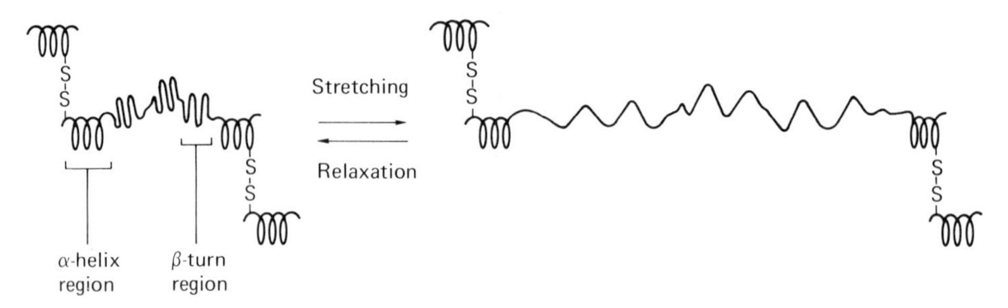


**Figure 17.1** Schematic representation of a polypeptide subunit of glutenin within a linear concatenation. Reproduced from Schofield (1986) with permission.

Somewhat surprisingly, Hoseney, Zeleznak and Lai (1986) have been unable to demonstrate any denaturation process by differential scanning calorimetry (DSC) (as might be expected if α-helical structures are present), and therefore propose that the glutenin is present as an amorphous random polymer. Slade indeed had reported in 1984 (Levine and Slade, 1987) that isolated native wheat gluten with 6% moisture possessed a $T_g$ of 66 °C, and subsequently Hoseney, Zeleznak and Lai (1986) confirmed this conclusion and published a graph of Ig versus wt% water which is a smooth curve from ~ 160 °C for the glassy polymer at < 1 wt% moisture to below 20 °C at a water content above 13 wt% moisture. The results show the extent of plasticization (about 10 °C/wt%) of gluten by water which is typical of many other water-compatible, amorphous polymers.

Gluten, of course, has reactive side groups and the effect of water and heat is to promote initially a time-dependent hydration and expansion of the molecule followed by intra- and intermolecular interactions which lead to aggregation and gelation. The effect of heat is, however, important not simply because of the baking process but also because gluten is extracted from flour, dried (almost invariably through the application of heat) and used as a fortifying additive in a variety of commercial flours. Schofield *et al.* (1983) have shown that there is a progressive decrease in the extractability of the gluten proteins in an SDS buffer which parallels the decline in baking performance; this falls to zero on heating to 75 °C. Other evidence points to changes in the glutenin proteins being responsible for the reduction in functionality (Pence, Mohammad and Mecham, 1953).

The mechanical work introduced during mixing may also be a potent factor in developing gluten structure. An appropriate speed of mixing may be necessary to give a gluten of the desired properties (MacRitchie, 1986) while over-mixing may lead to shear degradation (MacRitchie, 1975).

### *Starch*

The starch present as granules constitutes about 70% of raw wheat flour and consists of semi-crystalline polymer spherulites which are birefringent to plane polarized light. The structure and gelatinization behaviour of starch granules has been the subject of a recent review (Blanshard, 1987). The composition of the granules is approximately 75% of amylopectin (a highly branched, racemose poly α-glucan), 25% of the linear α-(1,4)-polyglucan amylose and 0.5–1.2% of lipid. X-ray diffraction measurements indicate that approximately 30% of the granule is crystalline and 70% amorphous. The architecture of the granule is shown in *Figure 17.2*. Overall the polymer chains are disposed radially while the endogenous lipid is primarily complexed with the amylose.

Exposure to excess water and increased temperatures leads to the process of gelatinization over the temperature range 50–65 °C which is accompanied by loss of birefringence and X-ray crystallinity and by a visible degree of swelling. The actual gelatinization temperature is affected both by water content and additives such as sucrose or other oligosaccharides. Their effects have been the subject of theories attempting to rationalize the observed behaviour. The first is based on a Flory–Huggins-type treatment of the gelatinization phenomenon as a melting process where the solvent is acting as a diluent. This treatment can be extended to three-component systems (Lelievre, 1973, 1976). More recently, free volume theory has been employed to rationalize the effects of both water (visualized as a plasticizer) and sugar/water systems where the sugar is regarded as a cosolvent (Levine and Slade, 1987).

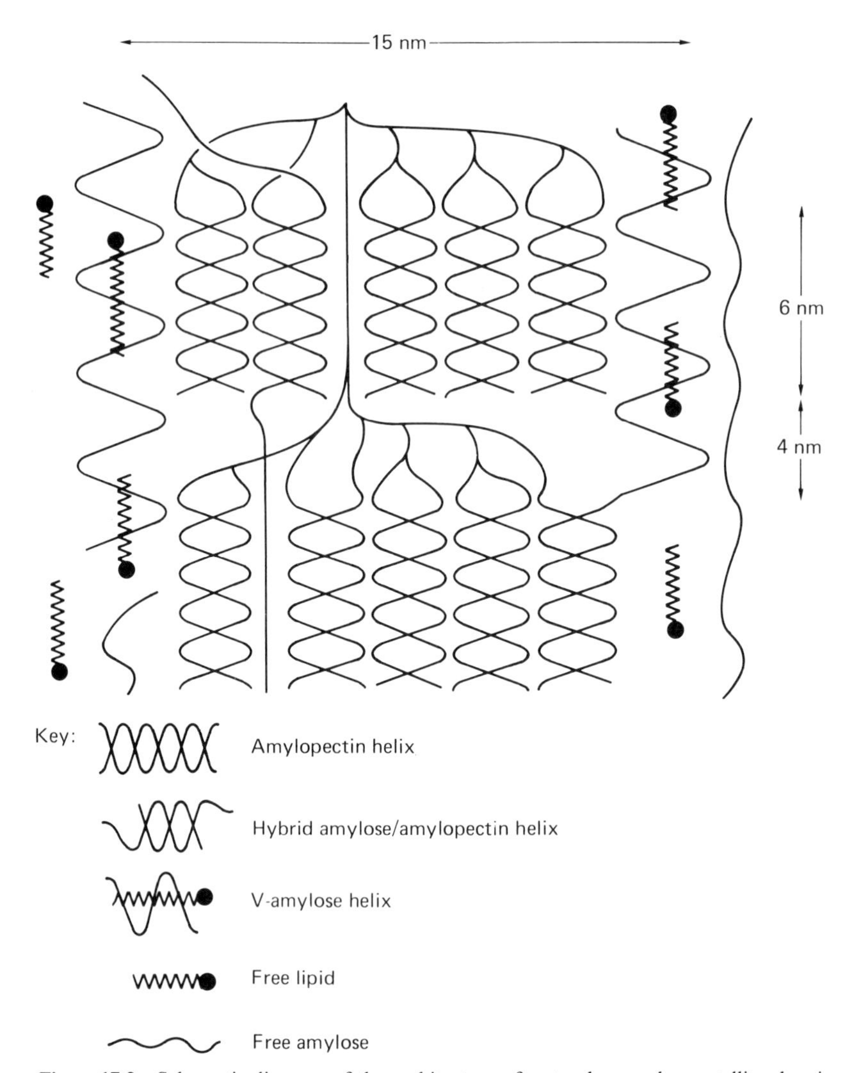


**Figure 17.2** Schematic diagram of the architecture of a starch granule crystallite showing the possible organization and distribution of different compounds.

In practice, of course, gelatinization of the starch granule occurs in a dynamic, limited water system where the other components compete with the starch for water and, indeed, where the competition from these other constituents changes as the baking process progresses and, possibly, at a rate different to starch itself. One such competitor is gluten. This aspect will be discussed at greater length later.

A consequence of this competition is that products vary in the degree of gelatinization of the starch after baking, depending not only on the initial components of the system but also on the heating regime that is adopted which may involve an associated dehydration. At one extreme we have conventional bread in which the majority of starch granules within the body of the loaf have lost their birefringence. At the other extreme are products of low initial moisture/high sugar contents where the loss of birefringence is marginal (*Table 17.1*). In general agreement with these

**Table 17.1** PERCENTAGE OF STARCH GRANULES GELATINIZED AND PERCENTAGE LOSS OF BIREFRINGENCE OF STARCH IN VARIOUS BAKED PRODUCTS. FROM LINEBACK AND WONGSRIKASEM (1980)

| *Product* | *Percentage gelatinized starch (enzymic method on isolated starch)* | *Percentage loss of birefringence (microscopic method)* |
|---|---|---|
| Angel food cake (fatless) | $97 \pm 4$ | 100 |
| White bread | $96 \pm 2$ | 100 |
| Cake doughnuts | $93 \pm 2$ | 98 |
| Pie crust | $9 \pm 0.4$ | 50 |
| Sugar cookies | $4 \pm 0.4$ | 9 |

results Abboud and Hoseney (1984) have shown using DSC that cookies have a substantial proportion of the starch which is not gelatinized. S. E. Gedney and G. Wheeler (1986, personal communication) have, however, examined a short dough biscuit formulation baked both in conventional convective and microwave ovens and shown by DSC, enzymic and X-ray methods that the inhibition of gelatinization is in large part due to the dehydration process accompanying baking in a convectively heated oven.

The exact nature of the starch granule and its contribution to the final product architecture varies. Structurally they may range from being substantially unchanged from the native state and function as building blocks, via partial to complete loss of order, with more or less swelling and variable degrees of exudation of granular polysaccharide which, if present, may contribute with one or more of gluten, sugar and fat to the intergranular matrix. The exuded polysaccharide may, depending on the water content and other contingent materials, behave physically as a glass or rubber while the amylopectin internal to the granule may likewise be in the glassy or rubbery state. At water contents greater than 20–30% w/w this may lead to retrogradation.

### *Sugar*

The principal sugar that is used is sucrose which may be presented either predissolved as a syrup or in the solid state but with varying particle sizes. Glucose syrups are also extensively employed while fructose syrups are being widely evaluated for possible advantages associated with texture, sweetening power or moisture retention.

From what has been said already, it will be evident that the presence of sugars affects the structural changes induced by heat/water in both proteins and starch. However, it is also undoubtedly true that the manner of presentation of the sugars has a significant effect. For example, in short dough biscuits, although all the sugar is dissolved by the baking process as can be demonstrated by X-ray diffraction, nevertheless the products are different depending on whether solid sucrose or sucrose syrups are initially used. It appears that textural development of the product during mixing/baking takes place along largely irreversible and non-interacting reaction paths.

Similarly, in the rather different system of cake and sponge batters, it is the interacting shortening fats and sugar crystals which are critical in the conventional two-stage method of mixing to the generation of the gas nuclei which contribute to the ultimate foamed structure.

In the final baked product, therefore, sugar may have modified the physical development of the protein and starch, may have contributed to the efficiency of development of the foam structure or even, in some products, actually be the dominating substance in the matrix. For example, in ginger biscuits the matrix cementing the flour components together is a sugar glass. The sugar/water relations are therefore critical not only for the stability of the glass, but also for the retention of flavour components.

### *Fat*

Fat is used in small amounts in bread, in larger proportions in sweet doughs (up to 15%) and of course in substantial proportions (15–25%) in cakes such as high ratio and Madeira cakes, in biscuits and pastry (15–30%). Considerable emphasis has been placed on the desirability of having the fat in the β′-form, particularly in the case of cake batters, but also in biscuit doughs. However, during the baking process the fats are totally melted. Therefore, the advantage of any distinct polymorphic form must be mediated during the dough preparation state. It is conventionally suggested that the β′-form facilitates the formation of stable air bubbles which act as nuclei in the development of the subsequent foam structure. After baking and cooling the solid/liquid ratio of any fat present may also have an important effect on the texture that is perceived by the consumer.

The actual rôle of the fat appears to be varied. For example it may, by reason of its hydrophobic properties and admixture with flour (particularly where this occurs at an early stage in the dough assembly process), limit the access of proteins and starch to the water thereby controlling gluten formation and starch gelatinization. Undoubtedly much of the art of baking resides in this area. In other instances the fat exercises a structural rôle. For example, in digestive biscuits the fat is believed to be responsible, with sugar, for 'spot welding' the starch granules together while in vol au vents the fat provides a mechanism by which a laminated structure may be generated. The layering of the fat within the pastry by the lamination process both prevents the isotropic movement of water vapour as well as providing a plane of mechanical weakness which permits the ready separation of sections of pastry.

### *Water*

The significance of water to both the baking process and the quality of the final products cannot be over-rated. Two viewpoints assist in understanding the rôle of water. Firstly, during baking itself water may be visualized as a catalyst, the physical properties of materials being particularly sensitive to the heat/moisture conditions prevailing. These conditions frequently promote plasticizing of the components and profound modification of their macromolecular order and interactions. Secondly, after baking water has an important rôle in providing the final texture. In some instances it is the absence of water and the preservation of the glassy state which is the objective; in other cases a rubbery texture is desirable and the appropriate combination of protein, starch and water give a foam matrix of the desired rheological properties; examples include bread (Wynne-Jones and Blanshard, 1986) and cake (Mizukoshi, 1985).

Flour doughs (of flour and water) where gluten formation has occurred contain

0.6–0.8 g water/g dry flour and both NMR and calorimetric measurements (Ablett, Attenburrow and Lillford, 1986) suggest that approximately half of this is associated with the flour macromolecules and hence unfreezable at –10 °C. Leung *et al.* (1976) did not observe any distinct differences in the water signals either on changing the flour type or mixing time, but did conclude that there is no simple relation between the state of the water and the viscosity of the dough.

During the baking process itself, as starch gelatinizes it picks up excess water but Nagashima and Suzuki (1981) have shown that there is no increase in the amount of non-freezable water. If so, the most significant change during starch gelatinization must be a redistribution of the water within the dough matrix.

The variation in initial and final water contents between different products and also the relative loss during baking are shown in *Table 17.2*.

**Table 17.2** PERCENTAGE MOISTURE CONTENTS OF TYPICAL AMERICAN BAKED PRODUCTS BEFORE AND AFTER BAKING. FROM DERBY, MILLER AND TRIMBO (1975)

| *Product* | *Before baking* | *After baking* |
|---|---|---|
| Cake | 37 | 30.0 |
| Bread | 43 | 37.7 |
| Pie crust | 19 | 1.3 |
| Biscuits | 41 | 30.1 |
| Russian teacake cookies | 11 | 6.6 |

## Development of structure in the baking process

Although there are significant changes in the rheological properties and interactions of macro- and micromolecules during the early stages of the baking process (including dough and batter formation and even in fermentation where this occurs), the changes occurring during the heating process are far more pronounced and are accompanied by major developments in product structure. However, it is the complexity of the system which poses a challenge to any investigator. Not only are there components which are only now being characterized and understood as individual constituents, but the interactions which occur do so in a constantly changing milieu of temperature and moisture which, furthermore, differs at any one time throughout the product. This is particularly true of conventional convective heating where the transmission of thermal energy within the product, largely by a process of conduction, is inevitably accompanied by pronounced mass transfer leading to dehydration of the product and rapid drying of the exterior surfaces. In addition there is the development of a porous structure which may be either (i) through the generation of $CO_2$ by fermentation or by the thermally promoted decomposition of baking powder; (ii) through the expansion of carefully engineered, pre-existent air bubbles as in cake batters; or (iii) by the vaporization of the water present, as in biscuits and pastry. It is obvious that with such complexity a systematic understanding will only emerge through a more analytical approach. A number of investigative avenues are possible.

Firstly, the method of heating may be modified to eliminate the thermal gradients and the variable time/temperature histories of different parts of a product. Hoseney has attempted to do this with a resistance oven (Junge and Hoseney, 1981) and has had some success in understanding the rôle of shortening in enhancing loaf volume

(Hoseney, 1986). A somewhat less sophisticated approach is the use of microwave heating.

Secondly, the complexity of the system may be reduced by examining the response of individual components to a time/temperature regime typical of a baking process. Subsequently reconstituted mixtures of varying composition and increasing complexity can be examined to gain further insight into the processes involved.

Within the compass of this chapter only a brief indication can be given of the general properties and behaviour of such systems.

### *Gluten*

Davies (1986) has reported that in the absence of starch the elastic modulus decreases first of all with increasing temperature, but at ~80 °C a substantial increase occurs reflecting the formation of a protein network by protein/protein aggregation; there is clear evidence that cysteine residues are implicated in these changes. However, relatively little work has been reported on the impact of a heating regime similar to baking processes on pure gluten as such. A developed gluten may be prepared by admixing dried gluten (Stadis Brand X wheat flour, kindly supplied by RHM and containing 8% moisture) with water in the ratio of 1:0.6 gluten:water in a pinmill for 8 min and thereafter standing for 6 min. When a 10 g sample of such a developed gluten was heated in a microwave oven (0.65 kW) at full power for 6 min, there was considerable expansion of the product giving a giant vacuous structure composed of irregularly-sized bubbles which tend to collapse on cessation of heating. Although the resultant product is difficult to quantify and relate to the heating process, it does clearly demonstrate the surface viscoelastic properties but poor aeration qualities of a gluten film. A more satisfactory form of heating involves compressing a 10 g sample of the above gluten between Perspex plates 5 mm apart and again submitting to the same microwave heating schedule. The resultant material was both highly expanded and friable and inevitably a flat structure but of moderately reproducible (± 5%) surface area. The size of bubbles was highly irregular. It is interesting that when a 10 g sample of hydrated gluten was heated between metal plates separated by a distance of 5 mm at a temperature of 210 °C for 18 min in a conventional convective oven, the resultant structure was generally comparable in texture but with more uniform bubble size. It was, however, less expanded than where microwave heating had been used. It is not unreasonable to infer that the slower rate of thermal input feasible by convective heating has permitted some surface dehydration and also structural setting of surface protein, both of which would tend to limit the expansion process on the one hand by reducing the residual water available for expansion, and on the other by enhancing the mechanical resistance of the surface of the product to such changes.

### *Gluten/starch*

Davies (1986) has reported that whereas the elastic modulus of a gluten sample shows a pronounced increase at ~80 °C, this increase occurs at ~60 °C when starch is present; this clearly reflects the process of gelatinization of the starch.

It is not unexpected, therefore, that when varying amounts of wheat starch are added to a 1:3 gluten to water system prepared as described later and heated in a microwave oven according to the prescribed schedule, pronounced differences are observed. The addition of increasing amounts of starch has been found to decrease the amount of spread and form a more regularly expanded structure which persists when the product is removed from the oven. Stress–strain measurements using the Instron texturometer under tension have shown that the stress at the yield point is markedly reduced with increasing starch (i.e. decreasing gluten) content (*Figure 17.3*). It was also evident that the yield point occurred at greater strain with the lower levels of starch.

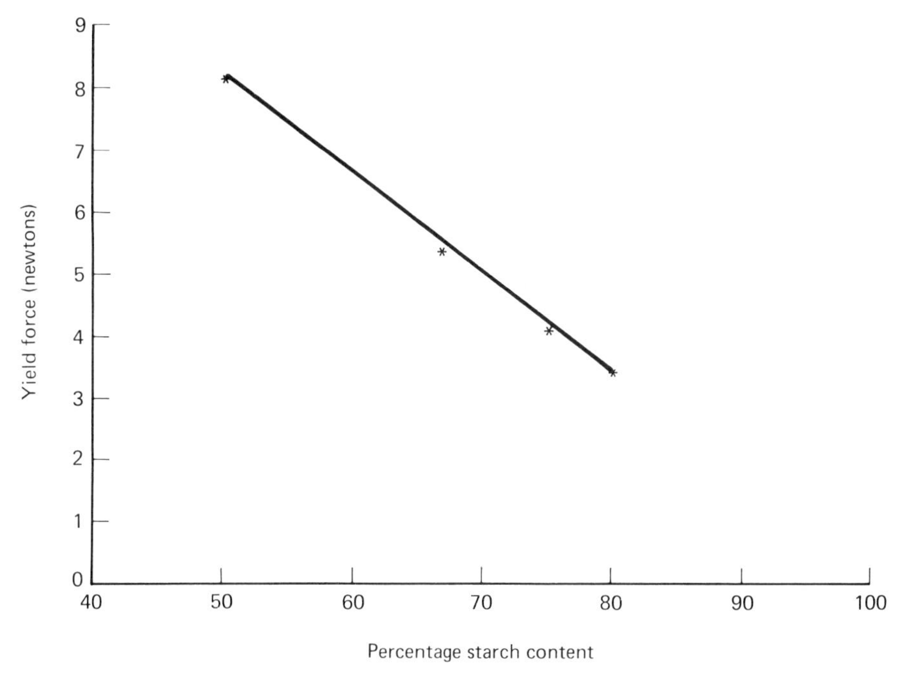


**Figure 17.3** Variation of yield force (stress) with starch content of a gluten/starch system.

A number of possible explanations can be proposed for this phenomenon. Firstly, the starch granules may be acting as a filler and consequently the overall elastic response of the dough is reduced. Secondly, the starch granules may provide weak points in the gluten films permitting the escape of water vapour at the expense of further expansion of the heated dough. Thirdly, the presence of the starch granules, particularly under conditions where gelatinization occurs, may lead to a temporary diversion of water from the expansion process. Fourthly, the gelatinization of starch at 60 °C in comparison to the structure development at ~ 80 °C for gluten, occurs at a temperature when the vapour pressure is less than half (19.94 kPa) the value at 80 °C (47.39 kPa) leading to reduced expansion. In fact all these processes may contribute to the final effect.

It is also possible to vary the rate of microwave heating by reducing the power input by a factor (dividing by $x$) while extending the time of heating by the same factor (multiplying time by $x$). The results have shown that a reduction in the heating rate leads to a reduction in expansion. There is also evidence from X-ray studies that a slower rate and extended time of heating leads to a more pronounced formation of V-amylose structures within the product.

The effect of extending the time of standing prior to baking of a gluten:starch:water (1:3:3) dough system has shown that there is a distinct increase in spread as the standing time is enhanced from 10–50 min. On the one hand it is possible that the increased time permits a more thorough hydration and development of the gluten or, more likely, that a near complete relaxation of the dough occurs prior to its forced expansion on baking.

### *Gluten/starch/other components*

#### *Fat*

The effect of fat as a third component (always presuming the presence of water) is varied but the mass of experimental evidence suggests that fat exerts its primary influence during the baking process by 'lubricating' the gluten extension.

Hoseney and coworkers have demonstrated using a resistance oven (Junge and Hoseney, 1981) that shortening extends the temperature range to 80 °C over which expansion of a dough with shortening continues unhindered, whereas in the absence of shortening, expansion was more limited above 55 °C (Hoseney, 1986; Moore, 1984). There was no evidence that starch gelatinization was responsible for this change at 55 °C or indeed the difference in composition between the two samples, nor was there significant difference in the permeability of the doughs to $CO_2$. The mechanism by which the shortening affects dough rheology was elucidated from the finding that flour, defatted with petroleum ether, did not show the rheological change at 55 °C. Further, the defatted flours did not give an increased volume when baked with shortening in the formula (Pomeranz, Shogren and Finney, 1968). Thus it appears that the free lipids in the flour, which bind to protein during mixing,

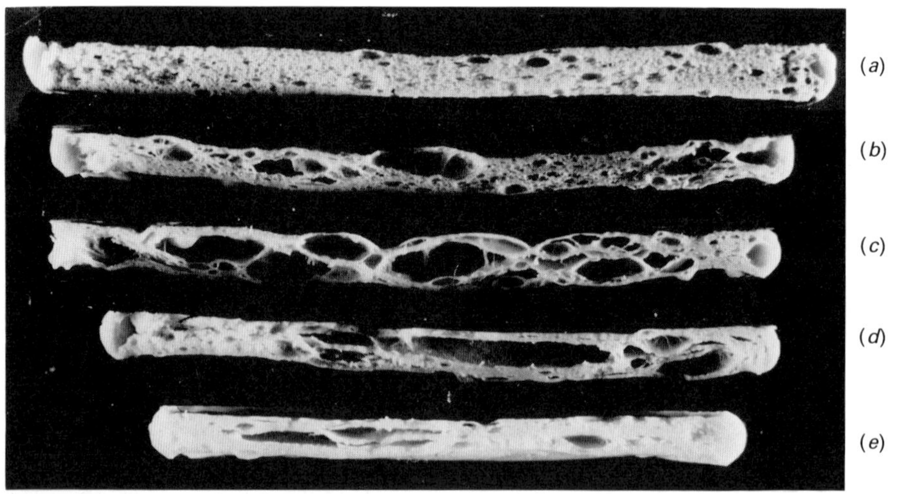


**Figure 17.4** Cross-section of a 1:3:4:1 gluten:water:starch:fat product heated in a microwave oven at varying rates from (a) 6 min at 100% power, (b) 8.6 min at 70% power, (c) 12 min at 50% power, (d) 20 min at 30% power to (e) 60 min at 10% power.

destabilize the protein so that it undergoes a rheological change at about 55 °C. The addition of shortening to the system prevents those lipids binding to the protein, probably by competing for hydrophobic binding sites. In consequence, the protein is more stable during heating and permits continued expansion up to ~ 80 °C.

In line with these conclusions, G. Wheeler and S. E. Gedney (1987, personal communication) have found that when a 1:3:4:1 gluten:water:starch:fat dough was mixed and heated in a microwave oven between Perspex plates in the fashion previously described, the spread was uniformly greater than when fat was absent. However, the spread was decreased by employing a low rate of microwave energy input, in exactly the same way as for the gluten/water/starch system. The internal structures of these materials also varied and the results can be seen in *Figure 17.4*. The sample prepared at full (100%) heating rate has a fine, honeycomb structure, but as the heating rate is reduced, the air pockets progressively increase in size. The sample heated at only 10% of the full power input had a rubbery texture with very little expansion.

Tamsdorf, Jonsson and Krog (1986) report that in high fat products (sweet doughs) with up to 15% fat, pure lard or tallow have a negative influence on the dough properties and crumb structure. These negative properties may be overcome by the use of emulsifiers.

### *Sugars*

As has already been indicated the overall effect of sugar is to elevate the temperature of starch gelatinization and protein coagulation. However, there are associated effects which arise through the competition of sugars with, for example, protein for water during the early stages of mixing of cake batters and biscuits leading to a minimization of gluten formation and an overall tenderizing effect (Shepherd and Yoell, 1976). Sucrose concentrations in the range 30–60 wt% have been shown to limit the exudation of amylose from untreated starch granules during gelatinization, whereas exudation does occur from chlorinated starch granules, the extent being linearly related to the degree of chlorination. It is not surprising that products with such chlorinated starch (e.g. high ratio cake mixtures) display a corresponding increase in paste viscosity and crumb firmness which alleviates the problem of collapse so often observed in high ratio cakes which do not use chlorinated flours.

Mizukoshi, Maeda and Amano (1980) have examined the relation between expansion and heat set of cake batter during baking. In some previous work Mizukoshi, Kawada and Matsui (1979) had established that starch gelatinization caused an increase in light transmission whereas the subsequent decrease in light transmission was the result of protein coagulation. The temperature at which light transmission was at maximum (which is a consequence of these two competing processes of gelatinization and coagulation) increased, as might be expected, for a series of cake batters with progressively increasing sucrose concentrations. The extreme of the range was 67 °C at zero sucrose concentration rising to 88 °C at 120% sucrose concentration relative to the flour. These workers found that there was almost an exact parallel between the effect of sucrose upon the cessation of bubble expansion prompted by starch gelatinization and protein coagulation and the initiation of gas release, i.e. volume extension continued to higher temperatures with progressively higher sucrose concentrations.

***Porous structure of baked cereal products***

Undoubtedly a foam is one of the most common types of structure observed in cereal products. However, only a very limited number of studies have been conducted on such systems. In terms of the process Handleman, Conn and Lyons (1961) have studied semi-empirically, using the Laplace equation, the relationship between bubble-to-bubble gas diffusion and bubble mechanics during baking and the ensuing cake quality but no thorough theoretical analysis appears to have been made of the effect of the non-Newtonian cake batter on bubble motion and deformation. Turning to the actual product, Mizukoshi (1985) has sought to relate the porosity of cake to its shear modulus using the Mackenzie (1950) equation with some degree of success while Ablett, Attenburrow and Lillford (1986), building on the generalized treatment of foams discussed by Ashby and Gibson (1983), have examined the effect of compression. When a foam is compressed the stress–strain curve shows three regions (*Figure 17.5*). At low strains the foam deforms in a linear elastic fashion, leading to a plateau of deformation at almost constant stress, and finally a region of densification as the cell walls crush together. Ablett, Attenburrow and Lillford (1986) have confirmed that food foams respond in the same way and that as with foams of synthetic polymers, the most important aspect of the structure in terms of the mechanical properties is the relative density $\rho/\rho_s$ where $\rho$ is the density of the foam and $\rho_s$ that of the solid matrix. Using the expression given by Ashby and Gibson (1983)

$$E/E_s = K_1\ (\rho/\rho_s)^2$$

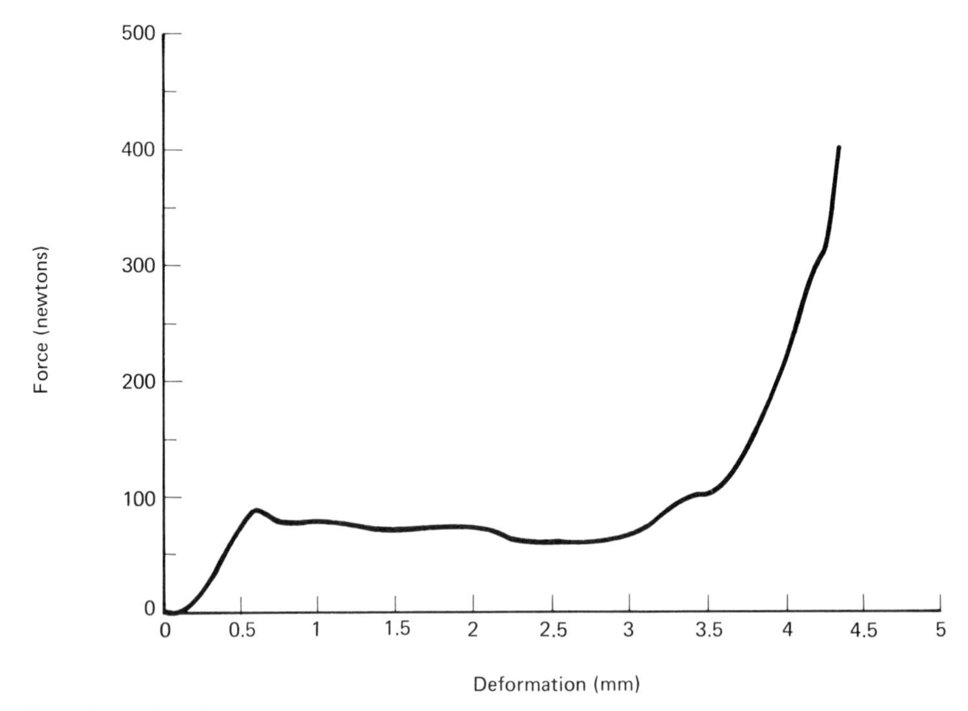


**Figure 17.5** Plot of stress versus strain for a 'baked' 1:4 gluten:starch system showing behaviour typical of a foam.

where $E$ is the observed bulk elastic modulus, $E_s$ is the modulus of the cell wall material and $K_1$ is a constant, they demonstrated with dried sponge cakes of varying densities that, as anticipated, a linear relation with a slope of 2 was observed between log $E$ and log $\rho$. G. Wheeler and S. E. Gedney (1987, personal communication) have likewise shown that gluten/starch systems, expanded to different degrees (i.e. of different densities) and produced as described on p. 321, also followed the above relation. Visual observation suggested that breakdown occurs both by brittle fracture and elastic buckling.

## Time-dependent changes occurring post-baking

### *General aspects*

The changes which take place after baking are of considerable interest since, if they are deleterious, they limit the shelf life of the product and acceptability to the consumer. Such changes include microbiological attack and deterioration in flavour and aroma but for many products it is the textural changes which are of dominating importance in the minds of both manufacturer and consumer.

Over the years various components have been considered as possible culprits, but undoubtedly the weight of evidence points to starch as being primarily responsible for the textural changes, though other components such as the proteins and pentosans are known to modify the perceived final texture or the rate at which it is attained. Early studies suggested that the changes in texture were broadly matched by increasing crystallinity in the starch. This process of starch crystallization is known as retrogradation. Further, the perceived textural changes and apparent enhanced dryness were not related to a gross loss of water from the product. Longton and Le Grys (1981) have, however, shown that the rate of retrogradation is controlled by the amount of water contained in the system. There is also clear evidence that while in dilute starch solutions it is the linear amylose which is responsible for retrogradation, in more concentrated systems typical of foodstuffs such as bread the amylopectin is the offending component.

The changes in texture can be followed by the usual static or dynamic rheological methods while the accompanying changes in crystallinity have been extensively investigated by DSC and, less commonly, by X-ray diffraction. These more precise measurements, when used to compare the rates of firming (staling) and retrogradation, have not given over good correlation (Hoseney, 1986). These rate processes have, however, been widely interpreted and analysed by means of the Avrami equation irrespective of whether the process was a textural change in bread, a rheological change in a starch gel or a calorimetric or diffraction-type measurement of crystallinity. Such an analysis does not, of itself, give any clear indication of the mechanisms responsible for such behaviour.

### *Application of polymer crystallization theory*

Some insight has been obtained through the application of the concepts of polymer physics. Many theories have been developed in this field to explain the energetics and kinetics of polymer crystallization, whether from the melt or from solution (Sanchez,

1974). All of the theories are based on classical nucleation theory which postulates that the nucleation rate for a phase transformation is the consequence of two processes: (i) the actual initiation and building of a nucleus of critical size (which is described by a term, $\exp(-\phi^*/kT)$ where $\phi^*$ is the free energy required to build the nucleus) and (ii) a transport process through a viscous medium (given by the term $\exp(-\Delta G/kT)$ where $\Delta G$ is the free energy barrier opposing transport of material across the interface). The actual nucleation rate $s^*$ is then given by

$$s^* = s_0 \exp(-\Delta G/kT) \exp(-\phi^*/kT)$$

where $s_0$ is a nearly temperature-independent constant.

The form of this expression has important consequences. At temperatures near the melting point of the system, the nucleation term is dominant and is inversely proportional to the undercooling ($T_m - T$) where $T_m$ and $T$ are the melting and crystallization temperatures respectively. Hence, near the melting point the nucleation rate decreases as the critical size of the nucleation increases. The nucleation rate therefore shows a negative temperature coefficient as the size of the critical nucleus decreases with decreasing temperature. However, at temperatures near to the glass transition temperature, the transport term is dominant and the nucleation rate depends upon the microscopic viscosity of the system. Thus, as the microscopic viscosity increases with decreasing temperature, the nucleation rate likewise decreases.

It is evident, then, that if crystallization is considered to be a process of secondary nucleation on a pre-existing substrate, then the positive and negative temperature coefficients which have been widely reported for polymer crystals can be explained.

### *The polymer crystallization theory of Lauritzen and Hoffman*

The theory of polymer crystallization developed by Lauritzen and Hoffman (1973) introduces a modified nucleation term in which the steady state net nucleation flux over the barrier to nucleation is integrated over all possible values of crystal thickness to give the total flux, $s_T$.

$$s_T \propto \exp(-K_g/T\Delta Tf)$$

where $K_g$ is a constant, $\Delta T$ is the undercooling and $f = 2T/(T_m + T)$. This term dominates the growth rate at small undercoolings.

The term governing growth rate at large undercoolings ($\simeq 100$ K) is deduced by considering the jump rate of local motions at temperature $T$. This is usually interpreted as representing the segmental jump motion of a polymer chain and may be expressed as (Hoffman, 1964):

$$(kT/h)J_1 \exp[-U^*/R(T - T_\infty)]$$

where $T_\infty$ is a hypothetical temperature at which viscous flow ceases and approximately equals $T_g$–30 K, while $U^*$ is the activation energy for steady state reptation, i.e. 'reeling in' the polymer chain.

If we gather together all the nearly temperature-independent terms into a constant factor $G_0$, then the growth rate can be expressed as

$$G = G_0 \exp[-U^*/R(T - T_\infty)] \exp(-K_g/T\Delta Tf)$$

The form of this equation permits zero growth rate at $T_m$ and $T_g$ and a maximum in the rate at some intermediate temperature.

Marsh and Blanshard (1987) have applied this approach to the study of the retrogradation of 50% starch gels (prepared from the equivalent of 50 g water and 50 g totally dry starch). The gels were prepared by a mild extrusion process and stored at varying temperatures under conditions to minimize mould growth. The crystallinity of the samples was determined by X-ray diffraction, the samples being examined periodically up to three weeks post-gelatinization by which time all the perceptible changes, as determined by X-ray diffraction, were complete. The absolute crystallinity was determined by the method of Hermans and Weidinger (1948). The raw crystallinity data were fitted to the Avrami equation and limiting crystallinities and rate

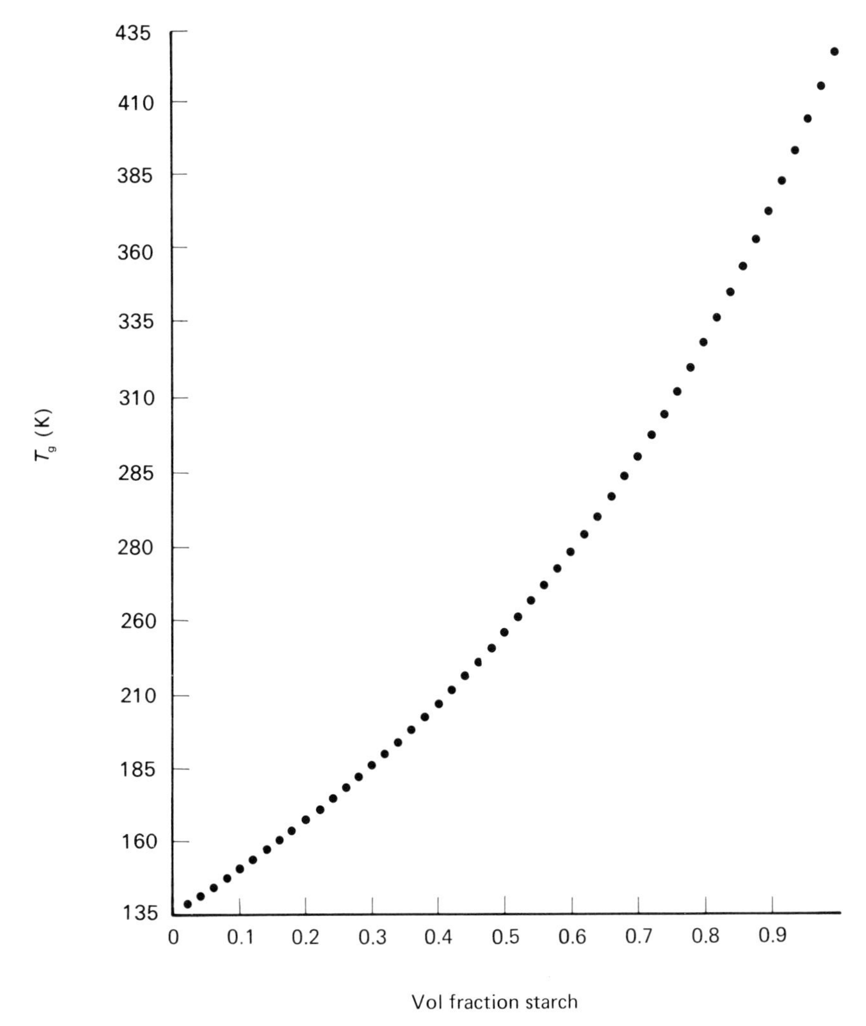


**Figure 17.6** Variation of the glass transition temperature ($T_g$) with the volume fraction of starch as calculated by free volume theory.

constants derived. Thereafter, by a process of fitting, trial values of $U^*$ and $T_\infty$ were substituted to obtain the best fit to the growth data. It was found (Marsh and Blanshard, 1987) that the optimum value of $K_g$ was in approximate agreement with those for synthetic polymers while the values of $U^*$ of 1.5 kJ mol$^{-1}$ and $T_\infty$ of 170 °K gave a best fit to the data. As $T_\infty = T_g - 30$ °K, the glass transition temperature ($T_g$) for a 50% starch and water gel is predicted to be approximately 200 °K.

A similar value of $T_g$ can also be predicted for a 50% gel using free volume theory (Marsh and Blanshard, 1987). Further, if these calculated values are plotted (*Figure 17.6*) it will be evident that no crystallization of starch/water systems which contain less than 20% moisture should occur at storage temperatures below room temperature. This behaviour is observed in practice.

### *The role of water*

Some reference ought to be made to the role of water in these systems, particularly since one of the notable features in the staling of bread is that there is a perceived increase in dryness. The subject has been investigated both by gravimetric (Willhoft, 1971) and more sophisticated spectroscopic (Lechert *et al.*, 1980; Leung, Magnuson and Bruinsma, 1983; Wynne-Jones and Blanshard, 1986) and calorimetric techniques (Bushuk and Mehrotra, 1977).

From the study of the individual starch and gluten components as well as in the combined state, Wynne-Jones and Blanshard (1986) postulate that on ageing of bread there is an increase in the water binding by the starch phase in bread as can be observed in simple starch gel systems but, at the same time, water is also transferred from the gluten phase and enters the starch phase as free water. The quantities of water and the rates involved in these processes are comparable and thus the overall effect is to produce a minimal change in the fraction of bound water and, consequently, little change in the relaxation behaviour of the system. This change, as shown by Willhoft (1971), is non-reversible. Thus, refreshening of the bread produces no change in the system, as seen by nuclear magnetic resonance.

## References


ABBOUD, A. M. and HOSENEY, R. C. (1984). Differential scanning calorimetry of sugar cookies and cookie doughs. *Cereal Chemistry*, **61**, 34

ABLETT, S., ATTENBURROW, G. E. and LILLFORD, P. J. (1986). The significance of water in the baking process. In *Chemistry and Physics of Baking* (Blanshard, J. M. V., Frazier, P. J. and Galliard, T., Eds), p. 30. London, Royal Society of Chemistry

ASHBY, M. F. and GIBSON, L. J. (1983). The mechanical properties of cellular solids. *Metallurgy Transactions*, **14**, 1755

BLANSHARD, J. M. V. (1987). Starch granule structure and function: a physicochemical approach. In *Starch: Properties and Potential* (Galliard, T., Ed.) p. 16. Critical Reports in Applied Chemistry. London, Society for Chemistry and Industry

BUSHUK, W. and MEHROTRA, V. K. (1977). Studies of water binding by differential thermal analysis. II. Dough studies using the melting mode. *Cereal Chemistry*, **54**, 320

DAVIES, A.P. (1986). Protein functionality in bakery products. In *Chemistry and Physics of Baking* (Blanshard, J. M. V., Frazier, P. J. and Galliard, T., Eds), p. 89. London, Royal Society of Chemistry

DERBY, R. I., MILLER, B. S. and TRIMBO, H. B. (1975). Visual observation of wheat starch gelatinization in limited water systems. *Cereal Chemistry*, **52**, 702

EWART, J. A. D. (1977). Reexamination of the linear glutenin hypothesis. *Journal of the Science of Food and Agriculture*, **28**, 191

EWART, J. A. D. (1979). Glutenin structure. *Journal of the Science of Food and Agriculture*, **30**, 482

HANDLEMAN, A. R., CONN, J. F. and LYONS, J. W. (1961). Bubble mechanics in thick foams and their effects on cake quality. *Cereal Chemistry*, **38**, 294

HERMANS, P. H. and WEIDINGER, A. (1950). Quantitative investigation by X-ray of 'amorphous' polymers and some other non-crystalline substances. *Journal of Polymer Science*, **5**, 269

HOSENEY, R. C. (1983). Gas retention in bread doughs. *Cereal Foods World,* **29**, 305

HOSENEY, R. C. (1986). Component interactions during heating and storage. In *Chemistry and Physics of Baking* (Blanshard, J. M. V., Frazier, P. J. and Galliard, T., Eds), p. 216. London, Royal Society of Chemistry

HOSENEY, R. C., ZELEZNAK, K. and LAI, C. S. (1986). Wheat gluten: a glassy polymer. *Cereal Chemistry*, **63**, 285

JUNGE, R. C. and HOSENEY, R. C. (1981). A mechanism by which shortening and certain surfactants improve loaf volume in bread. *Cereal Chemistry*, **58**, 408

LAURITZEN, J. I. and HOFFMAN, J. D. (1973). Extension of theory of growth of chain-folded polymer crystals to large undercoolings. *Journal of Applied Physics*, **44**, 4340

LECHERT, H., MAIWALD, W., KOTHE, R. and BASLER, W. D. (1980). NMR study of water in some starches and vegetables. *Journal of Food Processing and Preservation*, **3**, 275

LELIEVRE, J. (1973). Starch gelatinization. *Journal of Applied Polymer Science*, **18**, 293

LELIEVRE, J. (1976). Theory of gelatinization in a starch–water–solute system. *Polymers*, **17**, 854

LEUNG, H. K., STEINBERG, M. P., NELSON, A. E. and WEI, L. S. (1976). Water binding of macromolecules determined by pulsed NMR. *Journal of Food Science*, **41**, 297

LEUNG, H. K., MAGNUSON, J. A. and BRUINSMA, B. L. (1983). Water binding of wheat flour doughs and breads as studied by deuteron relaxation. *Journal of Food Science*, **48**, 95

LEVINE, H. and SLADE, L. (1987). Water as a plasticizer: physicochemical aspects of low-moisture polymeric systems. In *Water Science Reviews* (Franks, F., Ed.). Cambridge, Cambridge University Press (in press)

LINEBACK, D. R. and WONGSRIKASEM, E. (1980). Gelatinization of starch in baked products. *Journal of Food Science*, **45**, 71

LONGTON, J. and LE GRYS, G. A. (1981). Differential scanning calorimetry studies on the crystallinity of ageing wheat starch gels. *Starch*, **33**, 410

MACKENZIE, J. K. (1950). The elastic constants of a solid containing spherical holes. *Proceedings of the Physical Society, London*, **B63**, 2

MACRITCHIE, F. (1975). Mechanical degradation of cereal proteins during high speed mixing of doughs. *Journal of Polymer Science*, Symposium No. 49, 85

MACRITCHIE, F. (1986). Physicochemical processes in mixing. In *Chemistry and Physics of Baking* (Blanshard, J. M. V., Frazier, P. J. and Galliard, T., Eds), p. 132. London, Royal Society of Chemistry

MARSH, R. D. L. and BLANSHARD, J. M. V. (1988). The application of polymer crystal growth theory to the kinetics of formation of the β-amylose polymorph in a 50% wheat starch gel. *Carbohydrate Polymers*, (in press)

MIFLIN, B. J., FIELD, J. M. and SHEWRY, P. R. (1983). Cereal storage proteins and their effect on technological properties. In *Seed Proteins* (Doussant, J., Mosse, J. and Vaughan, J., Eds), p. 255. Phytochemical Society of Europe, Symposium Series No. 20. London, Academic Press

MIZUKOSHI, M. (1985). Model studies of cake baking. V. Cake shrinkage and shear modulus of cake batter during baking. *Cereal Chemistry*, **62**, 242

MIZUKOSHI, M., KAWADA, T. and MATSUI, N. (1979). Studies of cake baking. I. Continuous observations of starch gelatinization and protein coagulation during baking. *Cereal Chemistry*, **56**, 305

MIZUKOSHI, M., MAEDA, H. and AMANO, H. (1980). Model studies of cake baking. II. Expansion and heat sit of cake batter during baking. *Cereal Chemistry*, **57**, 352

MOORE, W. R. (1984). PhD Dissertation. Kansas State University

NAGASHIMA, N. and SUZUKI, E. (1981). Pulsed NMR and the state of water in foods. In *Water Activity: Influence on Food Quality*. (Rockland, L. B. and Stewart, G. F., Eds), p. 247. London, Academic Press

PENCE, J. W., MOHAMMAD, A. and MECHAM, D. K. (1953). Heat denaturation of gluten. *Cereal Chemistry*, **30**, 115.

POMERANZ, Y., SHOGREN, M. D. and FINNEY, K. F. (1968). Functional bread-making properties of wheat flour liquids. I. Reconstitution studies and properties of defatted flours. *Food Technology*, **22**, 324

SANCHEZ, I. C. (1974). Modern theories of polymer crystallisation. *Journal of Macromolecular Chemistry*, **C10**, 113

SCHOFIELD, J.D. (1986). Flour proteins: structure and functionality in baked products. In *Chemistry and Physics of Baking* (Blanshard, J.M.V., Frazier, P.J. and Galliard, T., Eds), p. 14. London, Royal Society of Chemistry

SCHOFIELD, J. D., BOTTOMLEY, R. C., TIMMS, M. F. and BOOTH, M. R. (1983). The effect of heat on wheat gluten and its involvement of sulphydryldisulphide interchange reactions. *Journal of Cereal Science*, **1**, 241

SHEPHERD, I.S. and YOELL, R.W. (1976). Cake emulsions. In *Food Emulsions* (Freiberg, S., Ed.), p. 215. New York, Marcel Dekker

TAMSDORF, S., JONSSON, T. and KROG, N. (1986). The role of fats and emulsifiers in baked products. In *Chemistry and Physics of Baking* (Blanshard, J. M. V., Frazier, P. J. and Galliard, T., Eds), p. 75. London, Royal Society of Chemistry

TATHAM, A. S., SHEWRY, P. R. and MIFLIN, B. J. (1984). Wheat gluten elasticity: a similar molecular basis to elastin. *FEBS Letters*, **177**, 205

WILLHOFT, E. M. A. (1971). Bread staling. I. Experimental study. *Journal of the Science of Food and Agriculture*, **22**, 176

WYNNE-JONES, S. and BLANSHARD, J. M. V. (1986). Hydration studies of gels and bread by $^{1}H$ NMR. *Carbohydrate Polymers*, **6**, 289

## 18

# EXTRUSION AND CO-EXTRUSION OF CEREALS

R.C.E. GUY and A.W. HORNE
*Flour Milling and Baking Research Association, Chorleywood, Hertfordshire, UK*

## Introduction

The manufacture of products based on cereal-rich formulations by extrusion cooking involves a wide range of processing techniques. Such products may be found in several sectors of the food industry including breakfast cereals, snacks, pregelatinized flours and starches, pet foods and animal feedstuffs (Harper, 1981; Linko, Colonna and Mercier, 1981).

The earliest form of extrusion processing concerned low temperature (< 90 °C) forming processes for pasta and cereal doughs where the cereal dough was compressed, possibly deaerated, before shaping at a die to form an intermediate for pasta, flakes, etc.

These early forms of extrusion processing are still used today but in addition the extruder has become a high temperature, short time (HTST) cooking unit. An early form of extrusion cooker, based on a single screw design, was operated by the Adams Company in the mid-1940s to manufacture a snack product from maize grits. Since that time extrusion cookers have been developed and improved so that they can be operated with a wide range of raw materials under well controlled processing conditions. Today there are many types of extruders available either with single or twin screws, each with individual designs and modes of operation.

In order to make some sense of the processing of cereal-rich formulations on such a range of machines it is necessary to examine the processing patterns on one type of machine at a fundamental level. Once the basic changes in the raw materials, caused by the processing in the machine, are understood and are related to product characteristics, the differences in processing due to machine design may be understood more easily. For this review research studies with twin screw extruders are mainly used to represent the extrusion process because of the simpler relationships between the process variables compared with single screw machines.

In the general range of extruded products structure may be created by two different mechanisms. Products may be formed at the die of the extruder by direct expansion. Alternatively unexpanded 'half products' as 'pellets' may be produced at the die which after careful drying may be expanded to form products by rapid application of heat as in frying, baking or the use of microwaves. A schematic representation of the two types of processes is shown in *Figure 18.1*. The production of 'half products' requires the use of special extruders with long barrels, or two extruders running in tandem, if cooking is performed during the process. However it is also possible to

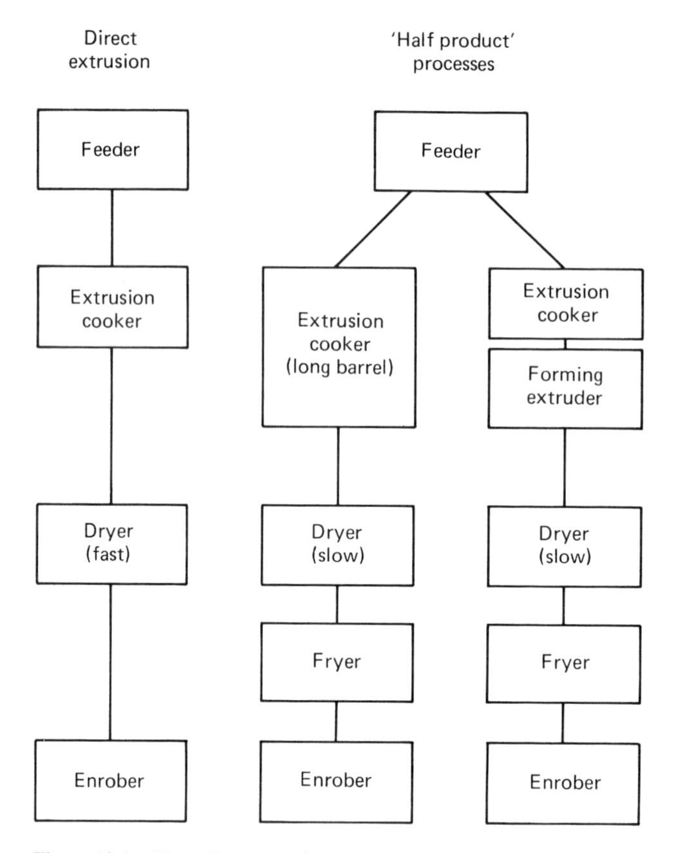


**Figure 18.1** Flow diagrams of extrusion cooking processes for the manufacture of products by direct expansion or via the 'half product' method.

manufacture 'half products' from cereal blends containing pregelatinized flours or starches using a simple forming extruder.

## Manufacture of products by direct expansion at the die

In the direct expansion process the farinaceous mix is compressed and subjected to high temperatures and powerful shearing forces at moisture contents of $\gtrsim 30\%$ (wet wt basis). The raw materials are transformed from the original powder form into a dense viscous fluid which is extruded through a small die at high temperatures and pressures. At the die exit the pressure falls to atmospheric and puffing occurs due to the release of water vapour. Products are formed by cutting the expanded extrudate into individual pieces either at the die or later when the extrudate has cooled and become hard and brittle.

In this chapter the transformation of cereal-based raw materials into extruded products is examined in detail, both in terms of the fluid developed within the extruder and the expansion phenomena at the die exit.

***Evidence for the fluid state***

Several workers have used capillary or slit die viscometers attached to extrusion cookers to examine the physical state of the cooked extrudates (Cervone and Harper, 1978; Van Zuilichem, Buisman and Stolp, 1974). It was found that the extrudate behaved as a viscous fluid giving a linear pressure drop along the die from entrance to exit. Similar studies carried out at FMBRA, with the capillary viscometer shown in *Figure 18.2* attached to Baker Perkins MPF 50D twin screw extruder, showed a linear pressure drop with wheat flour extrudates of 16–20% moisture content for a wide range of process conditions.

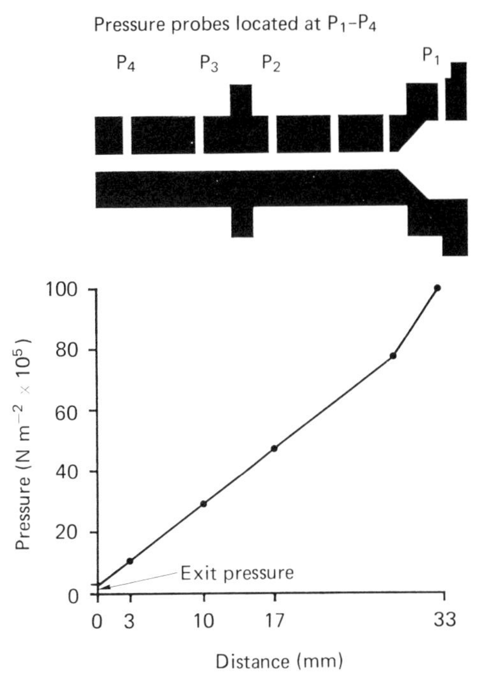


**Figure 18.2** A capillary die viscometer (12 × 300 mm) used on a Baker Perkins MPF 50D extruder with wheat flour processed at 300 rpm, 19% moisture, feed rate 800 g min$^{-1}$ at 150 °C. (Screw system as in *Figure 18.4*).

It has been shown that the viscosity of maize and wheat flours is non-Newtonian and may be described by a power law expression (Harper, 1981, *see* Eqn 1), where $\eta_{app}$ is the apparent viscosity (N s m$^{-2}$), $\gamma$ is the shear rate (s$^{-1}$) and $K$ and $n$ are constants sometimes referred to as the consistency index and the flow behaviour index, respectively. Results for a limited number of observations shown in *Table 18.1*, suggest that $n$ generally lies between 0.2 and 0.6 for well cooked fluid, indicating a shear thinning type of fluid viscosity.

$$\eta_{app} = K\gamma^{n-1} \qquad (1)$$

Studies at FMBRA (*Figure 18.3*) show how the viscosity of cooked wheat flour extrudates varies with moisture content and the level of applied shear forces as indicated by the specific mechanical energy input (SME). The fluid viscosity increases

**Table 18.1** VISCOSITY DATA FOR FLUID EXTRUDATES USING CAPILLARY OR SLIT VISCOMETERS FITTED TO EXTRUSION COOKERS

| *Type of cereal* | *Moisture* (% wet wt basis) | *Temperature* (°C) | η | *K* $(N\,s\,m^{-2})^n$ | *References* |
|---|---|---|---|---|---|
| Corn flour | 22–35 | 90–150 | 0.36 | 36 | Cervone and Harper (1978) |
| Corn grits | 13 | 177 | 0.45–0.55 | 28 | Van Zuilichem *et al.* (1974) |
| | | 193 | — | 17 | |
| | | 207 | — | 7.6 | |
| Wheat flour | 19 | 130 | 0.2 | 41 | Guy and Roberts (1986, personal communication) |

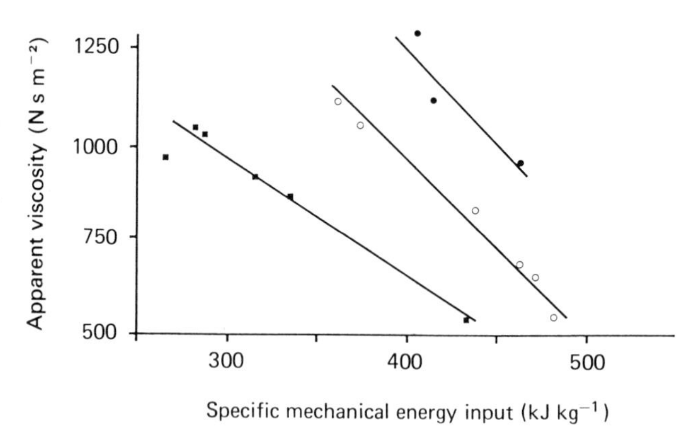


**Figure 18.3** Fluid viscosity of wheat flour at $60\,s^{-1}$ for ●, 16; ○, 18; and ■, 20% moisture contents versus SME inputs, measured on capillary die viscometer shown in *Figure 18.2*.

as the moisture content is reduced, but falls sharply as the fluid is subjected to increasingly high SME inputs.

A special feature of the results observed for wheat flour (*Figure 18.2*) was the small positive exit pressure. This tends to show the presence of elastic characteristics in the fluid extrudate (Han, 1976). In a recent review Harper (1986) also speculated on the presence of elastic characteristics which might give rise to die swell phenomena (Han, 1976) although no evidence was quoted to support this hypothesis. At FMBRA some practical studies with extrudates prepared from wheat flour, wheat starch and manioc starch were made on a screw system with cooling applied following the cooking process as in the 'half product' process. The extrudates were cooled to < 100 °C before reaching the die exit so that they were not expanded to an alveolar structure by water vapour, but remained as dense fluids ($1.4\,g\,ml^{-1}$ on cooling to 25 °C). All the extrudates showed large die swells with values of 50%, 127% and 200% for wheat flour, wheat starch and manioc starch, respectively. Since die swells of more than 10% under the test conditions indicated the presence of elastic characteristics, it is fairly certain that during the extrusion cooking of cereals a viscoelastic fluid is formed whose elastic characteristics are related to the starch component.

### *Structure formation in directly expanded products*

The most important property of the fluid developed within the extruder is its ability to expand at the die. For fluids cooled to below 100 °C there is no obvious change in density or production of any porous cellular structures. At higher temperatures water vapour is released at the die exit. This water vapour serves to nucleate bubbles in the fluid extrudate, some of which may expand and form the gas cells visible in the final product.

During the expansion stage the extrudate reaches its maximum size and is formed into the general shape of the product. The cellular crumb structures are formed by the gas cell walls of the water vapour bubbles after they have ruptured and cooling has taken place. Directly extruded products cut at the die may be characterized for the purpose of this chapter, in terms of their size, shape and cellular crumb texture.

The extrusion cooking process is designed to cook the cereal raw materials and develop them into a fluid which has suitable properties for expansion. This fluid should expand under controlled conditions to give the required size, shape and crumb texture. In order to achieve the desired effects it is necessary to understand how the independent variables, both for processing and raw materials, act together in the cooking process to develop suitable conditions at the die and the necessary physical properties in the fluid extrudate. Meuser, Van Lengerich and Kohler (1982) illustrated how the overall system could be studied in general terms.

At FMBRA detailed studies have been made on wheat flours and maize grits using a twin screw extruder. Some of the findings from these studies are used to illustrate the fundamental relationships between process variables and raw material factors and to show how these relationships affect the three major product characteristics of size, shape and crumb texture.

### *Factors affecting the size of products*

The size of extruded products is determined by independent process settings such as feed rate, die cross-sectional area and cutting speed, and the dependent variables related to expansion and collapse of the cellular structure of the extrudate.

The overall expansion of an extrudate is best measured by a displacement method employing fine grains of sand or glass beads to obtain the specific volume of individual product pieces. Normal commercial products tend to have specific volumes of between 3 and 10 ml $g^{-1}$ compared with 0.7–0.8 ml $g^{-1}$ for the fluid in the die channel. Once the die cutting speed and feed rate have been fixed so that the weight of each product piece is determined, the expansion and collapse must be controlled to ensure control over the size of the products. In all types of extrusion cookers it is necessary to apply both heat and shear forces to the cereal raw materials to achieve expansion of the extrudate. Raw flours may be extruded under low shear conditions but show little expansion with product specific volumes of 0.8–1.0 ml $g^{-1}$. Even if the mass temperatures are raised by barrel heating to give greater water vapour pressures at the die and to ‘melt’ all the crystalline regions in the starch granules, expansion remains fairly poor at 1.0–1.5 ml $g^{-1}$ specific volume.

For the Baker Perkins MPF 50D twin screw extruder the screw design illustrated in *Figure 18.4* incorporates a powerful shearing section, in which the raw materials may be cooked to give products ranging in specific volume from 1.5–25 ml $g^{-1}$ depending on the process settings and raw material composition used (Guy, 1986).

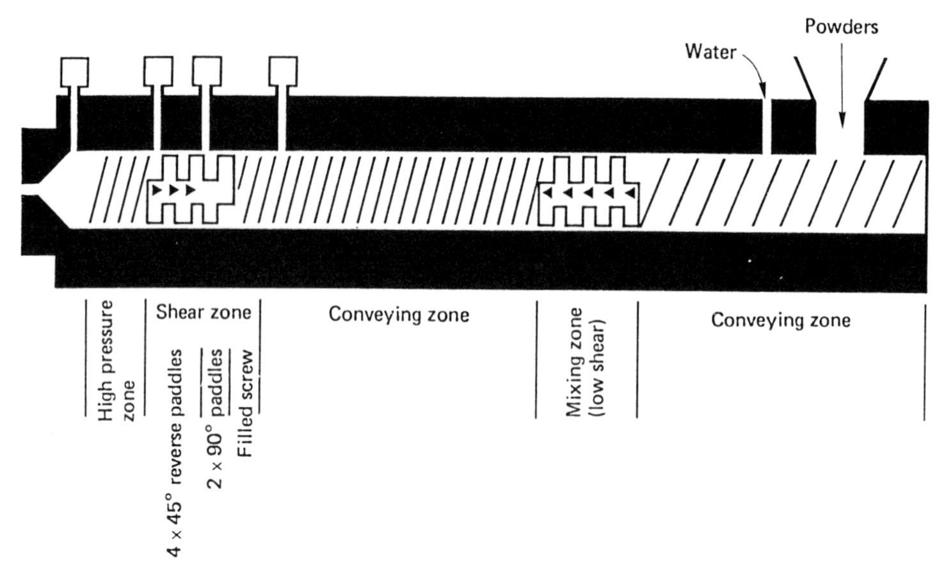


**Figure 18.4** Screw design for extrusion cooking on a Baker Perkins MPF 50D twin screw extruder.

The expansion of extrudates has been shown to increase with increasing SME input for moisture contents between 15 and 27%. SME input also correlated well with specific volume for changes in screw speed and feed rate. Park (1976) showed that the expansion of extrudates in a single screw machine correlated well with temperature and the degree of cook of the starch as represented by soluble starch determinations.

Examination of 'dead stop' shut downs of the Baker Perkins MPF 50D extruder (*Figure 18.5*) for a wide range of conditions and cereals (wheat, maize, triticale, rice, manioc and barley) has revealed similar processing patterns along the screws. This general pattern of processing provides the basis for a model of the extrusion cooking of all types of cereal mixtures and formulations with other ingredients. Therefore it is worthwhile to examine the physical changes taking in two typical examples such as a wheat flour with a soft endosperm texture and maize grits.

The cereal powder mix is conveyed largely unchanged in physical form up to the shear zone (*Figure 18.4*). At this point the screw system becomes completely filled due to the resistance to flow offered by the shear section which is made up of elliptical paddle elements arranged in a pseudo reversing screw formation. The filled flights of screw preceding the shear zone exert a powerful pumping action on the powders

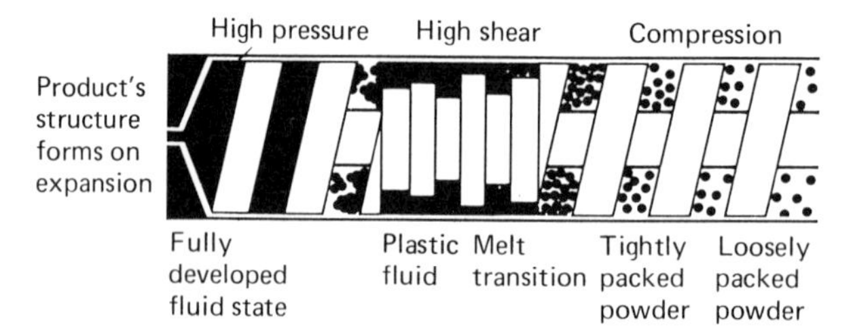


**Figure 18.5** Diagram of the changes observed in wheat flour on the screw system of a Baker Perkins MPF 50D extruder (16% moisture, 300 rpm, 800 g min$^{-1}$ feed rate).

compressing them to a dense mass (~ 1 g $ml^{-1}$) in the space between the barrel wall and the first pair of paddle elements. As the elements rotate they squeeze and shear the particles of flour causing frictional heating and mechanical damage to both particles and subparticles.

In flours and grits the frictional effects are concentration-dependent and decrease rapidly as the system becomes more dilute when the moisture content is raised as in *Table 18.2*. It can be seen that the peak mass temperature and SME input fall off as the moisture content is increased.

**Table 18.2** EXTRUSION COOKING OF MAIZE GRITS SHOWING VARIATION IN SME INPUT AND MASS TEMPERATURE IN SHEAR ZONE WITH MOISTURE CONTENT

| *Moisture* (% wet wt basis) | *SME input* (kJ $kg^{-1}$) | *Mass temperature in shear zone* (°C) |
|---|---|---|
| 15 | 506 | 180 |
| 17 | 478 | 173 |
| 19 | 392 | 159 |
| 21 | 327 | 146 |
| 23 | 307 | 144 |
| 25 | 266 | 140 |
| 27 | 247 | 139 |

Extrusion conditions: feed rate, 800 g $min^{-1}$; screw speed, 300 rpm; twin dies, 4 mm diameter; barrel temperature profile, 25/55/120/150/150 °C from feed port to die; for screw configuration as shown in *Figure 18.4*

For all the flours and grits extruded in a range of moisture contents below 25% an obvious transition occurs between compressed powder and a 'melt phase' within the shear zone. This transition may be at the start of the shear zone for wheat flours with hard endosperm texture, maize grits or starches, but takes place further into the zone for wheat flours with soft endosperm texture or formulations with cereal contents lower than ~ 80%. At the transition point the crystalline structures within the starch granules appear to 'melt' and cause a softening of the particles so that they can be compressed together (5–10 atmospheres pressure) to exclude the air trapped between the particles and form a dense viscoelastic fluid system. This fluid sets to a glassy state on cooling and has a density of ~ 1.4 g $ml^{-1}$ immediately following the transition compared with ~ 1 g $ml^{-1}$ for the compressed powder preceding the transition. It was observed that the majority of starch granules lost their birefringence, as viewed under crossed polarizing filters, at the transition point. Scanning electron microscopy revealed that the granules were flattened and slightly damaged on their surfaces after the initial change at the transition. Further passage along the shear section leads to some breakdown of the granules and the production of a continuous phase of starch polymers in the case of wheat flours.

Under similar processing conditions maize grits gave an earlier transition at the start of the shear zone and a more rapid breakdown of the granules so that they were almost completely dispersed at moisture contents ~ 19% for screw speeds of 300 rpm. These results were similar to those reported earlier for maize starch by Colonna *et al.* (1983) on a modified Creusot Loire extruder and for maize grits by Brenner, Richmond and Smith (1986) on a Baker Perkins MPF 50D extruder.

Examination of the melt phase for wheat flour using light microscopy revealed that the starch granules were detached from the endosperm proteins after passing through the transition. The protein phase in wheat flour and maize grits is broken down into small pieces (1–50 μm) and dispersed in the continuous phase (Guy, 1985). Similar results have been reported for rice flour by de Mosequada, Berg and Juliano (1986) which would indicate that in cereal systems in general the proteins do not form a developed continuous phase as in conventional baked products (Guy, 1986).

The starch granules in wheat flour were observed to be damaged and broken down to varying degrees during their passage through the shear zone of the extruder. At low SME inputs (< 250 kJ $kg^{-1}$) the granules were only partly dispersed into a continuous starch phase but at high SME inputs (> 600 kJ $kg^{-1}$) most of the structure of the starch granules had been broken down and dispersed.

Similar observations on the change in starch granules during extrusion cooking have appeared in many publications (see review by Linko, Colonna and Mercier, 1981), and it is generally agreed that starch granules are denatured from their native form and their polymers dispersed during processing (as described by Doublier, Colonna and Mercier, 1986). Many publications have appeared in recent years showing that the starch polymers are also degraded and reduced in molecular size by the extrusion process (Colonna and Mercier, 1983; Colonna *et al.*, 1983; Schweizer and Reimann, 1986; Schweizer *et al.*, 1986). The high molecular weight polymers of amylopectin and amylose are reduced to smaller molecules, averaging about half the original size of the native molecules.

### *Process variables affecting the denaturation and dispersal of starch granules in cereals*

The process variables which increase the level of starch granule denaturation and dispersal into a continuous polymer phase are similar to those which increase the level of applied shear forces as measured by the SME input (Meuser, van Lengerich and Reimers, 1984), such as screw design, screw speed or reduction in moisture content. In addition other factors which increase the time for which the shear forces are applied to the granules, such as feed rate, die size and screw design also tend to increase the development of starch into the dispersed state.

An example of two of these effects may be seen in *Figure 18.6* where the overall expansion as indicated by specific volume measurements of a series of products manufactured over a range of moisture contents from two different sizes of circular die, 3 mm and 4 mm diameter, both with similar land lengths (4 mm) is shown. The specific volume of products increases with decreasing moisture level and increasing shear input for moisture contents of 27–15%. It also increases as the die size is decreased from 4 mm to 3 mm diameter, a change which does not alter the fill in the extruder but causes an increase in SME input and possibly the retention time. For these maize samples the degree of cook assessed by a paste viscosity technique, as used by Launay and Lisch (1981), was medium to very high for samples from 27% to 15% moisture content.

The final product size was achieved by a combination of the expansion of the extrudate and a stabilization process which reduces the collapse on shrinkage of the delicate structures formed when the gas cells rupture at the point of maximum expansion. Those products manufactured with moisture contents ~ 19% had a good expansion and only a small amount of shrinkage (*Figure 18.7*). As the moisture

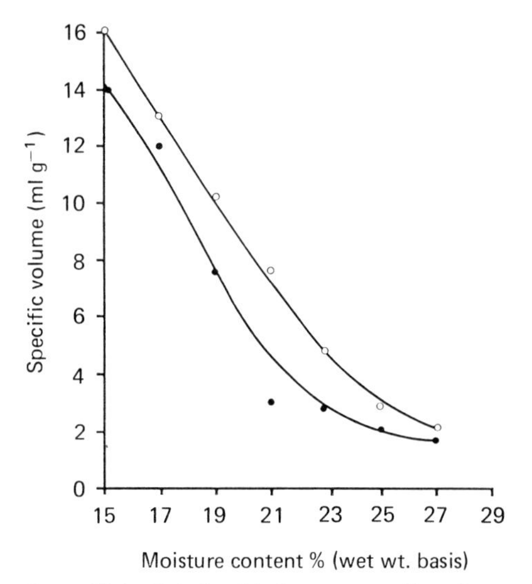


**Figure 18.6** Relationship between specific volume and moisture content for maize grits extruded at screw speed 300 rpm, feed rate 800 g $min^{-1}$ through ○, 3 or ●, 4 mm twin circular dies on screw system shown in *Figure 18.4*.

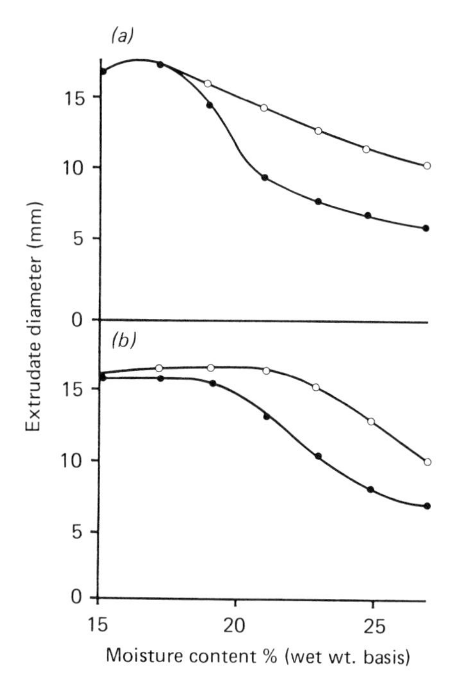


**Figure 18.7** Expansion and shrinkage for maize grits products extruded as in *Figure 18.6* showing the diameters of extrudates at their maximum when hot (○) and after cooling (●) to ambient temperature; (a) 4 mm diameter dies; (b) 3 mm diameter dies.

content increases to > 21% the collapse of the extrudate becomes very obvious with a decrease of up to 43% in the diameter of the extrudate on cooling to ambient temperature.

Therefore it would appear that the overall expansion of an extrudate as shown by specific volume measurements is achieved by expansion of bubbles of water vapour in an extensible continuous phase of starch polymers. At the time when the bubbles rupture at the maximum extensibility of the gas cell walls, the structure created may collapse under its own weight and possibly through some elastic effects, unless the viscosity in the fluid forming the cell walls is sufficiently high to balance these forces. Moisture evaporation effects both cooling and concentration which serve to increase fluid viscosity. As the temperature falls further over a short period of time the fluid system sets to a glassy state as shown by Mercier *et al.* (1980).

At high moisture levels the concentration of starch polymers in the fluid extrudate is too low to provide sufficient mechanical strength in the cell walls at the time of rupture. Therefore the porous structure tends to contract until it eventually stabilizes. There is a large excess of water vapour available to expand the fluid extrudate. Expansion appears to be limited only by the extensibility and gas holding properties of the cell walls of the gas bubbles. However, surface cooling effects may also play an important role in large extrudates.

The simple relationship between product specific volume and screw speed shown in *Figure 18.8* at low moisture contents (15–20% wet wt basis) where collapse is negligible, demonstrates that these desirable properties can be developed in a cereal flour or grit by increasing the level of applied shear forces. It would appear that as the starch granules are dispersed into a continuous polymer phase the expansion increases.

In the case of maize grits and wheat starch expansion levels off at high screw speeds due to surface cooling effects. The breakdown of the biopolymers themselves, as indicated earlier in this review, only tends to increase the potential expansion of an extrudate. Even small polymers found in glucose syrups < 20 DE give good expan-

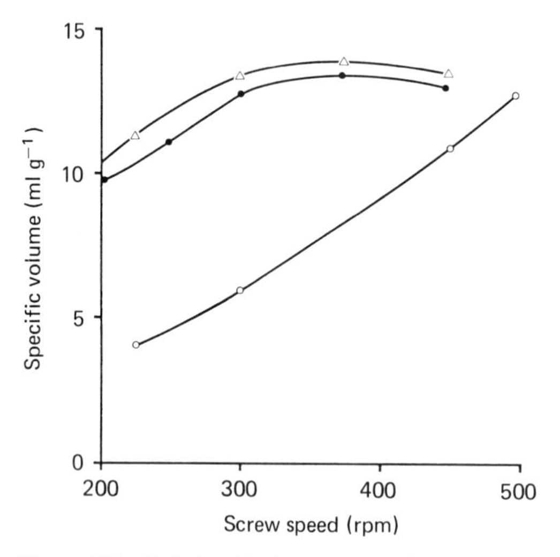


**Figure 18.8** Relationship between specific volume and screw speed for △, maize grits; ●, wheat starch, and ○, wheat flour at feed rate 800 g $min^{-1}$ and screw system as in *Figure 18.4*.

sion although they have a much lower viscosity in the fluid state and therefore such extrudates show extensive collapse and yield dense products.

Starch granules and other particles such as bran or protein pieces tend to reduce the extensibility of the cell walls. They tend to rupture the cells walls at a critical thickness related to their particle size, before the gas bubbles have expanded to reach their full potential. This effect may be observed in the manufacture of wholemeal products where up to 10% bran is added to the white flour (Andersson *et al.*, 1981; Guy, 1985). In coarsely textured extrudates, with a small number of gas bubbles available for expansion, bran causes a 50% reduction in expansion. However under conditions where a large number of gas cells are formed in the extrudate, the bran only causes a reduction of ~ 10% in expansion.

### *Raw material factors affecting product size*

The physical nature and the composition of raw materials are very important in the extrusion cooking of cereals. Raw materials help to generate frictional and shearing forces to act on the starch granules. In addition they also provide the biopolymers to from an extensible gas-holding bubble structure in the developed fluid and to stabilize the product on cooling.

Cereals may be used in the form of whole grains, grits, flours and starches in simple water systems or combined with other ingredients such as proteins, fibre, salt, oil, sugars, etc. If simple flour and water systems are considered, as in *Table 18.3*, the impact of different types of raw materials can be observed. Each material was extruded at similar process settings for screw speed, feed rate, moisture, die size and barrel temperature. Soft wheat (SW) flour had the lowest SME input and least expansion. Hard wheat (HW) flour, wheat starch and maize gave much greater SME input and expansion but not as high as the amylomaize starch. Wheat starch (WS) showed a high SME input indicating the effects of higher concentrations of starch

**Table 18.3** PROCESS VARIABLES AND PRODUCT CHARACTERISTICS FOR DIFFERENT TYPES OF RAW MATERIALS EXTRUDED AT 16% MOISTURE CONTENT, 450 RPM SCREW SPEED, 800 G MIN$^{-1}$ FEED RATE ON A BAKER PERKINS MPF 50D AS IN *FIGURE 18.4*, FITTED WITH TWIN CIRCULAR DIES (4 MM DIAMETER)

| *Variables* | *SW flour*[a] | *HW flour*[a] | *Wheat starch* | *Maize grits* | *Hylon VII amylomaize starch* |
|---|---|---|---|---|---|
| *Process:* | | | | | |
| SME (kJ kg$^{-1}$) | 458 | 512 | 535 | 568 | 603 |
| Die pressure (N m$^{-2}$ × 10$^5$) | 91 | 56.5 | 15.9 | 15.8 | 16.5 |
| Temperature in shear zone (°C) | 154 | 178 | 177 | 182 | 188 |
| *Product:* | | | | | |
| Specific volume (ml g$^{-1}$) | 9.2 | 14.9 | 13.1 | 14.0 | 25.0 |
| *L/D* ratio | 2.4 | 2.8 | 16.1 | 15.0 | 21.0 |
| Texture score for fineness (0–10) | 2.5 | 3.0 | 8.0 | 9.0 | 9.5 |

[a]SW = soft wheat flour; HW = hard wheat flour.

granules in the compressed powders. Starch granules in their native crystalline form create powerful frictional and shearing forces when compressed in the paddle system. This was illustrated in an extreme case with an amylomaize starch, Hylon VII (approximate amylose content, 70%). Hylon VII has a much higher 'melting point' for its crystalline regions than wheat starch. In high moisture conditions its crystalline regions melt at ~ 130 °C but this 'melt' temperature becomes much higher for low moisture conditions. In the extruder Hylon VII creates a high SME input and reaches a mass temperature of 198 °C in the shear zone to achieve a melt transition. The hard granules continue to create heat until the crystalline regions melt. Afterwards the granules disperse fairly easily to give an excellent gas-holding film with an enormous expansion of 25 ml $g^{-1}$ in the extrudate and no apparent collapse or shrinkage.

It was noted in the early reports (Feldberg, 1969) on extrusion that the expansion of extrudates fell as the level of flour used in the powder mix was reduced. This has been found to be the case for wheat flour and maize formulations at FMBRA. In the example for wheat flour shown in *Table 18.4* the raw materials were all processed under similar machine settings. Mix A of the wheat flour gave high SME input and good expansion. Addition of 7.5% (low level) minor ingredients to replace flour caused little change in SME input but some reduction in expansion whereas a 21.5% (high level) addition of minor ingredients caused a large decrease in both SME input and expansion. The dilution of the flour by replacement with other materials reduced the frictional effects in the shear zone as shown by lower mass temperatures and a reduction in the degree of dispersal of the starch.

Similar results were observed with maize grits as shown in *Table 18.5* where the reduction in SME input and expansion with maize dilution was even more striking than for wheat flour. At the lowest concentration of maize the shearing forces between particles were insufficient to break down many of the grits.

The general effect of replacing the cereal flours or grits by sugar and soya was to dilute the particles in the system reducing the level of cooking of the extrudates in the shear zone. This reduced the concentration of dispersed starch polymers in the cell walls leading to a lower expansion and more extensive shrinkage of the extrudates.

**Table 18.4** EFFECTS OF MINOR INGREDIENTS ON THE EXTRUSION COOKING OF WHEAT FLOUR AT 16% MOISTURE CONTENT USING THE CONDITIONS SHOWN IN *TABLE 18.2*

| *Dependent variables* | *Wheat flour mix A* | *Low level of minor ingredients*[(a)] | *High level of minor ingredients*[(b)] |
|---|---|---|---|
| SME input (kJ $kg^{-1}$) | 420 | 403 | 315 |
| Die pressure (N $m^{-2}$ × $10^5$) | 112.9 | 91.7 | 78.9 |
| Temperature in shear zone (°C) | 157 | 162 | 148 |
| Specific volume (ml $g^{-1}$) | 6.9 | 3.9 | 2.1 |
| *L/D* ratio | 2.3 | 2.7 | 3.3 |
| Texture score for fineness (0–10) | 3 | 3 | 2 |

[(a)]Low level: 5% sugar, 1% soya isolate, 1% salt, 0.25% groundnut oil and 0.25% distilled monoglyceride.
[(b)]High level: 15% sugar, 5% soya isolate, 1% salt, 0.25% GNO and 0.25% DGMS.

**Table 18.5** EFFECTS OF MINOR INGREDIENTS ON THE EXTRUSION COOKING OF MAIZE GRITS AT 16% MOISTURE CONTENT USING THE CONDITIONS SHOWN IN *TABLE 18.2*.

| *Dependent variables* | *Maize grits* | *Low level of minor ingredients*[(a)] | *High level of minor ingredients*[(a)] |
|---|---|---|---|
| SME input ($kJ\ kg^{-1}$) | 486 | 460 | 287 |
| Die pressure ($N\ m^{-2} \times 10^5$) | 32.8 | 17.0 | 22.0 |
| Temperature in shear zone (°C) | 178 | 169 | 141 |
| Specific volume ($ml\ g^{-1}$) | 14.2 | 11.3 | 1.7 |
| *L/D* ratio | 3.5 | 3.2 | 4.9 |
| Texture score for fineness (0–10) | 3.0 | 3.5 | 2.0 |

[(a)]Low and high levels of minor ingredients as in *Table 18.4*.

It can be seen therefore that the expansion process for cereals is balanced between the machine process settings and the composition and nature of the farinaceous mix. Both sets of factors contribute to the shearing forces applied to the starch granules and are of equal importance.

### *Factors affecting the shape of extruded products*

A fluid extrudate would be expected to expand from the die channel to give roughly the same cross-sectional shape as the die. Its overall shape would be determined by the cutting system employed to form individual pieces of products. At FMBRA the shapes of products manufactured with circular dies have been studied in great detail for wheat, maize, rice, triticale and other starch-based raw materials. The length to diameter ratio (*L/D*) for the cylindrical products was used to compare the shape of products and to assess radial and longitudinal expansion indices.

For a standard set of conditions, with mass flow rate of $800\ g\ min^{-1}$ from twin (4 mm diameter, 4 mm land length) dies, products were cut at $400\ cuts\ min^{-1}$ at each die. The products from coarsely textured extrudates with a relatively small number of gas cells had *L/D* ratios in the range 2–4. Under these conditions it can be estimated that the theoretical *L/D* ratio for isotropic expansion should be between 13 and 14 for fluids containing 16–20% moisture. Therefore it would appear that all the coarsely textured products had displayed anisotropic expansion with a strong bias in the radial direction.

As the numbers of gas cells increased in the expanded extrudates the *L/D* ratios of the products also increased. It was found that plots of *L/D* ratios against the ln (gas cell counts per unit weight) gave good correlations for individual wheat flours and maize grits as shown in *Figure 18.9*. The graphs of white wheat flour and maize grits had approximately similar slopes but wholemeal wheat flour gave a significantly greater slope than the other two raw materials.

It could be inferred from the relationship observed between shape, as defined by the *L/D* ratio, and the numbers of gas cells in the product that a fluid extrudate formed

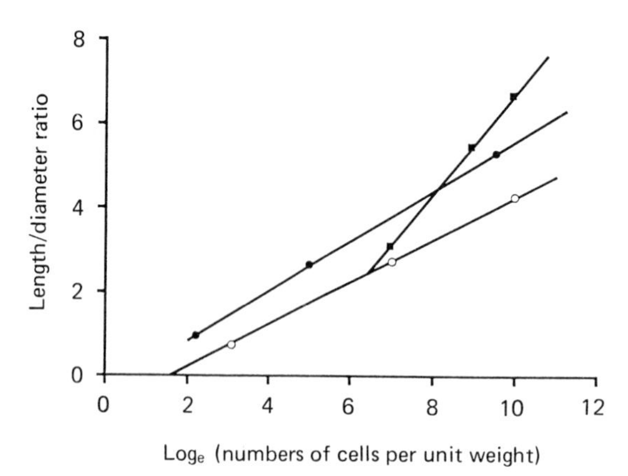


**Figure 18.9** Relationship between shape (*L*/*D* ratio) and numbers of gas cells per unit weight for product made from ●, maize grits; ○, white wheat flour; and ■, wholemeal wheat flour.

without gas cells should have a small *L*/*D* ratio and display a strong radial bias or die swell. The experiments with wheat flour and wheat and manioc starches described earlier confirmed this hypothesis. The starch-rich fluids showed a large die swell and gave *L*/*D* ratios of 0.7 to 0.8.

It would appear that this die swell of the fluid extrudate, created by the presence of elastic characteristics within the fluid, controls the overall expansion of the extrudate when the numbers of gas bubbles are relatively small ($\sim 600\ g^{-1}$). Its effect diminishes as the number of bubbles nucleated and expanded as the die increases. This might be due to a randomizing effect of the bubbles on the directional friction influence of die swell. Park (1976), who also recognized a similar effect on longitudinal expansion in his own studies, suggests that as the numbers of gas bubbles increase more expansion occurs within the die channel giving enhanced longitudinal expansion.

Die swell occurs in starch fluids with very low degrees of cook soon after the starch granules have lost their birefringence. Examination of a range of products with similar coarse textures ($\sim 600$ gas cells $g^{-1}$) but with specific volumes between 2 and 18 ml $g^{-1}$ for wheat flour and maize grits showed a fairly constant value for *L*/*D* ratios 2–3 over a wide range of degrees of cook (*Figure 18.10*). It would appear that the elastic characteristics developed in the starch phase at low levels of cook remain fairly constant over a wide range of development in which the starch granules are dispersed and some degradation of the polymers occurs. The elastic characteristic of the fluid produces a die swell as the fluid is forced through a die system with a converging throat or entrance region (Han, 1976). Energy is stored in the fluid in this region which can be released at the die exit to cause a swelling at right angles to the direction of flow, radial swelling in the case of simple circular dies. The size of the die swell effects are related to the design of the dies. In *Table 18.6* a comparison is made between two dies of 3 mm and 4 mm diameter respectively using similar raw materials and processing conditions. It can be seen that the smaller die causes an increase in expansion and an increase in radial bias by comparing the experimental *L*/*D* ratio with the theoretical value for isotropic expansion.

It is also possible for the size of the die swell effect to be influenced by the composition of the fluid. Lowering the starch concentration by replacing it with other materials such as bran proteins or by hydrocolloid gums at levels $\gtrsim 5\%$ may reduce

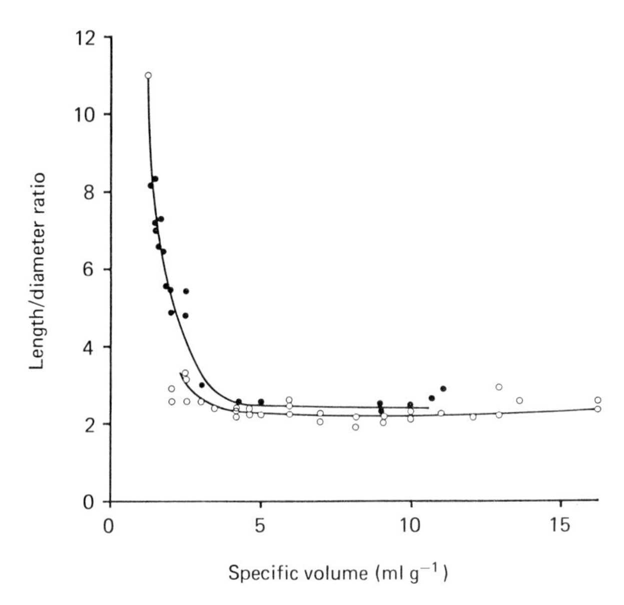


**Figure 18.10** Effect of increasing levels of starch development as shown by specific volume measurements and the shape (*L/D* ratio) of extruded products made from wheat flour (○) and maize grits (●) formulations.

**Table 18.6** EFFECTS OF CHANGES IN DIE CHANNEL DIAMETER FOR CIRCULAR DIES 4 MM LAND LENGTH

| | *Die channel diameter* | |
|---|---|---|
| *Product characteristic* | *3 mm* | *4 mm* |
| Specific volume ($ml\ g^{-1}$) | 10.2 | 7.7 |
| Diameter $D_1$ (in mm) | 15.8 | 16.6 |
| $D_1/D_0$ | 5.22 | 4.15 |
| Length $L_1$ (in mm) | 51.6 | 39.5 |
| $L/L_0$ | 0.39 | 0.69 |

Extrusion conditions as in *Table 18.2*, $D_0$ die diameter and $L_0$ is theoretical length of cylinder of same diameter as the die for the extrusion of unit mass of fluid.

the die swell. This is demonstrated (*Figure 18.9*) in the graphs of *L/D* ratio against ln (gas cell counts $g^{-1}$) where wholemeal flour is found to give a much greater slope than white flour or maize grits. The slope of this curve reflects the faster change of shape of the products with increasing gas cell counts due to a smaller die swell.

It was found that the overall expansion of wheat flours or maize grits in the range of moisture contents 14–20% was not influenced by the numbers of gas cells present until these numbers exceeded $60 \times 10^3\ g^{-1}$. For such extrudates an extremely fine cell structure was developed with some bias towards longitudinal expansion (*L/D* ratios 15–25). This extreme form of expansion caused exceptional surface cooling effects, revealed by distorted shapes, splitting and other defects which tended to reduce overall expansion.

Many papers in the past have reported results for product expansion in terms of a single dimensional change such as the radius for circular dies. Clearly such a measurement may have been misleading as often the shape of the extrudate changes due to changes in the numbers of gas cells or the other factors mentioned herein. The radius of a product may increase in the low to medium range of cook while the gas cell numbers are small and then decrease at high levels of cook, due to increased gas cell numbers, even though the overall expansion increases roughly in proportion to the level of cook.

### *Factors affecting the crumb structure of extruded products*

Little has been reported in the literature about the bubble formation in extrudates or the factors which influence the way the gas cell structure of the crumb is produced.

It has been observed (Park, 1976) that the nucleation of bubbles is dependent on a positive water vapour pressure compared to the total pressure. Thus expansion would tend to begin at ~ 100 °C at the die exit when the water vapour pressure is sufficient to exceed the small exit pressure for the die, the water being held in a liquid state within the extruder due to the high pressure exerted by the pumping action of the screws.

It is suggested that as the temperature of the fluid is increased by changing extrusion conditions, the potential for nucleation increases and theoretically the number of gas cells found in the expanded extrudate should also increase as the temperature rises. Experimental observations on many products extruded over a wide range of temperatures from 120 to 200 °C do not support this view. Other factors which influence the expansion of the bubbles were suggested by Park (1976), including the surface tension of the fluid and its rheological properties such as resistance to extension. There will be a tendency for the larger bubbles to grow at the expense of smaller ones and possibly only a small number of the bubbles grow into the large gas cells observed in the product.

Practical examinations of extrudates formed at simple circular dies shows that coarse product textures are associated with high viscosity fluids and high die pressures. Fine crumb textures are produced for lower viscosity fluids at lower die pressures. The viscosity of the fluid may be influenced both by starch concentration and the degree of cook. Low viscosities are obtained at low starch levels, as with high moisture contents, or for higher starch levels with high degrees of cook when most of the starch granules have been dispersed. Die pressure is determined by the resistance of the die channel, feed rate and fluid viscosity.

Examples of fine cell structures may be observed with high moisture levels for maize. As the moisture level decreases the texture becomes very coarse but below 17% moisture the frictional effects between the grits and the shearing elements increase so that a high degree of cook is obtained and the texture becomes extremely fine.

Similar effects can be observed with changes in screw speed for the same conditions of feed rate, die size and 16% moisture level. At low screw speeds the fluid has a low degree of cook and a relatively high viscosity. The extrudate texture is coarse but becomes finer as screw speed increases and the die pressure falls. A very fine texture is obtained for the well cooked fluids produced at high screw speeds.

However, if the feed rate is increased the die pressure rises and the texture again becomes coarser. An increase from 50 to 72 kg $h^{-1}$ greatly reduces the numbers of gas cells to give a fairly coarse-textured extrudate.

The relationship between the properties of the fluid and the physical conditions encountered in the die channel of temperature, pressure, shear, etc. are complex and require detailed studies to gradually build up a model to cover all the possible relationships and effects. However, the empirical evidence currently available has given some directions as to how this might be achieved.

## 'Half products' or 'pellets'

The extrusion cooking process for 'half products' may be performed in the shear zone of an extruder in the same way as for direct expansion processes. A low to medium level of development in the starch phase is sufficient for good expansion later in the process so high moisture contents of 25–30% are often used. The developed fluid is formed into precise shapes at the die either by cooling to below 100 °C in a long barrel extrusion cooker to prevent bubble formation or by using two extruders and forming the first expanded extrudate into a dense fluid in the second low temperature forming extruder. The cooling process also increases fluid viscosity and helps to give a good retention of the shape of the cross-section of the die in the 'half products'. Die swell may also affect this shaping process as in the direct expansion process.

The shapes cut from the continuous extrudate are dried very carefully to produce a stable 'half product' or 'pellet' which may be stored for several months before use. An extruded shape may be dried from 30% to ~10% moisture to form a stable intermediate. If the moisture is removed quickly the 'half product' is said to be 'case hardened' and gives irregular expansion in the central region (Gerkens 1963, 1964, 1965; Harper, 1981). A long slow drying process under well controlled conditions is often used to form a product in which the moisture distribution is kept fairly even from centre to edge. Such 'half products' expand more uniformly to give precise control of shape.

### *Expansion of 'half products'*

The essential requirement for the expansion of 'half products' appears to be a rapid heating process such as frying. In this type of process the temperature of the water within the 'half product' is raised to well above 100 °C while its solid matrix is being softened by the heating process. Water vapour is nucleated to form bubbles which expand to form an expanded cellular structure in the products. Such a mechanism was described for the puffing of popcorn by Hoseney, Zeleznak and Abdelraham (1983) in which a temperature of 170 °C was obtained within maize kernels before explosive puffing took place. In popcorn the bubbles of water vapour were thought to be nucleated in the vitreous regions of the grain at the centres of starch granules. The granules were destroyed during expansion supplying polymers to contain the expanding gases. In 'half products' a vitreous matrix is formed from the extruded fluid on cooling to provide a similar system for explosive puffing.

No detailed studies on the effects of levels of development of starch, concentration, cereal types, etc. on the mechanism of expansion of 'half products' have been reported in the literature to date.

## Co-extrusion processes

The principles for co-extrusion processes are similar to those for direct expansion of cereals. A developed cereal fluid is extruded through a special annular die in which a central tube may be added to co-extrude a filling or second extrudate inside the first extrudate as it is formed into a hollow tube or more complex shape (Anon, 1984).

The second extrudate pumped into the central tube is usually a fat-based filling based on savoury flavourings such as cheese or sweet ones such as chocolate. A normal expanded cereal formulation of the required specific volume, thickness and shape is obtained by die design and control of the processing variables as described in the earlier section on direct expansion processes. Several die designs have been described to give simple co-extrusion and to minimize contact time in the high temperature regions for the filling (Vincent, 1986). The filling itself has to be balanced to give a combination of properties so that it can be pumped but has sufficient viscosity to set in the cereal extrudate.

## Acknowledgements

The author wishes to acknowledge the contributions made to the studies at FMBRA by Dr S.A. Roberts and would like to thank the Ministry of Agriculture, Fisheries and Food for their support in studies on raw materials performance in extrusion cooking.

## References

ANDERSSON Y., HEDLUND, B., JONSSON, L. and SVENSSON, S. (1981). Extrusion cooking of a high-fibre cereal product with crispbread character. *Cereal Chemistry*, **58,** 370–374

ANON (1984). Co-extrusion process creates unique snacks. *Snack Food*, February, 30–31

BRENNER, P.E., RICHMOND, P. and SMITH A.C. (1986). Aqueous dispersion rheology of extrusion cooked maize. *Journal of Texture Studies*, **17,** 51–60

CERVONE, N.W. and HARPER, J.M. (1978). Viscosity of an intermediate dough. *Journal of Food Processing*, **2,** 83

COLONNA, P. and MERCIER, C. (1983). Extrusion cooking of manioc starch. *Carbohydrate Polymers*, **3,** 88–107

COLONNA, P., MELCION, J.P., VERGNES, B. and MERCIER, C. (1983). Flow mixing and residence time distribution of maize starch within a twin screw extruder with the longitudinally split barrel. *Journal of Cereal Science*, **1,** 115–125

de MOSEQUADA, M.B., BERG, C.M. and JULIANO, B.O. (1986). Rice varietal differences in properties of extrusion cooking of rice flour. *Food Science*, **19,** 173

DOUBLIER, J.L., COLONNA, P. and MERCIER, C. (1986). Extrusion cooking and drum drying of wheat starch II. Rheological characterisation of starch pastes. *Cereal Chemistry*, **63,** 240–246

FELDBERG, C. (1969). Extruded starch-based snacks. *Cereal Science Today*, 14 June, 211–214

GERKENS, D.R. (1963). Process for Producing Expanded Foodstuffs. US Patent 3 076 711

GERKENS, D.R. (1964). Process for Producing in Expanded Foodstuff and in Intermediate Therefor. US Patent 3 220 852

GERKENS, D.R. (1965). Method for Producing a Crispy Expanded Formed Foodstuff. US Patent 3 131 063

GUY, R.C.E. (1985). The extrusion revolution. *Food Manufacture*, January, 26–29

GUY, R.C.E. (1986). Extrusion cooking versus conventional baking. In *Chemistry and Physics of Baking*, (Blanshard, J.M.V., Frazier, P.J. and Galliard, T., Eds). Royal Society of Chemistry Publication, No. 56

HAN, C.D. (1976). *Rheology in Polymer Processing*, pp. 111–126. London, Academic Press

HARPER, J.M. (1981). *Extrusion of Foods*, Volumes I and II. Boca Raton, Florida, CRC Press

HARPER, J.M. (1986). Extrusion texturisation of foods. *Food Technology*, March, 70–76

HOSENEY, R.C., ZELEZNAK, K. and ABDELRAHAM, A. (1983). Mechanism of popcorn popping. *Journal of Cereal Science*, **1,** 43–52

LAUNAY, B. and LISCH, J.M. (1983). Twin screw extrusion cooking of starches of low behaviour of starch pastes, expansion and mechanical properties of extrudates. *Journal of Food Engineering*, **2,** 259

LINKO, P., COLONNA, P. and MERCIER, C. (1981). High temperature short time extrusion cooking. In *Advances in Cereal Chemistry*, Volume IV, (Pomeranz, Y., Ed.) pp. 145–235. St. Paul, Minnesota, American Association of Cereal Chemists

MERCIER, C., CHARBONNIERE, R., GREBUT, J. and DE LA GUERIVIERE (1980). *Cereal Chemistry*, **57**, 4

MEUSER, F., VAN LENGERICH, B. and KOHLER, F. (1982). The effects of extrusion parameters on the functional properties of wheat starch. *Staerke*, **34,** 366–372.

MEUSER, F., VAN LENGERICH, B. and REIMERS, H. (1984). Extrusion cooking of starches. Comparison of experimental results derived from laboratory and full scale extruders by means of system analysis. *Staerke*, **36,** 194–199

PARK, K.H. (1976). Elucidation of the extrusion puffing process. PhD Thesis, University of Michigan

SCHWEIZER, T.F. and REIMANN, S. (1986). Influence of drum drying and twin screw extrusion cooking on wheat carbohydrates. Comparison between wheat starch and flours of different extraction. *Journal of Cereal Science*, **4,** 193–203

SCHWEIZER, T.F., REIMANN, S., SOLMS, J., ELIASSON, A-C. and ASP, N-G. (1986). Influence of drum drying and twin screw extrusion on wheat carbohydrates, II. Effects of lipids on physical properties, degradation and complex formation of starch in wheat flour. *Journal of Cereal Science*, **4,** 249–260

VAN ZUILICHEM, D.J., BUISMAN, G. and STOLP, W. (1974). Shear behaviour of extruded maize. Presented at 4th International Congress of Food Science and Technology, Madrid

VINCENT, N. (1986). Co-extrusion Die Assembly for a Cooker Extruder. UK Patent 2 162 788

19

# THE EVALUATION OF FOOD STRUCTURE BY LIGHT MICROSCOPY

F.O. FLINT
*Department of Food Science, University of Leeds, UK*

## Introduction

The use of a microscope of any type is only one of several experiments that can be done to learn more about a foodstuff. Usually these experiments involve chemical analysis but when food structure is being studied physical methods of testing become important. The light microscope (LM) can yield information that supports both physical and chemical areas of study, e.g. it can be used to show the distribution of a specific food constituent that has been assessed chemically and it can give a visual explanation as to why foods of similar chemical constitution have measurably different textures. This is because texture is a consequence of structural organization which is itself due to the combined effect of chemical composition and physical forces. This structural organization is termed microstructure since in most cases it is only visible in any detail through some type of microscope (Stanley and Tung, 1976).

In everyday life we use the unaided eye to assess the structure of things about us and this most sensitive of our senses can see features down to a fraction of a millimetre (0.2 mm = 200 μm). Two points 200 μm apart will appear as two points; closer together the points seem to merge, i.e. the eye has a resolving power of 200 μm. The LM has a resolution of rather less than 1 μm.

Compared with the electron microscope which can resolve detail down to 0.01 μm the magnification possible with the light microscope seems modest, but it spans the most useful range for many processed foods where the features present are measured in micrometres. Excessive magnification has its drawbacks; it needs to be remembered that as magnifications increase the area being examined at any one time decreases quite sharply. This is marked even for the small range covered by the light microscope, e.g. using a × 10 eyepiece the field seen with the × 10 objective has a diameter of about 1 mm so that the area seen is rather less than one square millimetre; moving up to the × 40 objective only one-sixteenth of this area is seen at any one time. An appreciation of this is why experienced food microscopists stress the importance of using a low power stereomicroscope before resorting to the compound instrument (Apling, 1980; Vaughan, 1979).

## Use of the stereomicroscope

The magnification obtained depends on the instrument chosen but is of the order × 5 to × 50; this is associated with a field of view of several millimetres in diameter. Stereomicroscopes are characterized by a large working distance; this and the erect image obtained make them the most 'user friendly' of all microscopes.

In the initial microscopical examination of animal feedstuffs stereomicroscopes are well established (American Association of Feed Microscopists, 1978; Vaughan, 1979). Many animal feed constituents can be identified simply by viewing a sieved sample with the stereomicroscope using incident light, e.g. rolled barley can be readily detected and distinguished from rolled oats, wheat or rye. A similar technique can be applied to the examination of the raw cereal material used in extruder cooking where variation in particle size and the presence of 'fines' can affect the microstructure of the final extrudate.

In observing finished products the prime requirement is a really flat surface that can be mounted parallel to the low power objective. The erect image obtained shows a good depth of focus and this can be enhanced by careful lighting. For this, the modern trend is to use a fibreoptic light source powered by a quartz halogen lamp. This provides an intense but cold light which does not warm the specimen as does the traditional tungsten lamp. Newer still are fluid light guides which are claimed to give 40% more light than fibre guides but these are more expensive and may not be as long lasting. Stereomicroscopes are ideal for the rapid examination of extruded products. The large field of view enables a number of air cells to be seen at the same time so that air cell size and shape at different sampling positions can be compared (*see Figure 19.1*).

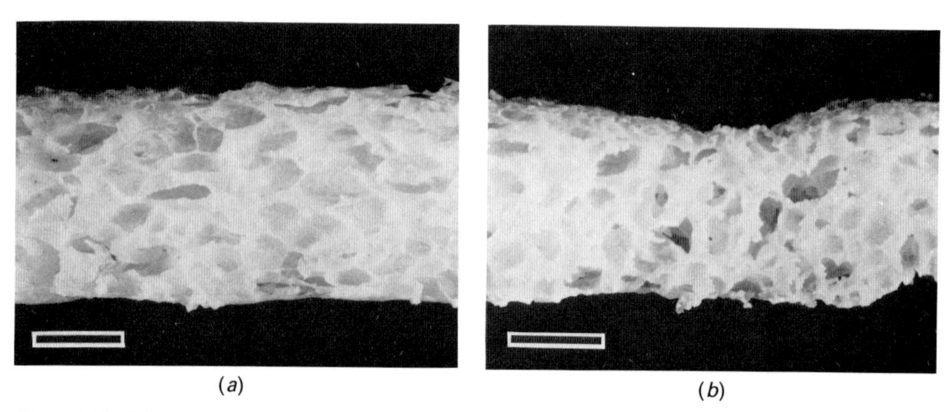


**Figure 19.1** Photomacrographs of extruded wheat and rye flat bread. Surface of levelled edges viewed by incident light: (a) edge parallel to direction of extrusion; (b) edge at right angles to direction of extrusion. Bar = 2 mm.

Although most specimens for the stereomicroscope rely solely on lighting for contrast a limited amount of staining is possible. *Figure 19.2* shows the effect of iodine vapour on a bread sample. The 'sharpening' of contrast in the stained specimen facilitates comparison of similar structured products and can be an aid in the measurement of air cell size.

*Figures 19.1* and *19.2* were taken at low magnification but as the magnification is increased the depth of focus is reduced which limits the information that can be obtained from surface preparations using light microscopy. For detailed surface work the scanning electron microscope (SEM) is a much more powerful tool; the depth of field is much greater, spatial resolution much higher (10 nm) and specimen prepara-

(a) (b)

**Figure 19.2** Photomacrographs of commercial sliced bread. Surface viewed by incident light: (a) unstained; (b) stained with iodine vapour. Bar = 5 mm.

tion is relatively easy. It is for these reasons the SEM has been used so extensively for studies of food microstructure (Bechtel, 1983; Cohen *et al.*, 1981). Several authors of SEM work also advocate the integration of LM and SEM (Angold, 1982; Stanley and Tung, 1976; Varriano-Marston, 1981). The instruments are seen as being complementary rather than in competition. This is a recognition of the range of techniques available to the light microscopist which yield information about the physical and chemical nature of the specimen at the submicroscopic level.

## Use of the compound light microscope

Most light microscopy is done with the compound microscope with the specimen being viewed by transmitted light. The most useful range of magnificaion for food structure studies is × 40 up to × 400. Even though useful magnification up to × 1000 is available on most instruments, this is only needed for work with subjects like emulsions where a high power objective may be needed to exploit the resolving power of the instrument. In observing most food structures it is important to first use the low power objectives with their large field diameters before looking for the detail given by the higher powered lenses. No microscopist studying food structure should be without a × 4 objective if only to aid in the orientation of a specimen and it is useful to have a × 20 objective as well as a × 10 and × 40. Often the × 20 lens reveals nearly as much detail as the × 40 lens and the area seen is double.

## Specimen preparation

For incident light microscopy the surface needs to be flat. For transmitted light work the preparation must be sufficiently thin for light to pass through but not so thin that important detail is lost. Foods of small particle size like starches and powders and those that easily form a thin layer (like purées and stable emulsions) can be simply spread or smeared on a slide. Solid foods, especially where there is an interest in structural orientation, should be carefully sampled noting the orientation before sectioning. This can be done most rapidly by using a cryostat, i.e. a microtome housed in a refrigerated cabinet. By freezing the water naturally present in the food the ice gives support to the specimen during sectioning and the frozen sections are collected on slides ready for staining and mounting. With dry samples, minimal hydration using water-based media enables sections to be taken without the swelling that full hydration might cause. This technique is appropriate for a wide range of processed foods including comminuted meats, oven and extruder cooked products, roller and spray dried powders and dried food mixes.

Before examination with the microscope all specimens need mounting. The choice of mountant depends on the nature of the specimen, e.g. where fats are present a water-miscible, rather than a solvent-based, mountant is appropriate and the mountant may contain a stain, e.g. a lipid-soluble dye which will colour fats. Stains are not always needed to demonstrate constituents. Physical methods of obtaining contrast in the microscope image need a minimum of preparation usually involving only a simple mountant.

## Physical methods of obtaining contrast

There are several methods of contrast which depend on the physical make-up of the specimen and these are useful for two reasons: they demonstrate structure at the molecular level and they require the minimum of preparation which limits the possibility of introducing artefacts.

### *Birefringence*

The most important method of obtaining physical contrast depends on the property known as birefringence. Uncooked starch, cellulose, fats, muscle and connective tissue are all birefringent, i.e. possess two refractive indices. This means that they have an orderly molecular structure which interacts with transmitted light waves so that light traverses the specimen in two sets of waves vibrating at right angles to one another and travelling at different speeds. With ordinary light this does not produce any visible effect. However, if the light used is vibrating in only one plane, i.e. it has been passed through a polarizing filter (the polarizer) before meeting the specimen, then birefringence can be shown by allowing the light emerging from the specimen to pass through a second polarizing filter (the analyser). Polarizer and analyser are set at a 'crossed' position which extinguishes all the transmitted light surrounding the specimen, whilst allowing some of the light which has traversed the specimen to pass through. The amount of light reaching the eyepiece of the microscope is, of course, greatly reduced but this is made up for by the high contrast between bright specimen and dark background.

Simple plastic 'Polaroid' filters can be fitted to any light microscope and if the microscope has a good light source the analyser can be permanently positioned within the body of the instrument without affecting bright field work.

Polarization microscopy is a useful technique for the study of starches and starch-based products, especially in combination with simple staining techniques, e.g. identity of a starch can be established by its size, shape and the position of the polarization cross (*see Figure 19.3*).

Damaged granules, whether mechanically or thermally damaged (gelatinized) lose their birefringence so that they are invisible when a preparation is viewed between fully crossed polars, but by slightly uncrossing the polars both intact granules

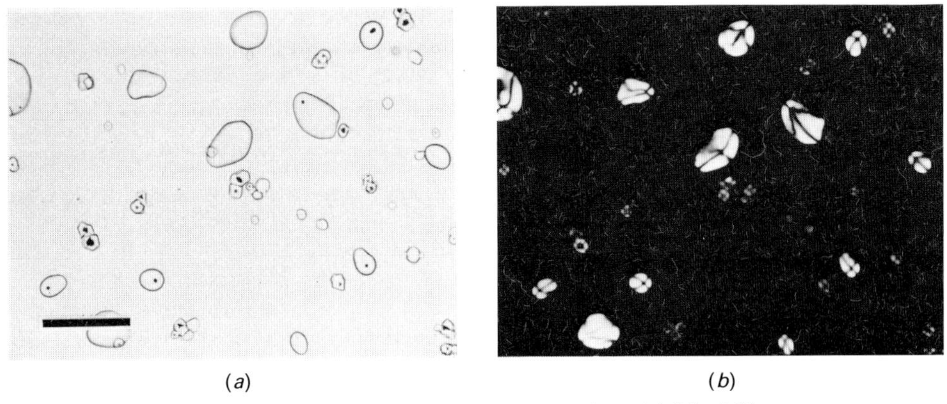

(a) (b)

**Figure 19.3** Photomicrographs of potato and corn starch granules: (a) bright field illumination; (b) same field viewed between crossed polars. Potato starch has large ovoid granules; in polarized light these appear bright with well-marked polarization crosses. Corn starch has small polyhedral and rounded granules; these show well-marked symmetrical polarization crosses. Bar = 100 μm

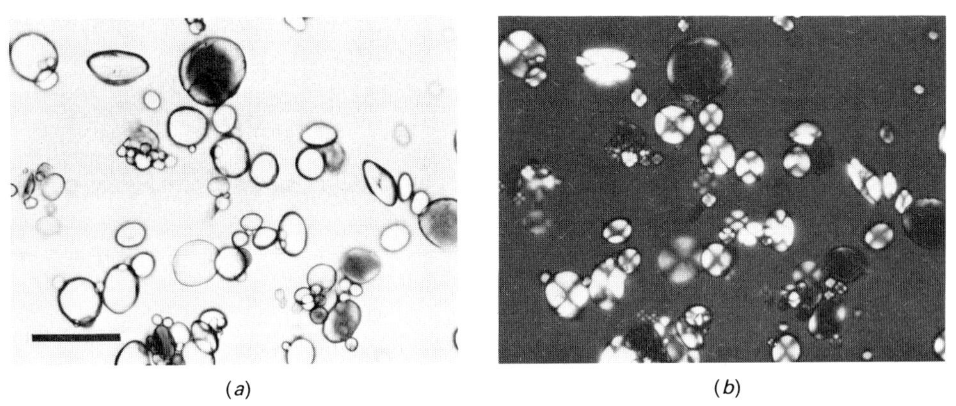

(a) (b)

**Figure 19.4** Photomicrographs of intact and mechanically damaged wheat starch mounted in 0.5% Chlorazol Violet R: (a) bright field illumination; (b) same field viewed between crossed polars. Only damaged area of granules are stained, undamaged starch is birefringent. The lens-shaped granules trapped 'on edge' show characteristic elongated 'cross'. Bar = 100 μm.

(birefringent or anisotropic) and damaged granules (isotropic) can be seen at the same time (*Figure 19.4*).

Polarization microscopy has been used in studies of the microstructure of a range of baked cereal products, most of which contain intact as well as gelatinized starch granules (Varriano-Marston, 1983).

## Interference microscopy

Like phase contrast microscopy to which it is related, interference microscopy relies on differences in thickness and refractive index of constituents to provide differences of light intensity in the microscope image. The image obtained with the interference microscope is much crisper than that given by phase contrast; of the several interference systems available the Nomarski Differential Interference Contrast (DIC) technique is recommended for those whose interest lies in visibility rather than in quantitative measurement of phase change (Hartley, 1979).

The Nomarski DIC image is one of high contrast, it is reminiscent of an SEM image and appears to have the raised relief associated with SEM pictures. It will be appreciated, of course, that the relief effect of the Nomarski image is due to optical path differences arising from differences in refractive index and thickness rather than from geometrical differences. The image is produced by light transmitted rather than reflected by the specimen and can form part of a very flexible viewing arrangement. Nomarski DIC uses polarized light with standard strain-free objectives. With the minimum of manipulation the same field of view may be seen using bright field, then polarized light and, finally, with the Nomarski prisms positioned, an interference image is obtained (*Figure 19.5*).

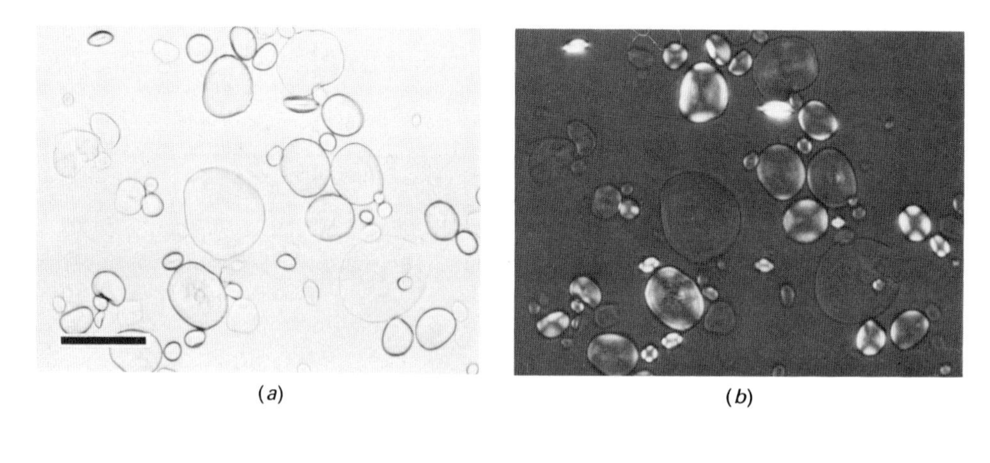

(*a*) (*b*)

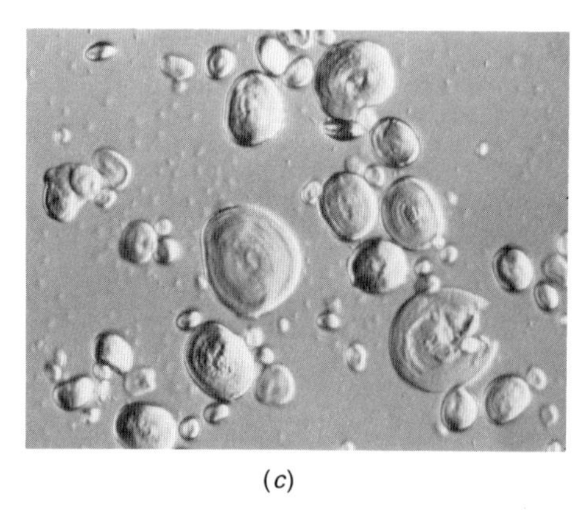

(*c*)

**Figure 19.5** Photomicrographs of wheat starch heated in water to 55 °C: (a) bright field illumination; (b) same field viewed between crossed polars; (c) same field viewed by Nomarski Differential Interference Contrast (DIC). Gelatinized granules have fainter outlines in bright field and have lost birefringence. Polarization crosses appear more diffuse than those in *Figure 19.4*. Nomarski DIC shows structure present in swollen granules (note striations). Bar = 50 μm.

## Chemical methods of obtaining contrast

Methods of staining which are firmly based on the chemistry of constituents are particularly suitable for food specimens. Such histochemical methods reveal the structure as well as the chemistry of constituents and have the advantage over histological staining of being more reliable when used with heat processed foods.

### *Demonstration of proteins and anionic polymers*

All basic dyestuffs, e.g. methyl violet and basic fuchsin stain proteins and other structures containing negatively charged groups but this is usually of limited value in distinguishing different proteins and anionic polymers such as pectin and agar from one another.

The basic dyestuff toluidine blue has metachromatic properties and is able to differentiate a range of anionic materials. This is due to the different dye binding properties of the constituents being reflected in the alignment of the dye that is bound. Where dye molecules are aligned closely, dye polymers are formed and these have a red colour whereas the monomer is blue. A comparatively strong solution of toluidine blue is used (Flint and Firth, 1981); normally this 0.2% solution would have a strong purple colour indicating the presence of toluidine blue polymers but the aqueous solution used contains 30% glycerol and a critical amount of phenol (0.66%) which restricts polymerization in solution. Polymerization occurs *in situ* when the dye mountant is put on a section yielding a differentially stained preparation within two minutes (*see Table 19.1*)

**Table 19.1** STAINING PROPERTIES OF TOLUIDINE BLUE STAIN MOUNTANT

| *Constituent* | *Staining reaction* |
|---|---|
| Muscle fibres | Normally pale blue (pale purple in presence of phosphates), muscle cell nuclei red violet |
| Raw collagen | Pale pink, cells purple |
| Cooked collagen (rind) | Paler than raw collagen, often lilac-grey, cells dark blue |
| Elastic tissue fibres | Turquoise |
| Soya protein | Dark blue-purple |
| Soya carbohydrate (cell wall material) | Deep pink |
| Commercial gluten | Pale blue-green |
| Pectin powder | Blue-purple |
| Agar powder | Magenta |

### *Demonstration of starch*

#### *Use of iodine*

Staining with aqueous iodine is the classical technique for starch identification. A 0.3% solution of iodine in 6% potassium iodide colours normal starches blue but the so-called waxy starches, e.g. waxy maize and waxy rice, which have a higher than normal amylopectin content stain a dull red colour (Evers, 1979). A much weaker solution of iodine (0.03%) in potassium iodide was used by Jones (1940) to demonstrate mill-damaged starch which swells in water allowing it to take up iodine more readily than does intact starch. Gelatinized granules respond in a similar manner but where the interest lies in structure this further swelling may be undesirable, e.g. in starch-containing emulsions or where a section of an extruded product would be disrupted by water. In these cases iodine vapour staining is recommended.

### *Iodine vapour staining for starch*

The technique is simple to set up needing only a closed glass staining jar on the base of which is a moist filter paper sprinkled with iodine crystals. Smears and sections are placed in the container and the progress of staining observed through the glass. The experiment is best done in a fume cupboard.

### *Use of direct cotton dyes*

Jones (1940) was the first to use a direct cotton dye (Congo Red) to colour damaged starch granules. Congo Red is an acid dye but the acid groups merely serve to solubilize the dye; like cotton dyeing, staining of mill- or heat-damaged starch is due to hydrogen bonding of the elongated dye molecule with the 'linear' starch molecules made accessible by the swollen condition of the damaged granules. Other direct cotton dyes have been used including Chlorazol Violet R, Chlorazol Violet N and Chlorazol Violet E (Flint and Moss, 1970). Of these Chlorazol Violet R (CI 22445) used as a 0.5% aqueous solution gives the best colour contrast between stained gelatinized and unstained intact starch.

## ***Histochemical demonstration of pectin***

Pectin can be distinguished from some gelling agents, e.g. agar and gelatin by the toluidine blue reaction described earlier but a more specific test for pectin in sections is the hydroxylamine ferric chloride method. Strictly this test, which was introduced by Reeve (1959), is a method for methylated pectin with which the hydroxylamine forms hydroxamic acid residues which are able to chelate with the iron added in the ferric form. A red-brown colour is formed *in situ*. The colour can be intensified by previously methylating the sections with hot, acidified absolute methanol.

## ***Demonstration of lipids***

### *Use of oil-soluble colours*

It is usual to demonstrate liquid fats with oil-soluble colorants, e.g. Sudan III, Sudan IV and the related Oil Red O; all these colour liquid fats red, Oil Red O giving the most intense colour. More sensitive is Sudan Black B which yields blue-black shades and which also reacts with phospholipids. All these oil-soluble colours require organic solvents for their solution but the dye solvent must contain water to minimize fat losses, 70% ethanol or 60% isopropanol being the usual choice. Despite this, the solvent may still affect the structure of some processed foods, oil-bearing emulsions being particularly vulnerable. An alternative method is to use a vapour staining technique (Flint, 1984).

*Osmium tetroxide vapour staining for lipids*

This technique is similar to that described for the iodine vapour staining of starch but osmium tetroxide (so-called osmic acid) is a more hazardous material because the vapour can permanently affect the eyes. All staining should be done in a fume cupboard, gloves should be worn and goggles for the non-spectacle wearer. The staining jar, which should have a drop-on lid, has its base lined with dry filter paper and two glass rods are positioned on this to prevent the slide touching the paper. A 5 ml vial of 2% osmium tetroxide is poured over the paper and the lid replaced. Slides holding smears or sections can then be put in and the progress of staining, which takes several minutes, observed through the glass. During staining the tetroxide is reduced to lower oxides and this is seen as a general 'browning' of the specimen.

*Combined vapour staining of lipids and starch*

Iodine vapour staining of starch can follow the osmium vapour staining of liquid fats but the lipid staining must be done first, otherwise the iodine will block the double bonds needed for osmium tetroxide reduction.

## The combination of physical and chemical methods of obtaining contrast

In any work on food structure it is wise to use more than one contrast technique. The combination of physical and chemical methods is particularly helpful because it enables more information to be gained than either technique could yield alone as the following examples illustrate:

### *Damaged and undamaged starch (Figures 19.4 and 19.6)*

Whether dilute iodine or a dye of the Chlorazol series is used to detect damaged starch, viewing the specimen between crossed polars is worthwhile. Intact granules show a polarization cross and, by slightly uncrossing the polars, the stained damaged granules can be clearly seen at the same time. Polars are particularly useful in demonstrating residual intact granules which could easily be overlooked in specimens which consist mainly of gelatinized starch, e.g. in cake and some extruded products.

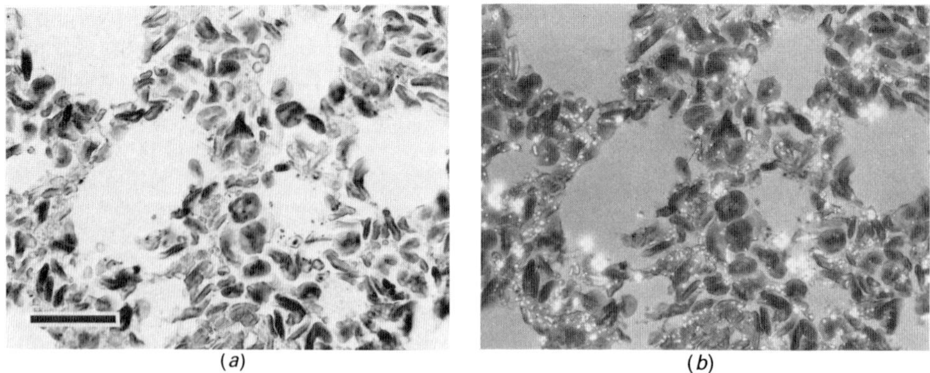


**Figure 19.6** Photomicrographs of 10 μm section of high ratio cake stained with Chlorazol Violet R: (a) bright field image, all granules appear to be gelatinized; (b) same field viewed between partially crossed polars, intact starch granules now visible (note polarization crosses). The tiny birefringent particles consist mainly of finely divided wheat bran. Bar = 100 μm.

### *Solid and liquid fats (Figure 19.7)*

In the staining of lipids the methods referred to do not colour solid fats but by viewing the specimen between partially crossed polars, the solid fats which are crystalline can be seen at the same time as the stained liquid fats. If there is doubt about whether the crystalline material is fatty this can be resolved by gently warming the slide. Solid fats can even be stained with lipid-soluble dyes if the staining solution and slide are both warmed.

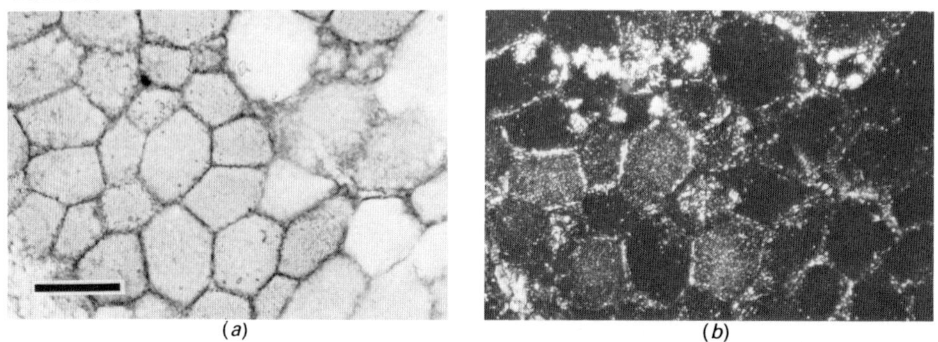


**Figure 19.7** Photomicrograph of 10 μm section of adipose tissue stained with Oil Red O: (a) bright field image, fat cells within connective tissue matrix are coloured red; (b) viewed between crossed polars the unstained crystalline fat is visible due to its birefringence. Bar = 100 μm.

### *Effect of Sirius Red staining on the birefringence of collagen (Figure 19.8)*

It might be expected that the staining of collagen would reduce the total light transmitted by a preparation and therefore make the birefringence of collagen more difficult to see. Sirius Red F3BA appears to be unique in that it not only colours collagen a more intense red than is found with the classical van Gieson method (Sweat, Puchtler and Rosenthal, 1964) but the Sirius Red stained collagen is itself more birefringent than the unstained collagen. This has been used to demonstrate collagen in meat products and to show the difference in structure between raw and cooked collagen (Flint and Pickering, 1984).

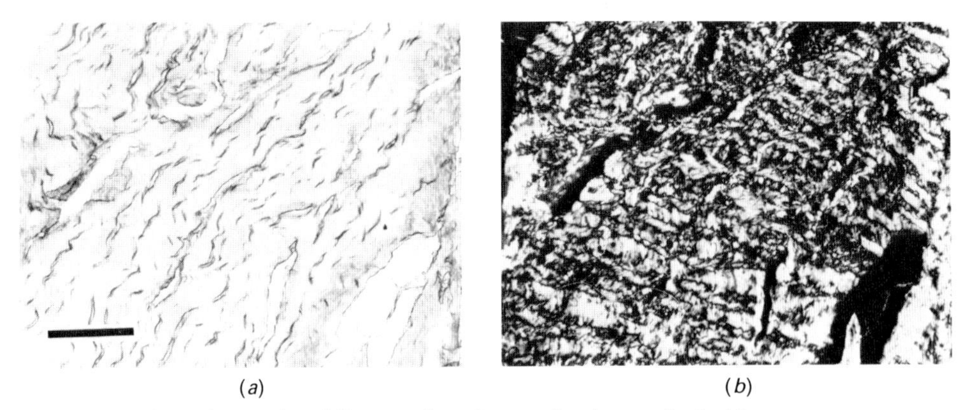


**Figure 19.8** Photomicrographs of 10 µm section of connective tissue stained with picro-Sirius Red F3BA: (a) bright field image, collagen is coloured red; (b) same field viewed between crossed polars, stained collagen is brightly birefringent with gold, orange and green polarization colours. Bar = 100 µm.

## Staining and interference microscopy (*Figure 19.9*)

Although the interference microscope is designed for the examination of unstained specimens and staining tends to degrade the image obtained, very light staining can be helpful with Nomarski DIC. It enables different constituents to be more readily recognized, e.g. starch granules and oil droplets. In emulsion work it can help to distinguish between oil-in-water and water-in-oil systems.

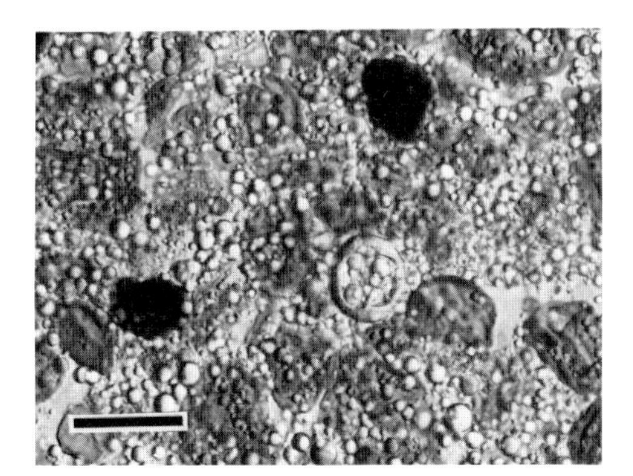

**Figure 19.9** Photomicrograph of low calorie salad cream smear stained with iodine vapour, viewed by Nomarski DIC. Shows stained swollen starch granules surrounded by unstained oil droplets. The spherical particle (centre) is undispersed spray dried egg. Bar = 50 µm.

## Estimation and measuring size with the optical microscope

In the evaluation of food structure it can be useful to know the size of features seen under the microscope. When only a rough estimate of size is needed this can be done by eye, using as a reference point material in the field of view whose dimensions are known, e.g. the presence of a recognized starch. Alternatively a knowledge of the field diameter for the particular objective being used can help in approximate sizing. Field diameters are simple to measure, needing only a stage micrometer. This consists of a microscope slide accurately engraved with a millimetre scale divided into tenths and hundredths. When in focus, the field diameter for each eyepiece–objective combination can be measured.

Where greater precision is needed, e.g. in assessing particle size distribution, the same stage micrometer can be used to calibrate an eyepiece graticule. The eyepiece graticule consists of a glass disc on which is engraved an arbitrary scale consisting of 100 equal units. This is best located in a special focusing eyepiece which has an adjustable eye lens so that the scale can be precisely focused by individual users. By carefully superimposing the image of the stage micrometer on the eyepiece graticule the value of the graticule units for each eyepiece–objective combination can be measured.

## Quantitative microscopy

An eyepiece graticule can also be used to assess the relative proportions of different constituents in a series of microsections. This time the graticule carries an array of points which are superimposed on the microscope image. The proportion of hits scored on each constituent by the points is a measure of the section area occupied by that constituent, and the area taken up by each constituent reflects the volume of that constituent in the initial sample. Even with very careful sampling the number of points that must be counted for close accuracy runs into thousands, e.g. 10 500 points are counted in a method for the determination of texturized soya protein in comminuted meats (Flint and Meech, 1978).

### *Automatic counting*

Automatic counting uses a television camera based system to scan the microscope image working line by line for 720 lines. The tiny unit length scanned at any one time is called a 'picture point' and rates approaching $10^7$ picture points per second are achieved. This makes the system much faster than even the best designed manual method but feature recognition is much poorer. Feature recognition with the television scanner depends on grey level detection and the scanner is not sensitive to the small differences in colour readily seen by the eye. For automatic image analysis it is particularly important that the image presented is one of high contrast. This can be achieved by both the staining method chosen and the use of suitably coloured filters to further increase image contrast.

## Conclusions

The optical microscope has a useful part to play in the assessment of raw materials and in the evaluation of the microstructure that processing imposes on those raw materials. With the preparative techniques now available, results can be achieved rapidly and can be related to other methods of analysis in a synergistic way.

Light microscopy techniques enable problems to be viewed from a different angle.

This other view is available to the non-specialist because when we use an optical microscope we are, in effect, extending our own optical powers. Apling (1980) advocates a revival in the application of light microscopy by both the food analyst and the researcher, partly because the instrument and techniques are ones directly available to the food scientist himself to be used whenever he needs them.

Learning to use the microscope is important. Unlike most scientific instruments which give obviously nonsensical results when incorrectly adjusted, the optical microscope will nearly always provide some sort of image (*see Figure 19.10*).

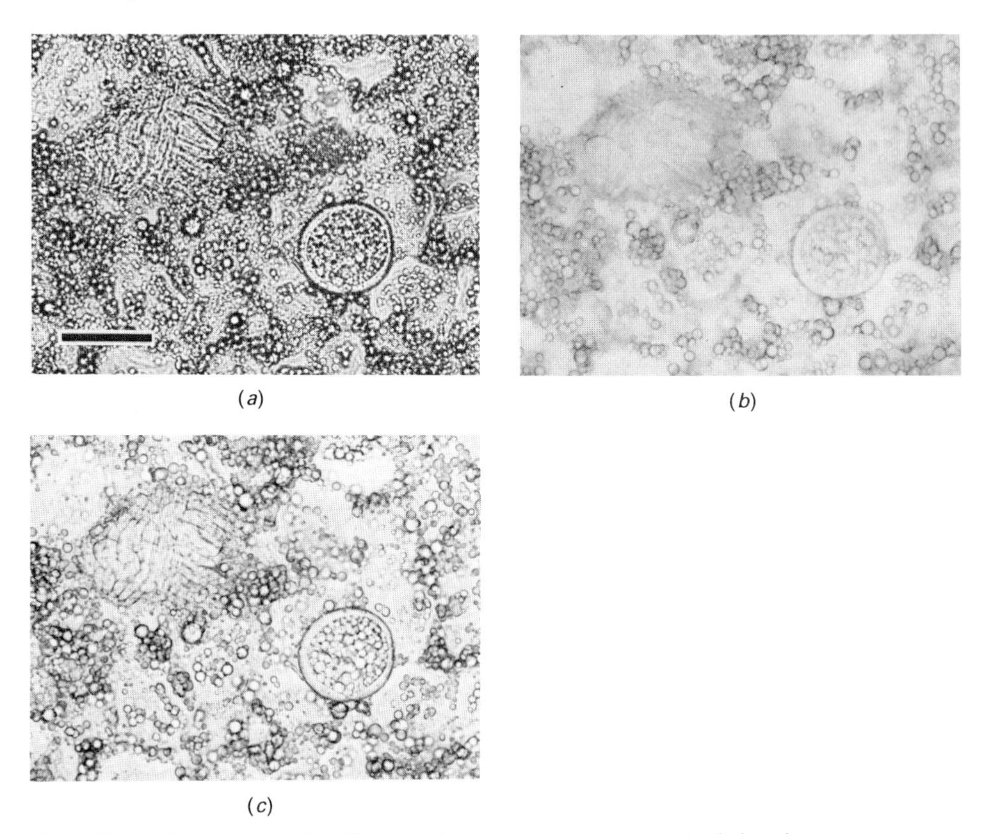

(*a*) (*b*) (*c*)

**Figure 19.10** Photomicrographs of low calorie salad cream smear, unstained viewed bright field with microscope incorrectly (a) and (b) and correctly adjusted (c): (a) substage condenser not focused; (b) condenser focused but aperture (iris) diaphragm fully open; (c) microscope correctly adjusted; (a) shows that the increased contrast obtained by diffraction effects is at the expense of resolution; (b) resolution is improved, oil droplets within the spray dried egg particle are now visible but the image is degraded by glare; (c) correctly adjusted detail within the spray dried egg particle and the mustard fragment (left) can be seen. Bar = 50 μm.

The improvement that correct setting up gives can be so marked and takes so little time that checking the instrument before starting a new task is always worthwhile. The text by Bradbury (1984) is a particularly good guide in this. One chapter is devoted to the practical use of the microscope and this provides both details of setting up the instrument and a trouble-shooting key dealing with some of the causes of poor image quality.

To obtain a good image, sample preparation is also important. For transmitted light work it is obvious that the specimen must transmit light, i.e. it must appear translucent if not transparent. For incident light work the emphasis shifts to obtaining a flat surface that can be mounted parallel to the objective lens.

Although formal training is not strictly necessary it can reduce the learning period. Practical courses such as those run regularly by the Royal Microscopical Society are helpful and these provide the opportunity to meet other microscopists as well as to use a wide range of equipment.

## Acknowledgement

Thanks are due to Barry M. Firth for technical help with the photographs illustrating this chapter.

## References

AMERICAN ASSOCIATION OF FEED MICROSCOPISTS (1978). *Manual of Microscopical Analysis of Feedstuffs*, 2nd Edition

ANGOLD, R. (1982). The structure of baked cereal products. *Proceedings of the Royal Microscopical Society*, **17**(Micro 82 Supplement), S28

APLING, E.C. (1980). Advances in the light microscopy of foods. In *Developments of Food Analysis Techniques—2*, (King, R.D., Ed.), pp. 151–185. London, Applied Science.

BECHTEL, D.B. (Ed.) (1983). *New Frontiers in Food Microstructure*. St. Paul, Minnesota, American Association of Cereal Chemists

BRADBURY, S. (1984). *An Introduction to the Optical Microscope*, Microscopy Handbook 01, Royal Microscopical Society. Oxford, New York, Oxford University Press

COHEN, S.H., DAVIES, E.A., HOLCOMB, D.N. and KALAB, M. (Eds) (1981). *Studies of Food Microstructure*. Illinois, Scanning Electron Microscopy Inc.

EVERS, A.D. (1979). Cereal starches and proteins. In *Food Microscopy*, (Vaughan, J.G., Ed.), pp. 139–191. London, New York, Academic Press

FLINT, F.O. (1984). Applications of light microscopy in food analysis. *Microscope*, **32,** 133–140

FLINT, F.O. and FIRTH, B.M. (1981). A toluidine blue stain mountant for the microscopy of comminuted meat products. *Analyst*, **106,** 1242–1243

FLINT, F.O. and MEECH, M.V. (1978). Quantitative determination of texturised soya protein by a stereological technique. *Analyst*, **103,** 252–258

FLINT, F.O. and MOSS, R. (1970). Selective staining of protein and starch in wheat flour and its products. *Stain Technology*, **45,** 75–79

FLINT, F.O. and PICKERING, K. (1984). Demonstration of collagen in meat products by an improved picro Sirius Red polarisation method. *Analyst*, **109,** 1505–1506

HARTLEY, W.G. (1979). *Hartley's Microscopy*, 2nd Edition, p. 200. Charlbury, UK, Senecio Publishing Co.

JONES, C.R. (1940). The production of mechanically damaged starch in milling as a governing factor in the diastatic activity of flour. *Cereal Chemistry*, **17,** 133–169

REEVE, R.M. (1959). A specific hydroxylamine–ferric chloride reaction for histochemical localization of pectin. *Stain Technology*, **34,** 209–211

STANLEY, D.W. and TUNG, M.A. (1976). Microstructure of food and its relation to texture. In *Rheology and Texture in Food Quality*, (de Man, J.M., Voisey, P.W., Rasper, V.F. and Stanley, D.W., Eds), pp. 28–78. Westport, Connecticut, AVI Publishers

SWEAT, F., PUCHTLER, H. and ROSENTHAL, S.I. (1964). Sirius Red F3BA as a stain for connective tissue. *Archives of Pathology*, **78,** 69–72

VARRIANO-MARSTON, E. (1981). Integrating, light and electron microscopy in cereal science. *Cereal Foods World*, **26,** 558–561

VARRIANO-MARSTON, E. (1983). Polarization microscopy: applications in cereal science. In *New Frontiers in Food Microstructure*, (Bechtel, D.B., Ed.), pp. 71–108. St. Paul, Minnesota, American Association of Cereal Chemists

VAUGHAN, J.G. (1979). Animal feeds—plant constituents. In *Food Microscopy*, (Vaughan, J. G., Ed.), p.395. London, Academic Press

20

# AN ELECTRON MICROSCOPIST'S VIEW OF FOODS

D.F. LEWIS
*Leatherhead Food Research Association, Leatherhead, Surrey, UK*

## Introduction

Microscopists generally see foods quite differently from other scientists. This extra viewpoint can often be of great benefit in considering a problem concerning the behaviour of foodstuffs. The structure of foods is invariably linked to properties such as texture, rheology, flavour release, appearance and stability on processing. Indeed an understanding of the microscopic structure of foods is often the only way to link the observed chemical and physical characteristics of a foodstuff. However, finding and understanding the structure of foods can be very difficult and persuading non-microscopists of the authenticity and relevance of microscopical findings is often even more difficult.

In fairly crude terms food structure can be considered on four levels:

1. Chemical structure is concerned with the molecules which make up the foodstuff and the way in which these molecules interact with each other.
2. The electron microscopic structure of foods deals with the aggregation of molecules and their assembly into components which typically measure between 2 nm and 1 μm.
3. The light microscope covers the range from about 0.6 μm to 1 mm.
4. The macroscopic level of structure considers those features which are perceived by human senses even though instrumental methods may be used to quantify them; firmness, coarseness, colour and wetness are examples of macroscopic structure.

In order to understand fully the behaviour of foods, consideration should be given to all levels of structure and an attempt to understand food based on only one structural level is likely to give misleading results. Hence, whilst this chapter considers the electron microscopist's view of food it is important always to combine the results of electron microscopy with other observations.

Electron microscope studies have now been carried out on a wide range of foods and food-related materials and, in considering these findings, a general model for food structures is suggested. The model is that foods consist of a mixture of continuous matrices and discontinuous inclusions. The precise nature of the model will vary from food to food but if the general model is adopted then it is possible to relate eating and handling properties to changes in the nature of the continuous matrices or the inclusions. Examples of continuous matrices in food systems include gel structures, sugar syrups and liquid lipids. Examples of inclusions are crystals, air, water or oil droplets, compact solids, fibres and membranes.

The concept of matrices and inclusions are illustrated in two ways. Firstly, the microscopic structure of a wide range of foods can be examined with a view to demonstrating the continuous and discontinuous phases. Secondly, the effect of making changes to the matrices and inclusions by processing can be studied in relation to change in the other properties of foods.

The concept of matrices and inclusions can be seen at both the light and electron microscope level and is common to natural and 'manufactured' foods as the examples in the following sections demonstrate.

## Plant tissues

A wide range of plant tissues are eaten as foods. Leaves, petioles, stems, roots, fruits and seeds are all used as foodstuffs. Despite the variety of structures on a macroscopic scale, common cell types are present. One of the most common cell types in plant tissue is the parenchyma cell. At the light microscope level (*Figure 20.1*) parenchyma tissue can be seen as consisting of a continuous network of cell walls with cell contents and air cells forming the inclusions. The cell contents themselves are also a mixture of continuous and discontinuous phases. The membrane-bound cytoplasm forms the continuous phase whilst nuclei and vacuoles form the inclusions. Depending on the plant tissue, starch grains, protein bodies, fat droplets, chloroplasts and pigment granules can be seen as cellular inclusions by light microscopy. At the electron microscope level inclusions such as mitochondria and small vesicles become apparent. Also, the cell walls themselves are revealed as rather more complex structures (*Figure 20.2*). Combining electron microscopy with chemical analysis indicates that the cell walls consist of fibrous inclusions of cellulose trapped in a mixed network of pectin, hemicelluloses and protein. The role of the constituents of plant tissue is illustrated by the softening that occurs on cooking. Most tissues lose crispness early on in the cooking process and this is caused by the cell membranes breaking down and

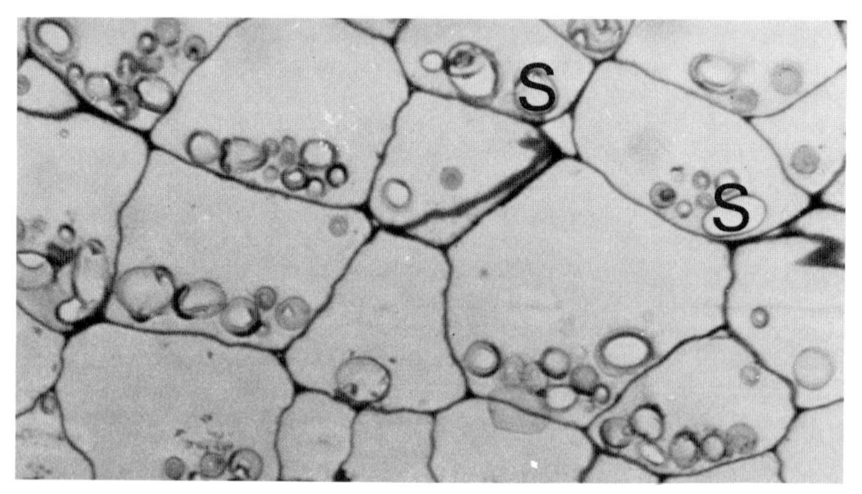


**Figure 20.1** Light microscopy of potato parenchyma; S = starch gain. (Magnification × 100)

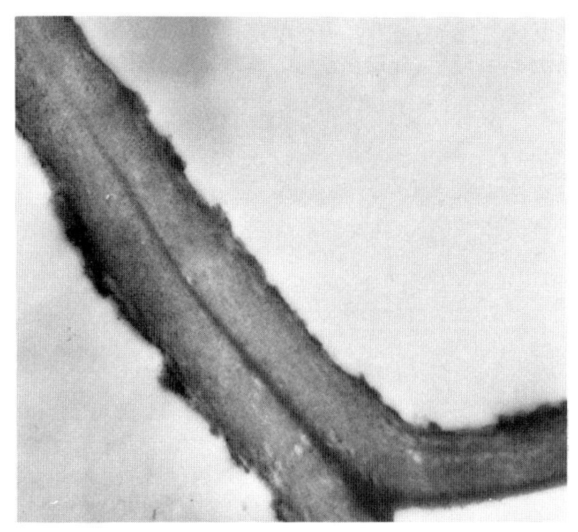

**Figure 20.2** Transmission electron microscopy of raw beetroot parenchyma cell wall. (Magnification × 25 000, reduced to 70% in reproduction.)

destroying the turgor pressure. In raw tissue the turgor pressure keeps the cell contents pressing against the cell walls. Loss of turgor pressure occurs in all processed plant foods. Longer cooking produces more softening and this is brought about by changes in the cell wall matrices, the pectin component being particularly affected. The cell walls start to swell and separate into layers (*Figure 20.3*), the swelling commonly starting from the air spaces. The changes in the cell wall matrix alter the fracture process on cutting or biting: in raw tissue the fracture breaks across cell walls leaving open cells (*Figure 20.4*), whilst in cooked tissue the cells separate along the pectin-rich junctions known as middle lamellae (*Figure 20.5*). Other processes can also cause swelling and softening of the cell walls, although the precise nature of the change depends on the process and probably on the type of tissue involved.

This description represents a fairly simple view of the relationship of matrices and

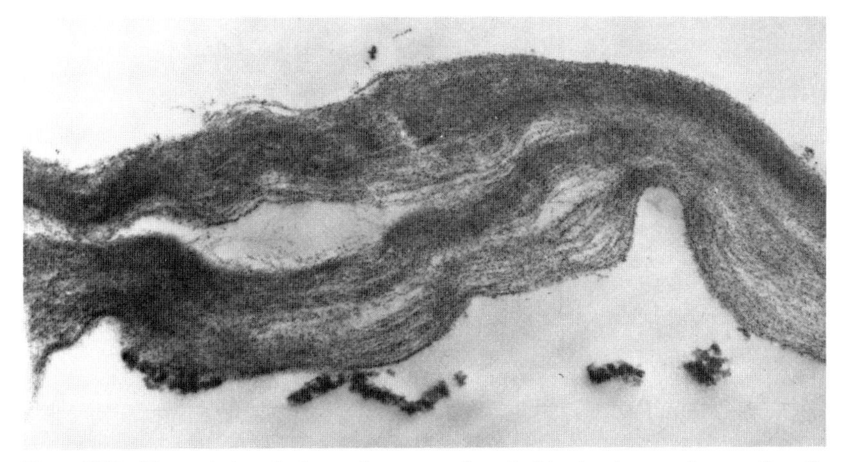

**Figure 20.3** Transmission electron microscopy of cooked beetroot parenchyma cell wall. (Magnification × 25 000, reduced to 80% in reproduction.)

**Figure 20.4** Scanning electron microscopy of raw beetroot, cut surface; U = fractured edge of cell wall. (Magnification × 150, reduced to 80% in reproduction.)

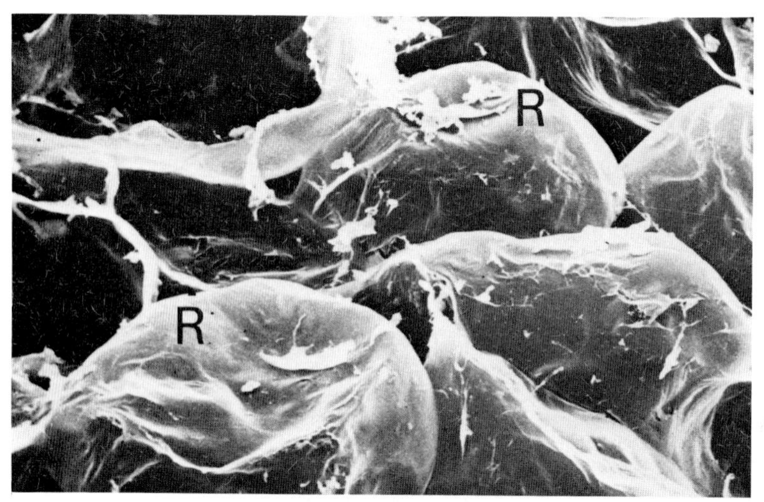


**Figure 20.5** Scanning electron microscopy of cooked beetroot, cut surface; R = separated face of cell wall. (Magnification × 150, reduced to 80% in reproduction.)

inclusions with softening on cooking and when starch grains, protein bodies and fat droplets are also included within the cells then the relationships are more complex.

## Meat

Like plant tissue, meat is made up of cells. In this case, the cells are considerably elongated and the interstitial connective tissue is the equivalent of the cell wall matrix in plants. At the light microscope level (*Figure 20.6*) the cell contents are seen to be proteinaceous structures arranged in a regular manner to give a periodic banding. Nuclei, mitochondria and fat droplets can also be seen as cellular inclusions by light

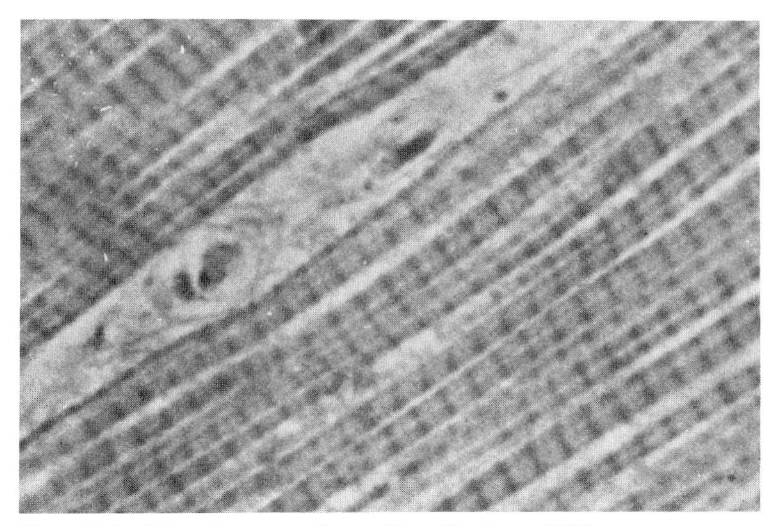

**Figure 20.6** Light microscopy of meat. (Magnification × 2000)

microscopy. The connective tissue contains blood vessels, nerves, fat cells and other cells as inclusions as well as the muscle fibres.

Electron microscopy and chemical analyses show both the connective tissue and the muscle cell contents to be complex structures. The connective tissue comprises banded collagen fibres surrounded by a complex glycoprotein matrix called ground substance (*Figure 20.7*). Proteins inside the cells are found to be ordered in a very precise way to give the regular banding pattern seen in the light microscope (*Figure 20.8*). The banded fibrils are made up of myofibrillar proteins, mostly myosin and actin, and are surrounded by a continuous matrix of sacroplasmic proteins. The sarcoplasmic proteins also surround mitochondria, nuclei, fat droplets and membrane-bound vesicles.

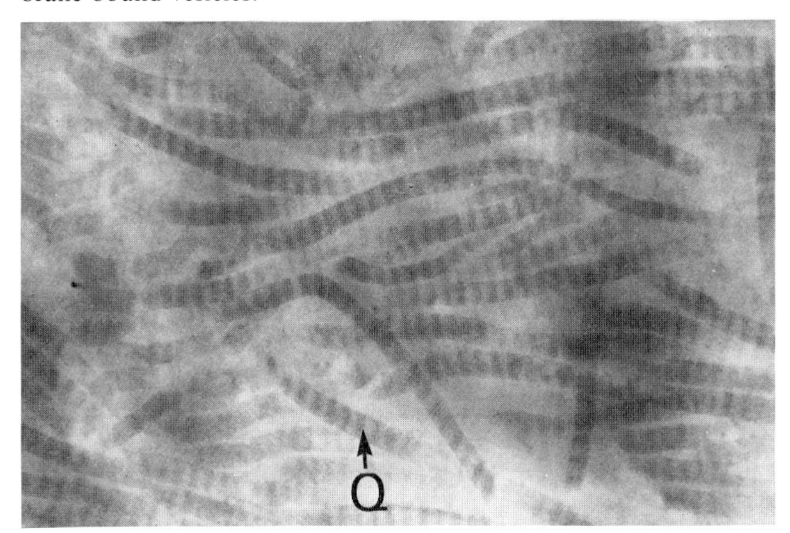


**Figure 20.7** Transmission electron microscopy of connective tissue in meat; Q = banded collagen fibre. (Magnification × 30 000, reduced to 80% in reproduction.)

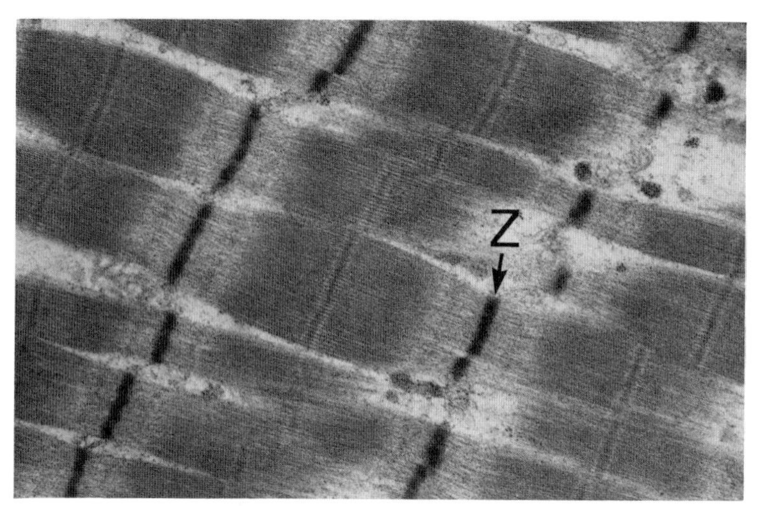


**Figure 20.8** Transmission electron microscopy of myofibrillar proteins in meat; Z = line. (Magnification × 25 000, reduced to 80% in reproduction.)

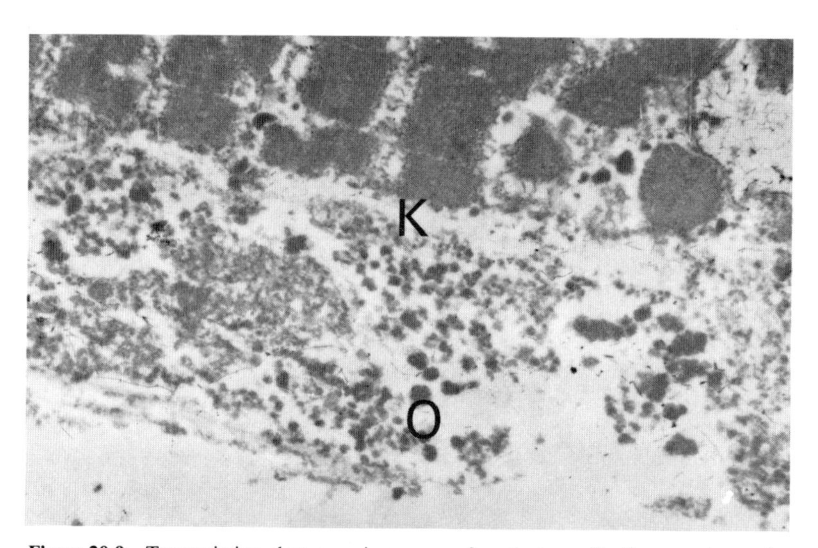


**Figure 20.9** Transmission electron microscopy of cooked myofibrillar proteins and connective tissue in meat; K = coagulated myofibrils; O = coagulated collagen fibres. (Magnification × 15 000, reduced to 80% in reproduction.)

When meat is cooked it normally loses water and gradually becomes more tender. These processes are related to changes in the connective tissue matrix and the meat cell contents. On heating to 60–70 °C, the meat cell contents coagulate (*Figure 20.9*), the fibrous nature of the myofibrillar proteins is lost and the sarcoplasmic proteins largely precipitated. These processes are accompanied by shrinkage, loss of water and some softening. On more prolonged or intensive heating the collagen fibres in the connective tissue shrink and collapse (*Figure 20.9*), which results in considerable softening. Eventually, the collagen fibres may form a gelatin gel, which will help to retain some water. The amount of water lost from meat on cooking can be controlled by the addition of salt and pH regulators such as polyphosphates (*Figure 20.10*). In this case the myofibrillar and sarcoplasmic proteins in the meat are to some extent dispersed and set into a water-holding gel on heating.

In meat products such as sausages, burgers or luncheon meats the myofibrillar proteins along with any added proteins are dispersed to form a continuous matrix enmeshing intact meat fibres, fat cells, starch and seasoning. The extent to which the protein network is continuous controls the stability of the product on cooking.

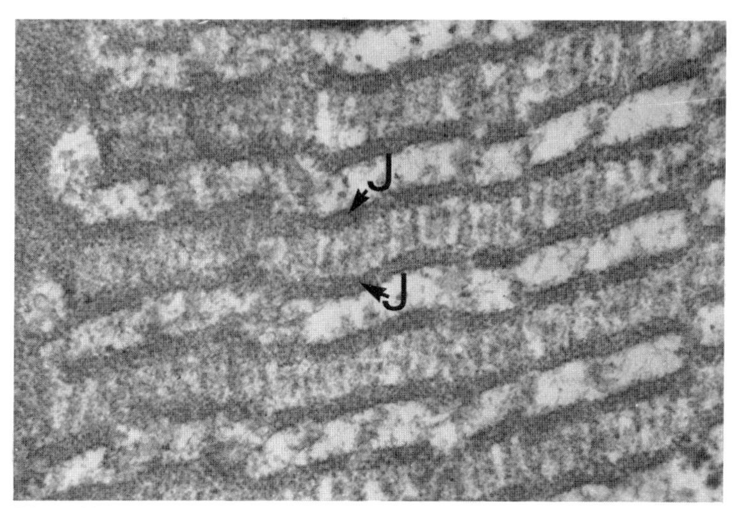


**Figure 20.10** Transmission electron microscopy of ham type product; J = dark staining bands at actomyosin-rich interface between A and I bands. (Magnification × 15 000, reduced to 80% in reproduction.)

## Fats

Animal fatty tissues, like meat, consist of cells embedded in a matrix of connective tissue. In this case the cell contents consist almost entirely of fat with just a thin ring of proteinaceous cytoplasm around the edge of each cell. At chill temperatures the fat inside the cells contains crystals surrounded by liquid. The cellular nature of fatty tissue and the crystalline nature of the fat within the cells can be observed using light microscopy and polarized light (*Figure 20.11*). Electron microscopy shows clearly the fat crystals embedded in liquid fat (*Figure 20.12*) and the arrangement of collagen fibres within the connective tissue (*Figures 20.13* and *20.14*). Differences in the arrangement of collagen fibres affect the behaviour of the fatty tissue; in pork jowl fat, where the collagen fibres are tightly packed in the ground substance, little cell breakage occurs during processing and consequently cooking losses are low. In flare fat, however, where the collagen fibres are loosely packed, considerable cell breakage and fat loss are produced on processing. The nature of the fat within the cells also influences the behaviour of the fat cells during comminution processes; thus cells with high levels of crystalline fat tend to break more readily than cells with low levels of crystallinity.

Fats extracted from animals or plants also show a structure consisting of a liquid oil phase with crystalline inclusions. The performance of the fat depends on the ratio of crystal to oil and on the size and shape of the crystals. For example, fats with small crystals will tend to produce a more aerated cake batter when creamed with sugar.

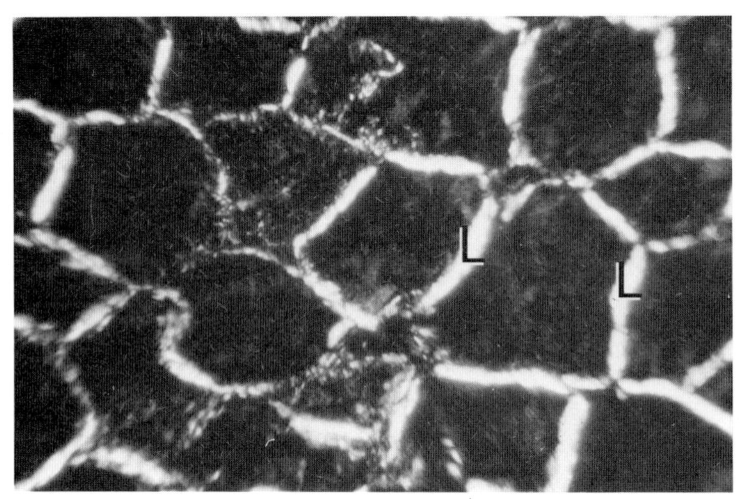


**Figure 20.11** Light microscopy of fatty tissue; L = crystalline layer at edge of cells. (Magnification × 150, reduced to 80% in reproduction.)

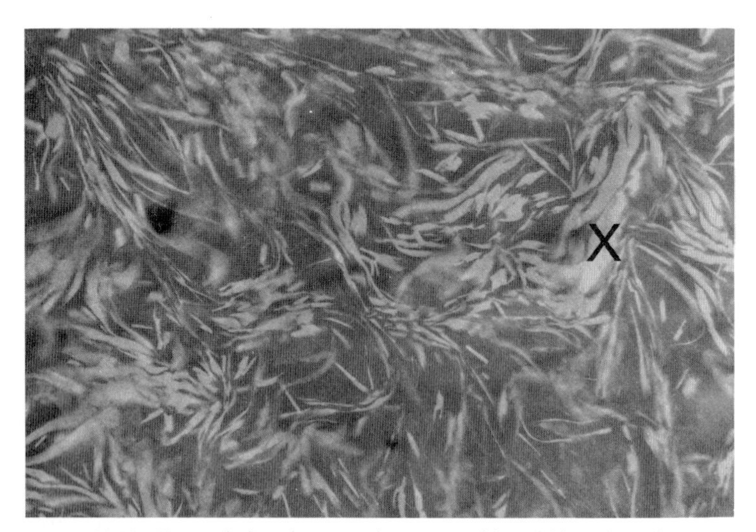


**Figure 20.12** Transmission electron microscopy of fat within a fat cell; X = crystals within fat cells. (Magnification × 10 000, reduced to 80% in reproduction.)

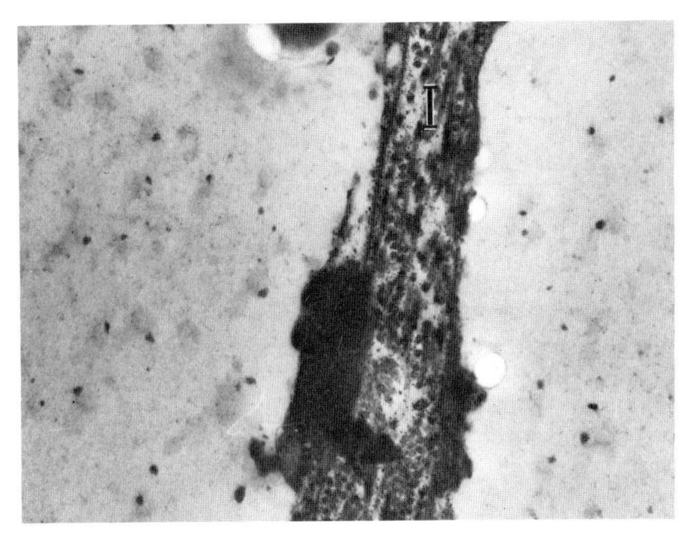


**Figure 20.13** Transmission electron microscopy of connective tissue in pork jowl fat; I = interstitial connective tissue. (Magnification × 15 000, reduced to 80% in reproduction.)

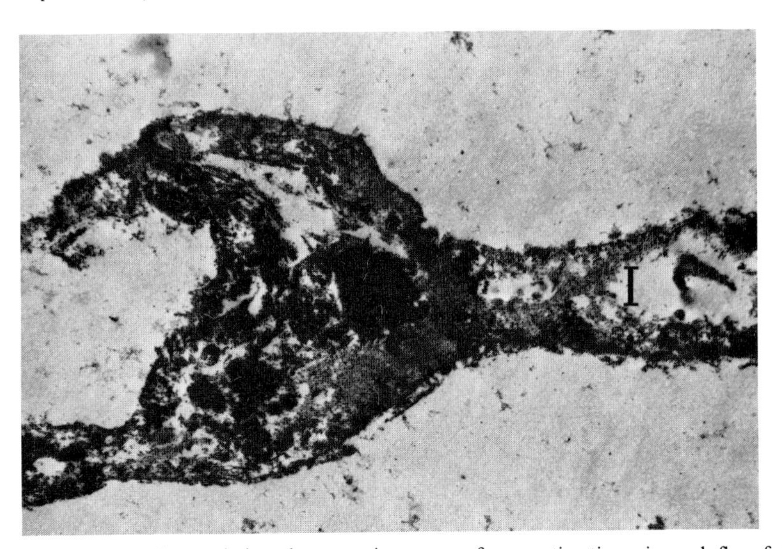


**Figure 20.14** Transmission electron microscopy of connective tissue in pork flare fat; I = interstitial connective tissue. (Magnification × 10 000, reduced to 80% in reproduction.)

## Foams and emulsions

Foams and emulsions are obvious examples of systems containing inclusions in a continuous matrix. These dispersions are clearly seen by light microscopy (*Figures 20.15* and *20.16*). Using electron microscopy it can be shown that the edge of the air or oil inclusions is often a concentrated protein layer (*Figures 20.17* and *20.18*). In general, stable foams and emulsions rely on having the air or oil inclusions as small as possible and this in turn requires that the protein layer at the edge of the inclusions is

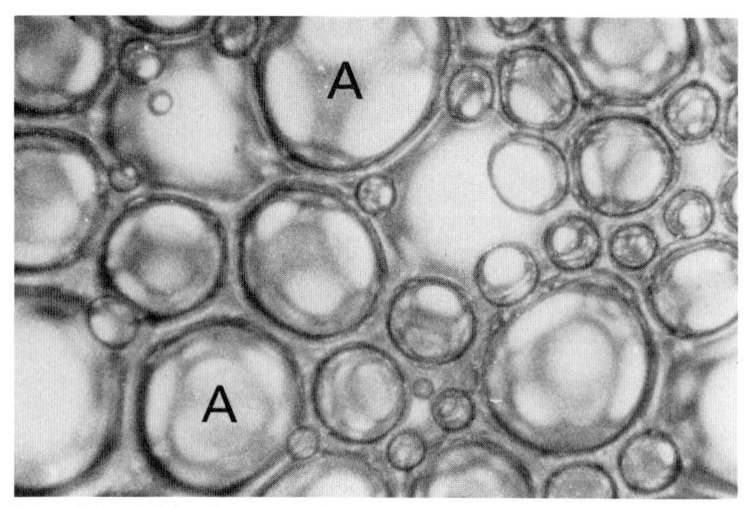


**Figure 20.15** Light microscopy of protein foam; A = air bubble. (Magnification × 100, reduced to 80% in reproduction.)

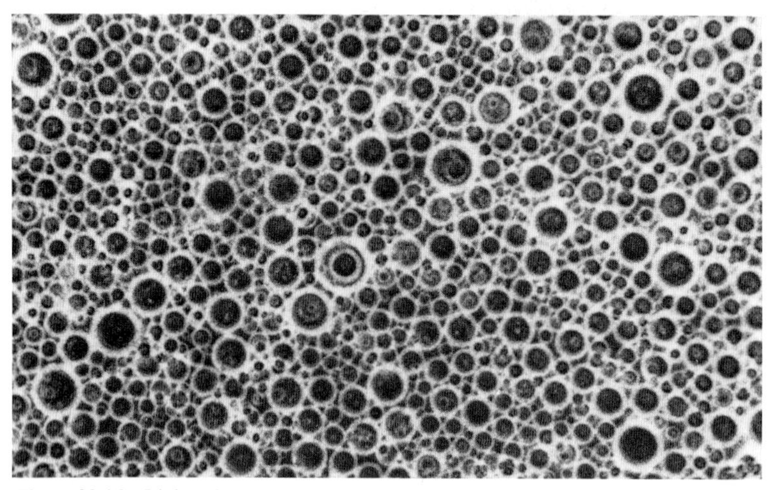

**Figure 20.16** Light microscopy of a good emulsion. (Magnification × 100)

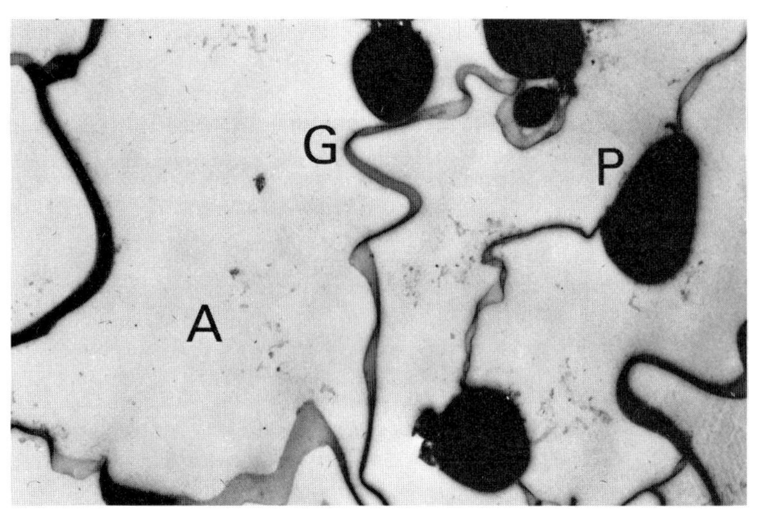


**Figure 20.17** Transmission electron microscopy of protein foam; A = air bubble; G = membrane; P = protein aggregate. (Magnification × 10 000, reduced to 80% in reproduction.)

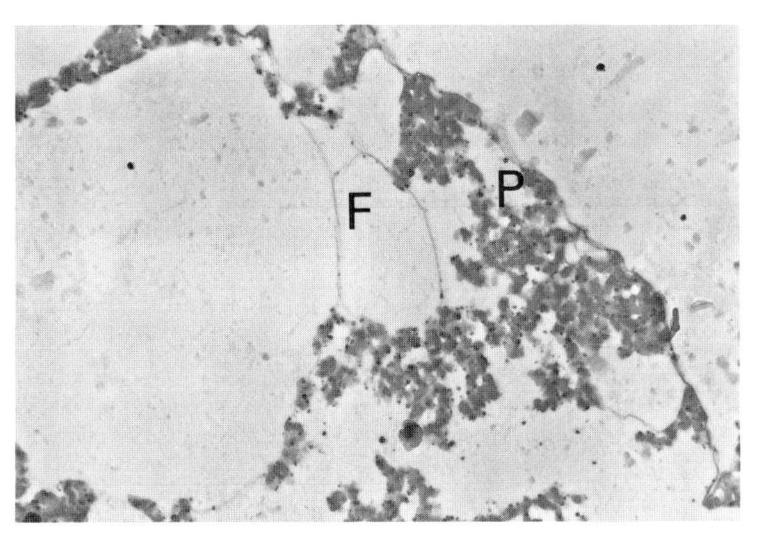


**Figure 20.18** Transmission electron microscopy of emulsion showing protein aggregation; F = fat droplet; P = protein aggregate. (Magnification × 10 000, reduced to 80% in reproduction.)

both strong and flexible. In toffee manufacture, the nature of the final product is greatly influenced by the ability of the protein to form a thin layer at the fat droplet surface. In normal toffee, casein associates with whey in the fat droplet membrane to produce a slightly brittle membrane which allows some fat release during cooking. A small amount of fat droplet breakdown helps to produce the toffee flavour but if too much breakdown occurs then the toffee becomes unsightly and greasy. The stability of the emulsion can be controlled by altering the protein content of the mixture. Thus a toffee made with only whey protein will have a very stable emulsion and this will affect the flow properties of the toffee and its flavour.

Some foods are mixtures of foams and emulsions and may even contain other inclusions. For example, ice cream (*Figure 20.19*) contains air, fat droplets, sugar crystals and ice crystals all embedded in a frozen gel matrix. In this case the relationship between the state of the various inclusions and the properties of the ice cream are quite complex.

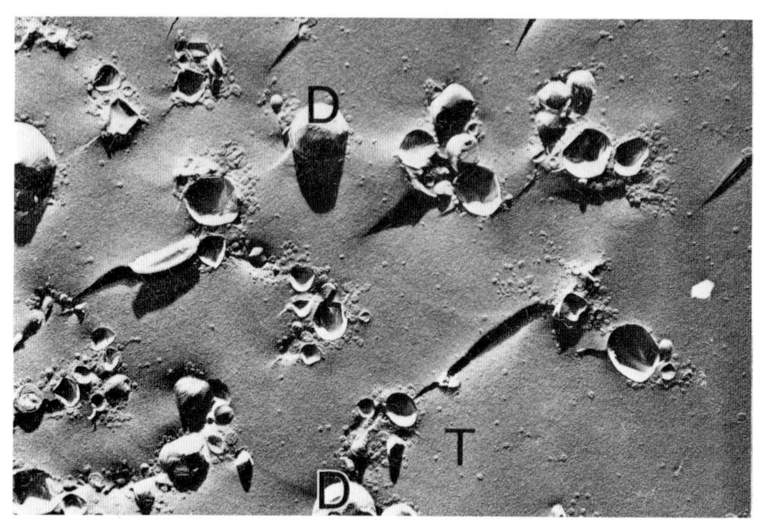


**Figure 20.19** Transmission electron microscopy of ice cream pre-mix; D = droplet; T = frozen matrix. (Magnification × 10 000, reduced to 80% in reproduction.)

## Chocolate

Chocolate is a dispersion of cocoa particles, sugar and milk solids in a fat matrix. These components can be demonstrated by light microscopy (*Figure 20.20*). Electron microscopy shows that the fat phase itself is made up of a continuous liquid phase with fat crystal inclusions (*Figure 20.21*). A combination of the techniques of electron microscopy, X-ray diffraction and differential scanning calorimetry has shown that crystalline fat in chocolate can exist in six different polymorphic forms, each with a different melting point. Chocolate is tempered in order to produce a fat phase with small crystals with a reasonably high melting point. If the crystals are too large or in a low melting point form then the chocolate bar is likely to be unstable and large, finger-like crystals may develop on the surface, producing a white powdery appearance known as bloom (*Figure 20.22*).

The structure of the cocoa particles, sugar and milk solids has a marked effect on the flow behaviour of the molten chocolate; in general the more very fine particles that are present the more viscous the chocolate. Microscopy can be used to observe the breakdown of particles. If the milk solids are added as a dense spray-dried powder such as calcium caseinate (*Figure 20.23*), the particles remain intact during the refining process and a low viscosity couverture results. However, if a highly aerated

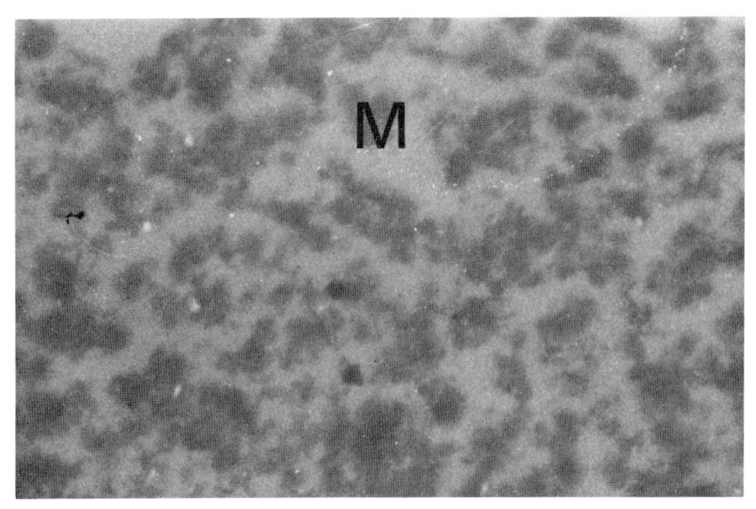


**Figure 20.20** Light microscopy of chocolate; M = milk protein particle. (Magnification × 100)

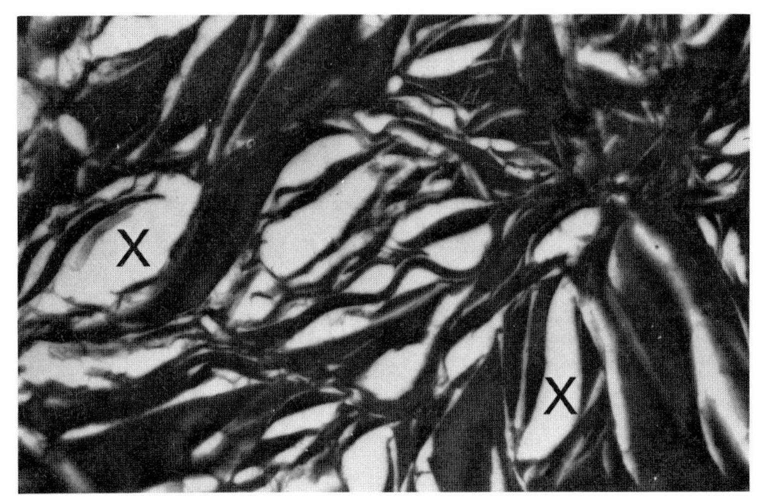


**Figure 20.21** Transmission electron microscopy showing crystals in cocoa butter; X = crystals within fat cells. (Magnification × 10 000, reduced to 80% in reproduction.)

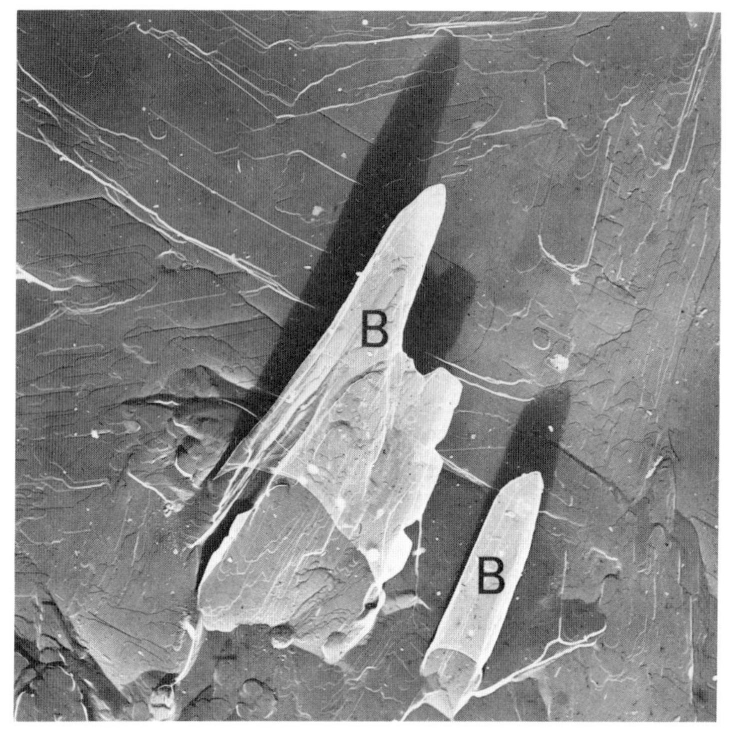


**Figure 20.22** Transmission electron microscopy of bloom on chocolate; B = bloom crystal on chocolate. (Magnification × 10 000, reduced to 80% in reproduction.)

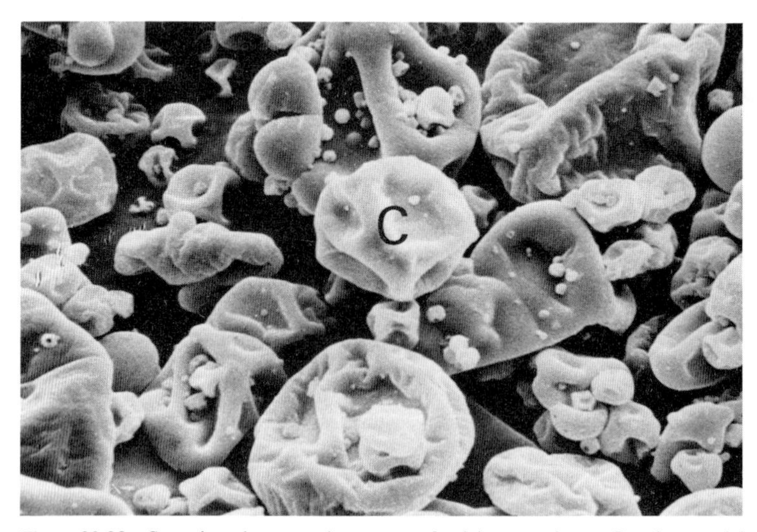


**Figure 20.23** Scanning electron microscopy of calcium caseinate; C = dense calcium caseinate particle. (Magnification × 500, reduced to 80% in reproduction.)

**Figure 20.24** Scanning electron microscopy of calcium-reduced ultrafiltered spray-dried milk; F = fractured aerated particle. (Magnification × 500, reduced to 60% in reproduction.)

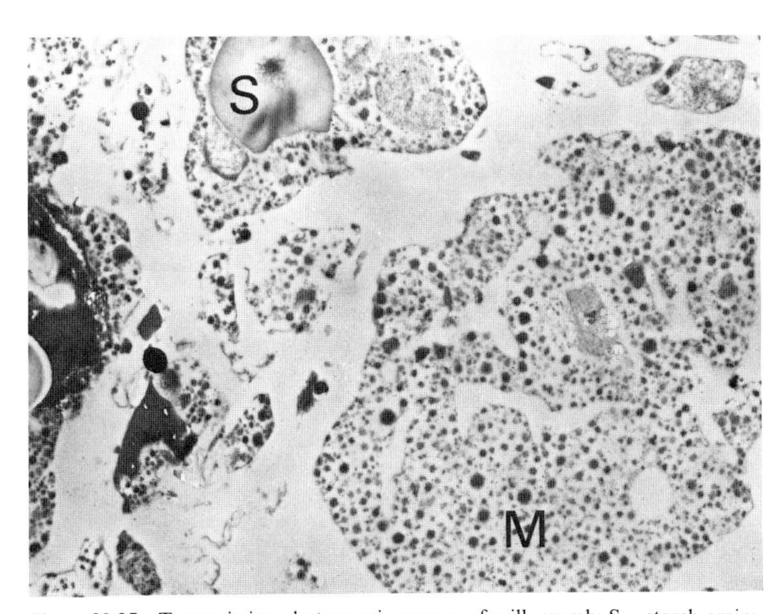


**Figure 20.25** Transmission electron microscopy of milk crumb; S = starch grain; M = milk protein particle. (Magnification × 10 000, reduced to 80% in reproduction.)

powder such as a spray-dried ultrafiltered milk is used (*Figure 20.24*), considerable breakdown is produced on processing and a high viscosity couverture results.

In the UK most milk chocolate is prepared from milk crumb. Milk crumb is produced by mixing liquid milk, sugar and cocoa mass and drying before crushing and milling. The final crumb is a continuous matrix of milk protein containing sugar and cocoa particle inclusions (*Figure 20.25*). The precise manner of production affects the distribution and size of the inclusions in the crumb and consequently the extent to which it breaks down on refining. As with the milk powders, those crumbs which are fragile and break into many small pieces give rise to more viscous chocolates.

## Gel systems

Gels form the continuous phase of a number of manufactured foods. They generally form by aggregation of molecules into a three-dimensional network. In some cases the aggregation may be into a fibrous network such as pectin (*Figure 20.26*), gelatin or agar and in other cases the network may be produced from clumps joined together, such as in sarcoplasmic protein gels (*Figure 20.27*), yoghurt, bovine serum albumin, whey protein and egg white gels. Fibrous gels are more usually encountered where the gel sets on cooling, presumably as this allows time for the molecules to aggregate in an ordered form.

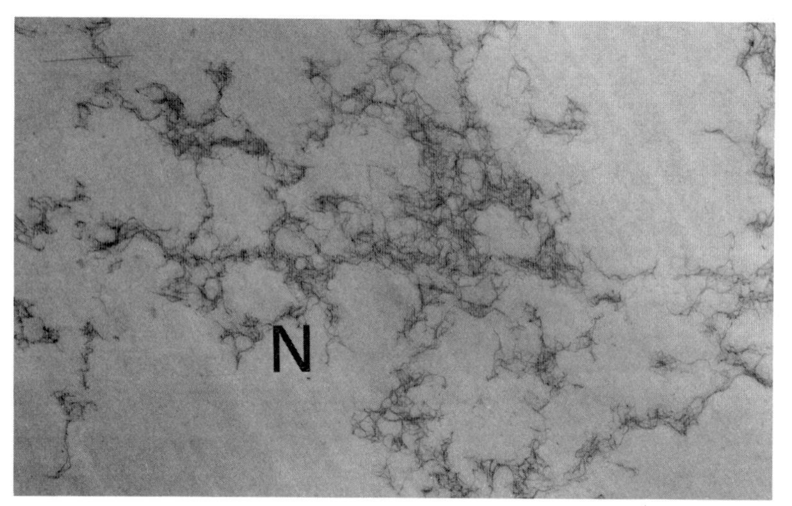


**Figure 20.26** Transmission electron microscopy of pectin gel; N = pectin fibrils in gel network. (Magnification × 25 000)

Some gels develop normally as a mixture of continuous and dispersed phases. Examples of this type of gel are starch gels, where the starch grains remain embedded in an extracted amylose matrix (*Figure 20.28*) and soya isolate gels, where many of the spray-dried particles are found enmeshed in a network of extracted protein (*Figure 20.29*). Starch gels are a good example of a system where the dispersed inclusions and the continuous matrix both contribute to the overall properties of the gel. During cooking of a starch slurry, the starch grains swell and amylose is extracted from the

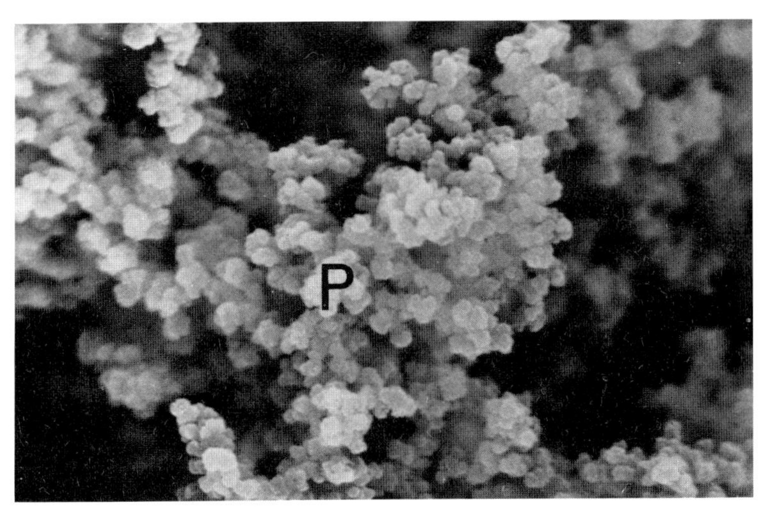


**Figure 20.27** Scanning electron microscopy of sarcoplasmic protein gel; P = protein aggregate. (Magnification × 1000)

**Figure 20.28** Scanning electron microscopy of starch gel; S = starch grain. (Magnification × 250, reduced to 80% in reproduction.)

grains; on cooling the amylose sets into a gel network. The viscosity of the starch pastes increases as the starch grains swell and then decreases as the starch grains disintegrate. Gels can be formed by cooling the mixtures at various stages of cooking. Early on in the cooking cycle, the starch grains are relatively compact and only a small amount of amylose 'cement' is available to form the gel and this results in a weak, brittle gel. At the point of maximum viscosity in the cooking cycle, the gel formed on cooling has very swollen but intact starch grains and a fair amount of amylose to act as cement and this results in a reasonably firm gel. Prolonged cooking

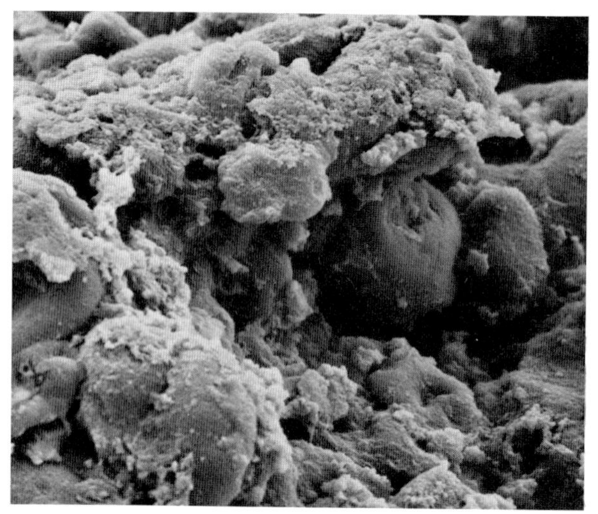

**Figure 20.29** Scanning electron microscopy of soya isolate gel. (Magnification × 500, reduced to 60% in reproduction.)

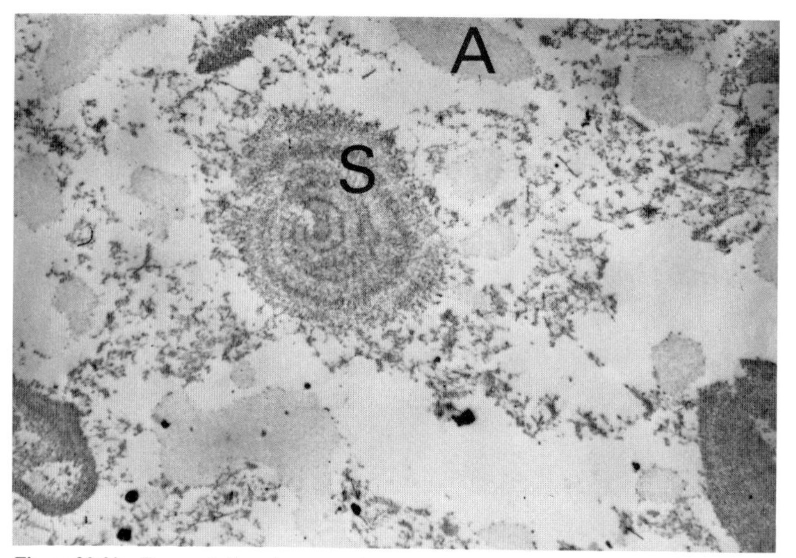


**Figure 20.30** Transmission electron microscopy of starch/agar mixture; S = starch grain; A = agar particle. (Magnification × 5000)

causes the breakdown of the starch grains and this results in a soft but fairly elastic gel.

Gelling agents are often used as mixtures; in some cases discrete phases are formed whilst in other cases a combined gel network results. An example of a mixture which forms discrete phases is agar and acid-thinned corn starch (*Figure 20.30*). In the example shown the starch has formed the continuous phase whilst the agar is present as discrete areas. Varying the proportions of agar and starch makes it possible to reverse the phases and to change the distribution of the dispersed phase, and consequently to modify the texture of the overall system.

In other mixtures of gelling agents, a combined gel is produced. The example shown in *Figure 20.31* is a commercial egg custard in which casein has been incorporated into a carrageenan network, producing junction points for the carrageenan.

The whole area of interactions between different gelling agents and between gelling agents and other food constituents is one which food science is only just beginning to understand and considerable advances in this area should be seen in the near future.

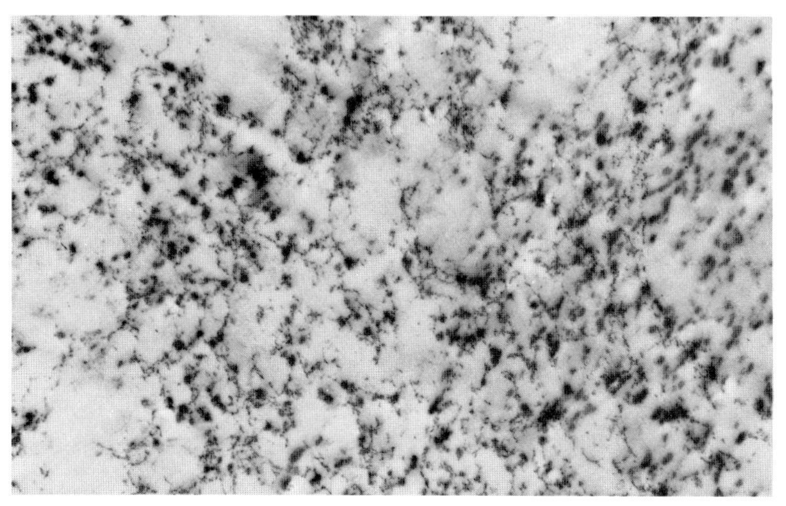

**Figure 20.31** Transmission electron microscopy of egg custard (casein/carrageenan). (Magnification × 10 000)

## Conclusions

This chapter has set out to give a broad outline of the electron microscopic structure of a wide range of foodstuffs and to illustrate the basic relationships between structure and function in foods. A wide range of microscopical techniques is available and the microscopist has to be careful in his methods of sample preparation. However, the essential feature of food microscopy is that the microscopic structure should be related to as many other evaluations as possible. In this way microscopy adds a unique facet to any investigation of food behaviour.

## Acknowledgment

All features are reproduced by courtesy of the Leatherhead Food R.A. and I am grateful to the Director, Dr A.W. Holmes, for permission to present this paper.

21

# SMALL DEFORMATION MEASUREMENTS

S.B. ROSS-MURPHY
*Unilever Research, Colworth House, Sharnbrook, Bedfordshire, UK*

## Introduction

The process of masticating and ingesting food materials involves subjecting the food to a range of deformations, flows, enzymic and thermal treatments, whose purpose is to break down the structure into a suitable form for swallowing and predigestion. It is for these reasons that the study of the failure of food materials has been historically of such importance, and several chapters in this volume discuss such aspects in detail. More recently the physical study of food materials and concurrent improvements in rheological equipment have resulted in a much deeper understanding of structure–property relationships in model food systems, and the application of such ideas to real food products (Ross-Murphy, 1984).

For example, just as NMR spectroscopy enables a detailed structural probe of food materials to be made at the 'molecular' level (i.e. over distances of, say, 0.5–5 nm), so the technique of 'mechanical spectroscopy' enables materials to be probed over supramolecular distances (say 1 μm–10 mm). From this, relationships between levels of structure and of structural organization can be made. This has resulted in the growth of a new approach to food rheology. The traditional 'phenomenologist' is being substituted by the 'structuralist'.

It is important at this stage to emphasize that mechanical spectroscopy like other spectroscopic techniques does not essentially perturb the system. It maintains its association with other spectroscopic techniques such as IR and CD/ORD which have also been applied to foods and biopolymers. However, it does not necessarily relate to the large deformation (cutting, spreading or chewing) regime which applies to the use of food materials in practice. The present chapter illustrates the small deformation technique, applied both to model and real food systems.

## Small deformation measurements

Any rheological experiment can, in principle, be carried out in two ways: either one can impose a small force (*stress*) and measure the deformation of the sample (*strain*) or conversely one can impose a fixed amount of movement (strain) and measure the stress developed in this sample. Clearly, if we are to restrict ourselves to small deformation measurements alone, only the latter regime is appropriate, since then the strain is the control variable and the stress which is developed is dependent upon the nature of the sample and its intrinsic material properties.

Stress and strain are defined formally in a number of texts, e.g. Ferry (1980); for the present purposes we shall only consider the strain applied in a small deformation, parallel plate experiment, as illustrated in *Figure 21.1*. The *maximum* shear strain is given by:

$$\gamma = \frac{\theta R}{d} \tag{1}$$

where $\theta$ is the angle of displacement (radians), $R$ is the radius and $d$ the distance of separation of the plates. (N.B. since $\theta R$ is itself a distance of arc, $\gamma$ is, as it should be, dimensionless.)

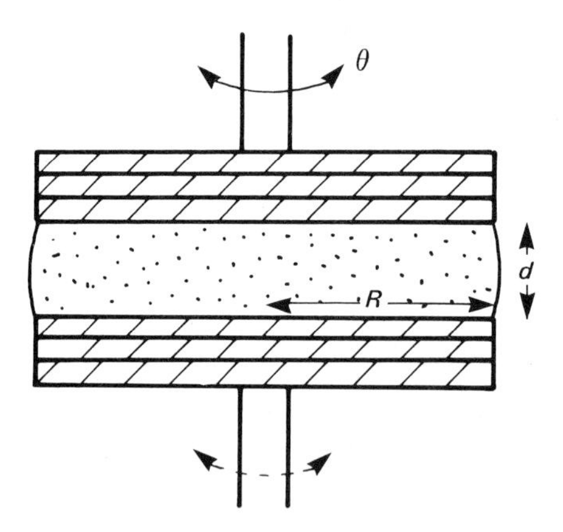


**Figure 21.1** Illustration of parallel plate geometry, with plates of radius $R$ and separation $d$. The upper plate is driven through a maximum strain $\theta$ radians, and the oscillatory stress transmitted through the sample is measured with a stress transducer coupled to the bottom plate. In practice $R$ would be 25–50 mm, $d$ 0.5–2 mm and $\theta \sim 5 \times 10^{-3}$.

Normally in mechanical spectroscopy, a sinusoidal *strain* amplitude (of fixed maximum strain) is applied to one of the plates (*Figure 21.2*), whilst the second is connected to a force transducer. In practice this usually consists of a torsion bar and a linear voltage displacement transducer (LVDT); the stress exerted on the opposite parallel plate by the slab of sample itself oscillates sinusoidal. In general, however, both the phase and amplitude of this shear stress wave are different to that of the original strain wave.

The ratio of the total stress to the strain is given by:

$$G^* = \sigma^*/\gamma \tag{2}$$

where $G^*$ is the complex shear modulus, and $\sigma^*$ is the complex shear stress.

$$\sigma^* = \frac{2g}{\pi R^3} \times m^* \tag{3}$$

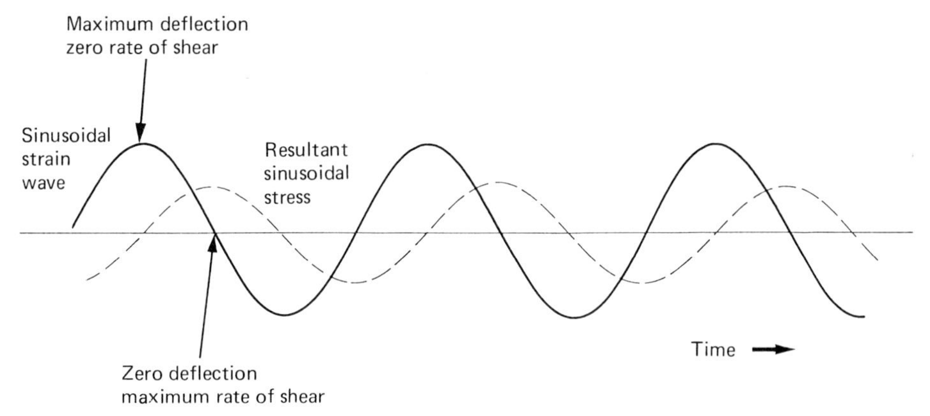


**Figure 21.2** Oscillatory (sinusoidal) shear. At the extremes of the oscillatory cycle the shear strain (—) is maximum, but shear rate (- - -) is zero, whilst at the null position where shear strain is zero, shear rate is at its maximum value. For a solid, the resistance to deformation (stress, $\sigma^*$) increases with increasing strain (i.e. in phase with the imposed deformation), whilst for a Newtonian liquid, $\sigma^*$ is greatest at the maximum shear rate (90° out of phase with the imposed deformation). Reproduced with permission from Morris and Ross-Murphy (1981).

where $m^*$ is the torque measured by the transducer (e.g. in kg m, or g cm) and $g$ is the acceleration of gravity. In practice, just as with alternating current theory, $G^*$ is separated into real and imaginary parts, $G'$, the storage modulus, and $G''$, the loss modulus, respectively. The inverse ratio of these two is the loss tangent tan $\delta$. In other words $G^* = G' + iG''$ and $G''/G' = \tan\delta$. In the technique of mechanical spectroscopy itself, it is usual to vary say the frequency ($\omega$) of the applied strain wave, or alternatively to keep $\omega$ constant and vary temperature. This chapter will consider both experiments, but in either case the dependence of $G'$ and $G''$ (or tan $\delta$) on the change in temperature or of frequency furnishes valuable information about the nature of the measured sample.

To illustrate this, at a fixed but low frequency, say 1 rad $s^{-1}$, for a fluid system $G'' \gg G'$. Conversely for a solid material $G' \gg G''$. In practice this means that for a material such as a protein solution at room temperatures, tan $\delta$ is very large, i.e. the phase angle $\delta \sim 1.57$ rad ($\sim 90°$), and the stress wave is almost completely out of phase with the applied strain wave. On heating the protein solution at high enough concentrations at or above the denaturation temperature, the sample begins to gel and $G'$ increases (as does $G''$ but to a more limited extent), and consequently tan $\delta$ decreases. The change in value of tan $\delta$ and that of the absolute magnitudes of $G'$ and $G''$ with temperature help us to monitor the protein gelation (Egelansdal, Fretheim and Harbitz, 1986; Richardson and Ross-Murphy, 1981). All this can be carried out under such small oscillations that the process is essentially unperturbed by this applied strain, so we can monitor the formation of structure under non-destructive conditions.

Recently a number of oscillatory constant *stress* rheometers have appeared on the market (Carrimed, Rheotech Viscoelastic Analyser, Deer Rheometer, etc.) and all of these are essentially very similar in design and specifications. Note that if the stress is oscillating, but the maximum stress $\sigma$, is constant, then we are measuring, not a modulus as given by Eqn (2), but a compliance, defined by:

$$J^* = \gamma^*/\sigma \tag{4}$$

and then just as before we can define real and imaginary components of this $J'$ and $J''$. It is important to appreciate three points here. Firstly $G'$ and $J'$, for example, are not in general reciprocally related (although their magnitudes $J^*$ and $G^*$ are); rather the exact relationship between the two is given by:

$$G' = \frac{J'}{(J'^2 + J''^2)} = \frac{1/J'}{1 + \tan^2\delta} \tag{5}$$

Secondly since the strain becomes the independent variable, there is no independent control to constrain this strain to be small—for fluids, for example, the strain will tend to increase inversely proportional to the applied frequency; this effect is, of course, less serious for more solid materials, but should still be carefully examined. Finally it is an inherent feature of most of this generation of constant stress instrumentation that the natural (resonant) frequency, $f_r$, of the measurement system lies in the range 0.5–5 Hz ($= 2\pi f_r$ in rad $s^{-1}$). This puts an upper limit on the accessible frequency range of the instrumentation since no measurements can be made within, say, half a decade of $f_r$. The remaining descriptions will apply only to instruments in which the strain is maintained constant and small, and the resultant stress is monitored.

## Constitutive equations and time–temperature superposition

Applications of the technique of mechanical spectroscopy are only now becoming more widespread in the areas of food science or biophysics, although the application in synthetic polymer science is extremely widespread. Nevertheless there have been a number of isothermal studies including investigations of blood clotting, and the formation of mucin and fibrin gels (Bell *et al.*, 1984; Hartert and Schaeder, 1962; McIntire, 1980; Nelb, Kamykowski and Ferry, 1980). As far as food science is concerned the coagulation of milk by proteolytic enzymes, including the effect of added $CaCl_2$, have been examined by Garnot and Olson (1982). The effect of heat on a cake batter, on a wiener emulsion and a commercial whey concentrate have also been published (Beveridge and Timbers, 1985; Dea, Richardson and Ross-Murphy, 1984; Ngo and Taranto, 1986) and the application to some of these systems will be described below.

In a number of the above studies very simple apparatus based upon vibrating reeds or sinusoidally oscillating bar vibrators has been used. All of these, in principle, perform the same experiment as illustrated in *Figure 21.2*, but usually the 'geometry' is not defined as formally as is this, so there is no simple relationship between geometry, stress and strain corresponding to Eqns (1) and (2). Despite this, the ratio tan δ depends only upon the phase angle, and if this is the only parameter required then such apparatus is quite acceptable.

Conversely there is a feeling that commercial apparatus to make good measurements of mechanical spectra are extremely expensive. Up to a point this is true, but compared, for example, to NMR equipment the cost is still competitive. Given machining and electronics expertise, apparatus may be constructed without excessive expenditure (*cf.* Richardson and Ross-Murphy, 1981). In succeeding sections the response of model and real food systems in small deformation measurements as a function of frequency, strain and/or temperature is described. However, before considering these a few details appropriate to the investigation of viscoelastic materials are given.

As suggested earlier the modulus of a material is defined as the ratio of stress to strain. For a given applied strain it is more correct to write

$$\sigma = f(\gamma, \omega, \dot{\omega}, \ddot{\omega}, \ldots, T, P, \ldots) \tag{6}$$

This 'constitutive' equation implies that the 'constitution' of a given material depends, in general, not just on $\gamma$, but on oscillatory frequency, $\omega$, strain rate, $d\gamma/dt$ ($=\dot{\gamma}$), $d\omega/dt$ ($=\ddot{\gamma}$), and higher terms, temperature ($T$) and pressure ($P$), etc. In practice it is obviously advantageous to hold, say, temperature, pressure and strain constant and to vary the frequency. Moreover, the terms in Eqn (6) are, in general, not independent, and frequency and temperature may be linked by the principle of time–temperature superposition.

This principle was originally introduced by workers in synthetic polymers, and is best introduced qualitatively by considering the familiar experiment of plunging a piece of rubber tubing into liquid nitrogen. The rubber becomes hard and 'glassy', and exactly the same phenomenon is detected if the rubber is subjected to an extremely high frequency oscillatory strain (say $> 100$ kHz). Such high frequencies can be generated by ultrasonic shear instruments, and thus at these frequencies the material behaves as a glassy solid. An example more appreciable to workers in the food area is the measurement of water at extremely high frequencies. Under these circumstances the shear modulus of water ($\sim 10^{-3}$ Pa at room temperature and at $\omega = 1$ rad s$^{-1}$) approaches that of ice at 0 °C, viz $\sim 10^{10}$ Pa. Qualitatively we can say that at very high frequencies the movement of a water molecule is restricted almost as if it were in an ice crystal lattice. By varying the temperature then we can access a much wider effective frequency range than a single oscillatory instrument can access at any one temperature. At this juncture we should point out that this superposition does not always succeed, and obviously if a material is modified physically or chemically by temperature then we would not expect it to do so; later, however, we shall illustrate its apparent success when applied to an ice cream in the range $-5$ °C to $-36$ °C.

## Mechanical spectra for model systems

In many texts on elementary viscoelasticity, the behaviour of ideal 'spring and dashpot' models is described in some detail—some workers have spent considerable effort in modelling real systems with arbitrary combinations of springs and dashpots in series and parallel. Unfortunately this is not always terribly helpful since most experimental data can be fitted with two exponential decay terms, but this does not mean that the fitted decay constants have, in general, any physical significance. By contrast, in this text, examples of behaviour are represented by measured spectra for typical polysaccharide solutions and gels, an approach originally inspired by the excellent illustrative monograph of Ferry (1980). *Figure 21.3* shows mechanical spectra for a 'dilute solution', (dextran at 5%), a gel (agar at 1%) and a 'concentrated solution' (lambda carrageenan at 5%). The dextran is typical of a fluid and shows $G'$ and $G''$ both increasing with frequency, with $G'$ approaching the value of $G''$ at the highest accessible frequency. By contrast the gel spectrum is that of a solid material,

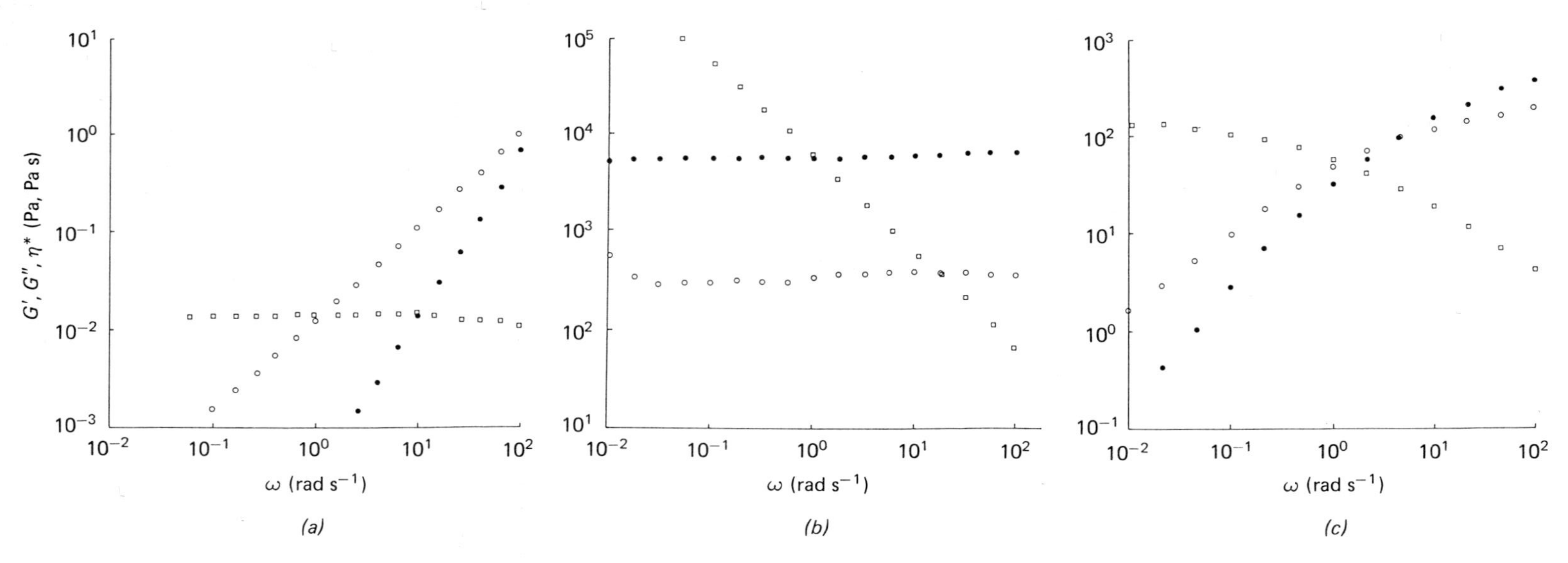


**Figure 21.3** Typical mechanical spectra for polysaccharide systems showing the frequency dependence of (●) $G'$ and (○) $G''$ (in Pa) and (□) $\eta^*$ (in Pa s) for (a) a dilute solution; (b) a gel; and (c) a concentrated solution. Reproduced with permission from Morris and Ross-Murphy (1981)

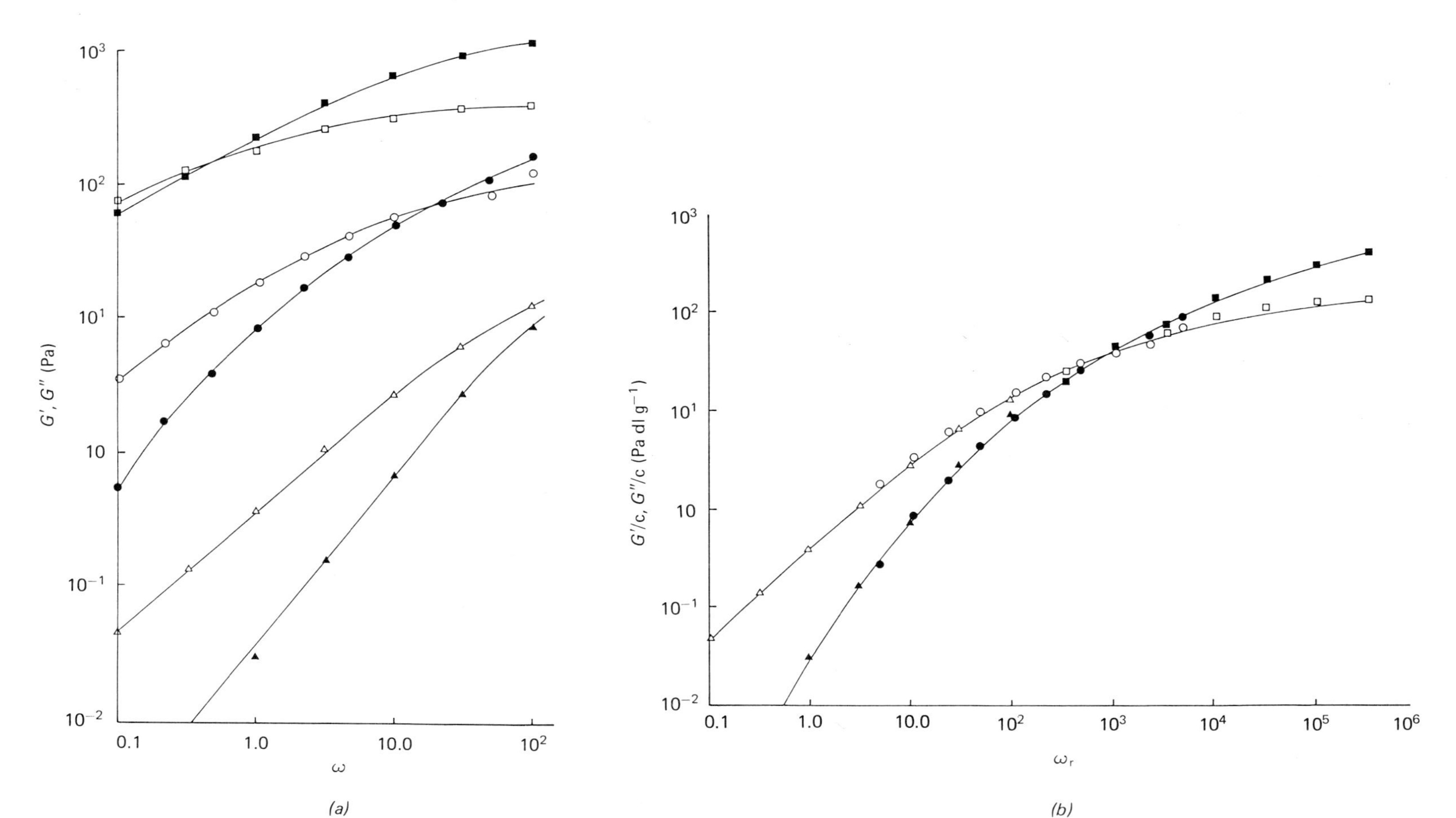


**Figure 21.4** (a) Mechanical spectra for three samples of aqueous guar galactomannan solution (1%, 2%, 3% w/w) (△, ▲ = 1%, ○, ● = 2%, □, ■ = 3%), open symbols $G''$, filled $G'$. (b) Frequency–concentration superposition for the above data obtained (i) by dividing observed moduli by concentration, and (ii) by a lateral shift of the data for the 2% and 3% solutions along the frequency axis. Reduced frequency ($\omega_r$) is identical to the applied frequency ($\omega$) for the 1% sample. Reproduced with permission from Robinson, Morris and Ross-Murphy (1982).

neither $G'$ nor $G''$ is very sensitive to frequency, and $G'' > G'$. Examination of the spectrum for the concentrated solution system shows that it has the features of a viscoelastic fluid in this frequency window ($10^{-2}$–$10^{2}$ rad s$^{-1}$); at the slow frequencies it is quite similar to that for the dextran, whereas at the highest frequencies $G'$ crosses over the trace of $G''$, and both traces become flatter. The indications are that at higher frequencies still, a more frequency-independent trace similar to that of the gel sample would be obtained.

Of course by changing the temperature, i.e. by cooling the sample, as mentioned earlier we might expect to access higher effective frequencies, but then for samples such as these we might also modify the complete system by doing so. An alternative strategy would be to measure samples at different concentrations. The equivalence of cooling the sample and/or increasing the polymer concentration to extend the effective frequency range follows, since both reduce the amount of 'free volume'. To illustrate this more graphically we reproduce *Figure 21.4* which shows data for guar polysaccharide solutions measured at 1%, 2% and 3% w/w; in this way the transition from a fluid to a gel-like spectrum can be seen more completely. It might appear from this that if we could measure the gel spectrum of *Figure 21.3(b)* at much lower frequencies it would also flow as does the dextran solution. This is not, however, always the case because if the gel crosslinks are permanent, the spectrum will extend in just the illustrated fashion to lower and lower frequencies, without ever showing the 'terminal flow' characteristics of *Figure 21.3(c)* (e.g. Ferry, 1980). For physical gels with crosslinks of long (but finite) lifetime, intermediate behaviour is to be expected (Clark and Ross-Murphy, 1987).

## Strain dependence

All of the above spectra have been collected in the linear viscoelastic regime where Eqn (2) holds, i.e. where stress is proportional to strain. The value of this strain varies considerably with the nature of the material, and this helps to characterize the materials just as does the frequency dependence above. For example, both rubber and

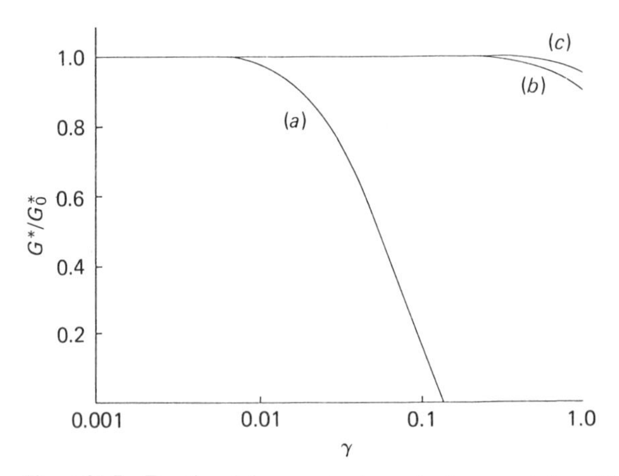


**Figure 21.5** Fractional decrease in dynamic shear modulus with strain for typical examples of (a) a colloidal dispersion, (b) a biopolymer gel and (c) a biopolymer solution. Reproduced with permission from Ross-Murphy (1984), copyright 1984 Society of Chemical Industry

margarine can produce very similar mechanical spectra, since at low strains both behave as solids. The spectrum in question will appear qualitatively similar to that of the gel of *Figure 21.3(b)*, but compared with this the absolute moduli would be $\sim 10^4$ times as great. However, the real difference between the two systems is clear from the maximum strain limit for this linear viscoelastic behaviour—for the margarine it is typically $\sim 10^{-3}$–$10^{-4}$ strain units, whereas for the rubber it is $\sim 1$ strain unit. *Figure 21.5* illustrates this strain dependence more graphically by charting typical responses for a dispersion, a gel and a solution. All of these systems show the reduced modulus $G^*/G^*_0$ decreasing monotonically with increasing strain, although the opposite effect ('strain hardening') is occasionally seen. Further discussion on strain dependence will be limited, since large deformation properties are discussed in more detail in Chapter 22 of this volume.

## Application of small deformation measurements to food precursors

In the remainder of this chapter the application of the technique of mechanical spectroscopy to multicomponent, multiphasic food systems will be discussed in more detail. The discussion will be restricted to only a few systems although there are many other examples in the literature. Nevertheless the examples chosen are meant to represent typical regimes of frequency, temperature, etc. and the extension of these principles to other systems should be obvious.

Much of the current academic interest in the structure and rheological properties of food and food-like materials is based upon the analogy between the development of mixing rules for composite materials based upon a knowledge of the properties of its components. The 'structuralist' approach to food rheology has been pursued in a number of other chapters in this volume, but it is now accepted that this extension of materials science has profound implications for biophysical understanding, and consequently has been the subject of a number of treatises (Dea, Richardson and Ross-Murphy, 1984; Ross-Murphy, 1984). In the longer term, the traditional phenomenological view of food rheology, as measurement followed by parameterization, must be tempered by an understanding of the underlying physicochemical principles of 'structuring'. Just as no one would attempt to characterize the gelation of gelatin simply by measuring the mechanical spectra of the system just at, say, 15 °C i.e. where it is a solid gel and 45 °C when it has 'melted' out, so we hope to illustrate that an understanding of the properties of individual components of a composite help us to comprehend its overall behaviour.

### *Cake batter mix*

The first example of this 'synthetic' approach is the trace obtained (*Figure 21.6*) for a typical cake batter system over the temperature range 30–120 °C (Dea, Richardson and Ross-Murphy, 1984).

The batter was introduced into a parallel plate (Rheometrics) instrument and then heated at 3 °C $\text{min}^{-1}$; the oscillatory strain frequency applied was 10 rad $\text{s}^{-1}$ and $\gamma_{max}$ was 0.01. At least initially both $G'$ and $G''$ fall steadily as the 'fluidity' of the batter is increased, but at 76°C there is small increase in $G'$ and $G''$ followed by a much more pronounced increase at $\sim 82$ °C. The water/concentration profile of the formulation

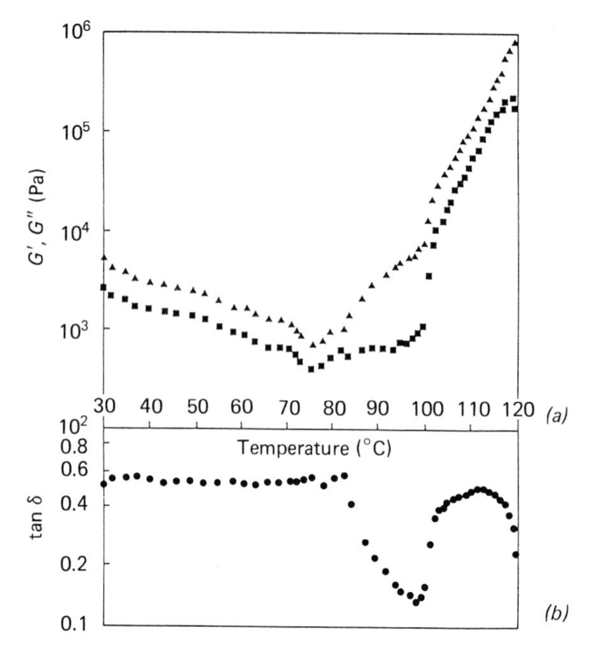


**Figure 21.6** Cake structure formation (a) $G'$, $G''$ plotted against temperature; (b) as (a), but tan δ ($= G''/G'$). Heating rate 3 °C min$^{-1}$, $\omega = 10$ rad s$^{-1}$. ▲, $G'$; ■, $G''$; ●, tan δ. Reproduced with permission from Dea, Richardson and Ross-Murphy (1984).

is difficult to ascertain precisely under the above heating regime, but it is reasonable to assign the small peak to the gelatinization of starch and the much more pronounced increase to the formation of protein gel (particularly ovalbumin, unfolding temperature ~ 80 °C). The modulus of the composite gel system increases steadily as the temperature is increased, but note that ≃ 100 °C the latent heat of vaporization ensures that the sample cannot respond to the oven heating profile, so that the temperature of the *sample* remains nearly constant for ≃ 10 minutes. Finally all available water must be evaporated from the system and the sample temperature rises to its final value of 120 °C; the ratio $G''$ to $G'$ then begins to decrease as a more 'elastic' product is formed. Experiments at a lower heating rate (1 °C min$^{-1}$) allowed more of the water to evaporate before the 'gel' structure was set up, producing a system with $G'$ even higher still—the principle is, of course, crucial to the production of materials with the overall texture of biscuits rather than sponge cakes!

Very recently Ngo and Taranto (1986) published a more systematic study of the effect of sucrose levels on the properties of a range of cake batters. They also used a Rheometrics instrument, and charted cure curves of $G'$, $G''$ against temperature for both heating and cooling cycles. Amongst their interesting observations was that tan δ increased with the increase of sucrose level, just as did the cake volume. This they attributed to the delay in starch gelatinization on increasing the sucrose content.

### *Ice cream mix*

Rather than a heat-set gel, the second example also reported by Dea, Richardson and Ross-Murphy (1984) may be regarded as a cold-set system, in that it is a typical ice cream mixture. When this is obtained ex-votator, introduced into the parallel plate geometry and then cooled, the results of *Figure 21.7* are obtained. Here the initial temperature was − 5 °C, the cooling rate was − 1 °C min$^{-1}$, the strain $\gamma_{max}$ was $10^{-3}$ (or less) and the frequency was again 10 rad s$^{-1}$. Note that $G'$ increases steadily to ~ − 26 °C and then remains approximately constant. Further, on reheating this trace is almost exactly reversible; any discrepancy in this reversibility merely reflects the slight temperature drop hysteresis over the test geometry. $G''$ shows a pronounced maximum centered about − 16 °C; on either side of this $G''$ decreases to very small values (and the results are limited by the phase angle resolution of the rheometer). Again $G''$ temperature traces are reversible. The curves closely reflect the temperature dependence of the volume fraction of ice in the overall system, and suggest that this dominates the small deformation behaviour.

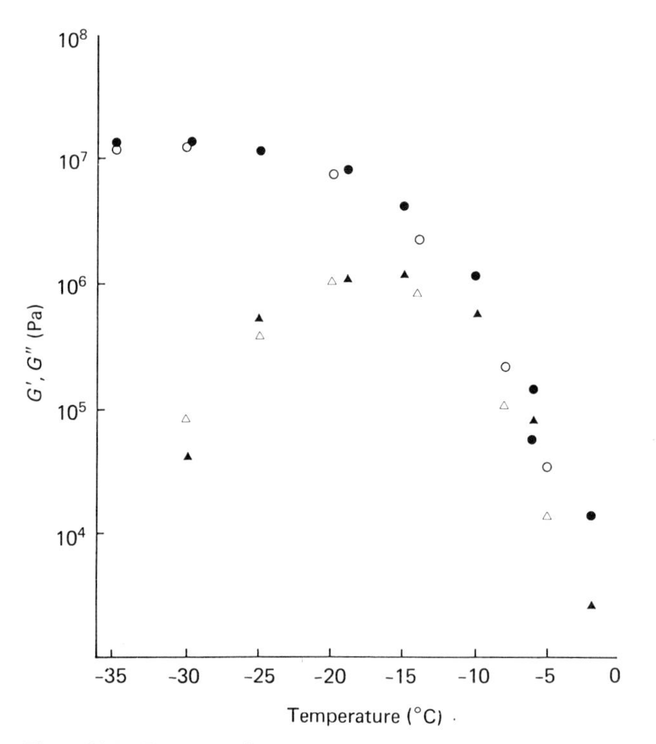


**Figure 21.7** Ex-votator ice cream, cooling and heating profile. $G'$, $G''$ vs. temperature (1 °C min$^{-1}$). $G'$: ○ ●, ; $G''$: △, ▲. Open symbols: cooling − 5 °C → − 35 °C; closed: reheating − 35 °C → − 2 °C. Reproduced with permission from Dea, Richardson and Ross-Murphy (1984).

To confirm this, in separate experiments frequency-dependent mechanical spectra were measured for small cylinders of pre-stored material over the same temperature range. For illustration, at − 8 °C there is still some frequency dependence although $G' > G''$; at the lower temperatures (− 24 °C) much flatter spectra are obtained. These traces are given in *Figure 21.8*. By applying a time–temperature superposition to all the data collected and plotting frequencies relative to those measured at − 5 °C, a trace is obtained which extends over 20 decades of effective frequency (*Figure 21.9*). Two features of this are immediately apparent: the trace is qualitatively the mirror

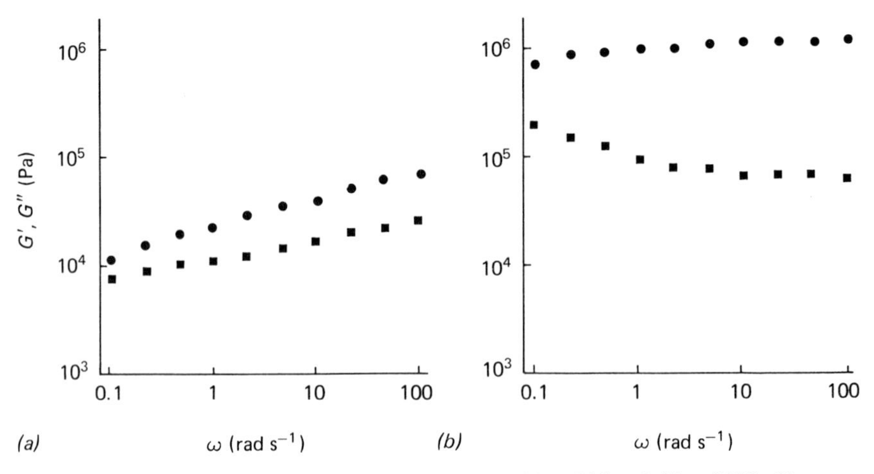


**Figure 21.8** Frequency dependence of ice cream disc at (a) − 8 °C and (b) − 24 °C. $G'$: ●; $G''$: ■. Reproduced with permission from Dea, Richardson and Ross-Murphy (1984).

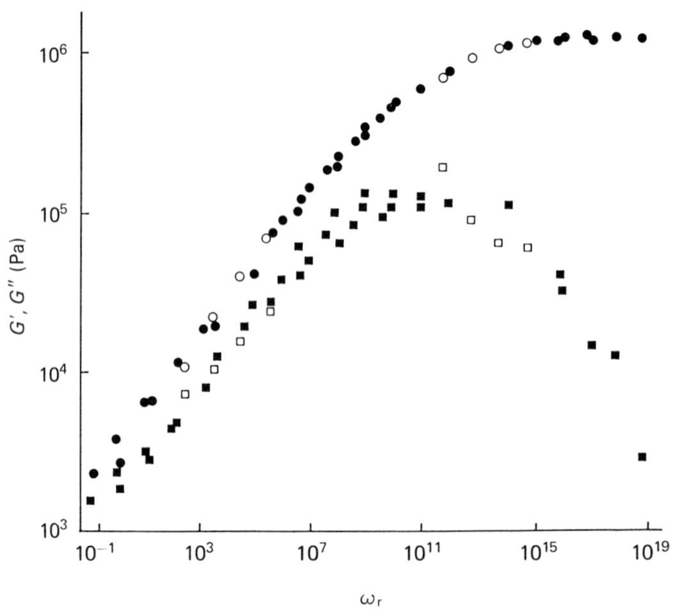


**Figure 21.9** Time–temperature superposition of system in *Figure 21.8*; $\omega_r$ = frequency relative to − 5 °C. Symbols as in *Figure 21.8* except data of *Figure 21.8* shown as open symbols. Reproduced with permission from Dea, Richardson and Ross-Murphy (1984).

image of the single frequency/temperature trace of *Figure 21.7* (as it should be), and the shape, although much broadened, is not dissimilar to that of a simple spring and dashpot (Maxwell element) model (see, for example, Ross-Murphy, 1984). The loss peak is centered around $\simeq 3 \times 10^{11}$ rad s$^{-1}$, i.e. the relaxation time is $\simeq 2 \times 10^{-11}$ s, which is quite close to that measured for the dielectric relaxation of water in water/ice mixtures; this again suggests that the overall small deformation properties are governed by the ice/water component of the mixture.

***Wiener batter***

Beveridge and Timbers (1985) have illustrated how comparatively simple instrumentation may be used to obtain information which, although not as detailed as that discussed above, is still valuable in monitoring physicochemical changes by small amplitude oscillatory changes. In their recent paper application to a number of systems is discussed, including a whey protein concentrate, an egg albumin preparation, and a wiener emulsion. This latter consists of animal matter, fat, water and salt (9% protein, 26% fat, 65% moisture) dispersed at 4 °C.

*Figure 21.10* illustrates the results obtained, in which temperatures, total output torque and input amplitude are plotted together. The different regions, A, B, C, D and E correspond to the different external temperature regimes. For example, heating is started at A (after overnight equilibration from 4 °C to 20 °C), and there is an initial decrease in the torque (and since the strain is constant, of $G^*$) as the fat melts. At B the proteinaceous material forms a heat-set gel (*cf. Figure 21.6*), which then levels off; recooling from 85 °C to 20 °C produces still further increases in stress development. Apart from the presence of the fat phase, the trace is quite similar to published spectra for pure thermally-treated protein solutions. A particularly interesting example of this is the recent work by Egelansdal, Fretheim and Harbitz (1986) using a Bohlin instrument. This has illustrated how the unfolding of regions of the myosin macromolecule, previously investigated by traditional thermal methods, can also be followed rheologically.

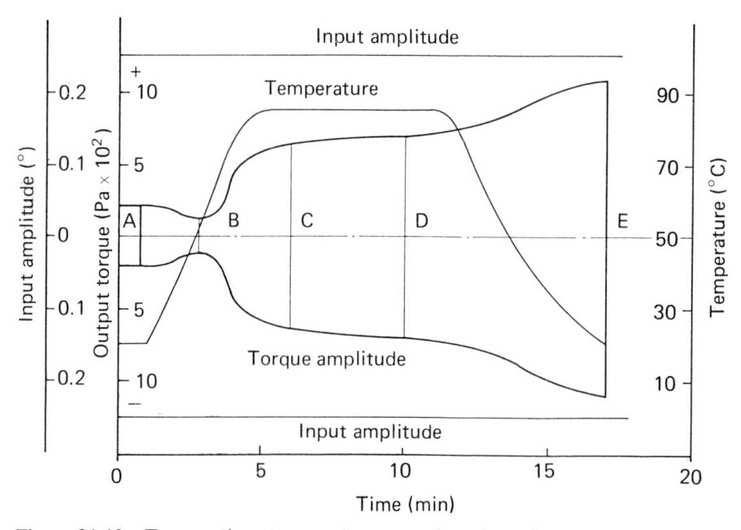


**Figure 21.10** Torque–time–temperature trace for wiener batter at 85 °C. The letters refer to the different processes in the curing: A = recovery of initial deformation; B = fat melting; C = gelation of meat proteins; to final value at D (see temperature trace), and E = further final increase on recooling. Reproduced from Beveridge and Timbers (1985), copyright 1986 Food and Nutrition Press Inc., Westport, Connecticut

## Conclusion

This chapter has briefly illustrated how the technique of small deformation oscillatory testing can be used to monitor the structure formation during thermally-induced changes to both model systems and more realistic food precursors. Future applications of the technnique should be very widespread; in this connection it is interesting that in the related sphere of medical research, small amplitude measurements are now replacing more traditional steady shear instrumentation to monitor blood clotting. This technique is now being applied as a routine screening aid. We suspect that in the future the same change from traditional steady shear rheometry to small deformation oscillatory testing will be true in the area of food sciences.

## References


BELL, A.E., ALLEN, A., MORRIS, E.R. and ROSS-MURPHY, S.B. (1984). Functional interactions of gastric mucus glycoprotein. *International Journal of Biological Macromolecules*, **6,** 309–315

BEVERIDGE, T. and TIMBERS, G.E. (1985). Small amplitude oscillatory testing. *Journal of Texture Studies*, **16,** 333–349

CLARK, A.H. and ROSS-MURPHY, S.B. (1987). Structural and mechanical properties of biopolymer gels. *Advances in Polymer Science*, **83**, 57–192

DEA, I.C.M., RICHARDSON, R.K. and ROSS-MURPHY, S.B. (1984). Characterisation of rheological changes during the processing of food materials. In *Gums and Stabilizers for the Food Industry 2*, (Phillips, G.O., Wedlock, D.J. and Williams, P.A., Eds), pp. 357–366. Oxford, Pergamon Press

EGELANSDAL, B., FRETHEIM, K. and HARBITZ, O. (1986). Dynamic rheological measurements on heat-induced myosin gels. *Journal of the Science of Food and Agriculture*, **37**, 944–954

FERRY, J.D. (1980). *Viscoelastic Properties of Polymers*, 3rd Edition. New York, Wiley

GARNOT, P. and OLSON, N.G. (1982). Use of oscillatory deformation technique to determine clotting times and rigidities of milk clotted with different concentrations of rennet. *Journal of Food Science*, **47,** 1912–1915

HARTERT, H. and SCHAEDER, J.A. (1962). The physical and biological constants of thrombelastography. *Biorheology*, **1,** 31–39

MCINTIRE, L.V. (1980). Dynamic materials testing: biological and clinical applications in network-forming systems. *Annual Review of Fluid Mechanics*, **12,** 159–179

MORRIS, E.R. and ROSS-MURPHY, S.B. (1981). Chain flexibility of polysaccharides and glycoproteins from viscosity measurements. *Techniques in the Life Sciences* **B310**, 1–46

NELB, G.W., KAMYKOWSKI, G.W. and FERRY, J.D. (1980). Kinetics of ligation of fibrin oligomers. *Journal of Biological Chemistry*, **255,** 6398–6402

NGO, W.H. and TARANTO, M.V. (1986). Effect of sucrose level on the rheological properties of cake batter. *Cereal Foods World*, **31,** 317–322

RICHARDSON, R.K. and ROSS-MURPHY, S.B. (1981). Mechanical properties of globular protein gels: I. Incipient gelation behaviour. *International Journal of Biological Macromolecules*, **3,** 315–322

ROBINSON, G., MORRIS, E.R. and ROSS MURPHY, S.B. (1982). Viscosity–molecular weight relationships, intrinsic chain flexibility and dynamic solution properties of guar galactomannan. *Carbohydrate Research*, **107**, 17–32

ROSS-MURPHY, S.B. (1984). Rheological methods. In *Biophysical Methods in Food Research*, (Chan, H.W.-S., Ed.), pp. 138–199. Oxford, Blackwell Scientific

## 22

## BEHAVIOUR OF FOODS IN LARGE DEFORMATION

E.B. BAGLEY and D.D. CHRISTIANSON
*US Department of Agriculture*, Peoria, Illinois, USA*

## Introduction

Large deformations can be important both in the creation and evaluation of food structure. The creation of structure will require processing and many processes, e.g. extrusion, will impose large deformations on the food material. In the evaluation of food structure, measurements made under conditions of large deformation can provide information complementary to that obtained in the small deformation range. Specifically, time and non-linear viscoelastic effects will often be particularly significant at large strains and stresses. Fracture for many foods will occur at high deformation, though fracture may be observed at low strain levels with brittle foods or even with soft foods tested at high deformation rate.

It might seem at first glance that extending measurements on foods to strain levels at which these complications can occur is merely confusing an already complicated situation. However, experience in characterizing synthetic polymer behaviour has shown the value of extending measurements to strain and stress levels where these complications occur (Bagley and Schreiber, 1969). In the molten state polymers show long-range interactions on a supramolecular scale reflected in properties such as normal stresses or 'die swell' in extrusion. Properties depend not only on molecular weight and molecular weight distribution but on how molecules interact, perhaps through chain entanglement. While there is disagreement as to the reality or significance of chain entanglements in polymer systems, there is no doubt that there are long-range structural or molecular organizational effects that influence the measured properties of polymer systems. These structural interactions in polymers are manifested particularly well under large deformations. The same considerations will be even more significant with foods for which the complex composition can lead readily to time and strain dependency of many measured physical properties. A good example of a food in which such long-range structural interactions are important would be Mozzarella cheese, as discussed recently by Masi and Addeo (1984). Fat globules are suspended in a water and whey protein matrix with a three-dimensional crosslinked structure of casein molecules. The crosslinks between casein chains involve ionic and hydrogen bonds, which break and reform if the stress and strain levels are high enough, leading to stress relaxation and creep and to irreversible

* The mention of firm names or trade products does not imply that they are endorsed or recommended by the US Department of Agriculture over other firms or similar products not mentioned.

deformation, i.e. to flow. Other food systems, such as gelatin gels, also may undergo breaking and reformation of bonds, crosslinks or other structure, leading to permanent deformation and to complex time effects that make it difficult to characterize the material. As Mitchell (1979) comments, 'It is a feature of polysaccharide and gelatin gels that the creep compliance and stress relaxation functions change slowly over long periods of time.... These time changes make it impossible to measure equilibrium parameters such as the rigidity modulus.' Nevertheless, as he goes on to comment, for other gels the situation may be different.

There is a large body of literature dealing with the rheology of polymers. The five volumes of *Rheology—Theory and Applications*, edited by Eirich, are a rich source of information including the chapter by Blatz (1969) on large deformations. Polymer mechanics texts include that by Long (1961), Janeschitz-Kriegl (1983) and Williams (1973). Fracture in polymers is the subject of the recent work by Kinloch and Young (1983). Treloar (1975) provides a very readable summary of information on theory and practice of large deformations. Rheology texts more directly related to foodstuffs include De Man *et al.* (1976) and Sherman (1979). Additional references are given by Bagley (1983).

In this chapter attention will first of all be drawn to considerations involved in definition of strain and in the use of the stored energy function. Examples of problems arising in study of large deformations will then be given, with reference to viscoelastic doughs and to the concept of elongational flows. This leads naturally to discussion of compressional experiments as applied both to viscoelastic fluid and viscoelastic solid foodstuff behaviour. The need to examine material behaviour in more than one deformational mode will be emphasized. Finally, problems involved in characterizing the fracture properties of foods will be discussed.

## Stress, strain and the stored energy function

While there is no ambiguity in the definition of stress, one troublesome aspect of the treatment of large deformations to which particular attention should be given is that there are numerous definitions of strain in the literature (Darby, 1976). Peleg (1984) has compared various strain measures and shows the resultant stress/strain plots associated with some of these different definitions. To the extent that only a simple description of experimental data in one deformation mode is required, the choice of strain definition can be a matter of convenience. A more fundamental phenomenological description of large deformations, however, considers the stored energy function, $W$, which is expressed in terms of the strain invariants, $I_1$, $I_2$ and $I_3$, rather than strain alone. These invariants are given in terms of the three principal extension ratios, $\lambda_1$, $\lambda_2$ and $\lambda_3$ as:

$$I_1 = \lambda_1^2 + \lambda_2^2 + \lambda_3^2 \tag{1}$$

$$I_2 = \lambda_1^2\lambda_2^2 + \lambda_2^2\lambda_3^2 + \lambda_3^2\lambda_1^2 \tag{2}$$

$$I_3 = \lambda_1^2\lambda_2^2\lambda_3^2 \tag{3}$$

If the system volume does not change during deformation, $I_3$ is constant and unity. The stored energy function is then dependent only on $I_1$ and $I_2$, so

$$W = W(I_1, I_2) \tag{4}$$

but the phenomenological theory does not specify the form of $W$. Further, measured stress deformation data yield the derivatives with respect to $I_1$ and $I_2$ of the strain energy function. As Kearsley and Zapas (1980) note, the determination of the dependence of strain energy on the strain invariants from experimental data, while tedious, is practical. Further, they note that for some elastomers, the strain energy can be expressed in the Valanis–Landel form, which simplifies the analysis. Kearsley (1980) emphasizes that the results obtained from different deformational modes (uniaxial extension/biaxial extension; torque/normal force in torsion; uniaxial extension/torque in torsion) can be combined to determine the strain energy function.

Treloar (1975, Chapter 10) presents an excellent discussion of the physical reasons for the advantages of using a combination of experiments in different deformational modes to obtain a reliable description of the strain energy function, $W$. A commonly used two-constant equation, due to Mooney, relates $W$ to $I_1$ and $I_2$ as:

$$W = C_1(I_1 - 3) + C_2(I_2 - 3) \tag{5}$$

Treloar notes that 'as long as we restrict ourselves to a limited range of variables, the Mooney equation *necessarily* provides a fair approximation to the behaviour of the material. . . . Experiments involving any particular strain, e.g. simple extension or simple shear, provide too restricted a basis for the derivation of the true form of $W$.' Thus the relations $(\partial W/\partial I_1) = C_1$ and $(\partial W/\partial I_2) = C_2$ require a linear Mooney plot in simple extension. However, the converse is not true and a linear Mooney plot could also result from an appropriate variation of $(\partial W/\partial I_1)$ and/or $(\partial W/\partial I_2)$ with strain. This possibility has often been disregarded, particularly in food sytems. To assess the true picture of the elastic properties of a material with any hope of reliability, as Treloar comments, requires experiments covering as wide a range of strain as possible.

Superimposed on these considerations is the fact that many foods will show effects of time as well as strain. These time effects may arise because of structural changes with applied stress or deformation as in an ostensibly solid material such as a gelatin gel (Mitchell, 1979). Other food materials may actually be viscoelastic fluids rather than viscoelastic solids, as in doughs referred to in the next section. This also raises the question as to whether a given food is 'solid' or 'liquid'. This issue is best resolved in terms of the Deborah Number, $D$, defined by Reiner (1964) as the ratio of relaxation time of a material to the time of observation. If $D$ is small (short relaxation time, long observation time) the material flows and behaves more like a liquid; if $D$ is large (long relaxation time, short observation time) the material appears to be a solid. Thus the slow rearrangements in the gelatin gels noted by Mitchell (1979) will result in 'flow', i.e. irreversible deformation, so that on a long time scale of observation these gels would appear to be fluid-like in their behaviour. In the discussion of large deformation effects given below, foods are considered as viscoelastic materials; the terms 'solid' or 'liquid' are reserved for use in connection with specific experimental time scales or testing rates.

## Large deformation behaviour of viscoelastic doughs

Doughs are particularly interesting materials to examine in large deformation because they show, in exaggerated form, all the problems which can arise in rheological measurements of viscoelastic materials. Doughs are time and strain dependent. They are fluid enough to sag in extensional experiments [and thus must be supported, for example in a buoyant medium, for experiments in tension (Tschoegl, Rinde and Smith, 1970a,b)], but at the same time are so viscous and elastic that measurements in standard viscometers are difficult, if not impossible. In a cone-and-plate viscometer a dough will often 'roll out' of the gap before steady-state viscosity values can be established. This problem has been discussed by Bloksma and Nieman (1975) who note for their systems that roll out occurs at a total shear of the order of 20. The critical factor for roll out was not shear rate alone but the product of shear rate and time. Baird (1983) has discussed briefly methods for measuring dough rheological properties and has commented also on experimental difficulties involved in obtaining accurate and repeatable results for food doughs. He includes some dynamic data on defatted soy doughs.

Dynamic measurements have proved informative, particularly for small amplitude studies, as in the work of Hibberd and Wallace (1966) who examined the linear aspects of wheat flour dough behaviour. These studies were extended to larger deformations where non-linear behaviour is observed (Hibberd and Parker, 1975). The question of 'large' or 'small' deformation is, however, relative. Thus Smith, Smith and Tschoegl (1970) also examined response of doughs in dynamic testing and found that even at peak-to-peak displacements corresponding to strains as low as $2 \times 10^{-2}$ the modulus of the dough depended on strain amplitude. In this sense, then, even these small strains can be considered 'large deformations'. The dynamic shear modulus also depended on frequency, and on time at constant amplitude and frequency. Normally, dynamic methods are more relevant to small than to large deformation behaviour.

The most direct way to measure foodstuffs in large deformation is to stretch a sample in uniaxial extension, as done by Schofield and Scott Blair (1932) for doughs. This experiment is complicated by problems of sag, already alluded to, and the problem of grasping the dough sample to extend it. Schofield and Scott Blair resolved the sag problem by floating the dough on mercury. Subsequently Tschoegl, Rinde and Smith (1970a) applied a technique from methods designed to examine large deformation and fracture of rubbers to doughs. In this method a ring of material is stretched to fracture at constant rate. In applying the procedure to doughs they found it necessary to suspend the dough ring in a fluid of matching density and, since the dough did not slip freely on the hooks, it was necessary to determine an effective circumference, $C_e$, of the entire loop. They also found that a 'race track' specimen, rather than a circular ring, was the preferred shape for the sample.

A significant problem in the extensional testing mode is the need to determine this effective circumference, $C_e$. The dough in extension does not draw down uniformly and the calculation of true stress is critically dependent on the determination of $C_e$. The effective circumference is affected to a certain extent by the dough character, moisture level, improver concentration, etc. (Rasper, 1975). Unhappily, the correction can also depend on temperature and crosshead speed, although in the Lemhi and Kansas flour doughs examined by Tschoegl, Rinde and Smith (1970a,b) the value of the correction was independent of crosshead speed, temperature or water absorption. Nevertheless, the correction has to be checked for every system, since, as Rasper (1975) notes, this is a very important factor in stress and strain calculations for extensograph measurements.

A scientifically interesting but relatively unexplored aspect of extensional behaviour of doughs is the interpretation of the data in terms of extensional viscosity. Trouton (1906) showed for very viscous materials, such as pitch and shoemaker's wax, which do not lend themselves readily to the usual viscometric measurements, that extensional deformation would yield an elongational viscosity, $\eta_{el}$, which for Newtonian materials should be three times the viscosity, $\eta$, measured in shear. Comparison of shear viscosity with elongational viscosity by Ballman as shown in *Figure 2.22* of Middleman (1968) shows agreement between the shear viscosity and one-third of the elongational viscosity for molten polystyrene when the data were extrapolated to zero shear stress and zero tensile stress, respectively. However, the situation has turned out to be more complex than this result indicates, as will be described below.

The general subject has been recognized as a matter of considerable importance in polymer processing because of the various elongational or extensional flows of commercial interest such as fibre spinning. The importance of such measurements was also recognized (as early as 1944) in the food area by Scott Blair (1944) who provided an extensive review of experimental techniques applied up to that time to bubble inflation in the baking industry, as well to butter, cheese, casein and flour doughs. The subject has been reviewed more recently by Petrie (1979).

## Comparison of viscosities in shear and in elongational flow

The behaviour of dough at large deformation in simple extension can be illustrated by reference to the work of Rasper (1975) who examined the response of doughs from Canadian Hard Red Spring wheats both untreated and chemically improved. The experimental procedures and associated calculations are described by Rasper, Rasper and De Man (1974). Data are plotted as stress versus strain curves and different plots are obtained at different testing rates. The higher the extensional rate the higher the observed stress at a given strain level. This is physically reasonable, reflecting the relaxation processes occurring in the viscoelastic fluid dough as the dough is deformed. The higher the extension rate the less time is available for decay of the imposed stresses. When the data at various rates and deformation are plotted as sample length versus time of extension a family of curves is obtained, each curve being associated with a constant stress level. These plots are not linear but are concave down approaching linearity beyond times of about 100 s. For this linear region a slope, $dL/dt$, can be computed for each stress level. A coefficient of absolute viscosity is then calculated from the relation:

$$\text{Coefficient of absolute viscosity} = \text{stress}/(3\, dL/L_0dt) \qquad (6)$$

where $dL/L_0dt$ is the relative velocity of extension. Results from Rasper's *Table 7* for flour B are tabulated in *Table 22.1* which lists stress, extensional rate and viscosity. The extension rate was calculated from viscosity and stress data of Rasper's *Table 7* using Eqn (6).

As Rasper comments, the viscosity values obtained in this experiment fall within the expected range of dough viscosities reported in the literature, which is between $10^4$ and $10^6$ Pa s. However, the fascinating feature of *Table 22.1* is that the viscosity *increases* with both stress and extension rate whereas the expected behaviour for a viscoelastic non-Newtonian dough would be viscosity *decreasing* with increasing stress and extension rate, or possibly, at the low extension rates shown in *Table 22.1*, a

**Table 22.1** VISCOSITY DATA FOR FLOUR B. FROM RASPER (1975)

| *Stress* (Pa) | *Extension rate* ($dL/L_0dt$) ($s^{-1}$) | *Viscosity* (Pa s) |
|---|---|---|
| 158 | $2.73 \times 10^{-3}$ | $5.79 \times 10^4$ |
| 251 | $4.09 \times 10^{-3}$ | $6.13 \times 10^4$ |
| 398 | $4.36 \times 10^{-3}$ | $9.12 \times 10^4$ |
| 631 | $5.32 \times 10^{-3}$ | $11.86 \times 10^4$ |
| 1000 | $6.45 \times 10^{-3}$ | $15.50 \times 10^4$ |
| 1580 | $7.23 \times 10^{-3}$ | $21.86 \times 10^4$ |
| 2510 | $8.18 \times 10^{-3}$ | $30.70 \times 10^4$ |

viscosity *independent* of either stress or rate! Constant extensional viscosity was observed for polyisobutylene in *Figure 2.22* of Middleman (1968).

That doughs are non-Newtonian, shear thinning materials is demonstrated by the data of Bloksma and Nieman (1975) who measured dough viscosity in a cone-and-plate viscometer. Their *Figure 4* shows that viscosity does decrease with increasing shear rate, dropping from about $2 \times 10^5$ to $2 \times 10^4$ Pa s as shear rate increases from $0.65 \times 10^{-3}$ to $20 \times 10^{-3}$ reciprocal seconds at 25 °C. This conflicts with the results of *Table 22.1* and thus the interpretation of extensional data in terms of an extensional viscosity needs to be reconsidered.

These extensional results reported by Rasper should be considered relative to the remarks of Petrie (1979, p. 77) who noted: 'We obtain two types of information from measurements of stress, strain and strain rate. If steady flow is attained, with stress and strain rate reaching constant values then their ratio gives the elongational viscosity, which is a material property. This may then be measured and recorded as a function of elongational stress or strain rate. The second type of information is obtained from unsteady flows where the stress or the strain rate (or both) are functions of time. In this case we do not obtain a material property directly, since the ratio of the time-dependent stress to strain rate depends on the mode of stretching as well as on the material and, as well as on the time since the start of the experiment, it may depend on the deformation history of the material before the experiment. It is possible to obtain a function of time which is a material property if we define the experiment completely.'

Extension experiments using instruments such as the Instron are normally carried out at constant crosshead speed. For such an experiment the strain rate, $\dot{\varepsilon}$, is given as (Middleman, 1968):

$$\dot{\varepsilon} = \frac{1}{L}\left(\frac{dL}{dt}\right) \tag{7}$$

where $L$ is the length of the sample at time $t$. If $(dL/dt)$ is constant and equal to the crosshead speed, $\dot{\varepsilon}$ decreases as $L$ increases. This is in contrast to the relative velocity of extension used in Eqn (6). Such results yield the second type of information referred to by Petrie (1979). The flow is unsteady and stress and strain rate are both functions of time. It is possible to devise experiments in which the strain rate defined by Eqn (7) is held constant by changing crosshead speed continuously as the sample elongates. It is also possible to devise experiments in which the stress is held constant by continuously adjusting the force as the cross-sectional area of the sample decreases during extension. The central question is whether a constant strain rate or constant stress experiment can be carried out over a sufficiently long period of time to attain a steady state. For the constant rate experiment a time long enough to reach a constant stress would be required; for the constant stress experiment the time would need to be long enough to reach a constant strain rate level.

In general, the times required to achieve a steady state are not accessible experimentally. Doughs, like many polymer systems, are characterized by a wide distribution of relaxation times. The distribution of relaxation times in doughs and the use of the Central Limit Theorem has been discussed by Grogg and Melms (1958) and Shelef and Bousso (1964). Long times are therefore needed to reach steady-state flow behaviour and this may not be feasible, in general, based on results by Demarmels and Meissner (1985). They developed a multiaxial elongational rheometer in which polyisobutylene was deformed at constant strain rate and comment with surprise that 'It is remarkable that within the range of strain rate and total strain investigated so far no steady state of the stress could be found.' Rasper's results, then, are in harmony with those of Demarmels and Meissner and undoubtedly reflect the effect of a very broad distribution of relaxation times which prevents attainment of a steady state in the experimental time range accessible in their extensional experiment.

Using the nomenclature recommended by the Society of Rheology (Dealy, 1984) the results of tensile (simple) extension flow should be expressed in terms of a tensile stress growth coefficient:

$$\eta_E^+(t,\dot{\varepsilon}) = \sigma_E/\dot{\varepsilon} \tag{8}$$

where $\sigma_E$ is the net tensile stress and with $\varepsilon$ defined as the Hencky strain, $\ln(L/L_0)$, and $\dot{\varepsilon}$ as $(d\varepsilon/dt)$. As $t$ goes to infinity the limit of the stress growth function is the tensile viscosity given in Eqn (9).

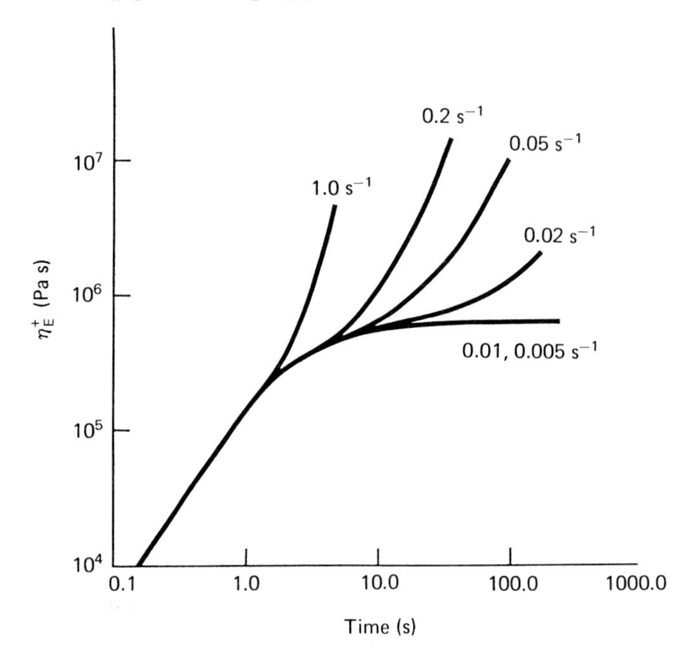


**Figure 22.1** Stress growth coefficient in extension as a function of time for polyethylene at 160 °C at various elongation rates. After Ide and White (1978).

$$\eta_E(\dot{\varepsilon}) = \lim_{t \to \infty} [\eta_E^+(t,\dot{\varepsilon})] \tag{9}$$

The stress growth coefficient should thus be plotted as a function of time. Results of such an experiment, obtained by Ide and White (1978) for a polyethylene, are shown in *Figure 22.1*. It is only at the lowest extension rates (0.005 and 0.01 $s^{-1}$) that a limit to $\eta_E^+$ appears to be reached at long times. At higher rates (0.05, 0.2 and 1 $s^{-1}$) the stress growth coefficient increases without bound and no limit is evident at experimentally accessible times.

## Compressional experiments

The uniaxial compressional experiment is an attractive one for examining the response of doughs and other foodstuffs to large deformations. The experiment can be carried out using equipment commonly employed in food laboratories for texture profile analysis. There is no need to calculate an effective mass. The evaluation of stress is simply the applied force divided by the cross-sectional area of the platens in contact with the sample. Thus if a cylindrical sample of initial radius $R_0$ and initial height $h_0$ is deformed to a cylindrical sample of radius $R$ and height $h$, then

$$R_0^2 h_0 = R^2 h \tag{10}$$

and $R^2$ is readily calculated. The stress $\sigma$, when the sample has attained the height $h$ is:

$$\sigma = (\text{Force}) h / \pi R_0^2 h_0 \tag{11}$$

A potential difficulty with the compressional experiment is that the cylindrical sample will stick to the compressing platens during the experiment. This would cause the sample edge to bulge and the sample would assume a barrel shape. This problem can be treated quantitatively (Christianson, Casiraghi and Bagley, 1985) and the effects of this sample bulging can be surprisingly large. The problem is, however, best avoided by eliminating or minimizing frictional losses by lubrication of the platen/sample interface (Bagley, Christianson and Wolf, 1985; Bagley, Wolf and Christianson, 1985).

The uniaxial compression experiment is equivalent in terms of sample deformation to the biaxial extension experiment as carried out by Demarmels and Meissner (1985). Biaxial extension is certainly relevant to doughs and bread baking because this is the type of deformation the dough undergoes both in processing and in bubble growth during baking.

*Figure 22.2* shows the stress growth coefficient obtained in lubricated uniaxial compression as a function of time for a 50%–50% corn flour–water dough. Crosshead speeds ranged from 0.2 to 20 cm $min^{-1}$. The total time scale covered is more than three decades, but the results for a particular crosshead speed cover only about one decade. All curves appear to approach the same limit at lower times, a straight line with a slope of approximately unity. However, for each curve there is a second region observed at longer times where the stress growth coefficient increases relatively slowly with time, the magnitude of the coefficient being higher the lower the crosshead speed. This result is qualitatively similar to that reported by Demarmels and Meissner (1985) for polyisobutylene. Certainly, the biaxial extension result for both the polyisobuty-

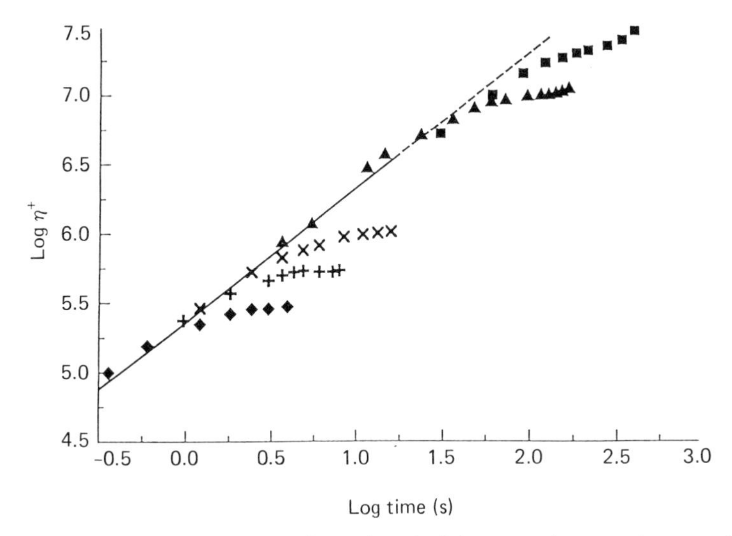


**Figure 22.2** Stress growth coefficient in uniaxial compression at various crosshead speeds for a 50% corn flour dough. Crosshead speeds are: ◆, 20 cm min$^{-1}$; +, 10 cm min$^{-1}$; ×, 5.0 cm min$^{-1}$; ▲, 0.5 cm min$^{-1}$; ■, 0.2 cm min$^{-1}$; $\eta^+$ measured in Pa s.

lene and corn flour dough is quite different from the result in uniaxial extension (*Figure 22.1*). The curves of *Figure 22.2* certainly describe the material response over a time scale pertinent to times involved in processing and should serve two purposes:

1. The curves characterize the dough and should correlate with composition and method of preparation.
2. The values of the stress growth coefficient, at times appropriate to processing times, should correlate with processing behaviour.

The comparison of material behaviour both in extension and in compression should prove more valuable in fully understanding the material than measurement of either extensional or compressional response alone. The theoretical advantages of such comparisons have been discussed by Kearsley (1980), Kearsley and Zapas (1980) and Treloar (1975). The problem for elastic materials comes down to evaluation of the stored energy function, $W$, in terms of the strain invariants, $I_1$, $I_2$ and $I_3$ which, in turn, are expressed as functions of the deformations $\lambda_1$, $\lambda_2$, $\lambda_3$ as summarized in Treloar (1975). Treloar points out the 'experiments involving any particular strain, e.g. simple extension or simple shear, provide too restricted a basis for the derivation of a true form of $W$'. While the situation is even more complex with viscoelastic fluids (such as doughs) than with viscoelastic solids (crosslinked gels or rubber) because of the irreversible processes which occur, a systematic investigation of dough response in two deformation modes would be most valuable. These remarks are equally pertinent to the following section on more solid-like foods.

## Elasticity of foods and normal stresses at large deformations

The application of a force to deform a material may result in the generation of new forces, termed 'normal stresses'. Thus when a shear stress is applied to a viscoelastic fluid in a cone-and-plate viscometer, the resultant normal stress perpendicular to the plate will tend to force the cone and plate apart. These normal stress effects are especially significant in characterization of polymer melts (Bagley and Schreiber, 1969) and the measurement and interpretation of normal stresses have been extensively discussed in the polymer literature.

Judging from experience with synthetic polymers, normal stress measurements on foodstuffs should yield results very sensitive to commercially important properties (Bagley and Schreiber, 1969). In particular, normal stress phenomena appear to relate to long-range structure in polymers and probably do so too in food systems. Nevertheless, little has been done to investigate normal stress in foodstuffs. Prentice (1984, p. 182) has pointed out that the measurement of these normal stresses is difficult and that 'only a few such measurements have been made on food materials, not sufficient to make a comprehensive survey of them'.

Viscoelastic liquid or semi-solid foods will show normal stress effects in shearing flows, as in a cone-and-plate viscometer. For such materials, techniques developed for examination of normal stresses in polymer solutions and melts may be applicable. However, some of the experimental methods used with polymers have difficulties which occur also when applied to highly viscoelastic foods, such as 'roll out' from the gap of the cone-and-plate viscometer alluded to earlier. One procedure avoiding this problem that has been found useful for polymer melts is measurement of 'die swell' in extrusion. When a viscoelastic material flows through a capillary die of radius $R_0$ the extrudate radius, $R$, can be as much as six times greater than $R_0$ due to the normal stresses generated during shear in the capillary itself. For many foods, however, this approach is not feasible because of the 'puffing' of a food extrudate which may result from vaporization of water when the extrudate emerges from the capillary tube, where it was under high pressure, into the atmosphere.

The investigation of elastic effects, as manifested by the generation of normal forces, is thus certainly hampered by experimental difficulties. One procedure that seems to have some promise is to examine the behaviour of foods in torsion. This method has been described by Treloar (1975). If a cylinder of radius $a$ and height $h$ is twisted through an angle $\theta$ the torsional strain, $\psi$, is given as:

$$\psi = \theta/h \tag{12}$$

The total couple ($M$) and the axial load ($N$) are then given by:

$$M = 4\pi\psi \int_0^a r^3 \left(\frac{\partial W}{\partial I_1} + \frac{\partial W}{\partial I_2}\right) \mathrm{d}r \tag{13}$$

$$N = -2\pi\psi^2 \int_0^a r^3 \left(\frac{\partial W}{\partial I_1} + 2\frac{\partial W}{\partial I_2}\right) \mathrm{d}r \tag{14}$$

where $W$ is the stored energy function and $I_1$ and $I_2$ are the strain invariants as defined in Treloar (1975). For a Mooney material such as a crosslinked rubber, the derivatives of the stored energy function are both constants, designated commonly as $C_1$ and $C_2$ and Eqns (13) and (14) reduce to:

$$M = \pi\psi a^4(C_1 + C_2) \tag{15}$$

$$N = -\frac{\pi}{2}\psi^2 a^4(C_1 + 2C_2) \tag{16}$$

$M$ is then linear with torsional strain, while the normal stress is linear with the square of the torsional strain. $C_1$ and $C_2$ can be evaluated separately. $C_1$ is a function of network structure, the crosslink density; $C_2$ has, in Treloar's words 'some entirely independent origin'.

Christianson *et al.* (1984) applied the torsional method to a variety of gels and dispersions and there were no difficulties in evaluation of $C_1 + C_2$, the value so obtained being confirmed by independent measurements in simple shear. Unfortunately there was variability in the determination of $C_1 + 2C_2$ from the normal stress measurement, and the values of $C_1$ and $C_2$ evaluated separately did not make physical sense, as noted by Christianson *et al.* (1984). More extensive consideration of these results and approach is necessary.

An alternative approach to evaluation of $C_1$ and $C_2$ separately is to measure the material response in simple tension (or uniaxial compression). For the Mooney material the force, $f$, on the unstrained cross-section varies with strain, $\lambda$, as:

$$f = 2\left(\lambda - \frac{1}{\lambda^2}\right)\left(C_1 + \frac{C_2}{\lambda}\right) \tag{17}$$

where $\lambda = L/L_0$. A plot of $f/2(\lambda - 1/\lambda^2)$ versus $1/\lambda$ should be linear with the intercept at $\lambda = 1$ giving $C_1 + C_2$ and the slope giving $C_2$. The difficulty, as noted by Treloar (1975, p. 225), is that 'while the relations $(\partial W/\partial I_1) = C_1$, $(\partial W/\partial I_2) = C_2$ necessitate a linear Mooney plot ... the converse is not true and the apparent consistency of the data with the Mooney equation could equally well be attributed to a variation of $(\partial W/\partial I_1)$ or $(\partial W/\partial I_2)$ (or both) with strain'. This fundamental problem is less severe in torsion because, as Treloar points out with reference to work by Rivlin and Saunders, the range covered experimentally for the two strain invariants is small so that consistency with the Mooney equation is to be expected.

It is evident that extreme care must be exercised in obtaining and analysing force/deformation response of foodstuffs. Combinations of different procedures would certainly seem warranted, examining the food not in one deformational mode only, but in several modes. Simple shear, torsional and compressional response of a variety of gels have been reported by Christianson *et al.* (1984), Bagley, Christianson and Wolf (1985) and Bagley, Wolf and Christianson (1985) using the Mooney equation. Values of $C_1 + C_2$ from torsion and simple shear were in agreement when the compressional response was obtained under lubricated conditions. For a wheat starch gel the values of $C_1$ and $C_2$ separately also agreed but the physical interpretation, as noted above, was not clear, with $C_1$ being zero or slightly negative while $C_2$ was large. There is thus a need for considerably more research, both into the mechanics of the behaviour of foods under large deformations and the interpretation of the various mechanical parameters describing the material response.

## Fracture

Fracture in solid foods has been reviewed by Hamann (1983) and Jowitt (1979). Fracture properties relate to sensory texture perception and thus are of interest *per se*. In addition, because fracture must relate to breaking of some bonds of the strong interactions between some portion of the food and a neighbouring portion on the macroscopic level, knowledge of the fracture behaviour should provide insight into these bonds and interactions. There is, unfortunately, a paucity of fundamental data on the fracture properties of foods and this represents another area of opportunity for the food scientist.

In studying fracture behaviour of food gels Hamann and co-workers have compared fracture behaviour as measured by various techniques, as reviewed by Hamann (1983). Specifically, he and his colleagues examined surimi, turkey, beef and pork gels (Montejano, Hamann and Lanier, 1984) in torsion and in compression. For the beef and pork gels good agreement was observed between true shear stress at fracture, appropriately calculated, for the two methods. Agreement was poorer for the turkey gel which the authors attributed to the large strain at failure which caused gross sample shape change before failure occurred. Surimi gels did not fail in compression though all gels failed in torsion.

More recently Montejano, Hamann and Lanier (1985) examined eight different heat-induced protein gels by torsional failure testing and Instron texture profile analysis (TPA) and compared results to sensory ratings by a trained profile panel. True shear strain at failure and TPA cohesiveness were the instrumental parameters which had the highest correlation with sensory perception. There appear to be myriad opportunities for more investigations into food properties/sensory perception along the lines being followed by Hamann and co-workers.

Fracture behaviour in uniaxial compression is determined by the extent to which the sample sticks or slips on the instrument platens. As noted earlier, if the sample sticks to the platens, a cylindrical sample will deform to a barrel shape, bulging at the free edge; if the sample/platen interface is lubricated, the sample will retain a cylindrical shape during deformation. *Figure 22.3* compares the fracture behaviour of a starch gel obtained under bonded and lubricated conditions. The cylindrical samples were all of the same initial radius (77 mm diameter) but of different initial

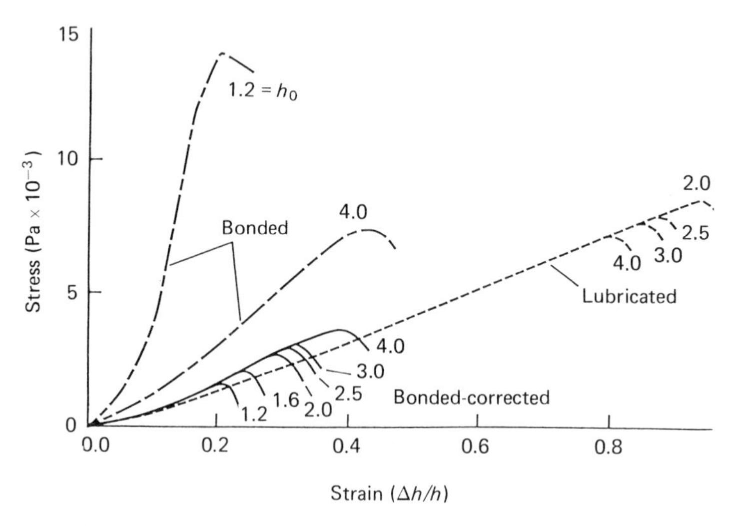


**Figure 22.3** Comparison of stress–strain response of wheat starch gels in bonded and lubricated uniaxial compression. The initial sample heights, $h_0$, range from 1.2 to 4.0 cm and the initial sample diameters were all 77 mm. From Christianson, Casiraghi and Bagley (1985).

heights (1.2–4.0 cm). The stress/strain curves obtained under lubricated conditions were independent of sample dimensions but the stress at fracture increased as the initial sample height decreased.

The stress/strain curves obtained with the samples bonded to the instrument platens were not independent of sample dimensions but depended strongly on $h_0$, the larger $h_0$ giving the lower stresses at a given strain. When these results were corrected for the sample bulge, the corrected stress/strain curves were in reasonable agreement with the lubricated results, independent of $h_0$. However, the stress at fracture for the bonded sample was much lower than for the lubricated samples. Further, the stress at fracture under bonded conditions increased with increasing $h_0$ in contrast to the $h_0$ dependence of fracture observed in lubricated compression. This result is not unexpected, since the bonded samples are fracturing in shear while the lubricated samples are fracturing in an extensional deformation. The magnitude of this difference in stress at fracture in bonded and lubricated compression would be expected to be significant, both in characterization of the mechanics of foodstuffs and in correlating with sensory evaluation.

Mitchell (1984) has commented that relatively little fundamental work has been carried out on large deformation and rupture behaviour of foodstuffs. In particular the rupture behaviour, i.e. the stress and strain at break, will depend on the rate of testing, increasing as the rate of strain increases. The results of rupture experiments can be given in terms of a failure envelope, as discussed by Smith (1969). Smith notes that 'when time–temperature reduction is applicable . . . data at different strain rates and temperature must then superimpose on a plot . . . to form a single curve—the failure envelope'. This has been done for biopolymer gels by McEvoy, Ross-Murphy and Clark (1984). They were particularly interested in relating observed maximum extension at break to the theoretical maximum extension for a network chain. They conclude that theories of extensibility and failure of rubber networks are adaptable (with limitations) to biopolymer gels and that the failure envelope approach is an informative way of presenting fracture data. However, as Mitchell comments, there is room for much more fundamental work in this general area.

## References


BAGLEY, E.B. (1983). Large deformations in food testing and processing. In *Physical Properties of Foods* (Peleg, M. and Bagley, E.B., Eds), pp. 325–342. Westport, Connecticut, AVI Publishers

BAGLEY, E.B. and SCHREIBER, H.P. (1969). Elasticity effects in polymer extrusion. In *Rheology, Theory and Applications*, Volume V (Eirich, F.R., Ed.), pp. 93–125. New York, Academic Press

BAGLEY, E.B., CHRISTIANSON, D.D. and WOLF, W.J. (1985). Frictional effects in compressional deformation of gelatin and starch gels and comparison of material response in simple shear, torsion and lubricated uniaxial compression. *Journal of Rheology*, **29**, 103–108

BAGLEY, E.B., WOLF, W.J. and CHRISTIANSON, D.D. (1985). Effect of sample dimensions, lubrication and deformation rate on uniaxial compression of gelatin gels. *Rheologica Acta*, **24**, 265–271

BAIRD, D.G. (1983). Food dough rheology. In *Physical Properties of Foods* (Peleg, M. and Bagley, E.B., Eds), pp. 343–350. Westport, Connecticut, AVI Publishers

BLATZ, P.J. (1969). Application of large deformation theory to the thermomechanical behavior of rubber-like polymers—porous, unfilled, filled. In *Rheology, Theory and Applications*, Volume V (Eirich, R.F., Ed.), pp. 1–55. New York, Academic Press

BLOKSMA, A.H. and NIEMAN, W. (1975). The effect of temperature on some rheological properties of wheat flour doughs. *Journal of Texture Studies*, **6**, 343–361

CHRISTIANSON, D.D., CASIRAGHI, E.M. and BAGLEY, E.B. (1985). Uniaxial compression of bonded and lubricated gels. *Journal of Rheology*, **29**, 671–684

CHRISTIANSON, D.D., NAVICKIS, L.L., BAGLEY, E.B. and WOLF, W.J. (1984). Rheological characterization of starch, starch–hydrocolloid and protein dispersions and gels in simple shear and torsion. In *Gums and Stabilizers for the Food Industry 2* (Phillips, G.O., Wedlock, D.J. and Williams, P.A., Eds), pp. 123–134. New York, Pergamon Press

DARBY, R. (1976). *Viscoelastic Fluids*. New York, Marcel Dekker

DEALY, J.M. (1984). Official nomenclature for material functions describing the response of a viscoelastic fluid to various shearing and extensional deformations. *Journal of Rheology*, **28**, 181–195

DE MAN, J.M., VOISEY, P.W., RASPER, V.F. and STANLEY, D.W. (1976). *Rheology and Texture in Food Quality*. Westport, Connecticut, AVI Publishers

DEMARMELS, A. and MEISSNER, J. (1985). Multiaxial elongation of polyisobutylene with various and changing strain rate ratios. *Rheologica Acta*, **24**, 253–259

GROGG, B. and MELMS, D. (1958). A modification of the extensograph for study of the relaxation of externally applied stress in wheat dough. *Cereal Chemistry*, **35**, 189–195

HAMANN, D.D. (1983). Structural failure in solid foods. In *Physical Properties of Foods* (Peleg, M. and Bagley, E.B., Eds), pp. 351–383. Westport, Connecticut, AVI Publishers

HIBBERD, G.E. and PARKER, N.S. (1975). Dynamic viscoelastic behaviour of wheat flour doughs. Part IV—Non-linear behavior. *Rheologica Acta*, **14**, 151–157

HIBBERD, G.E. and WALLACE, W.J. (1966). Dynamic viscoelastic behaviour of wheat flour doughs. Part I—Linear aspects. *Rheological Acta*, **5**, 193–198

IDE, Y. and WHITE, J.L. (1978). Experimental study of elongational flow and failure of polymer melts. *Journal of Applied Polymer Science*, **22**, 1061–1079

JANESCHITZ-KRIEGL, H. (1983). *Polymer Melt Rheology and Flow Birefringence*. New York, Springer-Verlag

JOWITT, R. (1979). An engineering approach to some aspects of food texture. In *Food Texture and Rheology* (Sherman, P., Ed.), pp. 143–155. New York, Academic Press

KEARSLEY, E.A. (1980). Determining an elastic strain energy function from torsion and simple shear. *Journal of Applied Physics*, **51**, 4541–4542

KEARSLEY, E.A. and ZAPAS, L.J. (1980). Some methods of measurement of an elastic strain energy function of the Valanis–Landel type. *Journal of Rheology*, **24**, 483–500

KINLOCH, A.J. and YOUNG, R.J. (1983). *Fracture Behavior of Polymers*. New York, Applied Science

LONG, R.R. (1961). *Mechanics of Solids and Fluids*. Englewood Cliffs, New Jersey, Prentice-Hall

MASI, P. and ADDEO, F. (1984). The effect of composition on the viscoelastic properties of Mozzarella cheese. In *Advances in Rheology*, Volume 4 (Mena, B., Garcia-Rejon, A. and Rangel-Nafaile, C., Eds), pp. 161–168. Universidad Nacional Autonoma de Mexico

McEVOY, H., ROSS-MURPHY, S.B. and CLARK, A.M. (1984). Large deformation and failure properties of biopolymer gels. In *Gums and Stabilizers for the Food Industry 2* (Phillips, G.O., Wedlock, D.J. and Williams, P.A., Eds), pp. 111–122. New York, Pergamon Press

MIDDLEMAN, S. (1968). *The Flow of High Polymers.* New York, John Wiley–Interscience

MITCHELL, J.R. (1979). Rheology of polysaccharide solutions and gels. In *Polysaccharides in Foods* (Mitchell, J.R. and Blanshard, J.M.V., Eds), pp. 51–72. London, Butterworths

MITCHELL, J.R. (1984). Rheology of gels. In *Proceedings of the International Workshop on Plant Polysaccharides, Structure and Function* (Mercier, C. and Rinaudo, M., Eds), pp. 313–339. Institut National de la Recherche Agronomique, Nantes, and Centre National de la Recherche Scientifique, Grenoble

MONTEJANO, J.G., HAMANN, D.D. and LANIER, T.C. (1984). Thermally induced gelation of selected comminuted muscle systems—rheological changes during processing, final strengths and microstructure. *Journal of Food Science*, **49**, 1496–1505

MONTEJANO, J.G., HAMANN, D.D. and LANIER, T.C. (1985). Comparison of two instrumental methods with sensory texture of protein gels. *Journal of Texture Studies*, **16**, 403–424

PELEG, M. (1984). A note on the various strain measures at large compressive deformations. *Journal of Texture Studies*, **15**, 317–326

PETRIE, C.J.S. (1979). *Elongational Flows.* San Francisco, Pitman

PRENTICE, J.H. (1984). *Measurements in the Rheology of Foodstuffs.* London, Elsevier Applied Science

RASPER, V.F. (1975). Dough rheology at large deformations in simple tensile mode. *Cereal Chemistry*, **52**, 24r–41r

RASPER, V.F., RASPER, J. and DE MAN, J. (1974). Stress–strain relationships of chemically improved unfermented doughs. *Journal of Texture Studies*, **4**, 438–466

REINER, M. (1964). The Deborah Number. *Physics Today*, January, 62

SCHOFIELD, R.K. and SCOTT BLAIR, G.W. (1932). The relationship between viscosity, elasticity and plastic strength of soft materials as illustrated by some mechanical properties of flour doughs I. *Proceedings of the Royal Society of London*, **A138**, 707–718

SCOTT BLAIR, G.W. (1944). *A Survey of General and Applied Rheology.* New York, Pitman

SHELEF, L. and BOUSSO, D. (1964). A new instrument for measuring relaxation in flour dough. *Rheologica Acta*, **3**, 168–172

SHERMAN, P. (1979). *Food Texture and Rheology.* New York, Academic Press

SMITH, J.R., SMITH, T.L. and TSCHOEGL, N.W. (1970). Rheological properties of wheat flour doughs. III. Dynamic shear modulus and its dependence on amplitude, frequency, and dough composition. *Rheologica Acta*, **9**, 239–252

SMITH, T. (1969). Strength and extensibility of elastomers. In *Rheology, Theory and Applications* (Eirich, F.R., Ed.), pp. 127–221. New York, Academic Press

TRELOAR, L.R.G. (1975). *The Physics of Rubber Elasticity.* London, Oxford University Press

TROUTON, F.T. (1906). On the coefficient of viscous traction and its relation to that of viscosity. *Proceedings of the Royal Society of London*, **A77**, 426–440

TSCHOEGL, N.W., RINDE, J.A. and SMITH, T.L. (1970a). Rheological properties of wheat flour doughs. I. Method for determining the large deformation and rupture properties in simple tension. *Journal of the Science of Food and Agriculture*, **21**, 65–70

TSCHOEGL, N.W., RINDE, J.A. and SMITH, T.L. (1970b). Rheological properties of wheat flour doughs. II. Dependence of large deformation and rupture properties in simple tension on time, temperature and water absorption. *Rheologica Acta*, **9**, 223–238

WILLIAMS, J.G. (1973). *Stress Analysis of Polymers*. New York, Wiley

23

# THE SENSORY–RHEOLOGICAL INTERFACE

P. SHERMAN
*Department of Food and Nutritional Sciences, King's College, University of London, UK*

## Introduction

One of the main functions of food rheology is to develop instrumental methods which will facilitate prediction of consumers' evaluations of the textural properties of foods. Two lines of approach to this problem have been adopted over the years. In the first, one or more readily available instrumental methods have been used and the degree of statistical correlation of the data with subjective evaluations of the same textural property has been determined. Where a significant correlation has been demonstrated the method has been adopted as a reliable predictor of the sensory evaluation. With a few exceptions, such as the cone penetrometer or NIRD–BFMIRA extruder for evaluating the spreadability of butter or margarine, most of the instrumental methods utilized do not impose the same mechanical conditions on the food as those to which it is exposed in sensory evaluations. It is this fact which clearly distinguishes the first approach from the second. In the latter approach the aim has been to simulate as closely as possible the mechanical conditions which prevail during mastication. The need to do this cannot be overemphasized because most fluid foods exhibit non-Newtonian flow, i.e. their measured viscosity depends on the rate of shear applied, and most solid foods are non-linear viscoelastic, i.e. their rheological parameters are a function of the applied stress. Consequently, if instrumental measurements are made under conditions which do not resemble those prevailing during mastication the resulting data cannot be used to predict the sensory evaluation. In other words '... instrumental measurements ... can be expected to correlate significantly with sensory evaluation only if the measurements to be correlated are compatible and belong to the same zone in the shear stress–shear rate (for liquids) or stress–strain (for solids) diagram of the material' (Mohsenin and Mittal, 1977).

The general philosophy behind the second approach to the problem can be further explained by reference to Stevens' psychological law:

$$\psi = KS^n \tag{1}$$

where $\psi$ is the sensory evaluation of the textural property, $S$ is the stimulus on which consumers base their evaluation and K and n are constants (Stevens, 1960). The aim in the instrumental test is to simulate, as closely as possible, the mechanical conditions in the mouth which give rise to the stimulus $S$. If this aim is achieved a double logarithmic plot of $\psi$ against $S$, where $S$ is now the instrumentally simulated stimulus, gives a straight line with a gradient n and an intercept K on the log $\psi$ axis.

## The three phases of texture evaluation during mastication

The texture of food is a composite of many different characteristics which are identified sensorily in the mouth in three consecutive phases (Brandt, Skinner and Coleman, 1963). These are the first bite phase which characterizes the initial contact of the teeth, tongue, etc., with the food, the masticatory phase and, finally, the residual phase just before the food is swallowed. Two mechanical properties which are identified during the first bite phase will be reviewed in this chapter. They are the viscosity of liquid foods and the firmness of solid foods. Mechanical characteristics are evaluated sensorily by the pressures exerted on the teeth, tongue and roof of the mouth during mastication (Szczesniak, 1963). The time span of the first bite phase is so short that it is unlikely that the food, after entering the mouth, has yet had time to mix with saliva so that its constitution is modified.

## Sensory evaluation of fluid food viscosity and its instrumental simulation

### *Oral evaluation*

The viscosity of liquid foods is identified in the mouth by forcing it to flow in the space between the surface of the tongue and the roof of the mouth. The liquid flows under the influence of the stress exerted by the tongue. Viscosity is then evaluated either by perception of the stress registered by the sense organs located on the surfaces of the superficial structures of the mouth, by perception of the rate of flow of the liquid food, or by some complex of the two (Wood, 1968). This action can be simulated by shearing samples in the narrow gap between the stationary and mobile members of a coaxial cylinder or cone–plate viscometer. Movement of the mobile member resembles movement of the tongue and the stationary member represents the roof of the mouth.

The shear rate ($\dot{\gamma}$) and shear stress ($\tau$) associated with the oral evaluation of viscosity can be determined by a relatively simple procedure (Wood, 1968). *Figure 23.1* shows the viscometric $\tau$ versus $\dot{\gamma}$ data for three shear thinning fluid foods (A, B and C), which appear, visually, to have very similar viscosities. Also included are the data for a Newtonian fluid (X), which is used as a standard for comparison with A, B and C. The standard fluid should meet two requirements. First, it should appear, visually, to have a viscosity similar to that of the shear thinning samples. Second, its linear $\tau$ versus $\dot{\gamma}$ plot should intersect and cross over the $\tau$ versus $\dot{\gamma}$ curves for the other samples. Glucose syrup diluted with an appropriate volume of water is usually suitable. In addition to the viscometric $\tau$ versus $\dot{\gamma}$ measurements a standard volume of each sample is presented to a sensory evaluation panel. The members of the panel are asked to indicate which of the three shear thinning samples appears to have a viscosity in the mouth which is higher, lower or equal to the viscosity of the Newtonian standard. Using the samples for which viscometric data are quoted in *Figure 23.1*, the sensory responses indicated that A appeared to have a higher viscosity than X, while B and C had lower viscosities than X. The only $\tau$ versus $\dot{\gamma}$ region where the sensory

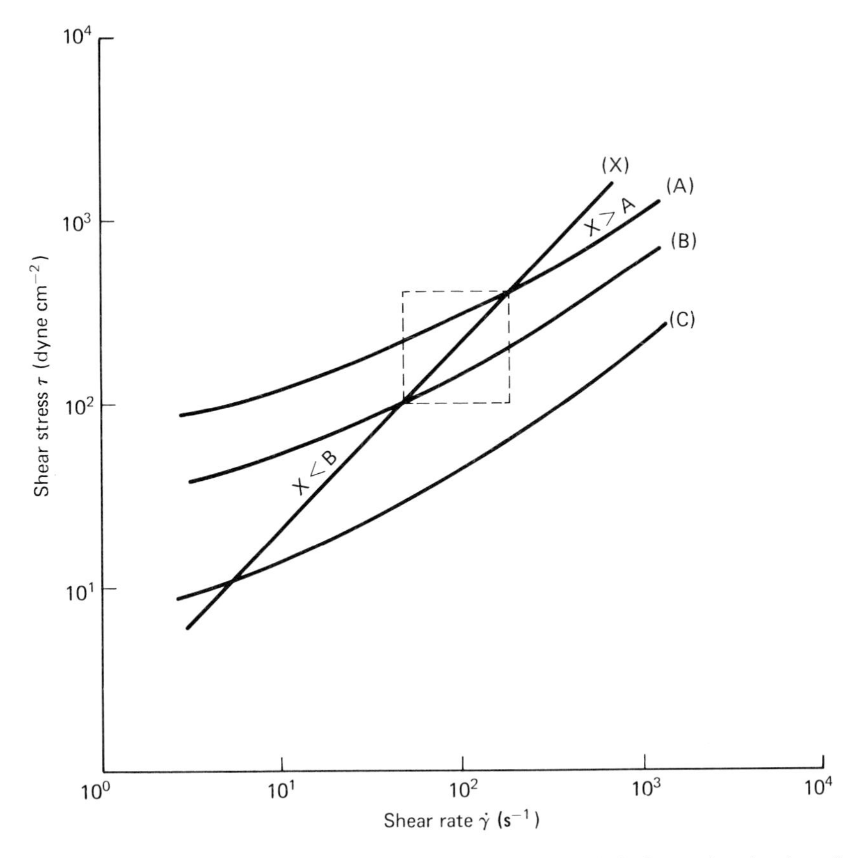


**Figure 23.1** Procedure for determining shear rate in the mouth during oral evaluation of viscosity.

and instrumental responses will agree is in the region enclosed within the broken line rectangle.

Similar tests can be made with other groups of liquid foods extending over a wide range of consistencies. The only limitations in the selection of samples are that within any one group all the samples should appear, visually, to have similar consistencies, and that there should be significant differences between the groups. A detailed study along these lines provided the data shown in *Figure 23.2* (Shama and Sherman, 1973a). Each broken line rectangle denotes the $\tau$ versus $\dot{\gamma}$ domain associated with the comparison of viscosities for a single group of samples. The two continuous line curves represent the approximate limits of $\tau$ and $\dot{\gamma}$ relating to oral evaluation of viscosity for a very wide range of foods extending from milk through to semi-solid chocolate spread. The broken lines superimposed on the data are viscometrically derived $\tau$ versus $\dot{\gamma}$ data for Newtonian fluids whose viscosities extend over six orders of magnitude.

Several important conclusions relating to the oral evaluation of viscosity can be drawn from *Figure 23.2*.

1. The continuous line curves indicate that the $\tau$ versus $\dot{\gamma}$ conditions are not constant for all foods irrespective of their flow properties. Very viscous foods are subjected

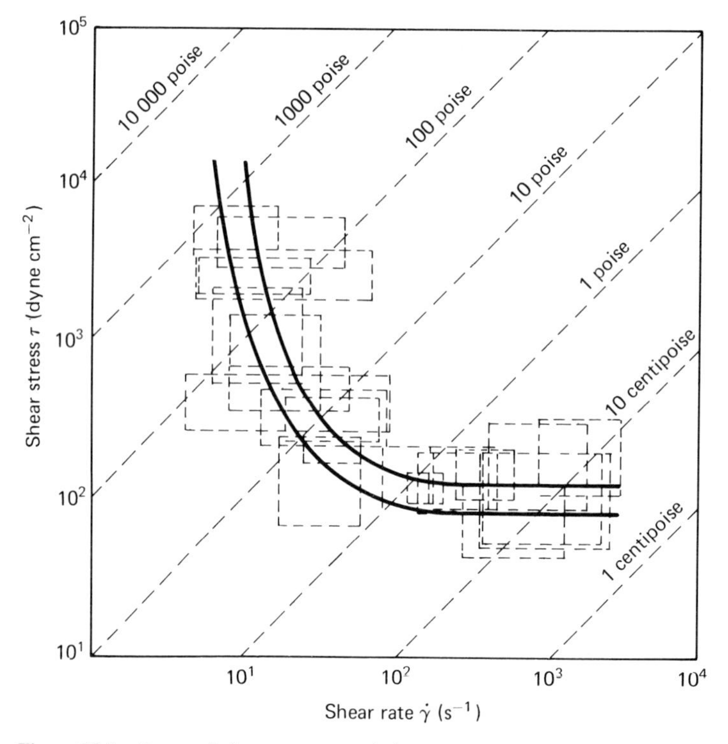


**Figure 23.2** Range of shear stresses and shear rates associated with the oral evaluation of viscosity.

to much lower $\dot{\gamma}$ and higher $\tau$ than very fluid foods during oral evaluation. The difference in $\dot{\gamma}$ can be two orders of magnitude or more.

2. The continuous line curves are distinctly flat in the region of high $\dot{\gamma}$ and low $\tau$, i.e. in the region associated with oral evaluation of very fluid foods. Many fluid foods deviate from laminar flow when they are subjected to high $\dot{\gamma}$ (Parkinson and Sherman, 1971) and the resulting turbulence produces an apparent increase in viscosity ($\eta$).

   For laminar flow $\tau$ is given by:

$$\tau = \eta\dot{\gamma} \tag{2}$$

   For turbulent flow $\tau$ is given by:

$$\tau_{\text{total}} = (\eta\dot{\gamma} + E_{\text{v}}) \tag{3}$$

   where $E_{\text{v}}$ is the eddy diffusivity of momentum.

   Turbulence is reduced by the mucin, serum albumin and globulin which are present in human saliva. However, as viscosity is evaluated almost immediately after the sample is introduced into the mouth it is unlikely that saliva has time to be secreted and so reduce the turbulence.

3. The nature of the stimulus associated with oral evaluation of viscosity differs for

fluid and very viscous, or semi-solid, foods. For the first named category the stimulus appears to be variable $\dot{\gamma}$ developed at a constant $\tau$ of $\sim 100$ dyne cm$^{-2}$, whereas for the latter category it is a variable $\tau$ developed at a constant $\dot{\gamma}$ of $\sim 10\,s^{-1}$.

The lingual pressures associated with oral evaluation of the viscosity of glucose syrup, determined by a capillary flow technique before and after dilution with water, ranged from $\sim 3.0 \times 10^5$ dyne cm$^{-2}$ for undiluted glucose solution to values at least two orders of magnitude lower for heavily diluted samples (Shama and Sherman, 1974). These values agree satisfactorily with the range of values quoted in *Figure 23.2*.

Statistical analysis of sensory evaluation data for a range of liquid foods indicated that 'thickness' was one of three attributes which best described the textural characteristics of liquid foods (Kokini, Kadane and Cussler, 1977). It was assumed that all sensory evaluations are proportional to specific forces on the tongue and a theoretical analysis suggested that thickness $\propto$ viscous force on the tongue

$$\propto \mathrm{m} V^{\mathrm{n}} \left[ h_0^{n/(\mathrm{n}+1)} + \left( \frac{W}{R^{\mathrm{n}+3}} \cdot \frac{\mathrm{n}+3}{2\pi \mathrm{m}} \right)^{1/\mathrm{n}} \left( \frac{\mathrm{n}+1}{2\mathrm{n}+1} \right) t \right]^{\mathrm{n}^2/\mathrm{n}+1} \tag{4}$$

where $V$ is the velocity with which the tongue moves relative to the roof of the mouth, $h_0$ is the original thickness of the sample on the tongue, $W$ is the load or normal force exerted by the tongue, $R$ is the effective radius of the tongue, $t$ is the time taken for the evaluation, n is the power law constant for non-Newtonian fluids and m is a consistency index. It was assumed that 'thickness' and viscosity are interrelated. From the sensory evaluation of 'thickness' data and viscometric measurements on a series of non-Newtonian fluid foods the $\tau$ versus $\dot{\gamma}$ conditions operative during the sensory evaluations were derived. The values derived agreed well with those quoted in *Figure 23.2* (see *Figure 23.3*). It is interesting to note that the only region where the two sets of data may deviate is at high $\dot{\gamma}$, i.e. in the region where the flow may be turbulent. Equation (6) does not allow for the possibility of turbulent flow.

The results of two other studies using aqueous solutions of gums and hydrocolloids have been used to question the validity of the conclusions drawn from *Figures 22.2* and *22.3* and to suggest that sensory evaluation of viscosity occurs over a range of shear rates rather than at an approximately constant shear rate. In the first of these two studies (Christiansen, 1979) sensory evaluation panelists were instructed to swish the solution around in the mouth as they would a mouth wash and then to expectorate. Consequently, the evaluation was not carried out instantaneously. A similar criticism, and others, can be levelled at the second study (Cutler, Morris and Taylor, 1983; Morris and Taylor, 1982) where the sensory data for viscosity evaluation (and no details are given of the criteria used) were correlated with viscosity measurements made in a rheometer after equilibrium had been reached. It would appear that these two studies describe the shear rate conditions associated with sensory evaluation of a consistency which is changing from the moment that the sample is introduced into the mouth until the time that the sample is in a condition suitable for swallowing, following its structural modification by being mixed with saliva. The data in *Figures 22.2* and *22.3* relate only to the first phase of the process, i.e. the evaluation of viscosity *immediately* after the sample enters the mouth. This, according to sensory evaluation panelists, is the way in which they evaluate viscosity orally. It may well be that we should evaluate two levels of viscosity: the initial and

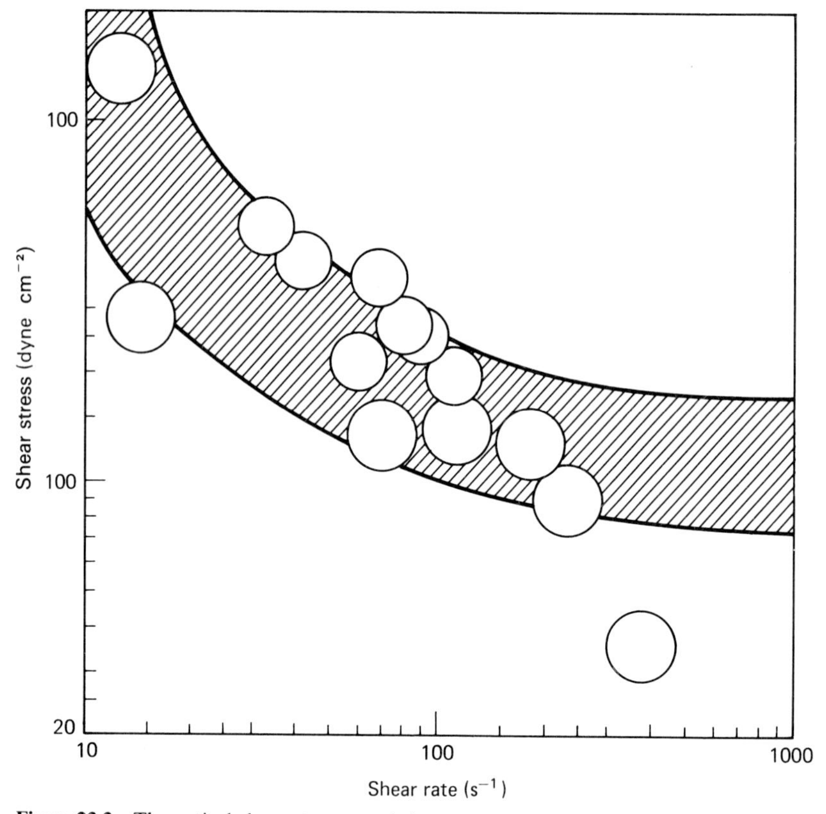


**Figure 23.3** Theoretical shear stresses and shear rates operative during oral evaluation of viscosity.

final viscosities. The shear rate range observation would then be relevant to the final evaluation process.

The continuous double line curve of *Figure 23.2* can be used to predict the $\tau$ and $\dot{\gamma}$ associated with the oral evaluation of viscosity for all liquid foods, and hence the conditions under which viscometric measurements should be made if the sensory assessment conditions are to be simulated. It can also be used to predict the rating of samples relative to each other. *Figure 23.4* reproduces the double line curve from *Figure 23.2* and superimposed on it are the rheometer $\tau$ versus $\dot{\gamma}$ data for many foods which are in general use. These data can be interpreted as follows. Condensed milk and tomato ketchup obviously have quite different flow characteristics. Nevertheless, their $\tau$ versus $\dot{\gamma}$ plots intersect within the limits of the two continuous line curves. This means that orally they will appear to have the same viscosity. The same conclusion can be drawn for condensed milk and St. Michael creamed tomato soup, and for chocolate spread and peanut butter. On the other hand, the intersection point of the $\tau$ versus $\dot{\gamma}$ plots for condensed milk and salad cream lies outside the two continuous line curves, and the latter are crossed at a much higher $\tau$ by the salad cream plot. Consequently, salad cream will be sensorily evaluated as having a much higher oral viscosity than condensed milk. The ratio of their viscosities can be derived from the $\tau$ and $\dot{\gamma}$ at which their respective plots intersect the two continuous line curves.

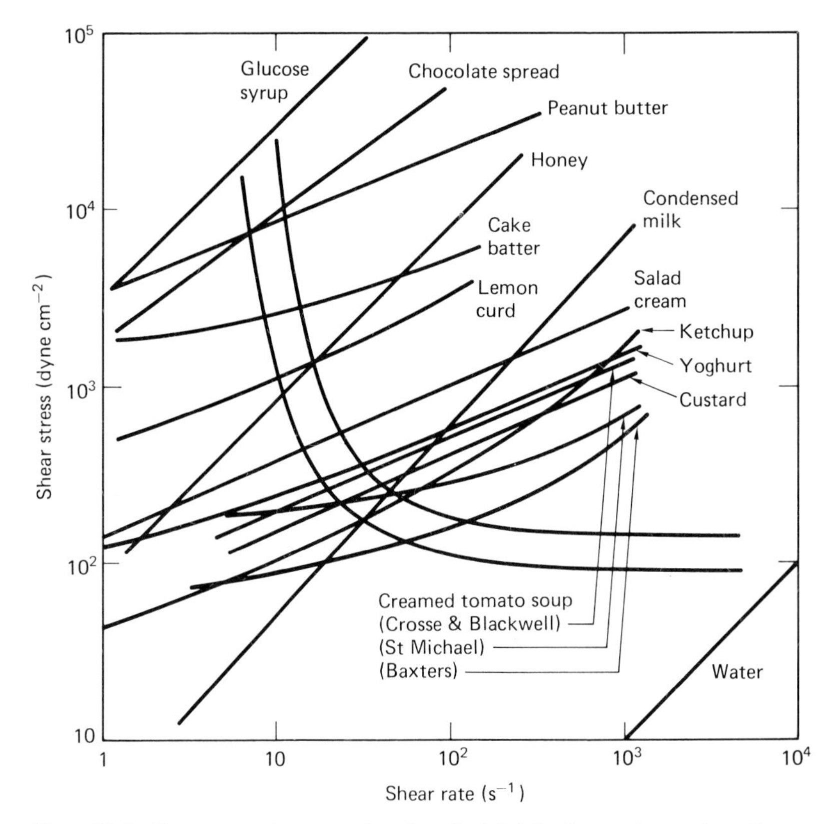


**Figure 23.4** Shear stress–shear rate data for scheduled foods superimposed on *Figure 23.2*.

### *Non-oral evaluation*

The viscosities of fluid foods may also be evaluated sensorily by non-oral methods, e.g. by stirring the contents of a container or by the rate at which the food flows along the wall of a container when it is tilted. These alternative procedures involve stimuli which not only differ from each other but they also differ from the stimuli associated with oral evaluation (Shama, Parkinson and Sherman, 1973). The stimulus associated with stirring is a variable $\tau$ ($10^2$–$10^4$ dyne cm$^{-2}$) developed at an approximately constant $\dot{\gamma}$ (90–100 s$^{-1}$), whereas for the tilting procedure, labelled 'visual conditions' in *Figure 23.5*, it is a variable $\dot{\gamma}$ (0.1–40 s$^{-1}$) developed at a widely variable $\tau$ (60–600 dyne cm$^{-2}$). Consequently, it is possible for different relative ratings of samples to be achieved by the three evaluation procedures depending on the $\dot{\gamma}$ dependency of viscosity for each sample.

## Sensory evaluation of firmness and its instrumental simulation

Consumers evaluate firmness in the mouth as the force required to achieve a given deformation (Szczesniak, 1963) or as the force required to compress the food between

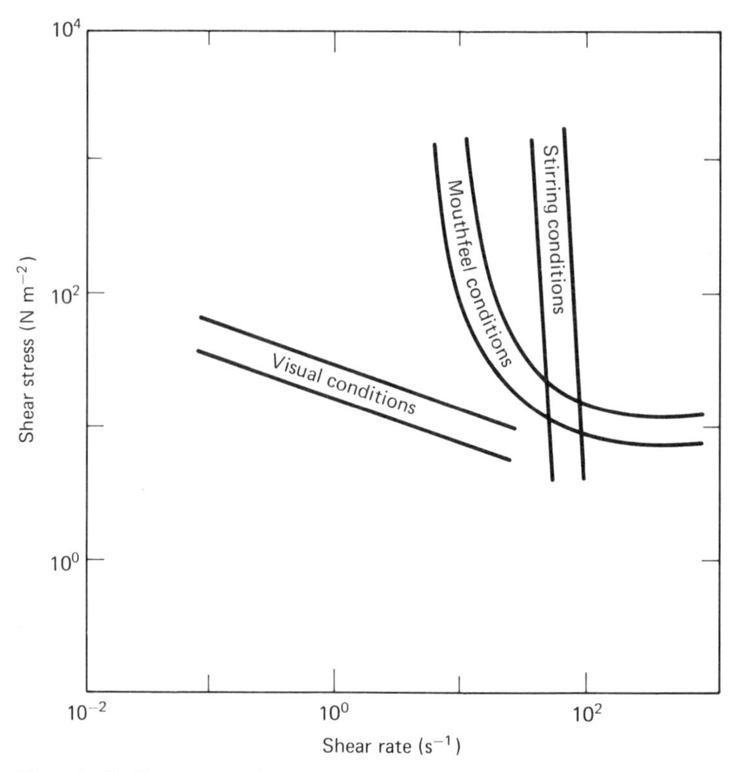


**Figure 23.5** Shear stress–shear rate conditions associated with oral and non-oral evaluations of viscosity.

the molar teeth (Abbott, 1973). The food's response to the application of this force is measured organoleptically by the pressure exerted on the teeth, tongue and roof of the mouth (Szczesniak, 1963).

When published chewing patterns (Howel and Brudevold, 1950) are analysed by Fourier's theorem it appears that during the first downstroke, i.e. when firmness is evaluated, there is an approximately linear increase in force up to the point where the maximum force is applied (Shama and Sherman, 1973b). The subsequent upstroke and the remainder of the chewing force pattern cannot be characterized in this way. This result suggests that sensory evaluation of firmness can be simulated rheologically by a compression test in which the compression force is increased linearly.

There is another factor which must be taken into account, however. Most people chew at a rate of 40–80 masticatory strokes per minute (Louridis, Demetriou and Bazapoulos, 1970) so that the downstroke of the first bite is completed in about 0.5 s. The actual time involved is rather longer for hard foods than for soft foods (Hedegard, Lundberg and Wictorin, 1970). Therefore, the compressive masticatory force is applied at a very fast rate. In spite of this fact instrumental compression tests have generally used low rates of compression (Shama and Sherman, 1973b) so that short time relaxation processes can develop (Shama and Sherman, 1973c) and a lower force is required to achieve a certain degree of compression than when a high compression rate is used.

The influence of compression rate, and how the use of the wrong compression rate can lead to incorrect conclusions, can be illustrated by reference to compression tests on White Stilton and Gouda cheeses (Shama and Sherman, 1973b). *Figure 23.6* shows the force–compression behaviour at compression rates of 5, 20, 50 and 100 cm min$^{-1}$. Sensory evaluation responses indicated that Gouda cheese was firmer than White Stilton. Yet, at a compression rate of 5 cm min$^{-1}$ the White Stilton required a higher force to achieve a given degree of compression than the Gouda, so that instrumental data did not agree with the sensory evaluation. When the rate of compression was increased to 20 cm min$^{-1}$ the force–compression curves for the two cheeses intersected and crossed over at two points. Only within the shaded region between the two intersection points was there agreement between the instrumental and sensory data. The region of agreement increased as the rate of compression was increased further to 50 and then 100 cm min$^{-1}$. The importance of selecting the correct compression rate

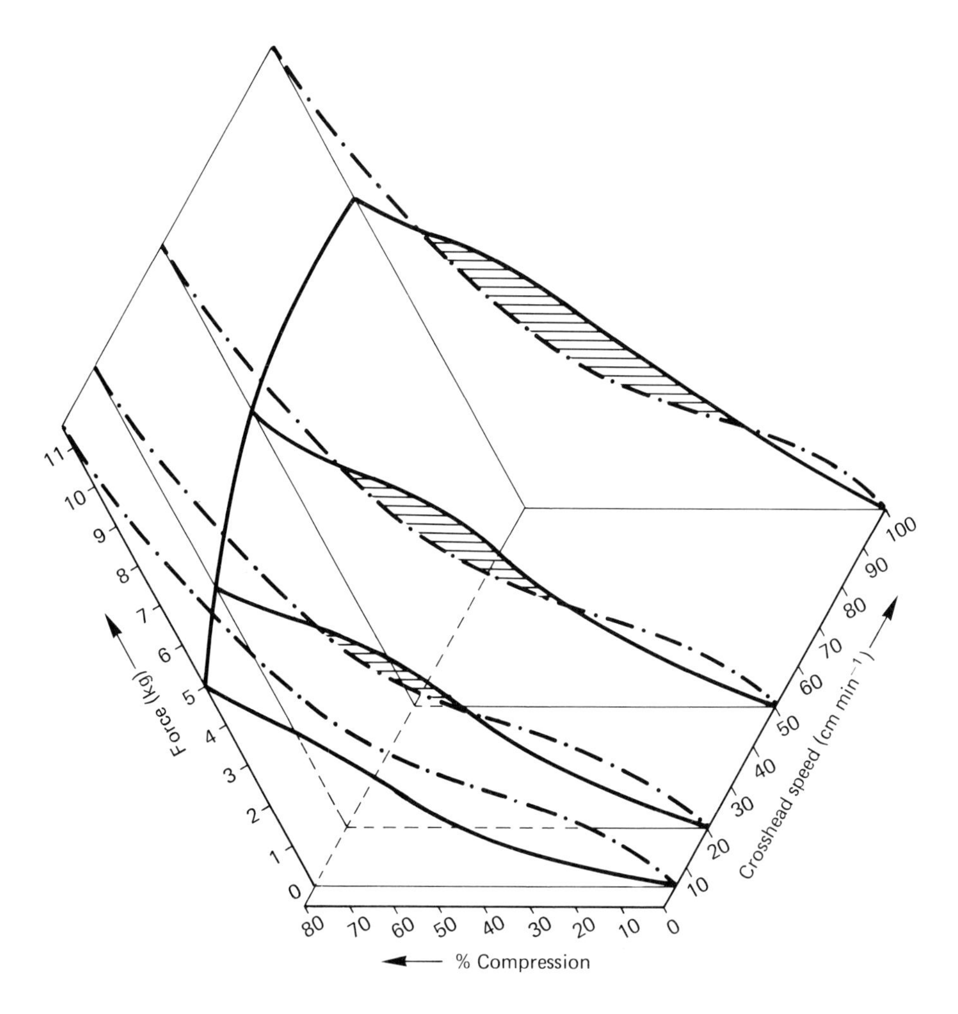


**Figure 23.6** Influence of compression rate on force–compression behaviour; –·–·–, White Stilton cheese; ——, Gouda cheese.

for simulating the sensory evaluation of firmness by squeezing samples between the fingers has also been demonstrated (Voisey and Crete, 1973).

A comparison of instrumental force–compression data obtained at various compression rates with sensory data from both oral and squeezing tests suggests that firmness evaluations by squeezing involves lower percentage compressions than oral evaluations (Boyd and Sherman, 1975). Squeezing does not normally involve a compression in excess of 50%, whereas in oral evaluations it can be 70% or more. The latter higher compression is well in excess of that at which the elastic limit is reached, whereas the compression achieved by squeezing is not. Consequently, it is possible that different stimuli are associated with sensory evaluation of firmness by squeezing and in the mouth. Evaluation by squeezing may depend on the food's compression behaviour within the elastic region, whereas orally it is also influenced by crack initiation and propagation.

In addition to the effect of compression rate there are other factors which influence firmness evaluation, including the dimensions of the samples being tested (Brinton and Bourne, 1972). This factor, which influences both sensory and instrumental procedures, appears to be related to the barrel deformation of samples when they are compressed. Cine films of the compression of Gouda (Culioli and Sherman, 1976) and Leicester cheeses (Vernon Carter and Sherman, 1978) indicated that barrel deformation was not initiated until 20% compression was achieved. Similar effects have been observed with bread (Abu-Shakra and Sherman, 1984). When the sample's upper and lower surfaces are lubricated with mineral oil barrel deformation does not occur when the compression force is applied. Now, the diameters of the surfaces increase at a faster rate than the central region so that the sides develop a concave shape. Samples squeezed between the fingers also deform into a barrel shape.

These observations suggest that the force identified in both instrumental and sensory evaluations of firmness is the sum of two components, i.e. the force required to overcome the friction at the sample's surfaces plus the force actually involved in compressing the sample. The influence of surface friction has been studied using German Loaf cheese (Atkin and Sherman, 1984), which is a reasonably homogeneous product. One to five cylinder-shaped samples were compressed at the same time. All samples tested at any one time had the same diameter and height, but different heights were used in different tests. The standard diameter adopted was 1.64 cm and the sample heights ranged from 1.1 to 3.3 cm.

The component of the measured force ($F_T$) which was involved in overcoming the friction was derived as follows. Compression of a single sample to a given level is given by:

$$F_{T(1)} = F_c + F_{f(1)} \tag{5}$$

where $F_c$ is the force component actually involved in compressing the sample and $F_{f(1)}$ is the force component required to overcome the surface friction.

In general, using $i$ samples:

$$F_{T(i)} = iF_c + F_{f(i)} \tag{6}$$

Dividing through by $i$:

$$\frac{F_{T(i)}}{i} = F_c + \frac{F_{f(1)}}{i} \tag{7}$$

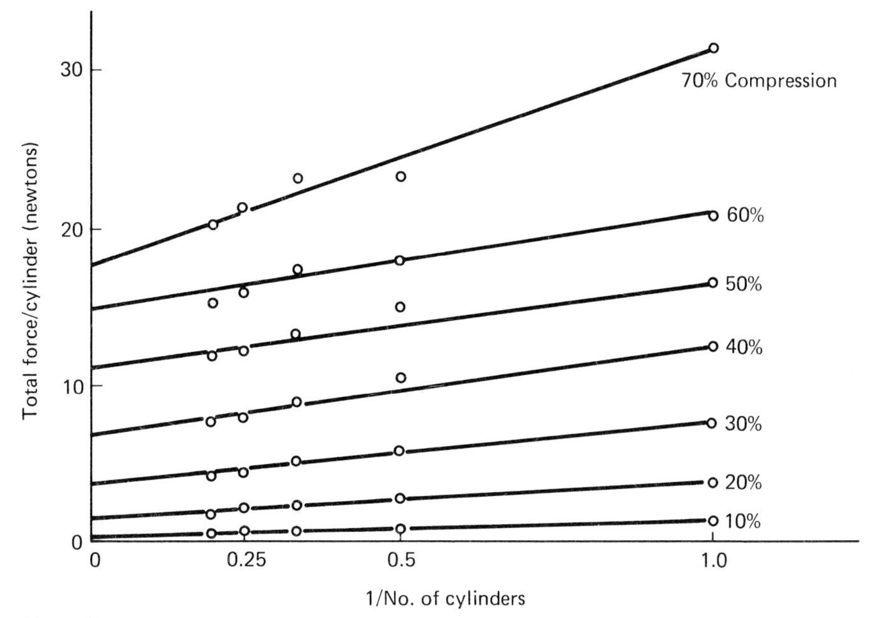


**Figure 23.7** $F_{T(i)}$ versus $1/i$ for German loaf cheese. Sample diameter/length ratio = 1.0; compression range 10–70%.

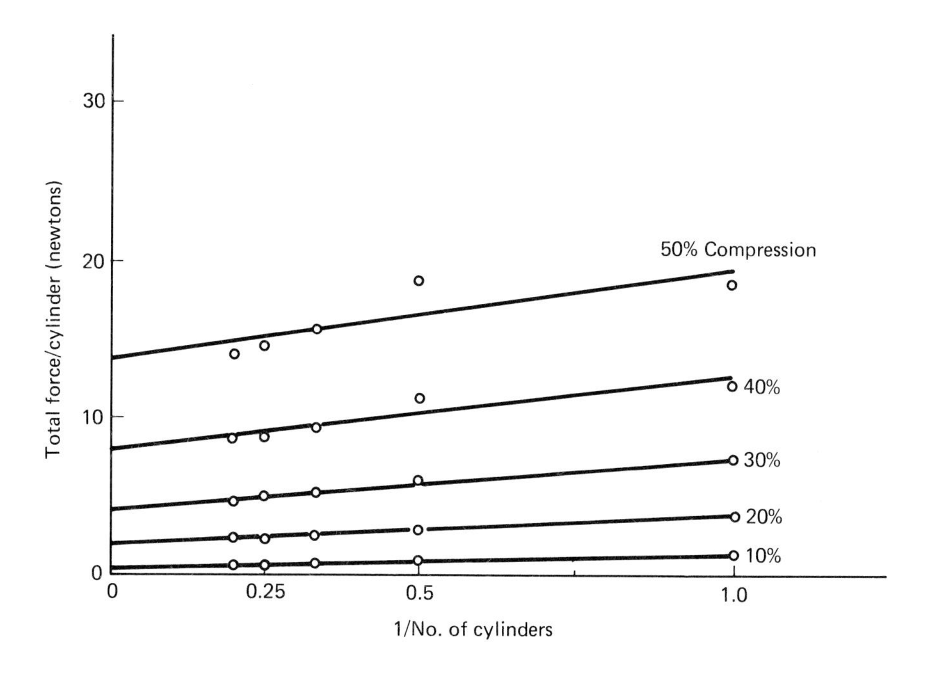


**Figure 23.8** $F_{T(i)}$ versus $1/i$ for German loaf cheese. Sample diameter/length ratio = 1.25; compression range 10–50%.

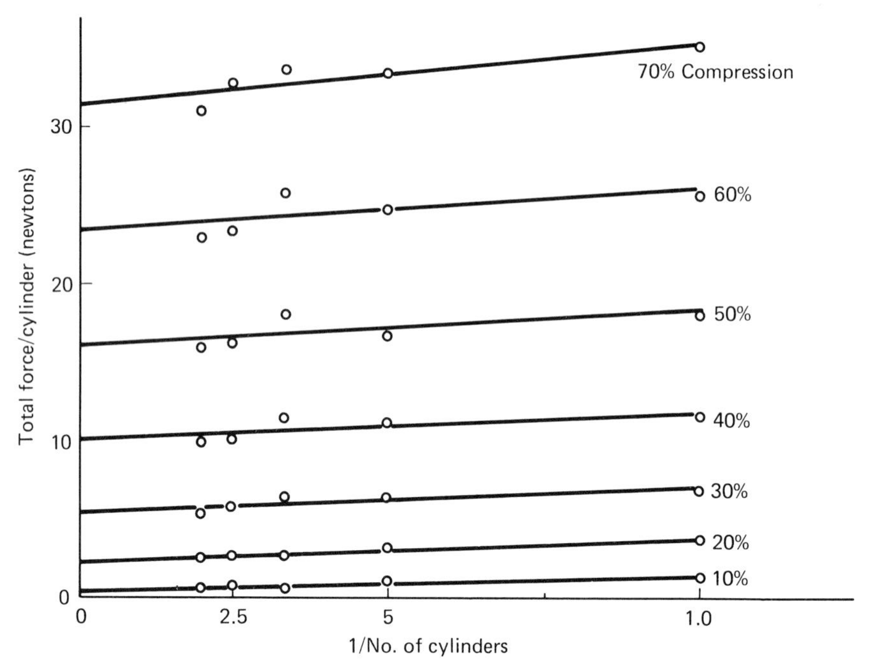


**Figure 23.9** $F_{T(i)}$ versus $1/i$ for German loaf cheese. Sample diameter/length ratio = 1.5; compression range 10–70%.

Therefore, if the friction effect is not a function of $i$ then a plot of $F_{T(i)}$ versus $1/i$, at a fixed compression, should be linear with a gradient $F_{f(i)}$ and an intercept $F_c$. The experimental data conformed to this analysis. By applying this analysis to the recorded force–compression curve the true compression curve can be derived.

*Figures 23.7–23.9* show $F_{T(i)}$ versus $1/i$ for diameter/length ratios of 1.0, 1.25 and 1.5, respectively. As the sample diameter/length ratio increases so does $F_c$ (*Figure 23.10*), but the friction decreases under these conditions. Two components of friction are involved. These are the coefficients of static friction ($\mu_s$) and of kinetic friction ($\mu_k$). The latter operates when the sample's surfaces move relative to one another. Especially at high percentage compressions and compression rates $\mu_k$ exerts the major effect. If $\mu_k$ is represented by the ratio $F_{f(i)}/F_{T(i)}$ then $\mu_k$ is highest at low compressions (*Figure 23.11*).

Previously it was stated that evaluating firmness by squeezing samples between the fingers involves lower forces and lower compressions than oral evaluation. Consequently, because the coefficient of friction is highest at low compressions, one would expect friction effects to have a greater effect on firmness evaluation by squeezing samples between the fingers than on oral evaluation.

Friction and lubrication effects have been reported also for gelatin and starch gels (Bagley, Christianson and Wolf, 1985) and for some natural and processed cheeses (Casiraghi, Bagley and Christianson, 1985). The friction effects can be eliminated by bonding the samples to the compression plates, but the instrumental evaluation of firmness does not then simulate the sensory evaluation.

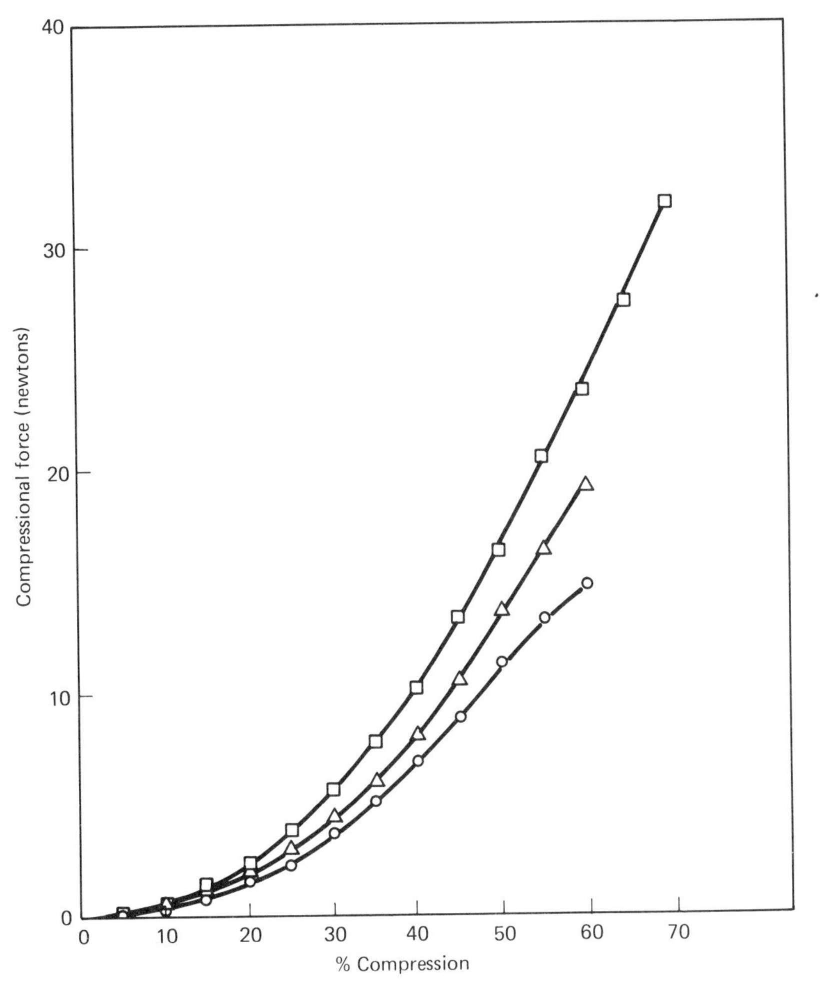


**Figure 23.10** True compression force ($F_c$) versus % compression for German loaf cheese; sample diameter/length ratios of 1.0, 1.25 and 1.5.

## Conclusions

Identical instrumental test conditions cannot be used for all foods irrespective of their 'viscosity' or firmness if the sensory evaluations of these textural parameters are to be simulated. Both instrumental and sensory evaluations are complex processes based upon two or more processes acting simultaneously. The precise way in which these processes manifest themselves depends on the magnitude of 'viscosity' or firmness exhibited. Before an instrumental method is adopted for predicting sensory evaluation of either textural property it is necessary to establish how the sensory evaluation is made and the criteria upon which it is based. This will involve detailed discussion

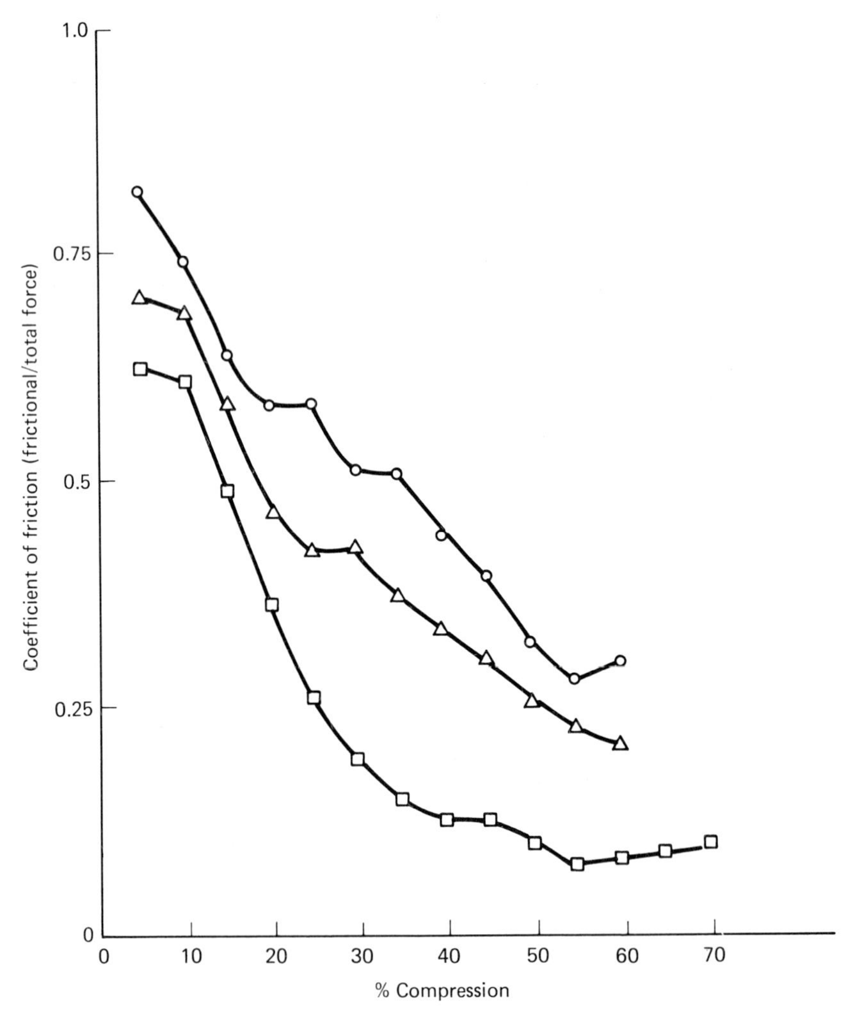


**Figure 23.11** Fraction of total force $F_{T(i)}$ required to overcome surface friction versus % compression for German loaf cheese, sample diameter/length ratios of 1.0, 1.25 and 1.5.

with sensory panelists so that an appropriate instrumental test can be formulated. Sensory data should then be correlated with instrumental data obtained with a variety of test conditions in order to ascertain which of the latter are most appropriate to the problem.

## References

ABBOTT, J.A. (1973). Sensory assessment of textural attributes of foods. In *Texture Measurement in Foods* (Kramer, A. and Szczesniak, A.S., Eds), pp. 17–32. Dordrecht, Reidel

ABU-SHAKRA, A.M. and SHERMAN, P. (1984). Evaluation of bread firmness by Instron compression tests. *Rheologica Acta*, **23**, 446–450

ATKIN, G. and SHERMAN, P. (1984). The influence of surface friction on road firmness evaluation by compression tests. In *Advances in Rheology*, Volume 4 (Mena, B., Garcia-Rejon, A. and Rangel-Nafaile, C., Eds), pp. 123–144. Universidad Nacional Autonoma de Mexico

BAGLEY, E.B., CHRISTIANSON, D.D. and WOLF, W.J. (1985). Frictional effects in compressional deformation of gelatin and starch gels and comparison of material response in simple shear, torsion and lubricated uniaxial compression. *Journal of Rheology*, **29**, 103–108

BOYD, J.V. and SHERMAN, P. (1975). A study of force–compression conditions associated with hardness evaluation in several foods. *Journal of Texture Studies*, **6**, 507–522

BRANDT, M.A., SKINNER, E.Z. and COLEMAN, J.A. (1963). Texture profile method. *Journal of Food Science*, **28**, 404–409

BRINTON, R.H. Jr. and BOURNE, M.C. (1972). Deformation testing of foods. III. Effect of size and shape on the magnitude of deformation. *Journal of Texture Studies*, **3**, 284–297

CASIRAGHI, E.M., BAGLEY, E.B. and CHRISTIANSON, D.D. (1985). Behaviour of Mozzarella, Cheddar and processed cheese spread in lubricated and bonded uniaxial compression. *Journal of Texture Studies*, **16**, 281–301

CHRISTIANSEN, C.M. (1979). Oral perception of solution viscosity. *Journal of Texture Studies*, **10**, 153–164

CULIOLI, J. and SHERMAN, P. (1976). Evaluation of Gouda cheese firmness by compression tests. *Journal of Texture Studies*, **7**, 353–372

CUTLER, A.N., MORRIS, E.R. and TAYLOR, L.J. (1983). Oral perception of viscosity in fluid foods and model systems. *Journal of Texture Studies*, **14**, 377–395

HEDEGARD, B., LUNDBERG, M. and WICTORIN, L. (1970). Masticatory function. A cineradiographic study. IV. Duration of the masticatory cycle. *Acta Odontologica Scandinavica*, **28**, 859–865

HOWEL, A.H. and BRUDEVOLD, F. (1950). Vertical forces used during chewing of food. *Journal of Dental Research*, **29**, 133–136

KOKINI, J.L., KADANE, J.B. and CUSSLER, E.L. (1977). Liquid texture perceived in the mouth. *Journal of Texture Studies*, **8**, 195–218

LOURIDIS, A., DEMETRIOU, N. and BAZAPOULOS, K. (1970). Chewing effects on secretion of stimulated mixed human saliva. *Journal of Dental Research*, **49**, 1132–1135

MOHSENIN, N.N. and MITTAL, J.P. (1977). Use of rheological terms and correlation of compatible measurements in food texture research. *Journal of Texture Studies*, **8**, 395–408

MORRIS, E.R. and TAYLOR, L.J. (1982). Oral perception of fluid viscosity. In *Progress in Food and Nutrition Science. Volume 6. Gums and Stabilisers for the Food Industry. Interaction of Hydrocolloids* (Phillips, G.O., Wedlock, D.J. and Williams, P.A., Eds), pp. 285–296. Oxford, Pergamon Press

PARKINSON, C. and SHERMAN, P. (1971). The influence of turbulent flow on the sensory assessment of viscosity in the mouth. *Journal of Texture Studies*, **2**, 451–459

SHAMA F. and SHERMAN, P. (1973a). Identification of stimuli controlling the sensory evaluation of viscosity. II. Oral methods. *Journal of Texture Studies*, **4**, 111–118

SHAMA, F. and SHERMAN, P. (1973b). Evaluation of some textural properties of foods with the Instron universal testing machine. *Journal of Texture Studies*, **4**, 344–352

SHAMA, F. and SHERMAN, P. (1973c). Stress relaxation during force–compression studies on foods with the Instron universal testing machine and its implications. *Journal of Texture Studies*, **4**, 353–362

SHAMA, F. and SHERMAN, P. (1974). Lingual pressure associated with oral evaluation of viscosity. *Biorheology*, **11**, 453–456

SHAMA, F., PARKINSON, C. and SHERMAN, P. (1973). Identification of stimuli controlling the sensory evaluation of viscosity. I. Non-oral methods. *Journal of Texture Studies*, **4**, 102–110

STEVENS, S.S. (1960). The psychophysics of sensory function. *American Scientist*, **48**, 226–253

SZCZESNIAK, A.S. (1963). Classification of textural characteristics. *Journal of Food Science*, **28**, 385–389

VERNON CARTER, E.J. and SHERMAN, P. (1978). Evaluation of the firmness of Leicester cheese by compression tests with the Instron universal testing machine. *Journal of Texture Studies*, **9**, 311–324

VOISEY, P.W. and CRETE, R. (1973). A technique for establishing instrumental conditions for measuring food firmness to simulate consumer evaluations. *Journal of Texture Studies*, **4**, 371–377

WOOD, F.W. (1968). *Rheology and Texture of Foodstuffs*, pp. 40–49. London, Society of Chemical Industry

# 24

# EVALUATION OF CRISPNESS

Z. M. VICKERS
*Department of Food Science and Nutrition, University of Minnesota, USA*

## Introduction

The importance of crispness, and the related sensation of crunchiness, was documented in the study by Szczesniak and Kleyn (1963) of consumer awareness of food texture. During a word association test, in which subjects were given a list of 79 food names and asked to list four descriptors they associated with each food, it was found that 'crisp' was mentioned more often than any other descriptor. Furthermore, foods having the greatest number of textural responses were either bland in flavour or possessed the qualities of crispness or crunchiness. A similar word association test using a different group of subjects (Szczesniak, 1971) confirmed those earlier results. Crisp was again used more frequently than any other textural descriptor.

Yoshikawa *et al.* (1970) conducted a study in Japan similar to that of Szczesniak and Kleyn. They used questionnaires with 97 food stimulus words to assemble a list of Japanese terms relating to texture. Kori-Kori (crisp) and Kari-Kari (crunchy) were among the 10 most frequently used words.

Crispness and crunchiness appear to be not only important and useful descriptors of food texture, but desirable qualities as well. Szczesniak and Kahn (1971) stated 'Crispness is synonymous with freshness; fresh vegetables, fruits and snacks are thought to be best when they are firm and crisp . . . Crispness is very important to the pleasure of substantial eating. It appears to hold a particular place in the basic psychology of appetite and hunger satiation, spurring one to continue eating . . . It is particularly good as an appetizer and as a stimulant to active eating . . . It appears to be universally liked and is often used as a very popular accent-contributing or dramatizing characteristic.' In their 1984 study of textural combinations the same authors added that crispness '. . . appears to be the most versatile single texture parameter. It is liked in nearly every combination . . . goes well with soft, smooth and creamy things when used as a central characteristic. It goes well with almost any other texture when used as an accessory or accent. It is very prominent in texture combinations that mark excellent cooking . . .'.

Iles and Elson (1972) had subjects rate both the crispness and their textural preferences for six foods, each altered in crispness to several degrees. They found a high correlation between textural preferences and sensory crispness in all six foods studied, indicating that the acceptability of a food's texture increased with increasing crispness. Vickers (1983) had 52 subjects evaluate the taped biting and chewing sounds of 16 different foods for pleasantness and nine other auditory descriptors. Of the nine auditory qualities, crispness was most closely associated with pleasantness.

The importance and desirability of crispness has spurred efforts to define and measure the sensation. Many scientists have worked towards this goal. This chapter examines some of these studies focusing primarily on the sensory perceptions. Our knowledge of the sensation of crispness has developed over many years and has been shaped by information from several sources including: clues from early literature and discussions, clues provided by experiments designed to test specific hypotheses about crispness and clues from partly successful instrumental measurements of crispness.

## Early sensory studies of crispness

Prior to 1975, several researchers had studied the sensory perception of crispness and developed sensory definitions or descriptions of the attributes. The Sensory Texture Profile technique developed in the early sixties (Brandt, Skinner and Coleman, 1963) does not specifically mention the term crispness. The technique regarded crispness as part of the brittleness scale [now called fracturability scale (Szczesniak, 1975)] and defined crispness as the ease or force with which a food crumbles, cracks, or shatters. Stier (1970) defined crispness in potato chips as 'brittle, crushable, or friable as far as the initial sensation was concerned'.

Brennan, Jowitt and Williams (1974) gave the following instructions for the sensory evaluation of crispness in biscuits. 'Place a small piece of biscuit between the molars and bite down slowly and evenly until a sudden and continuous breakdown of the biscuit structure occurs. Assess the rate at which this breakdown into small fragments occurs, using as nearly as possible the same biting rate for each sample. (The most crisp biscuit breaks down the most rapidly.)'

Bourne (1975) described the characteristics of a crisp food as follows: 'Crisp or crunchy foods are characterized by having a rigid, non-deformable, stiff structure that suddenly collapses with a brittle fracture and a rapid decay of the force after fracture, very low shear strength, breaks up under simple compression between the teeth with little or no grinding or tearing, rapid breakdown into small pieces, small number of chews per piece, not chewy, low work content required for mastication, sound effects associated with brittle fracture often desirable, structure usually comprised of cellular aggregates.'

## An auditory hypothesis of crispness

When we began our study of crispness, we found these definitions and means of assessing sensory crispness to be varied and inadequate (Vickers and Bourne, 1976a). Our goal at that time was to identify the sensation(s) common to all crisp foods, and the outcome of those studies was the hypothesis that crispness was an auditory sensation (Vickers and Bourne, 1976b).

All crisp foods produce noises when they are bitten and chewed. These noises can be characterized as having irregular amplitude versus time pictures (*Figure 24.1*). We suggested that the repeated breaking or fracturing of food samples during biting and chewing produce these acoustical characteristics. Sagara, Saeki and Yamaguchi (1969) show similar amplitude–time plots for vibrations of crisp foods.

The crisp sounds showed broad frequency spectra with most of the sound energy

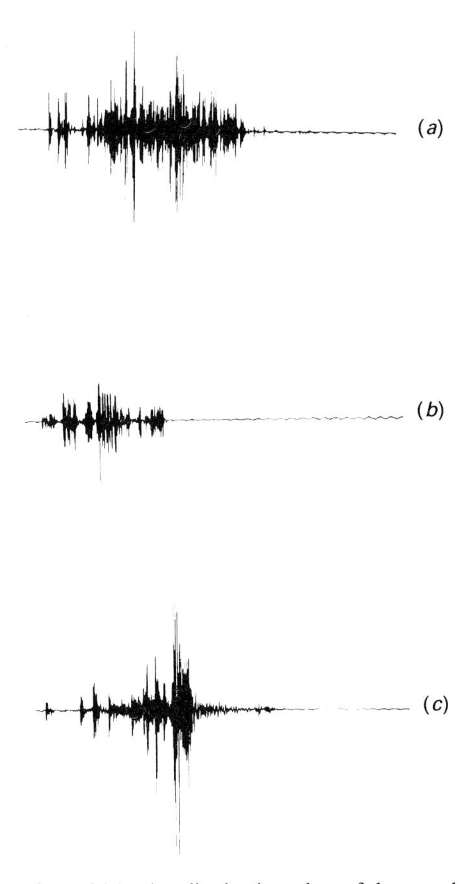


**Figure 24.1** Amplitude–time plots of the sound generated by the author biting: (a) Triscuit; (b) melba toast; and (c) gingersnap. The horizontal axis represents time; total time shown in each plot is a little more than 0.5 s. The vertical axis represents amplitude which is related to loudness.

falling into the 0–10 kHz range (*Figure 24.2*). Other than possessing a broad frequency range, there appeared to be no frequency characteristic unique to crisp sounds.

If crispness can be characterized by an irregular series of noises, what distinguishes a more crisp product from a less crisp one? The answer can best be seen by observing the differences in amplitude–time plots between the three green pepper samples shown in *Figure 24.3*. In the less crisp samples there is less total noise produced. On the other hand, louder sounds and/or a greater density of sound occurrences seem to characterize the crisper samples. From these observations, Vickers and Bourne (1976b) concluded that changes in either one or both of these two parameters, i.e. the number of sounds produced in a given biting distance and the loudness of the sounds produced, would probably cause a change in perceived crispness.

Since Vickers and Bourne first suggested that auditory sensations were involved in the perception of crispness, other investigators have contributed supporting evidence for that idea. Sherman and Deghaidy (1978) reported that their subjects used auditory information, particularly during the first bite, to determine crispness. Vickers and

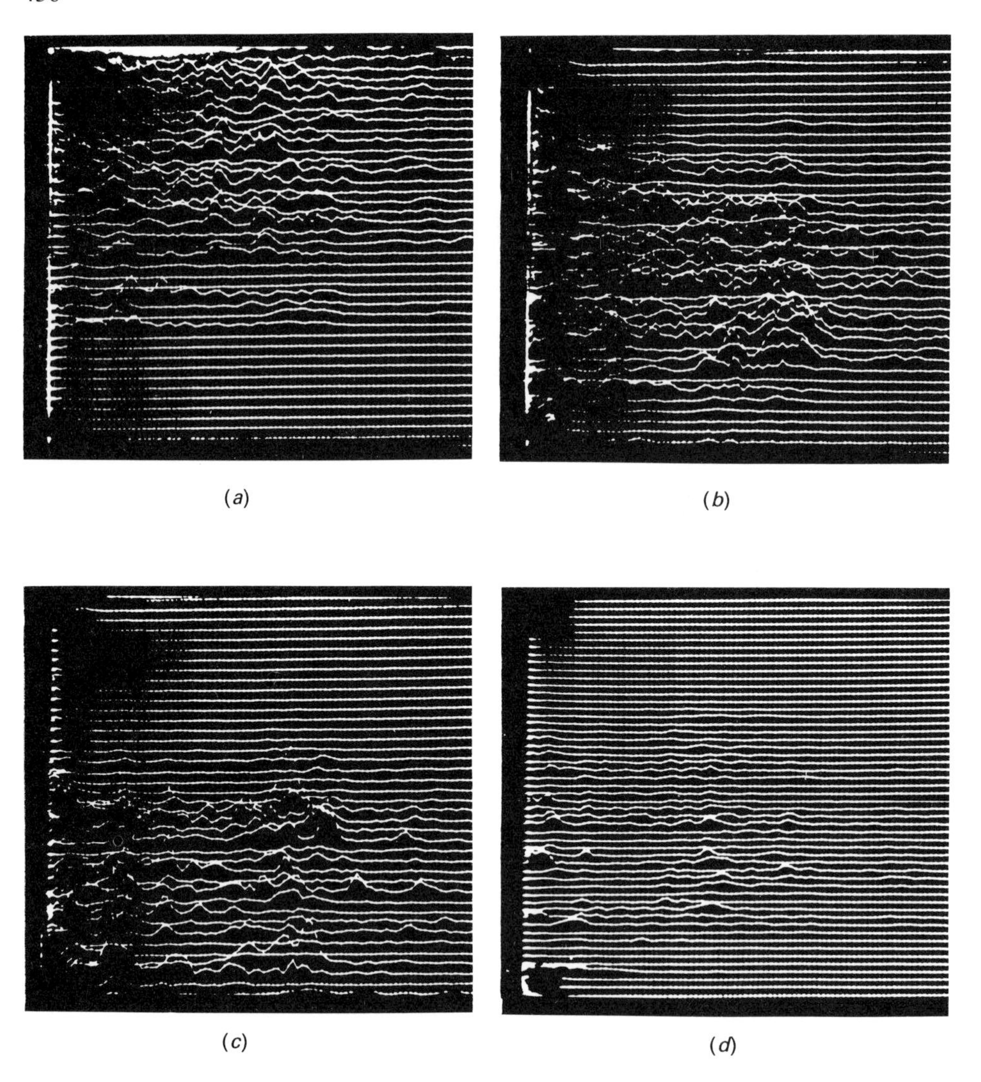


**Figure 24.2** Spectrum analyser displays of the sounds produced by the author biting: (a) Triscuit; (b) Pringles potato crisps; (c) fresh celery; and (d) a saltine $a_w = 0.27$. Each horizontal line represents a single 10 ms interval. For each 10 ms increment the amplitude of the sound (vertical axis) is plotted against the frequency (horizontal axis). Frequency range shown is from 0 to 10 kHz. Time proceeds from the lines at the bottom through to the lines at the top. Reprinted from Vickers and Bourne (1976b).

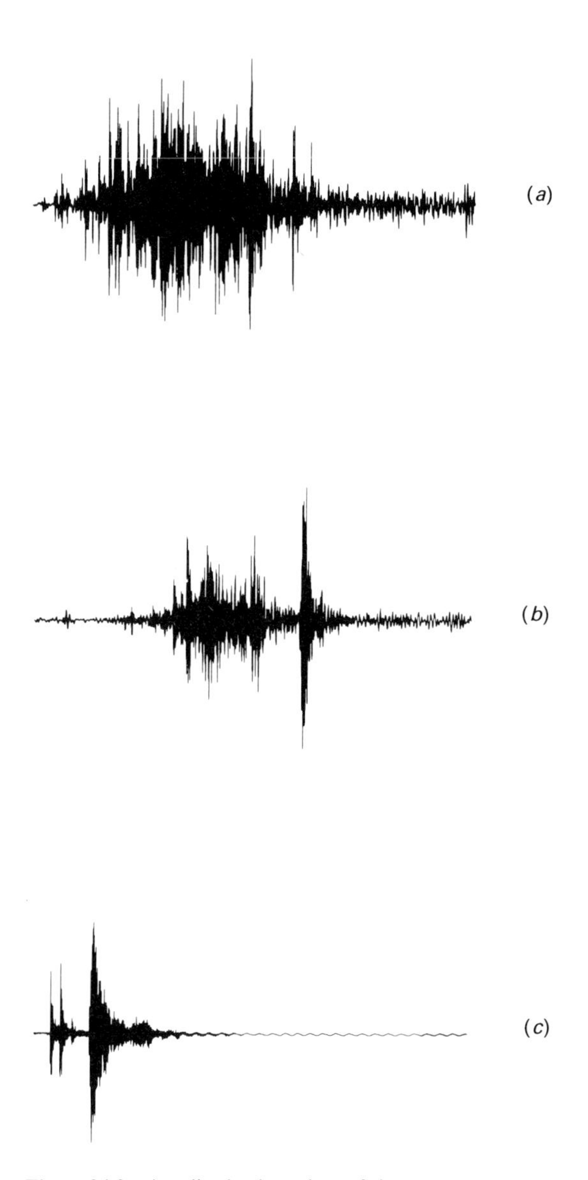


**Figure 24.3** Amplitude–time plots of the sound generated by the author biting: (a) water-soaked green pepper; (b) fresh green pepper; (c) blanched green pepper. Reprinted from Vickers and Bourne (1976b).

Wasserman (1980) demonstrated that subjects could successfully rate the sensation of crispness by listening to tape-recorded sounds of foods being crushed. Furthermore, there was a high positive correlation between the perceived loudness of the food crushing sounds and their perceived crispness. Stanley, Voisey and Swatland (1980), in a study of the texture–structure relationships in lean bacon, reported that their subjects evaluated crispness using the duration of the cracking sound during chewing. Mohamed, Jowitt and Brennan (1982) showed that the sensory judgements of the loudness of the sounds produced by biting a variety of biscuits had the highest correlation with sensory judgements of crispness. They found very poor correlations between sensory crispness and sensory hardness.

## Testing the hypothesis that crispness is a sound

Christensen and Vickers (1981) began to test the hypothesis that biting and/or chewing sounds were important for crispness. During one of their test sessions they had subjects evaluate separately the loudness and the crispness of 16 different food

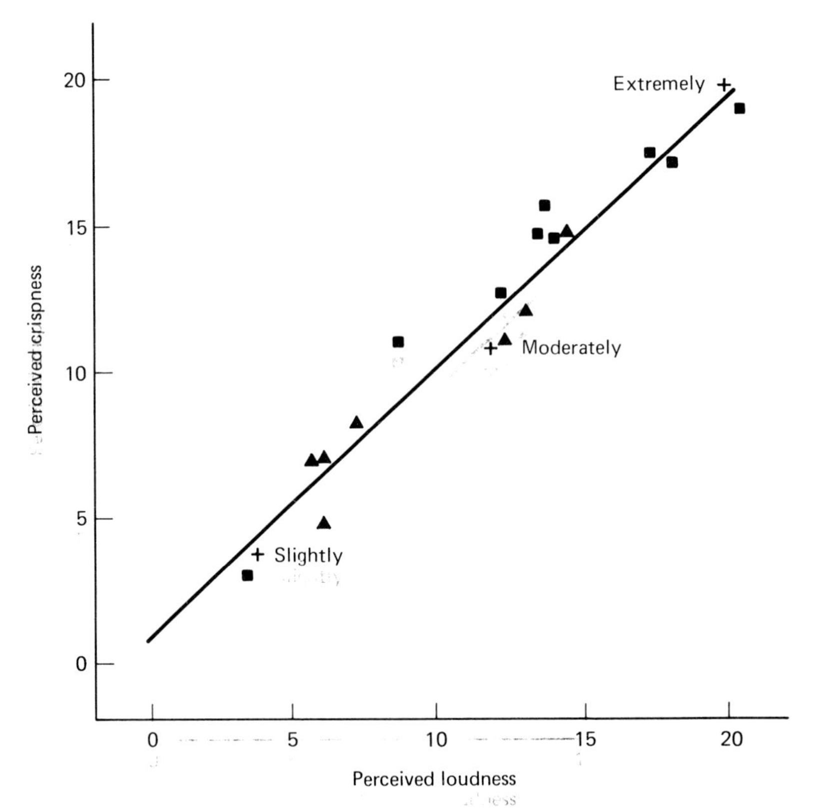


**Figure 24.4** Relationship between perceived loudness and crispness for wet (▲) and dry (■) food samples following biting and chewing each of the samples. Each point is the mean judgement for 20 subjects. The solid line is the function derived from the appropriate linear regression equation. Also included are the mean values assigned to the concept scale: slightly, moderately and extremely crisp or loud. Reprinted from Christensen and Vickers (1981)

samples by biting each food with the incisors and chewing twice with the molars. The judgements of crispness and loudness were highly correlated ($r=0.97$, *Figure 24.4*), and the slope of the regression equation was approximately 1.0, indicating that perceived loudness and crispness varied almost equally among the different food samples. Thus, the more crisp the food, the louder are the sounds. These results suggested, albeit indirectly, that biting and chewing sounds are an important cue for determining crispness.

A second experiment by Christensen and Vickers (1981) tested the hypothesis that auditory sensations are necessary for judgements and crispness. For this test, subjects judged the crispness of the same series of 16 foods while their biting and chewing sounds were blocked by a loud masking noise played through headphones. Subjects' judgements of crispness obtained during this auditory block session were compared with their judgements obtained under normal conditions. As shown in *Figure 24.5*, the subjects' judgements of crispness were largely unaffected by the presence or absence of an auditory block. The correlation between crispness judgements obtained

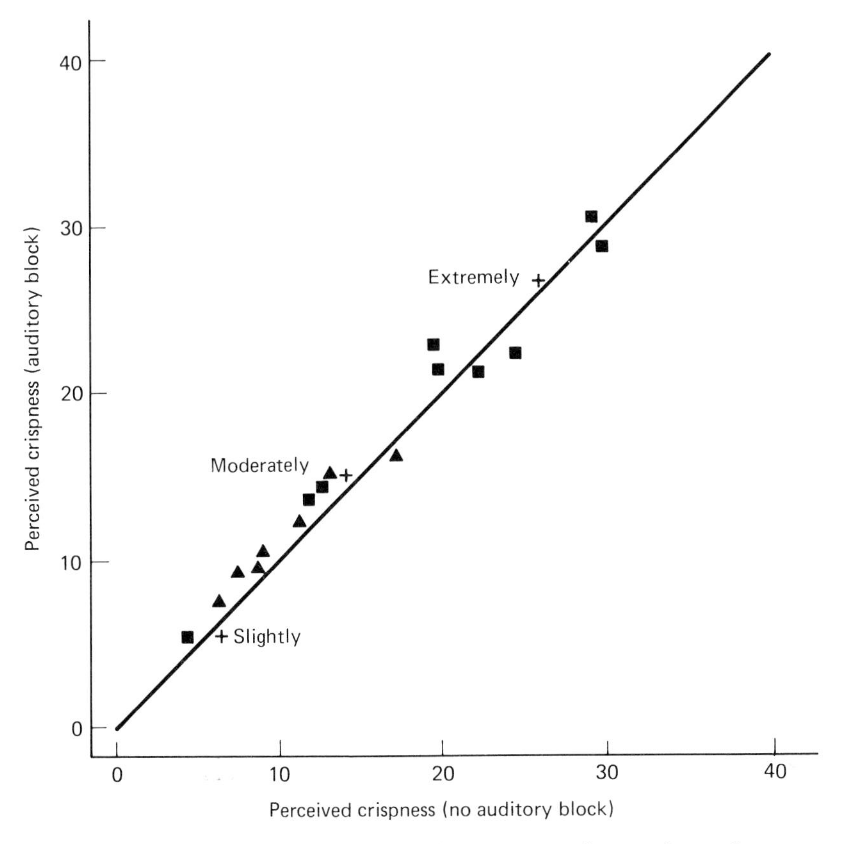


**Figure 24.5** Effect on perceived crispness of the presence or absence of an auditory block. Each point is the mean judgement of 16 subjects for each of the dry (■) and wet (▲) food samples. The solid black line represents the function X = Y. Also included are the mean values assigned to the concept scale: slightly, moderately and extremely crisp. Reprinted from Christensen and Vickers (1981)

during both conditions was high ($r = 0.99$). These data clearly showed that auditory cues are not essential for judgements of crispness.

In order to explain the data from both experiments, Christensen and Vickers suggested that a vibratory stimuli might form the basis for crispness determinations. Vibrations produced by biting and chewing a crisp food could be expected to produce both auditory and tactile sensations.

If the Vickers and Bourne (1976b) hypothesis was true, auditory sensations alone should be sufficient stimuli for making judgements of food crispness. Based on the Vickers and Bourne hypothesis and the findings of Christensen and Vickers, one might have predicted that auditory and non-auditory tactile crispness information are completely redundant, and, therefore, individuals can use either auditory or oral tactile cues to judge crispness. Vickers (1981) tested this prediction.

Subjects participating in Vickers' (1981) test were selected in the same manner and were served the same food samples as those in the Christensen and Vickers (1981) study. For this test, subjects judged the tape-recorded biting and chewing sounds of the 16 food samples for crispness, as well as evaluating the crispness of the foods by normal biting and chewing. The relationship between oral (normal biting and chewing) and auditory (sound only) judgements of crispness is shown in *Figure 24.6*.

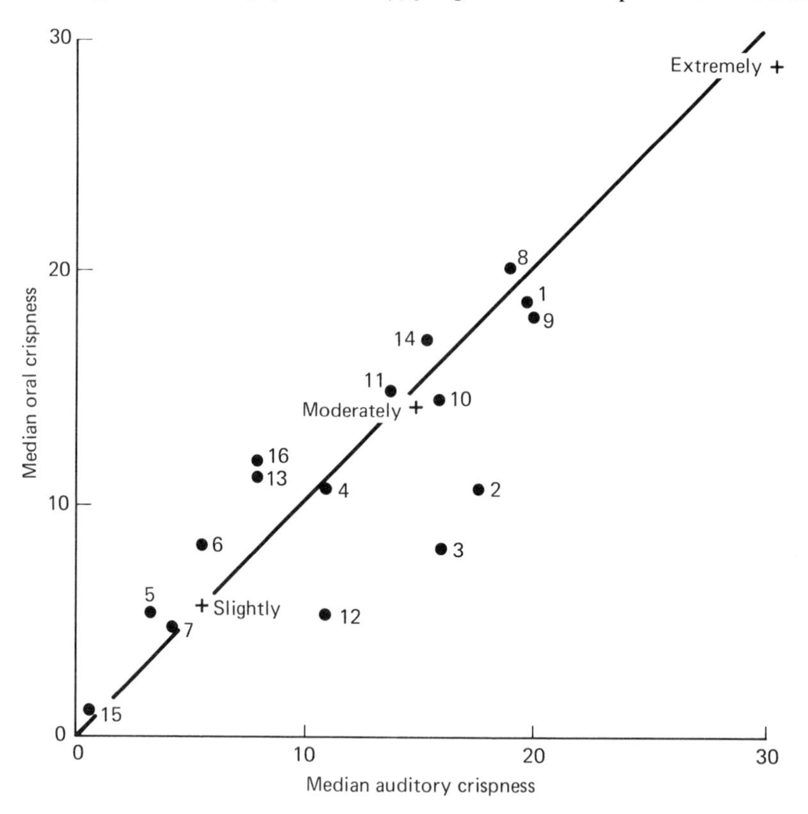


**Figure 24.6** Relationship between oral crispness and auditory crispness of a variety of wet crisp and dry crisp foods. Each point is the median judgement for 20 subjects. The solid black line represents the function X = Y. Also included are the median values assigned to the concept scale: slightly, moderately and extremely crisp. Reprinted from Vickers (1981).

The correlation coefficient for all data points on this plot is 0.82. Points 2, 3 and 12 on this plot were clearly outliers. If these points were omitted and the correlation coefficient recalculated, $r = 0.96$ (slope = 0.89). These outliers provided an interesting dilemma. They were judged more crisp when only the sound was evaluated than when the food was actually bitten and chewed. These three products (celery blanched 45 s, celery blanched 3 min and Triscuit $a_w = 0.75$) were familiar foods which had been treated to lower crispness. In the process of adjusting their crispness levels, several mechanical properties (e.g. rigidity, hardness, toughness, etc.) also changed. Whether the position of these three outliers should have been considered the result of an unfortunate confusion or the result of a valued part of the concept of crispness was unclear.

A similar test comparing crispness judgements was made on an assortment of 16 different crisp foods by Edmister and Vickers (1985). Each was tested at three different thicknesses and the biting and chewing sounds were recorded. A plot of their results is shown in *Figure 24.7*. The correlation coefficient for the 48 points on this plot is 0.65 (slope = 0.84). The scatter in *Figure 24.7* is much greater than that observed by Vickers (1981) and much of it cannot be explained by the above rationale for the outliers. These results suggested that judgements of auditory crispness are not completely redundant with oral crispness judgements. There are non-auditory stimuli that are sensed in the mouth that can influence people's perceptions of crispness.

More recently we have had people judge the crispness of a variety of potato crisps by normally biting and chewing them and also by listening to tape-recorded sounds of them being bitten and chewed (Vickers, 1987). Since the earlier studies showing that oral and auditory crispness were not highly correlated were based on studies of a

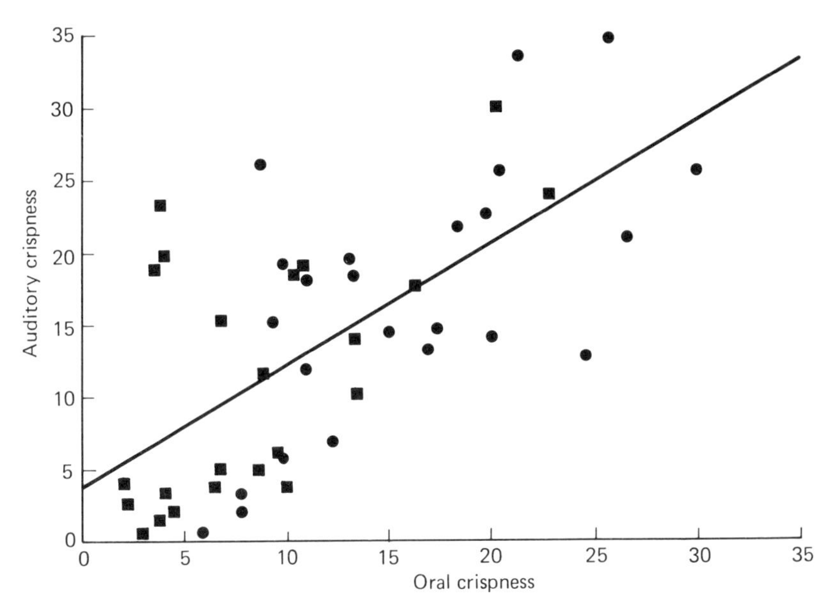


**Figure 24.7** The relationship between auditory and oral judgements of the crispness of 16 food samples, each at three different dimensions. Each point represents the mean normalized score for 20 subjects: ●, dry crisp foods; ■, wet crisp foods. The regression equation for this plot is: auditory crispness = 3.85 + 0.84 (oral crispness), $r = 0.65$. Reprinted from Edmister and Vickers (1985)

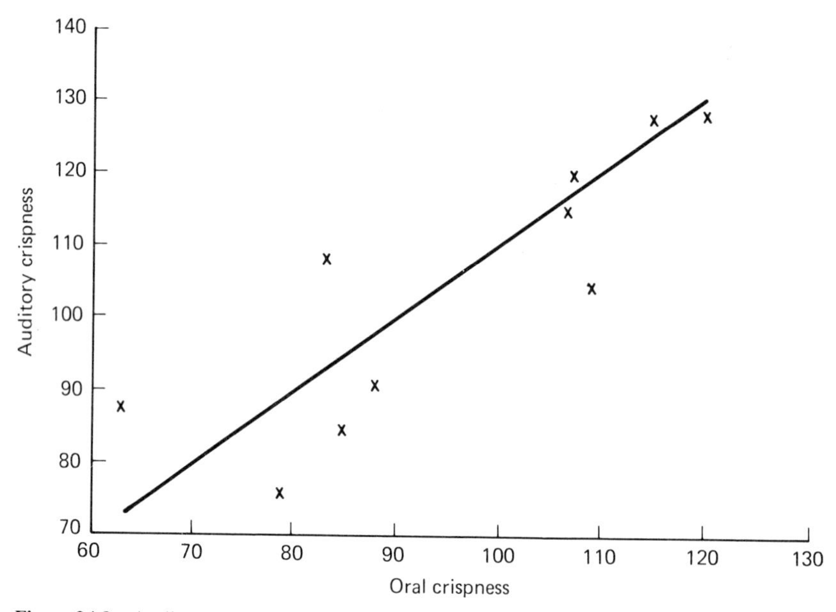


**Figure 24.8** Auditory versus oral crispness judgements of potato crisps. Each point is the mean judgement for 20 subjects. The solid black line is the regression line. From Vickers (1987)

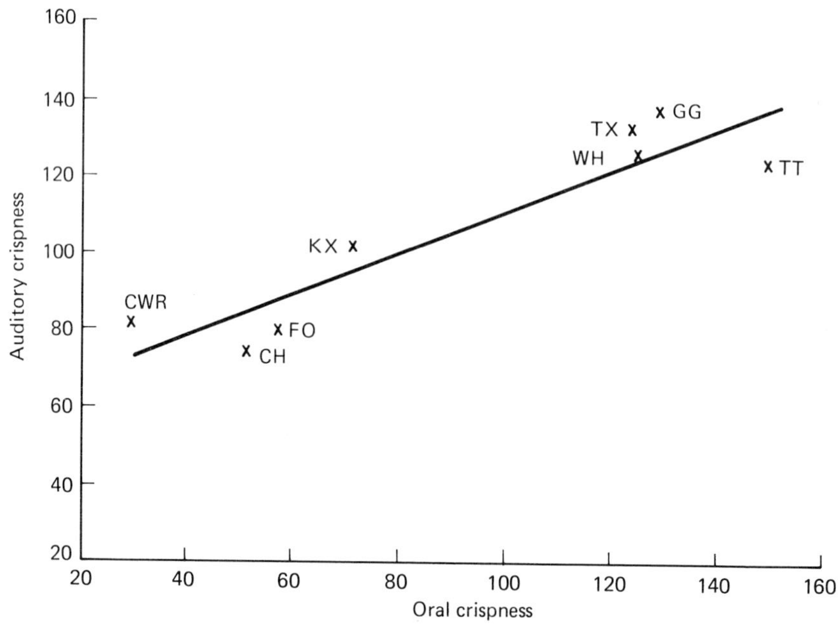


**Figure 24.9** The relationship between auditory and oral judgements of the crispness of eight ready-to-eat breakfast cereals. Each point represents the mean scores for 20 subjects. The line represents the following best fitting regression equation: auditory crispness = 57.3 + 0.54 (oral crispness), $r = 0.93$. (CWR: Crispy Wheats and Raisins, CH: Cheerios, FO: Fiber One, KX: Kix, WH: Wheaties, TX: Trix, GG: Golden Grahams, TT: Corn Total)

variety of food types, we thought that the correlation might be improved if only a single food type, potato crisps, was involved.

The auditory versus oral judgements from this study of potato crisps are shown in *Figure 24.8*. The correlation coefficient for this relationship is 0.85, in the same range as that observed by Vickers (1981). Thus studying a single food type did not improve the correlation. The auditory crispness judgements for potato crisps tended to be slightly higher than the oral judgements. This might have been due to a slightly different scale usage between the two parts of the test, or it may have been a real effect.

We have also examined both the oral and auditory crispness of eight ready-to-eat breakfast cereals. The relationship between the two types of judgements is shown in *Figure 24.9*. The correlation coefficient was 0.93, higher than observed in the studies cited above. This may mean that factors other than sound have less influence on the crispness of cereals than they do for other products. However, the correlation was not perfect so again other non-auditory factors are probably influencing crispness.

The tests of the hypothesis that crispness is an auditory sensation have shown clearly that the sensation of crispness has both auditory and non-auditory components. What specific acoustical stimuli and textural stimuli produce these sensations? Information about the stimuli have come from studies of the relationships between instrumental measurements and sensory judgements and from more general research on food crushing sounds.

## Acoustical stimuli and sensations

Drake (1963) was the first scientist to study food crushing sounds. Although he did not specifically study crispness, his data showed that sounds from crisp foods differ from those of non-crisp foods primarily in their amplitudes. Frequency or duration seemed to play a much less important role.

Sagara, Saeki and Yamaguchi (1969) used an adaptation of the General Foods Texturometer to study crispness in foods. By means of a piezoelectric element sandwiched into the plunger, vibrations produced by a food as it was crushed were sensed and analysed. The waveforms of the vibrations were rectified and integrated. Good correlations were found between this instrumental measurement and sensory evaluations of crispness in a traditional Japanese pickled vegetable.

Kapur (1971) recorded bone-conducted sounds by placing a very fine needle directly on human bone surfaces. He found the maximum resonance frequency of the jaw and skull bones to be around 160 Hz. Less than 50% of the noise level recorded at the mandible reached the mastoid process, which is quite close to the ear canal and would reflect the intensity of the sound actually heard. He found that the sounds produced by chewing a crisp vanilla wafer were markedly louder than those produced by chewing a soggy wafer.

Vickers and Bourne (1976b) studied some of the acoustical properties of tape-recorded biting sounds of wet and dry crisp foods. They suggested that the amplitude–time plots of more crisp sounds could be distinguished from those of less crisp sounds by the amplitude of the sounds and the number of sounds produced in a given bite distance (*Figure 24.3*).

Mohamed, Jowitt and Brennan (1982) recorded sounds produced by five varieties of dry crisp foods stored at five different relative humidities as they were fractured in a constant loading rate texture testing instrument. The equivalent sound level (average sound energy) correlated significantly ($r=0.71$) with sensory crispness. Edmister and Vickers (1985) evaluated several instrumental acoustical parameters for their ability to predict oral or auditory sensory crispness. Of the instrumental measurements they studied, the log of the product of the number of sound occurrences and their amplitude had the highest correlation ($r=0.89$) with auditory crispness judgements (*Figure 24.10*). This observation agreed with Vickers and Bourne's hypothesis that crispness was related to the number of sound occurrences times the loudness of the sounds. Vickers and Bourne had also specified that these two parameters be expressed in terms of biting distance. However, Edmister and Vickers' efforts to adjust for biting distance by dividing by the thickness of the food or the duration of the sound (neither of which were actually measuring biting distance) did not improve the relationship between auditory crispness and the instrumental parameters.

Seymour (1985) found that a variety of measures of sound pressure and sound intensity had large correlations with sensory crispness of several dry crisp foods.

In a recent study of the crispness of eight ready-to-eat breakfast cereals we found the best acoustical measure of crispness to be the amplitude of the sounds produced when the cereals were bitten ($r=0.57$). On the other hand Vickers (1987) found that the number of sounds produced during biting was the single best predictor of crispness in potato crisps ($r=0.92$). The amplitude of the potato crisp sounds was much less useful than observed in other studies.

Although the pitch of sounds is very important in other fields such as music and speech, its role as a stimulus for crispness is much less well defined. The few studies that have examined the relationship between crispness and the pitch of a sound have not found them to be highly related (Vickers and Wasserman, 1980; Vickers, 1983).

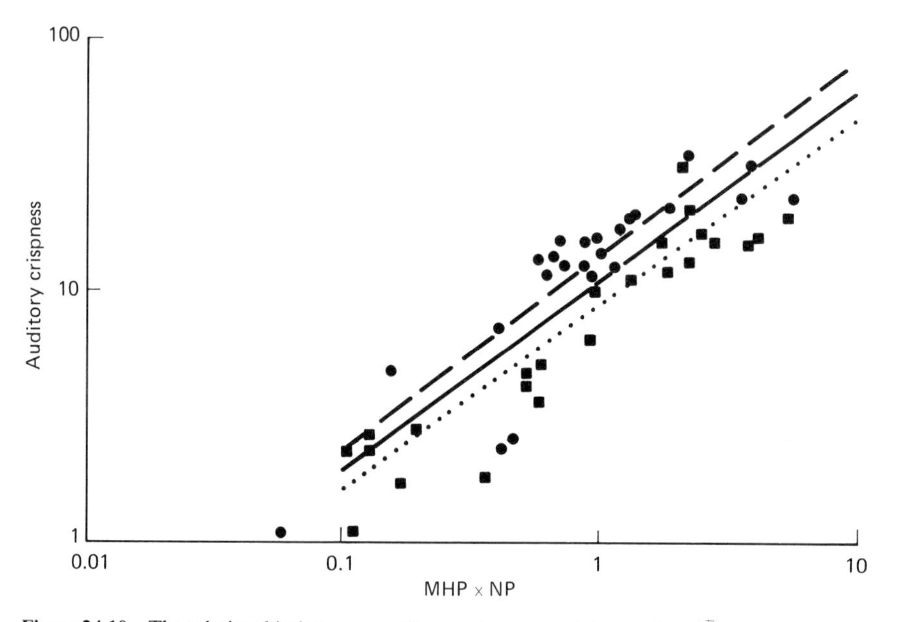


**Figure 24.10** The relationship between auditory crispness and the product of mean height of peaks (MHP) and number of peaks (NP) is shown in log–log coordinates. The line represents the linear regression equation for the 48 sounds (16 foods × 3 dimensions each). ●, dry crisp foods; ■, wet crisp foods; ———, wet and dry crisp foods combined; – – –, dry crisp foods only; ......., wet crisp foods only.

However, Vickers (1984, 1985) did find pitch to be useful for distinguishing between the sensations of crispness and crunchiness; crisp sounds tended to be higher in pitch than crunchy sounds.

Observations and results from all of the above studies support the thesis that the total amount of sound produced during biting and/or chewing is very important for crispness. There are probably variations among and within food products in the features of the sounds that make the greatest contribution to crispness.

## Non-auditory sensations that affect crispness

Prior to 1975 crispness was viewed primarily as a non-auditory textural sensation. Although definitions of crispness proposed at the time are typically inadequate for defining crispness, they are very useful when used to account for the non-auditory components of the sensation. One quality that emerges from nearly all of the early definitions is 'fracturability' or the tendency of crisp foods to break into many pieces. For example:

'ease or force with which a food crumbles, cracks or shatters' (Brandt, Skinner and Coleman, 1963—original texture profile)
'rate at which this breakdown into small fragments occurs' (Brennan, Jowitt and Williams, 1974)
'brittle, crushable, or friable' (Stier, 1970).

A second quality that has been used is biting or chewing force. Force has been incorporated into the texture profile definition of crispness (above) and has been mentioned by others as useful for judging or measuring crispness (Jeon, Breene and Munson, 1975; Seymour, 1985; Vickers and Christensen, 1980). However, the rôle of biting or chewing force in the sensation of crispness is not clear because others (Mohamed, Jowitt and Brennan, 1981; Seymour 1985; Vickers, 1987) have found force to be unrelated or inversely related to sensory crispness. Possibly the rôle of force is different in different foods.

A third non-auditory quality (which may be equivalent to fracturability) that appears to be closely related to crispness in dry crisp foods is compressive brittleness. Jowitt and Mohamed (1980) suggested that the terms 'crispness' and 'compressive brittleness' are equivalent for dry crisp foods. Using a compression test cell in a constant loading rate machine, Mohamed, Jowitt and Brennan (1982) showed that the ratio of work done during fracture to total work (a measure of compressive brittleness) correlated highly with sensory crispness ($r = 0.88$).

## Combination studies

If the sensation of crispness is produced by both auditory and non-auditory stimuli, studies examining both types of stimuli and their contributions to the crispness sensation should be most useful. A few such studies have been conducted.

Seymour (1985) used multiple linear regression to select combinations of two instrumental measurements that would best correlate with the sensory variable crispness. All of his combinations included one mechanical and one acoustical parameter. *Table 24.1* shows his best regression equations for four different products. In all cases except Crunch Twist, crispness was positively related to an acoustical

**Table 24.1** REGRESSION EQUATIONS FOR SENSORY CRISPNESS BASED ON THE BEST TWO INSTRUMENTAL PARAMETERS. FROM SEYMOUR (1985)

| *Product* | *Regression equation* | $r^2$ |
|---|---|---|
| Pringles Potato Crisps | Crisp = 13.6—0.19 Work + 0.03 MP4 | 0.96 |
| Crunch Twist | Crisp = 16.5—0.06 Force—0.11 SPL1 | 0.95 |
| O'Gradys Potato Crisps | Crisp = 6.1—0.08 Force + 0.21 SPL3 | 0.88 |
| Rippled Pringles Potato Crisps | Crisp = 9.1—0.004 Work + 0.003 ILT | 0.89 |

Force = Maximum force at failure in Newtons (stack of two potato crisps crushed in an Instron mounted Kramer shear cell).
Work = Work done to 1 cm deformation (mJ) for Pringles potato crisps.
Work = Work done to failure (mJ) for Rippled Pringles potato crisps.
MP4 = Mean sound pressure ($N\ m^{-2}$) in quadrant 4.
SPL1 = Mean sound pressure level (dBA) in quadrant 1.
SPL3 = Mean sound pressure level (dBA) in quadrant 3.
ILT = Acoustic intensity (watts $m^{-2}$) in 0.5–3.3 kHz.

parameter (one representing loudness) and inversely related to a force–deformation parameter.

This inverse relationship of crispness to force–deformation parameters and positive relation to acoustical parameters is also shown by Vickers (1987) in her study of potato crisp crispness. Her best regression equation was:

$$\text{Crispness} = -15.6 + 5.35\ \text{NP} + 133\ \text{MHP} - 6.21\ \text{Peak}$$

where NP = the number of sounds, MHP = mean height of peaks (both taken from oscilloscope displays of the bite sounds) and Peak = peak of force–deformation curve from a bite test cell. The correlation coefficient for this equation was 0.99.

In another recent study we have examined both the oral and auditory crispness of eight ready-to-eat breakfast cereals. The best combination of instrumental measurements from that study is described by the following equation:

$$\text{Crispness} = 538 + 539(\log \text{MHP}) - 222(\text{Peak force})$$

where MHP and Peak are as described above. The correlation coefficient for the above equation is $r = 0.86$. This was a considerable improvement over the best acoustical predictor alone (MHP, $r = 0.57$) or the best force–deformation predictor alone (slope, $r = -0.33$), but is not nearly as high as the correlations obtained by Seymour (1985) and Vickers (1987). Again we see that the acoustical measurement is positively related while the force–deformation parameters are inversely related to crispness.

Since the peak force measured in a bite test cell or the force or work measured in the Kramer shear test cell may indicate hardness or toughness, the negative relation of these parameters to crispness may mean that such hardness or toughness counteracts or suppresses the auditory sensations of crispness. Sensory measures of hardness and toughness support this idea. Seymour (1985) measured the sensory hardness of the products used in his test and found that hardness and crispness were inversely related. In our study of the crispness of the eight breakfast cereals we found that sensory crispness and sensory toughness were inversely related ($r = 0.87$). The idea that other textural sensations could diminish or enhance crispness is in line with observations in other senses such as taste or smell that show counteraction or synergism among different odorants and tastants.

One should note that the three studies discussed above deal only with a limited number of dry crisp foods. The relations between such measurements of hardness and toughness and the crispness of wet crisp foods have not been studied and may be very different from those of dry crisp foods.

Different crisp foods differ in their structural and mechanical properties; therefore, the exact mechanism by which their vibrations are generated may differ. The accompanying oral tactile sensations that enhance or counteract crispness will also vary from one food to another. More detailed analyses of the crispness of specific foods should provide a better understanding of the attribute crispness and its interactions with other textural attributes.

## Conclusion

Collectively, it appears that crispness can be determined largely, but not entirely, by biting or chewing sounds alone. Some additional information is provided by non-auditory cues. The sensations of toughness and hardness appear to counteract crispness in the dry crisp foods studied above. There are probably other non-auditory attributes that influence crispness judgements. Further studies are needed to identify these attributes and to determine the extent to which each of the attributes influences the crispness of different food products.

## References

BOURNE, M.C. (1975). Texture properties and evaluations of fabricated foods. In *Fabricated Foods*, (Inglett, G.E., Ed.), pp. 127–158. Westport, Connecticut, AVI Publishers

BRANDT, M.A., SKINNER, E.Z. AND COLEMAN, J.A. (1963). Texture profile method. *Journal of Food Science*, **28**, 404–410.

BRENNAN, J.G., JOWITT, R. AND WILLIAMS. A. (1974). Sensory and instrumental measurement of 'brittleness' and 'crispness' in biscuits. Presented at the Fourth International Congress of Food Science and Technology, Madrid, Spain, September 22–27.

CHRISTENSEN, C.M. AND VICKERS, Z.M. (1981). Relationships of chewing sounds to judgements of food crispness. *Journal of Food Science*, **45**, 574–578

DRAKE, B. (1963). Food crushing sounds. An introductory study. *Journal of Food Science*, **28**, 233–241

EDMISTER, J.A. AND VICKERS, Z.M. (1985). Instrumental acoustical measures of crispness in foods. *Journal of Texture Studies*, **16**, 153–167

ILES, B.C. AND ELSON, C.R. (1972). Crispness. BFMIRA Research Report No. 190

JEON, I.J., BREENE, W.M. AND MUNSON, S.T. (1975). Texture of fresh-pack whole cucumber pickles: correlation of instrumental and sensory measurements. *Journal of Texture Studies*, **5**, 399-410

JOWITT, R. AND MOHAMED, A.A.A. (1980). An improved instrument for studying crispness in foods. In *Food Process Engineering*, Volume 1, (Linko, P., Malkki, Y., Olkku, J. and Harinkari, J., Eds), pp. 292–300, London Applied Science

KAPUR, K. (1971). Frequency spectrographic analysis of bone conducted chewing sounds in persons with natural and artificial dentitions. *Journal of Texture Studies*, **2**, 50–61

MOHAMED, A.A.A., JOWITT, R. AND BRENNAN, J.G. (1982). Instrumental and sensory evaluation of crispness: I. In friable foods. *Journal of Food Engineering*, **1**, 55–75

SAGARA, T., SAEKI, R. AND YAMAGUCHI, M. (1969). Measurement and record of crispness of pickled vegetables. *Journal of Food Science and Technology (Japan)*, **16**, 350

SEYMOUR, S.K. (1985). Studies on the relationships between the mechanical, acoustical and sensory properties in low moisture food products. PhD Thesis. North Carolina State University

SHERMAN, P. AND DEGHAIDY, F.S. (1978). Force–deformation conditions associated with the evaluation of brittleness and crispness in selected foods. *Journal of Texture Studies*, **9**, 437–459

STANLEY, D.W., VOISEY, P.W. AND SWATLAND, H.J. (1980). Texture–structure relationships in bacon lean. *Journal of Texture Studies*, **11**, 217–238

STIER, E. (1970). Chew a chip and tell. *Food Technology*, **24**, 46

SZCZESNIAK, A.S. (1971). Consumer awareness of texture and of other food attributes, II. *Journal of Texture Studies*, **2**, 196–206

SZCZESNIAK, A.S. (1975). General Foods Texture Profile revisited—ten years perspective. *Journal of Texture Studies*, **6**, 5–17

SZCZESNIAK, A.S. AND KAHN, E.L. (1971). Consumer awareness of and attitudes to food texture. I: Adults. *Journal of Texture Studies*, **2**, 280–295

SZCZESNIAK, A.S. AND KAHN, E.L. (1984). Texture contrasts and combinations: a valued consumer attribute. *Journal of Texture Studies*, **15**, 285–301

SZCZESNIAK, A.S. AND KLEYN, D.H. (1963). Consumer awareness of texture and other food attributes. *Food Technology*, **17**, 4–77

VICKERS, Z.M. (1981). Relationships of chewing sounds to judgements of crispness, crunchiness and hardness. *Journal of Food Science*, **47**, 121–124

VICKERS, Z.M. (1983). Pleasantness of food sounds. *Journal of Food Science*, **48**, 783–786

VICKERS, Z.M. (1984). Crispness and crunchiness—a difference in pitch? *Journal of Texture Studies*, **15**, 157–163

VICKERS, Z.M. (1985). The relationship of pitch, loudness and eating technique to judgements of the crispness and crunchiness of food sounds. *Journal of Texture Studies*, **16**, 85–95

VICKERS, Z.M. (1987). Sensory acoustical and force–deformation measurement of potato chip crispness. *Journal of Food Science*, **52**, 138–140

VICKERS, Z.M. AND BOURNE, M.C. (1976a). Crispness in foods—a review. *Journal of Food Science*, **41**, 1153–1157

VICKERS, Z.M. AND BOURNE, M.C. (1976b). A psychoacoustical theory of crispness. *Journal of Food Science*, **41**, 1158–1164

VICKERS, Z.M. AND CHRISTENSEN, C.M. (1980). Relationships between sensory crispness and other sensory and instrumental parameters. *Journal of Texture Studies*, **11**, 291–307

VICKERS, Z.M. AND WASSERMAN, S.S. (1980). Sensory qualities of food sounds based on individual perceptions. *Journal of Texture Studies*, **10**, 319–332

YOSHIKAWA, S., NISHIMURA, S., TASHIRO, T. AND YOSHIDA, M. (1970). Collection and classification of words for description of food texture. I. Collection of words. *Journal of Texture Studies*, **1**, 437–442

25

# BEYOND THE TEXTURE PROFILE

E. LARMOND
*Agriculture Canada, Ottawa, Canada*

## Introduction

The importance of food texture is well recognized. This symposium itself provides ample evidence. However, texture has not always been considered important. In fact, many consumers are largely unaware of the texture of food, except when it doesn't live up to their expectations! Texture is taken for granted and will scarcely be noticed as long as it is the way people have learned to expect it. Flavour overshadows texture at the conscious level (Szczesniak and Kahn, 1971).

However, there is an underlying familiarity with texture. Texture plays an important role in the enjoyment of eating. Not only do consumers expect certain textures in foods that are in optimal condition, they unconsciously seek out texture contrasts which will contribute interest, variety and additional pleasure to eating. Szczesniak and Kahn (1984) reported that consumers enjoy texture contrasts within a meal, on the plate, within a multiphase food (e.g. cream-filled cookies) and within a single-phase food during consumption (e.g. chocolate changing phases in the mouth).

Schiffman (1986) has recently observed that overweight persons tend to have an excessive need for variety and intensity of texture, taste and odour. She cites the theory that each person has a setpoint that acts as a thermostat for flavour and texture. The setpoint for an overweight person is too high. Overweight subjects require more textures than thin subjects. Textures especially desired include crumbly, brittle, gummy, gooey, oily, greasy and creamy.

## Definition of texture

The International Organization for Standardization defines texture as 'all of the rheological and structural (geometric and surface) attributes of a food product perceptible by means of mechanical, tactile and where appropriate, visual and auditory receptors' (ISO, 1981). An examination of this definition reveals three important characteristics. Texture is physical in nature, quite unlike flavour which is recognized as being chemical in nature. It is also evident that texture is really a group of several properties rather than a single property. We describe foods as being soft, firm, crumbly, crunchy, crisp, brittle, tender, chewy, tough, thick, thin, sticky, tacky, gooey, oily, greasy, airy, chalky, flaky, fluffy, grainy, gritty, lumpy, pulpy, etc. The

ISO definition also indicates that texture is a sensory property, i.e. perceived by sensory receptors.

The definition of texture proposed by Szczesniak in 1963 encompassed the same three features. 'Texture is a composite of those properties which arise from the structural elements of food and the manner in which it registers with the physiological senses.'

One textural attribute is usually of prime importance in a particular food. We are concerned with the tenderness of meat, the flakiness of pastry, the smoothness of sauces, the crispness of crackers. Definitions of texture are often incomplete, being limited to the aspects of texture important in the product being studied (Harries, Rhodes and Chrystall, 1972). An important feature of food texture is that the food must be actively manipulated and masticated so that texture can be exposed and perceived (Jowitt, 1974). The terms 'kinesthesis' and 'rheology' are sometimes used instead of 'texture'. Texture is the accepted term for 'that one of the three primary sensory properties of foods which relates entirely to the sense of touch or feel, (Kramer, 1972). Kinesthesis is the corresponding psychological term and rheology is the physical term.

## Sensory evaluation

Since texture is a sensory property the ultimate way to measure it is by sensory evaluation which the Sensory Evaluation Division of the Institute of Food Technologists (1975) defines as 'a scientific discipline used to evoke, measure, analyze and interpret reactions to those characteristics of food and materials as they are perceived by the senses of sight, smell, taste, touch and hearing'.

Sensory evaluation has been criticized by many. The comments by Love (1983) are perhaps typical: 'Assessment (of texture) by means of a panel of trained tasters is expensive and less reliable than assessing flavours. Results from one laboratory are also not comparable with those of another using different personnel, and the values for texture given by the panel are not related mathematically (a score of four is not twice that of two except by chance).' Sensory panel tests are time consuming and the results are often difficult to analyse and interpret (Kiosseoglou and Sherman, 1983).

Many sensory tests deserve these criticisms but not because of inherent weaknesses, rather because of misuse of sensory procedures. Pangborn (1980) has listed the most common errors in using sensory evaluation as: incorrect selection and/or use of methods, poorly defined objectives, confounding or combining methods in an attempt to short-cut proper procedures, and failure to distinguish analytical sensory analysis from consumer-type tests.

Over the past several years a better understanding of the measurement of human behaviour has been gained and a more systematic and professional approach to sensory testing has been developed, so that results of sensory tests can be reliable and valid (Stone and Sidel, 1985).

When testing conditions are controlled so as to minimize or counterbalance extraneous factors that affect judgement, when the appropriate method is selected considering the objective of the study and when discriminating panelists are selected and trained to disregard personal likes and dislikes, sensory evaluations can be objective (Larmond, 1986).

## The texture profile

The General Foods texture profile method, defined as 'the organoleptic analysis of the texture complex of a food in terms of its mechanical, geometrical, fat and moisture characteristics, the degree of each present, and the order in which they appear from first bite through complete mastication', was published in 1963 (Brandt, Skinner and Coleman, 1963). The fundamental principles upon which it is based are that texture is composed of a number of different properties and that the sensory evaluation of texture is a dynamic process (Szczesniak, 1975). It is dependent on a classification system which groups textural characteristics into three main classes: mechanical, geometrical and others, referring mainly to moisture and fat content of the food (Szczesniak, 1963). Rating scales with reference materials to illustrate each point on the scales were developed for all the mechanical parameters (Szczesniak, Brandt and Friedman, 1963).

The panelists use standard scales to rate the intensity of each characteristic during training sessions and category scales in actual evaluations (Civille and Szczesniak, 1973). The original scales were intended to be expanded at any portion of particular interest. Panelists develop a texture profile ballot for the particular product during training sessions and during actual evaluations they work independently to rate the intensity of each characteristic. Their scores are then compared and a consensus score is arrived at following discussion, if necessary.

Since 1963, several modifications of the texture profile method have evolved. Major modifications to each component of the method will be described.

## Classification of terms

Any attempt to describe the texture of foods with commonly used terms will probably result in 40–50 texture terms. However, the terms will be imprecise, there will be confusion between them and overlap of definitions (Wedzicha, 1979). The usefulness of descriptive methods would increase substantially if the terms used were more accurately defined (Williams, 1975). Peleg (1983) has pointed out that words depend on their use for meaning and carry many different connotations. A word may have several definitions as well as a multitude of connotations.

In developing the classification system upon which the texture profile is based, Szczesniak (1963) started with a compilation of terms used in popular texture terminology, analysis of their meanings and definitions of rheological concepts. She was attempting to develop a classification system which would relate popular terms to fundamental rheological principles. Textural characteristics were grouped into three classes: mechanical characteristics, which are the ways that foods react to stress; geometrical characteristics, which are the arrangement of the constituents of food; and other characteristics, which are the mouthfeel qualities related to the perception of moisture and fat in food (*Table 25.1*). The classification was intended for use with both instrumental and sensory measurements of texture.

Sherman (1969) suggested modifications to this classification system, placing it on a more basic rheological foundation (*Figure 25.1*).

In an effort to classify texture terms, Yoshikawa *et al.* (1970a), using a controlled word association technique, collected words used to describe texture. These words were then used to construct texture profiles for 79 foods (Yoshikawa *et al.*, 1970b). Multivariate analysis of the texture profiles yielded seven orthogonal factors (Yoshik-

**Table 25.1** CLASSIFICATION OF TEXTURAL CHARACTERISTICS. REPRINTED FROM SZCZESNIAK (1963) WITH PERMISSION OF THE PUBLISHERS.

| *Primary parameters* | *Secondary parameters* | *Popular terms* |
|---|---|---|
| *Mechanical characteristics* | | |
| Hardness | | Soft, Firm, Hard |
| Cohesiveness | Brittleness | Crumbly, Crunchy, Brittle |
| | Chewiness | Tender, Chewy, Tough |
| | Gumminess | Short, Mealy, Pasty, Gummy |
| Viscosity | | Thin, Viscous |
| Springiness | | Plastic, Elastic |
| Adhesiveness | | Sticky, Tacky, Gooey |
| *Geometrical characteristics* | | |
| Particle size and shape | | Gritty, Grainy, Coarse, etc. |
| Particle shape and orientation | | Fibrous, Cellular, Crystalline, etc. |
| *Other characteristics* | | |
| Moisture content | | Dry, Moist, Wet, Watery |
| Fat content | Oiliness | Oily |
| | Greasiness | Greasy |

awa *et al.*, 1970c). The most important dimensions were: hard–soft, cold–warm, oily–juicy, elastic–flaky, heavy, viscous, smooth.

Jowitt (1974) proposed a systematic and comprehensive glossary of food texture terms. He grouped terms into three classes closely resembling those of Szczesniak (1963): 25 terms relating to the behaviour of the material under stress or strain; 20 terms relating to the structure of the material further divided into those terms relating to particle size or shape and those relating to shape and arrangement of structural elements; and 18 terms relating to mouthfeel characteristics. Definitions were also proposed for each term.

Szczesniak (1979) developed a classification of mouthfeel characteristics of beverages. There were 11 categories in this classification with viscosity listed as the first and most important sensation.

In spite of these and other attempts to develop a classification system, no single scheme has been adopted internationally. Little progress can be made in cooperative research in texture evaluation without standardization in textural terminology (Brennan and Jowitt, 1977).

## Reference scales

Standard texture rating scales are an important component of the texture profile method (Szczesniak, Brandt and Friedman, 1963). Standard rating scales with reference materials to illustrate the points on the scale were developed for all the mechanical characteristics. Familiar foods which exhibit a given characteristic as their most outstanding textural characteristic were selected. The scales could be expanded in any portion to allow for more precise ratings of similar products.

Although the standard rating scales have often been used exactly as originally described, they have at times been modified according to availability of products, and the range of textural characteristics being evaluated. Bourne *et al.* (1975), working in Columbia, developed standard scales for the mechanical characteristics using products readily available in that country. Vaisey and Shaykewich (1971) developed a 4-

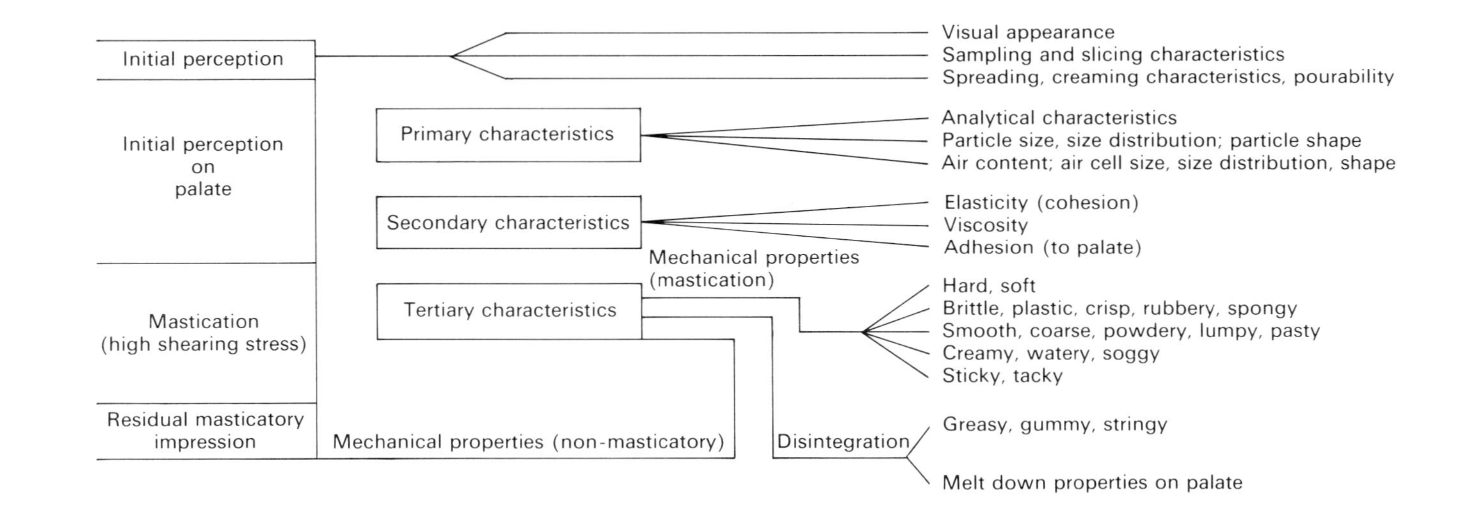


**Figure 25.1** Classification of textural characteristics. Reprinted from Sherman (1969) with permission of the publishers.

point hardness scale to assess the texture of fish. Cream cheese was the first point and gelatin gels of different strengths were used to illustrate points 2, 3 and 4. A 6-point scale for mealiness (a popular term meaning low degree of gumminess) was developed with hard cooked egg yolk as point 1 and undiluted condensed pea soup as point 6.

In the inaugural issue of the *Journal of Sensory Studies*, Munoz (1986) reported modifications to the standard scales of hardness, adhesiveness, fracturability, cohesiveness and denseness (gumminess). A texture profile panel selected new reference materials which better illustrated the textural characteristics and intensities or replace recommended food products which are no longer available. The panel also developed standard scales for other textural characteristics related to surface properties and to the behaviour of food material during mastication. These were wetness, adhesiveness to lips, roughness, self-adhesiveness, springiness, cohesiveness of mass, moisture absorption, adhesiveness to teeth and manual adhesiveness. Throughout this work the panelists used a 15 cm line to rate intensity and scale values reported represent the position on a 15 cm line scale (*Table 25.2*). Equidistance between points in the scale is not assumed.

**Table 25.2** FRACTURABILITY SCALE

| *Data of Szczesniak* et al. *(1963)*[a] | | *Data of Munoz (1986)*[a] | |
|---|---|---|---|
| *Rating* | *Product* | *Scale value* | *Product* |
| 1 | Corn muffin | 1.0 | Corn muffin |
| 2 | Angel puffs | 2.5 | Egg jumbo |
| 3 | Graham crackers | 4.5 | Graham cracker |
| 4 | Melba toast | 7.0 | Melba toast |
| 5 | Jan Hazel Cookies | 8.0 | Ginger snap |
| 6 | Ginger snaps | 10.0 | Thin bread wafer |
| 7 | Peanut brittle | 13.0 | Peanut brittle |
| | | 14.5 | Hard candy |

[a] Reproduced with permission of the publishers, Institute of Food Technologists and Food and Nutrition Press respectively.

## Intensity rating

In the original description of the texture profile method (Brandt, Skinner and Coleman, 1963) there is little mention of how the intensity of the characteristic was scaled. In his book Bourne (1982) shows a basic texture profile sheet obtained from Szczesniak which indicates that hardness was rated on a 9-point scale, gumminess on a 5-point scale, chewiness on a 7-point scale, etc. These numbers correspond to the points in the standard reference scales for those characteristics. In 1973, Civille and Szczesniak recommended the use of a 14-point intensity scale, going from 'not detectable' to 'strong' or 'large' for each characteristic.

Several different scales for measuring the perceived intensity of textural characteristics have been used. Scales may be classified as nominal, ordinal, interval or ratio scales, each type having different measuring power which determines the method of data analysis which may be used (Land and Shepherd, 1984). Interval scales and ratio scales are the types most commonly used for texture profiling.

Structured interval scales have a certain number of intervals, usually 7–9, which are

**Table 25.3** 8-POINT INTERVAL SCALE FOR TENDERNESS

| |
|---|
| — Extremely tough |
| — Very tough |
| — Moderately tough |
| — Slightly tough |
| — Slightly tender |
| — Moderately tender |
| — Very tender |
| — Extremely tender |

labelled with numbers or descriptive terms (e.g. *Table 25.3*). The descriptive terms must be carefully selected. The intervals on the scale are assumed to represent equal sensory distances.

The problems of selecting descriptive terms and ensuring equal sensory distances between points in the scale are overcome by using unstructured scales. These are usually horizontal lines, most commonly 15 cm long with descriptive terms near each end (*Figure 25.2*). Weiss (1981) reported the use of a diagonal line going up from left to right to represent the upward trend in the characteristic being evaluated.

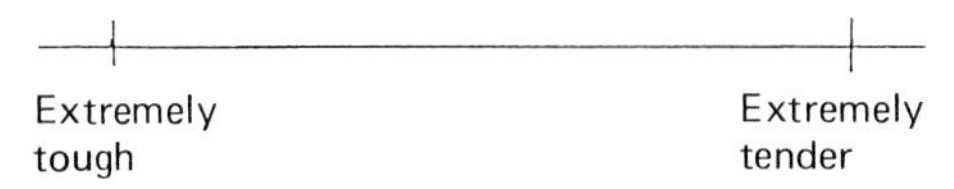


**Figure 25.2** Unstructured scale for measuring tenderness.

The method of magnitude estimation which yields results in a ratio scale has been described by Moskowitz (1983). According to this method the panelist assigns a number to indicate the intensity or magnitude of a specified characteristic, e.g. hardness in a sample. He then evaluates a second sample and assigns it a number which reflects the magnitude of the characteristic in comparison with the first sample. If the second sample is twice as hard as the first, the number he assigns is twice as high, e.g. 100 versus 50. Ratios of the measures can be calculated. Ratio scaling can be used to obtain power functions which relate the proportional change in measured physical properties to the resultant proportional change in the perceived sensation. The application of magnitude estimation to texture work has increased in recent years (Brennan, 1984).

Shand *et al.* (1985) compared the sensitivity and accuracy of category scaling, line scaling and magnitude estimation for measuring the intensity of attributes in the descriptive sensory analysis of beef steaks. They found category scaling to be the most sensitive method for detecting differences in steak quality attributes and the method preferred by the panelists. Line scaling was the least sensitive method for detecting differences even though the panelists found this method easy to learn and use. Magnitude estimation was as sensitive as category scaling to most treatment differences but the panel found that it required the most effort to use and preferred it least.

Davey (1983) found that a semi-structured line scale 50 mm long to rate the hardness of beef fat was preferred by the judges over any numerical or fully structured scale and permitted the judges to differentiate between two treatments.

Cardello, Matas and Sweeney (1982) rescaled the reference products which make up the six 'standard scales' of the texture profile using magnitude estimation. They plotted the category scale value for each item on the standard scales against the magnitude estimate for that item. Both scales can be used to measure intensity of attributes but the information they provide is not identical. Plots of the category scale data were concave downward which suggests greater discrimination at the lower end of the scale (*Figure 25.3*).

Bourne (1982) developed a comparative texture profile ballot for specific products to quantify the differences in textural properties between samples. The ballot is derived from a basic texture profile for a product and allows the panelists to focus on the degree of difference between samples in each textural characteristic. Experimental samples are rated against a control as having 1–5 points more or 1–5 points less of the characteristic than the control.

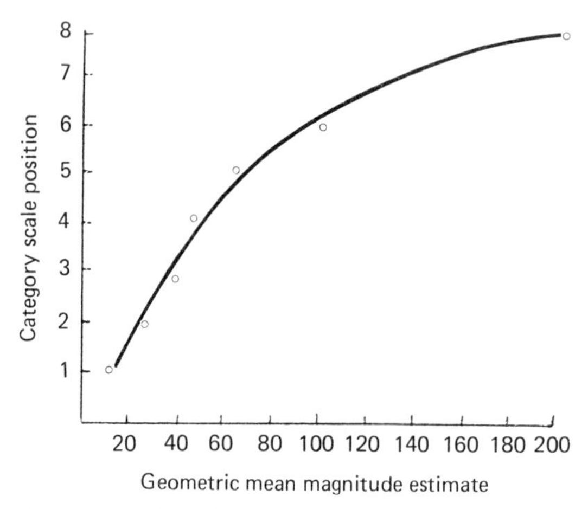


**Figure 25.3** Plots of the category scale position for each food item on the standard scale of hardness as a function of the geometric mean estimates for hardness of the same food items. Reprinted from Cardello, Matas and Sweeney (1982) with permission of the publishers.

The sensory characteristics of food may be compared without reference to specific characteristics using a method known as multidimensional scaling (Moskowitz, 1983). Samples are presented in pairs to the panelists who score the degree of similarity (or dissimilarity) using a category scale. The results are used to generate a similarity matrix, showing psychological distances among samples. The main advantage cited is that the panelists do not use descriptive terms, thus biases from word association are eliminated. Proponents of the method claim that the dimensions which emerge more truly represent the basic underlying psychological primaries. However, this method requires that large numbers of samples be assessed.

Chauhan and Harper (1986) compared descriptive profiling with direct similarity assessments of soft drinks. They converted profile assessments into a dissimilarity matrix between all pairs of drinks by computing the Euclidian distance between their respective ratings. The two methods yielded broadly similar results in terms of spatial representation of the stimuli.

### *Consensus versus mean scores*

The texture profile score for the intensity of each characteristic was a consensus score arrived at after discussion. The inherent problems of having a panel agree on a consensus are well known. In most laboratories panelists work individually and a mean score is calculated. Syarief *et al.* (1985) compared mean scores and consensus scores for profile ratings. They found that mean scores accounted for higher cumulative proportion of variance than the consensus scores when retaining the same number of factors.

### *Quantitative descriptive analysis*

A descriptive method which has had a major effect on profiling is the Quantitative Descriptive Analysis (QDA) which was introduced in 1974 (Stone *et al.*, 1974). This method is not limited to flavour or texture but includes all sensory characteristics. Panelists are carefully selected and trained for participation on a QDA panel. The panelists develop the descriptive terms for the set of products. They also define each term and where possible identify appropriate reference materials.

During product evaluations the judges work independently, replicating each evaluation at least four times (Stone, Sidel and Bloomquist, 1980). The replication permits monitoring of judge reliability and within-sample variation.

The intensity of each characteristic is rated on a line scale 6 in long with word anchors 0.5 in in from each end. The right end of the scale represents the greatest intensity. When first reported, there was a mid-point on the line scale, but dropping the mid-point was found to reduce variability by 10–15% (Stone and Sidel, 1985).

The data are submitted to analysis of variance and the results are expressed graphically as well as numerically. The graphic form shows in as simple a way as possible the degree of similarity or difference in a set of products. (*Figure 25.4*).

### *Multivariate analysis*

When describing the texture of a food product a sensory panel may identify several

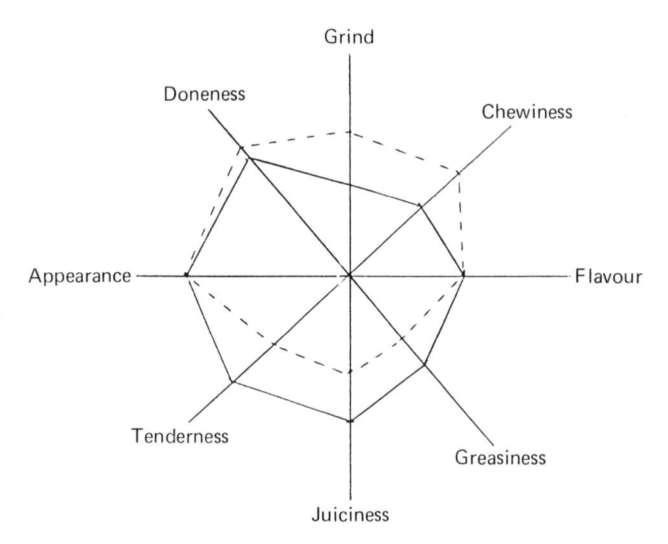


**Figure 25.4** Graphic representation of intensity of sensory characteristics of two hamburger patties.

characteristics which it feels are relevant to the description of that product. It is important to know if these characteristics are independent of one another or just variations of the same characteristic. Multivariate data analysis techniques permit simplification and interpretation of many different variables (Resurreccion and Shewfelt, 1985). Sensory analysts are using multivariate analysis with increasing frequency to understand the underlying properties and to obtain an insight to the relationship among variables which would not be evident from univariate analysis alone (Powers, 1984). Several multivariate analysis techniques have been used but the most frequently cited techniques are factor analysis, principal component analysis, cluster analysis and discriminant analysis (Ennis *et al.*, 1982).

### *Factor analysis*

Factor analysis is used to replace the original variables measured in an experiment with a smaller set of factors. The factors extracted are usually, but not always, independent of one another. Factor analysis can help to understand interplay among sensory sensations and to explore the main underlying structure of the variables. Factor analysis attempts to account for the variance of each attribute by partitioning it among the factors.

### *Principal component analysis*

Principal component analysis is another dimension-reducing technique whose purpose is to transform the original variables into a new set of principal components which are uncorrelated with each other and whose variances decrease (Smith, 1984). The new factors will not be any of the original attributes. The analyst must name the factors. This is done by examining the contribution of each of the attributes to the factors and applying a certain amount of intuition.

A panel used by Horsfield and Taylor (1976) identified 11 sensory parameters important in evaluating meat. Using principal component analysis the linear combinations, called principal components, of the 11 sensory parameters were calculated. Three principal components accounted for 95.8% of the total variance between the samples. In order to interpret the components they were rotated by the Varimax method to a new set of dimensions which they named succulence, flavour and toughness.

### *Cluster analysis*

By means of cluster analysis variables are grouped according to similarity. It can be applied to descriptive terms to see which ones are highly correlated so as to reduce the number of terms used routinely. The scores for the descriptors are examined for similarities and grouped.

### *Discriminant analysis*

Discriminant analysis is a statistical procedure which attempts to find which grouping of variables will permit the best discrimination among treatments with the lowest error.

Each of these techniques has a slightly different basis. For example, with principal component analysis the maximum variation is accounted for, while with discriminant analysis the maximum discrimination is sought (Williams, 1982). The selection of the form of multivariate analysis to use should not be limited to the method expected to be best. Palmer (1974) demonstrated this by examining the results of principal component analysis, stepwise discriminant analysis and cluster analysis to select the most important descriptors from a list of relevant ones.

Multivariate techniques are applied in an attempt to find simpler models for expressing and interpreting complex multidimensional data with minimum loss of information (Harries, 1973). If multivariate analysis of panel analysis results in fewer characteristics being evaluated in future tests, time and cost will be reduced and fatigue avoided. The use of multivariate analyses should also result in better understanding of the interrelationships among textural characteristics thus further developing texture analysis.

The results of consumer acceptance tests can be used to determine which textural characteristics are most important. In our laboratory a trained panel used firmness, chewiness, gumminess, adhesiveness, starchiness (as a component of flavour) and flavour to describe spaghetti. Eight brands of spaghetti were evaluated for these characteristics. Three hundred and seventy-five consumers of Italian descent rated the overall acceptability of the same brands. Correlation and regression analyses showed that firmness and gumminess ratings were sufficient to predict consumer acceptance. Consumers preferred spaghetti with low gumminess and high firmness scores (Larmond and Voisey, 1973). In all subsequent work on spaghetti firmness and gumminess were the characteristics evaluated.

### *Variations*

A survey of recent literature has shown that many variations of the classical texture profiling method are being used. Bourne (1982) pointed out that it may not always be necessary to use the entire texture profile. Measurements may be limited to parameters of interest.

Brady and Hunecke (1985) trained a panel to evaluate textural characteristics according to the texture profile method. The panelists then developed a ballot specifically for use in evaluating meat. They identified nine important characteristics which they evaluated on 15 cm unstructured scales.

In the sensory profiling of canned boned chicken, Lyon (1980) expanded the texture profile method and combined it with QDA. The panel identified characteristics common to the training samples, defined those characteristics with respect to generally established definitions and established the order of perception of relevant sensory attributes. Intensity of each attribute was rated on line scales which had been developed during training.

Cross, Moen and Stanfield (1978) identified tenderness, juiciness and connective tissue remaining at the end of mastication as the predominant attributes in meat texture. Intensity of these attributes was rated on an 8-point scale with some features drawn from flavour profiling and texture profiling. The procedures reported for panelist screening and qualifying, panelist feedback during training, the use of references and data analysis closely resemble the QDA method.

Horsfield and Taylor (1976) described a rather straightforward procedure which is very similar to one used by the author. 'Samples are given to the panel and the sensory parameters needed to describe the texture and flavour properties of the samples were listed. These parameters were then discussed by the panel to eliminate any ambiguities and equivalences. The reduced list of parameters was then passed to further group discussions with new samples. Where new parameters were needed, they were added and any apparent redundant parameters dropped. This procedure was repeated until a relatively stable unambiguous list of parameters was produced.'

Szczesniak, Loew and Skinner (1975) modified the texture profile so that it could be used by untrained consumers. They used a trained texture profile panel to compile a list of descriptive texture terms for the product and then got consumers to rate the intensity of each characteristic present in samples.

## Conclusion

When it was introduced in 1963, the texture profile method was the first systematic approach to the sensory assessment of texture. Since that time significant advances have been made in the development of sensory evaluation methods. It is now possible to measure with validity and reliability the amount of a given characteristic perceived in a sample. In most cases the textural characteristic is given its common term. The link with rheology is still weak. However, with the application of multivariate analysis techniques to sensory panel data, the underlying properties will become evident and a better understanding of texture will emerge. As a data base of this type is accumulated an acceptable classification system will evolve.

## References

BOURNE, M.C. (1982). *Food Texture and Viscosity: Concept and Measurement*. London, Academic Press

BOURNE, M.C., SANDOVAL, A.M.R., VILLALOBOS, M.C. and BUCKLE, T.S. (1975). Training a sensory texture profile panel and development of standard rating scales in Columbia. *Journal of Texture Studies*, **6**, 43–52

BRADY, P.L. and HUNECKE, M.E. (1985). Correlations of sensory and instrumental evaluations of roast beef texture. *Journal of Food Science*, **50**, 300–303

BRANDT, M.A., SKINNER, E.Z. and COLEMAN, J.A. (1963). Texture profile method, *Journal of Food Science*, **28**, 404–409

BRENNAN, J.G. (1984). Texture perception and measurement. In *Sensory Analysis of Foods*, (Piggott, J.R., Ed.), pp. 59–91. London, Elsevier Applied Science.

BRENNAN, J.G. and JOWITT, R. (1977). Some factors affecting the objective study of food texture. In *Sensory Properties of Foods*, (Birch, G.G., Brennan, J.G. and Parker, K.J., Eds), pp. 227–246. London, Applied Science

CARDELLO, A.V., MATAS, A. and SWEENEY, J. (1982). The standard scales of texture: rescaling by magnitude estimation. *Journal of Food Science*, **47**, 1738–1740, 1742

CHAUHAN, J. and HARPER, R. (1986). Descriptive profiling versus direct similarity assessments of soft drinks *Journal of Food Technology*, **21**, 175–187

CIVILLE, G.V. and SZCZESNIAK, A.S. (1973). Guidelines to training a texture profile panel. *Journal of Texture Studies*, **4**, 204–223

CROSS, H.R., MOEN, R. and STANFIELD, M.S. (1978). Training and testing of judges for sensory analysis of meat quality. *Food Technology*, **32** (7), 48–54

DAVEY, K.R. (1983). An instrument for the measurement of fat on sides of chilled beef. *Journal of Texture Studies*, **14**, 419–430

ENNIS, D.M., BOELENS, H., HARING, H. and BOWMAN, P. (1982). Multivariate analysis in sensory evaluation. *Food Technology*, **36** (11), 83–90

HARRIES, J.M. (1973). Complex sensory assessment. *Journal of the Science of Food and Agriculture*, **24**, 1571–1581

HARRIES, J.M. RHODES, D.N. and CHRYSTALL, B.B. (1972). Meat texture I. Subjective assessment of the texture of cooked beef. *Journal of Texture Studies*, **3**, 101–114

HORSFIELD, S. and TAYLOR, L.J. (1976). Exploring the relationship between sensory data and acceptability of meat. *Journal of the Science of Food and Agriculture*, **27**, 1044–1056

INSTITUTE OF FOOD TECHNOLOGISTS (1975). Minutes of Sensory Evaluation Division Business Meeting, Chicago, June 10

ISO (International Organization for Standardization) (1981). *Sensory Analysis Vocabulary*, Part 4, Geneva, Switzerland

JOWITT, R. (1974). The terminology of food texture. *Journal of Texture Studies*, **5**, 351–358

KIOSSELOGLOU, V.D. and SHERMAN, P. (1983). The rheological conditions associated with judgement of pourability and spreadability of salad dressings. *Journal of Texture Studies*, **14**, 277–282

KRAMER, A. (1972). Texture—definition, measurement and relation to other attributes of food quality. *Food Technology*, **26** (1), 34–39

LAND, D.G. and SHEPHERD, R. (1984). Scaling and rating methods In *Sensory Analysis of Foods*, (Piggott, J.R., Ed.), pp. 141–177. London, Elsevier Applied Science

LARMOND, E. (1986). Sensory evaluation can be objective. In *Objective Methods in Food Quality Assessment*, (Kapsalis, J., Ed.), pp. 3–14. Boca Raton, Florida, CRC Press

LARMOND, E. and VOISEY, P.W. (1973). Evaluation of spaghetti quality by a laboratory panel. *Canadian Institute of Food Science and Technology Journal*, **6**, 209–211

LOVE, R.M. (1983). Texture and fragility of fish muscle cells. Research at the Torry Research Station. *Journal of Texture Studies*, **14**, 323–352

LYON, B.G. (1980). Sensory profiling of canned bone chicken: sensory evalutaion procedures and data analysis. *Journal of Food Science*, **45**, 1341–1346

MOSKOWITZ, H.R. (1983). *Product Testing and Sensory Evaluation of Foods*. Westport, Connecticut, Food and Nutrition Press

MUNOZ, A.M. (1986). Development and application of texture reference scales. *Journal of Sensory Studies*, **1**, 55–83

PALMER, D.H. (1974). Multivariate analysis of flavour terms used by experts and non-experts for describing teas. *Journal of the Science of Food and Agriculture*, **25**, 153–164

PANGBORN, R.M. (1980). Sensory science today. *Cereal Foods World*, **25**, 637–640

PELEG, M. (1983). The semantics of rheology and texture. *Food Technology*, **37** (11), 54–61

POWERS, J.J. (1984). Current practices and applications of descriptive methods. In *Sensory Analysis of Foods*, (Piggott, J.R., Ed.), pp. 179–242. London, Elsevier Applied Science

RESURRECCION, A.V.A. and SHEWFELT, R.L. (1985). Relationship between sensory attributes and objective measurements of post harvest quality of tomatoes. *Journal of Food Science*, **50**, 1242–1245

SCHIFFMAN, S.S. (1986). Recent findings about taste: important implications for dieters. *Cereal Foods World*, **31**, 300–302

SHAND, P.J., HAWRYSH, R.J., HARDIN, R.T. and JEREMIAH, L.E. (1985). Descriptive sensory assessment of beef steaks by category scaling, line scaling and magnitude estimation. *Journal of Food Science*, **50**, 495–500

SHERMAN, P. (1969). A texture profile of foodstuffs based on well-defined rheological properties. *Journal of Food Science*, **34**, 458–462

SMITH, G.L. (1984). Statistical analysis of sensory data. In *Sensory Analysis of Foods*, (Piggott, J.R., Ed.), pp. 305–349. London, Elsevier Applied Science

STONE, H. and SIDEL, J.L. (1985). *Sensory Evaluation Practices*. Orlando, Florida, Academic Press

STONE, H., SIDEL, J.L. and BLOOMQUIST, J. (1980). Quantitative descriptive analysis. *Cereal Foods World*, **25**, 642–644

STONE, H., SIDEL, J., OLIVER, S., WOOLSLEY, A. and SINGLETON, R.C. (1974). Sensory evaluation by quantitative descriptive analysis. *Food Technology*, **28** (11), 24–34

SYARIEF, H., HAMANN, D.D., GIESBRECHT, F.G., YOUNG, C.T. and MONROE, R.J. (1985). Comparison of mean and consensus scores from flavor and texture profile analyses of selected food products. *Journal of Food Science*, **50**, 647–650, 660

SZCZESNIAK, A.S. (1963). Classification of textural characteristics. *Journal of Food Science*, **28**, 385–389

SZCZESNIAK, A.S. (1975). General Foods texture profile revisited—ten years perspective. *Journal of Texture Studies*, **6**, 5–17

SZCZESNIAK, A.S. (1979). Recent developments in solving consumer-oriented texture problems. *Food Technology*, **33** (10), 61–66

SZCZESNIAK, A.S. and KAHN, E.L. (1971). Consumer awareness of an attitudes to food texture I: Adults. *Journal of Texture Studies*, **2**, 280–295

SZCZESNIAK, A.S. and KAHN, E.L. (1984). Texture contrasts and combinations: a valued consumer attribute. *Journal of Texture Studies*, **15**, 285–301

SZCZESNIAK, A.S., BRANDT, M.A. and FRIEDMAN, H.H. (1963). Development of standard rating scales for mechanical parameters of texture and correlation between the objective and sensory methods of texture evaluation. *Journal of Food Science*, **28**, 397–403

SZCZESNIAK, A.S., LOEW, B.J. and SKINNER, E.Z. (1975). Consumer texture profile technique. *Journal of Food Science*, **40**, 1253–1256

VAISEY, M. and SHAYKEWICH, K. (1971). Training for measuring the taste of food. *Journal of the Canadian Dietetic Association*, **32**, 188–196

WEDZICHA, B.L. (1979). The measurement of food texture. *Nutrition and Food Science*, **58**, 8–11

WEISS, J. (1981). Rating scalings in the sensory analysis of foodstuffs, II: Paradigmatic application of a rating method with unstructured scale. *Acta Alimentaria*, **10**, 359–405

WILLIAMS, A.A. (1975). The development of a vocabulary and profile assessment method for evaluating the flavour contribution of cider and perry aroma constituents. *Journal of the Science of Food and Agriculture*, **26**, 567–582

WILLIAMS, A.A. (1982). Scoring methods used in the sensory analysis of foods and beverages at Long Aston research station. *Journal of Food Technology*, **17**, 163–175

YOSHIKAWA, S., NISHIMARU, S., TASHIRO, T. and YOSHIDA, M. (1970a). Collection and classification of words for description of food texture. I: Collection of words. *Journal of Texture Studies*, **1**, 437–442

YOSHIKAWA, S., NISHIMARU, S., TASHIRO, T. and YOSHIDA, M. (1970b). Collection and classification of words for description of food texture. II: Texture profiles. *Journal of Texture Studies*, **1**, 443–451

YOSHIKAWA, S., NISHIMARU, S., TASHIRO, T. and YOSHIDA, M. (1970c). Collection and classification of words for description of food texture. III: Classification by mutivariate analysis. *Journal of Texture Studies*, **1**, 452–463

26

# ORAL PERCEPTION OF TEXTURE

M. R. HEATH
*Department of Prosthetic Dentistry, London Hospital Medical College Dental School, London, UK*
and
P. W. LUCAS
*Department of Anatomy, National University of Singapore, Singapore*

## Introduction

The food industry aims to produce foods that are both pleasurable and not too difficult to eat. Pleasure is derived from taste, smell, temperature and vision. However, the perception of texture also gives pleasure and is the sensation that is the most relevant to a consideration of food structures. Assessment of the appreciation of food depends on subjective response. Considerable time and care has been given to taste panels but their results are inherently limited by the insensitivity of language. We hope that some physiological measurements might usefully complement these studies.

## Phases of oral processing

Three phases can be identified—an initial ingestion phase, a repetitive cyclical chewing phase and swallowing. Control of all three phases depends on sensory input. Ingestion demands assessment of the initial quality of food. If the food is ingested directly onto the tongue, then initial assessment is only a function of surface characteristics and small deformations. The tongue achieves this by pressing the food against the palate.

During incision and chewing, internal mechanical characteristics involving large deformations and fracture predominate. Usually only the detection of unexpected consistency or texture produces a conscious response to a previously unconscious pattern of activity but this serves to demonstrate that texture is being continually monitored. The now generally accepted concept of a central pattern generator for control of chewing movements (reviewed by Lund and Olsson, 1983), allows for peripheral input concerning the texture of food to modulate the pattern (Taylor, Appenteng and Morimoto, 1981). For example, Tornberg *et al.* (1985) showed that the duration of that part of the chewing cycle when the food is loaded depends on food texture.

During the final phase, texture is monitored for suitability of the bolus for swallowing and continued transport. The gag reflex protects the airway during swallowing by choking unsuitable food back into the mouth.

## Sources of sensation

Mastication is effected by voluntary muscles which can be consciously controlled but which normally act without much conscious control by using a wide range of sensory stimuli from the mouth. The cerebral cortex of the brain has a discrete area that is known to be highly important in the reception of sensory information from the body (somatic reception). This area has been mapped (Penfield and Rasmussen, 1957) so that the cortical regions to which information from different parts of the body are projected are known. The importance of the sensory reception from the facial and oral area is indicated by the proportion of the sensory cortex devoted to it (Penfield and Rasmussen, 1957) (*Figure 26.1*). However, much of the coordination of mastication by the central nervous system is subcortical, involving the basal ganglia and the brainstem (Lund and Olsson, 1983). The effectiveness of this subconscious control is indicated by those occasions when the unexpected triggers conscious attention.

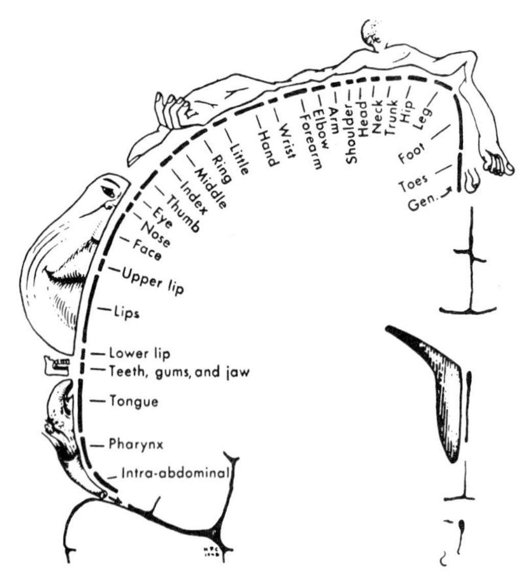


**Figure 26.1** Mapping of the somatosensory surface of the cerebral cortex demonstrates the disproportionate representation of the lips and tongue. Reproduced from Penfield and Rasmussen (1957) with permission of the publishers.

### *Oral mucosal receptors*

The oral mucosae are liberally innervated with sensory nerve endings which lie beneath the epithelium (*Figure 26.2*). The lips are particularly sensitive to temperature while the ability to make the finest two-point discrimination (1–2 mm) is found on the upper surface of the tip of the tongue (Ringel and Ewanowski, 1965). Superficial lingual sensory nerves are very fast adapting (*Figure 26.3*). Light mechanical stimulation produces a short discharge of pulses but none thereafter despite continued stimulation. It is easy to argue that this fast adaptation is essential for sensory perception of lingual contact during chewing and speech. The speed of this coordinated movement can be demonstrated during speech using a thin plate to cover the palate and upper teeth carrying electrodes that can detect the contact positions of the

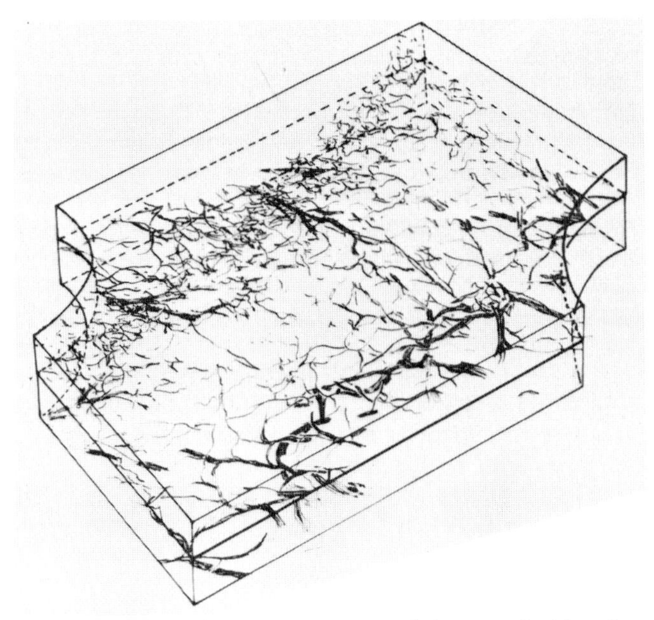

**Figure 26.2** The oral mucosae are generously innervated with endings sensitive to temperature and light touch. Reproduced from Dixon (1963) with permission of the publishers.

tongue to the plate (Heath, Ferman and Mollard, 1980). The plate may disturb the normal reflex production of precise speech initially, but adjustment to it appears rapid, as is also commonly observed following provision of new dentures.

The study of Laine and Siirilä (1971), in which they tested the intraoral and manual discrimination of different shapes, showed that man may be capable of far finer discrimination than he commonly consciously detects. This highlights the need for appropriate tests and accounts for some of the need for experienced tasters on taste panels.

**Figure 26.3** Superficial lingual sensory nerves are sensitive to light touch but adapt very quickly. Upper trace: electrical discharge in a single sensory fibre; lower trace: fine displacement of the tongue. Reproduced from Porter (1966) with permission of the publishers.

### *Periodontal mechanoreceptors*

The sensory nerve endings within the periodontal membrane, a soft tissue which supports the teeth and attaches them to bone, provide for very fine discrimination. Static tests show thresholds to displacements of 2–3 μm for single neurons (reviewed by Anderson, Hannam and Matthews, 1970), but for detection of foils between natural human incisors the average threshold is between 8 and 12 μm (Utz, 1983) whereas the threshold for denture wearers is 200–300 μm, about 20 times greater (Utz and Wegmann, 1985).

The dynamic threshold of natural teeth for detecting particles of aluminium oxide in yoghourt is also extremely sensitive, the absolute threshold being about 15 μm (Utz, 1983). However, the perception of steel balls hidden within peanuts is dramatically poorer being some 40 times less sensitive—about 700 μm (Owall and Vorwerk, 1974). This difference in perception is clearly dependent upon contrasts in texture. Again the distribution of thresholds by dentate subjects contrasts with the loss of sensitivity by denture wearers whose average threshold is 1.4 mm (*Figure 26.4*).

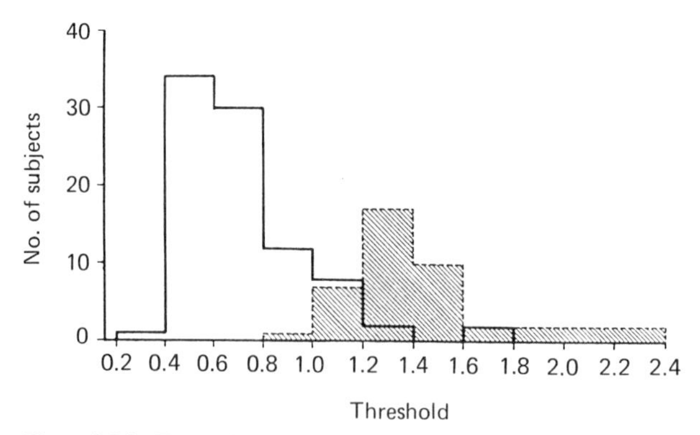


**Figure 26.4** Perception is much poorer without periodontal sensation. Owall secreted steel balls of various sizes within peanuts. The minimal size detected by subjects with full dentures (▨) was twice that detected by dentate subjects (□). Reproduced from Owall and Vorwerk (1974) with permission of the publishers.

The dynamic response of periodontal receptors is understandably difficult to study even in experimental animals. Different fibres fire maximally to forces in different directions (Cash and Linden, 1982) and the rate of application of force also seems to be important (Hannam, 1969). In man their stimulation is usually judged by the effect on some motor response with or without local anaesthetic. Hector (1984) showed that parotid salivary secretion is stimulated principally by ipsilateral periodontal mechanoreceptors during chewing, e.g. chewing on the right side of the mouth stimulates a flood of saliva from the right parotid gland.

### *Temporomandibular (jaw) joint receptors*

Animal experiments have shown that mechanoreceptors of the temporomandibular joints sense both mandibular position and movement (see review by Klineberg and Wyke, 1973). Although their importance in perception is uncertain, they would seem to be well situated to provide general information for control of the large and rapid movements of the mandible in contrast to the precision needed for the occlusal (tooth) contact for which the periodontal mechanism is ideally suited.

Part of the difficulty in designing experiments to assess the importance of any one source of sensation is the generosity of supply so that local anaesthesia of any one structure leaves sufficient other sensory pathways for motor control. For example, Owall (1978) attempted to assess the role of mechanoreceptors positioned around the jaw joint in the discrimination of gape. He failed to show any difference in discrimination with or without anaesthesia at gapes of less than 20 mm. One interpretation is that other sources of information may provide sufficient input below 20 mm. Alternatively, it is also possible that joint mechanoreceptors are poor sensors of small openings since the jaw is more or less a hinge in this range, with the jaw joint as its fulcrum. However, the results of Klineberg (1980), despite inherent limitations of his equipment, do indicate that anaesthesia of the capsule of the jaw joint enlarges the size of the chewing loop in all dimensions. Anaesthesia of the joints appears to indicate that their innervation is redundant for the control of swallowing movements (Ingervall *et al.*, 1971).

### *Muscle receptors*

All the above afferents which we have described are relatively straightforward compared with those from muscles. For any serious interest, the monograph by Matthews (1972) is recommended. Suffice for this review that most muscles are not only served by their principal motor (α) efferent nerves but also have elaborate spindle cells which are themselves capable of contracting. This fusimotor system, containing sensory and motor endings, provides sensitivity to tension, rate of change of tension and absolute length.

### *Sounds*

Intraoral sounds are, of course, monitored continuously. Studies by Drake (1965) and, more recently, Vickers (1981) suggest that the sounds of chewing food are likely to contribute to textural discrimination and should be included in a wider study. They appear to be implicated in protective inhibition of muscle activity (*see* below).

## Muscle activity

Some aspects of perception can be inferred from the modification of motor activity. The simplest example is the inhibition of muscle activity, evidenced by a silent period on an electromyography (EMG) recording. Thus during chewing some biscuits produce sounds with moderately fast rise times (*Figure 26.5*). The time to peak amplitude may be about 12 ms, but the corresponding activity of the masseter muscles precedes the initial sound by some 50 ms and shows complete muscle inhibition evident as a 'silent period', some 22 ms after the sound of fracture. Some chewing strokes show sequential fracture sounds and two separate periods of muscle inhibition may result. It is not conclusive that it is only sound that produces this inhibition (*see* also *Figure 26.9*), but a similar effect can be triggered by the sound of tapping any

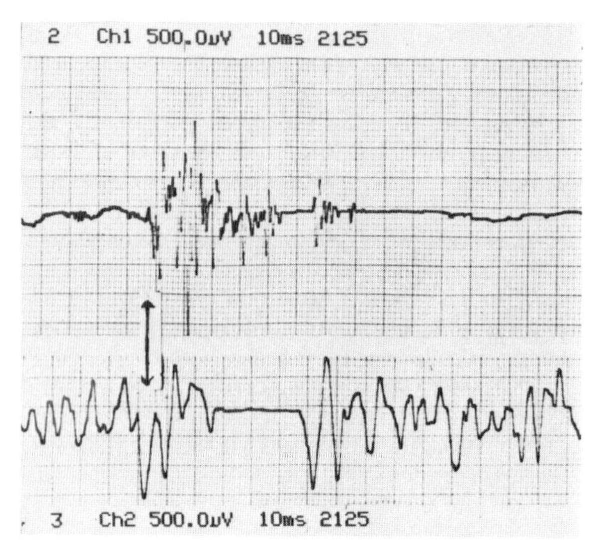


**Figure 26.5** The fracture of a biscuit during chewing followed by inhibition of muscle activity (a silent period) after a delay (latency) of some 22 ms. Upper trace: sound recorded from the skin over the forehead; lower trace: EMG of masseter muscle activity during part of a closing stroke, recorded with bipolar surface electrodes.

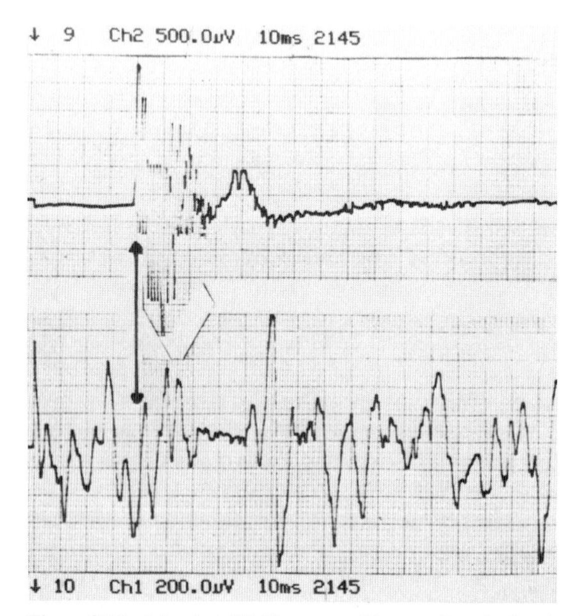


**Figure 26.6** Muscle inhibition caused by tapping the head, with a similar delay to biscuit fracture. Upper trace: the electrical signal from a microswitch mounted on a patella hammer; lower trace: EMG of masseter muscle activity during sustained clenching.

part of the skull (*Figure 26.6*) and has been extensively investigated in relation to tooth tapping (reviewed by De Laat, 1985).

This fast tooth contact in the absence of food is invariably followed by a silent period (*Figure 26.7*). It is not known whether these inhibitions are stimulated by vibration to the ear, by changes in the loading of the periodontal membrane or possibly by vibration to Golgi tendon organs (if these are present in the insertions of masticatory muscles).

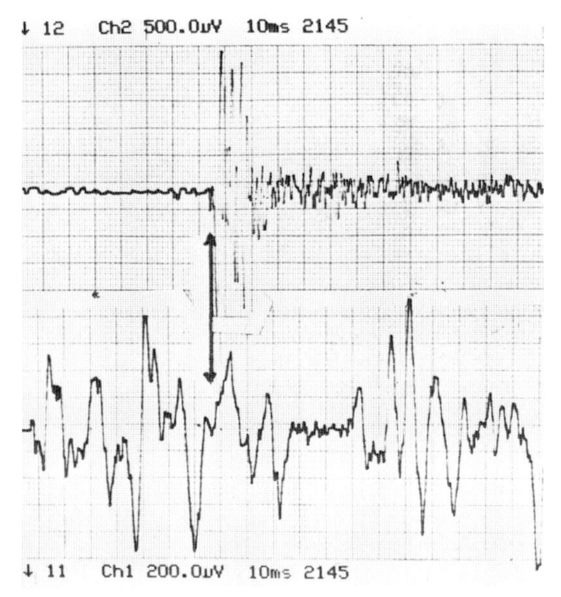


**Figure 26.7** Muscle inhibition caused by tapping the teeth together. Upper trace: as for *Figure 26.5*; lower trace: EMG of masseter muscle activity during fast closure of the mandible to clench the teeth together in the absence of food.

Inhibition of muscle activity occurs during some normal chewing cycles probably following cuspal contact between teeth but is dramatically more frequent with small hard particles in food (shown by experiments involving the detection of steel balls within peanuts; Owall and Vorwerk, 1974). However, despite the attractive simplicity of EMG it suffers from the grave limitation that extrapolation from the amplitude of the recorded activity to the force generated is only valid for isometric clenching forces (Möller, 1966). The difficulties of interpreting masticatory forces are peculiar because so many different muscles share the work at different phases of the chewing stroke and food is moved to different teeth between cycles.

Furthermore, different foods may be chewed principally by different teeth (*Figure 26.8*) but it is impossible to state which without careful, elaborate experiments such as those of Tornberg *et al.* (1985). Boyer and Kilcast (1986) showed interesting differences in the time course of EMG potentials generated during chewing two gels which had equivalent initial mechanical break loads as bench tested. Presumably these uncover intriguing real differences but there is no clear way of finding out the cause. Some effect may result from the use of different teeth for which a different mix

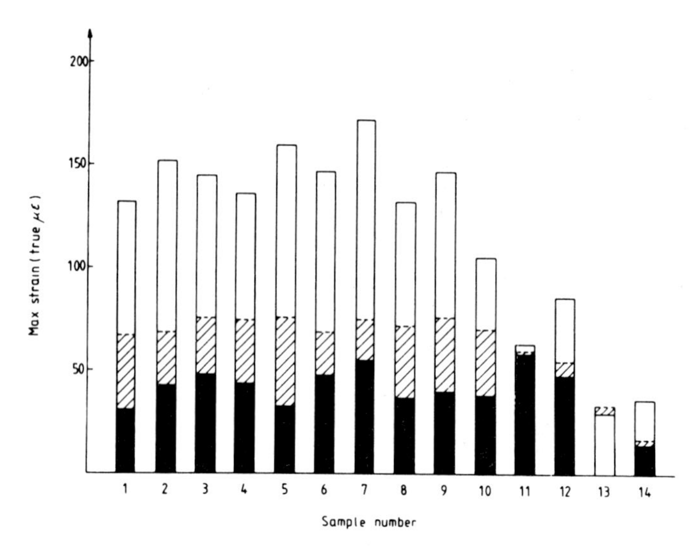


**Figure 26.8** Different foods are chewed principally on different teeth with different maximal forces; □, gauge 8; ▨, gauge 1; ■, gauge 9. Mean maximal loads on left and right molars (gauges 1 and 9) and a left premolar (gauge 8). Foods 1–9: whole meat; 10–12: hamburger; 13 and 14: sausage. For food 11, gauge 9 has recorded the largest strain, whereas for food 7 it is gauge 8 (which is a gauge on a premolar, a more anterior tooth). Reproduced from Tornberg *et al.* (1985) with permission of the publishers.

of muscle fibres will be active. Such selective control of food has been shown by Yurkstas and Curby (1953) and Tornberg *et al.* (1985). However, with due caution EMG is valuable for timing muscle activity (Olthoff, 1986).

## Forces

Since the classic work of Anderson (1953) there have been many attempts to measure forces exerted by teeth onto foods; data on maximal forces and rates of loading are of obvious value but dynamic records of normal chewing patterns are extremely difficult to obtain (Gibbs *et al.*, 1981). Quite apart from the complexity of mounting transducers within bridges or dentures, the transducers can only give accurate records of chewing at the site of placement. Tornberg *et al.* (1985) overcame this problem by mounting 14 strain gauges on a full arch bridge for one patient. He chewed nine different types of whole meat with similar mean maximal forces but sausages needed much lower loads. Restructured beef and hamburger were intermediate (*Figure 26.8*). The average chewing rate was also similar but within each cycle it appears that the rate of loading onto meats was slower than for hamburgers and sausages. However, the number of chewing cycles used was similar for all samples except the sausages. Their subject also gave sensory attributes for the meats which gave impressively high correlations with the number of cycles used, but no significant relationship with Warner–Bratzler tests. Even higher correlations were shown by multiple regression analyses.

One particularly successful study of forces was that of Bearn (1973) who constructed dentures with a wafer mounting of the teeth so that force was detected from all the cheek teeth on the side being investigated. This elegantly demonstrated the

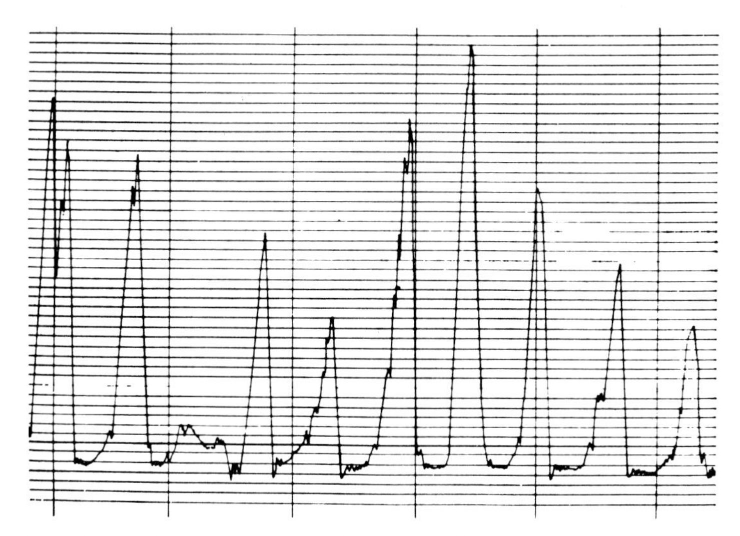

**Figure 26.9** Forces generated by denture teeth during chewing Marie biscuit. Note the double peak caused by the initial fracture. Reproduced from Bearn (1973) with permission of the publishers.

double sharp rise and fall of force that occurs within some early strokes when chewing biscuit (*Figure 26.9*).

## Food breakdown in the mouth

There is little doubt that particle size reduction is overwhelmingly important in the oral breakdown of natural unprocessed foods. However, foods developed by industry and in which food texture scientists are most interested bear little structural relation to natural foods. A significant aspect of human chewing behaviour is that much pleasure ensues from its taste and smell. The rate of release of taste may be a governing factor in the popularity of many foods and may have unconsciously been important in their 'design'. Taste is elicited by substances passing into solution in the saliva (e.g. the sugar from chewing gum, toffee), other foods are first hydrated by saliva (e.g. potato crisps, biscuits, breakfast cereals), others dissolve or melt (e.g. jellies). These events all involve continuously changing the internal characteristics of the foods to achieve a continuous source of taste and smell for the duration of chewing. Melting and significant hydration probably occur rarely in unprocessed foods.

For foods which possess a high moisture content, the expression of moisture may release the taste. An obvious class of natural food which behaves this way is juicy fruit. Moisture forced out by bursting turgid cells is probably far more significant than any particle size reduction. Critical for the release of moisture in foods in this manner is the position of the packets of moisture within the food, their individual size and their ease of release. The noise of fracture of crisp vegetables such as celery is probably associated with freshness and thus the perception of noise, texture and taste are interrelated in food selection.

Studies on jaw movement indicate that chewing involves perception of the food being chewed and therefore fundamental studies of the process should provide an understanding of the sensory basis for control. There are two processes involved in the fracture of foods:

1. Selection by the tongue and cheek of particle(s) into a bolus between the teeth giving rise to a *selection function* for the probability of fracture;
2. A *breakage function* representing the extent of fracture of the selected particles (Epstein, 1947).

Numerical solutions for these functions have been achieved with computers (Gardner and Austin, 1962; Lucas and Luke, 1983b; Lucas *et al.*, 1986; Olthoff, 1986; Voon *et al.*, 1986).

Measurements of the selection function with specific foods or materials are possible if the food can be shaped or coloured. These show that in non-sticky (non-bolus forming) foods, the selection function declines exponentially, to a power of between 1.5 and 3.0, as the size of particles is reduced (Lucas and Luke, 1983a; Van Der Glas *et al.*, 1985). The breakdown function cannot normally be measured on more than one chew unless the selection function takes an extremely low value (Gardner and Austin, 1962). Nevertheless, it forms an indication of the fragmentation pattern occurring in the early part of mastication.

## Jaw movement

The voluminous literature on jaw movement was reviewed by Bates, Stafford and Harrison (1975a,b) and more recently by Jemt (1984). Although little of this research has been oriented towards the influence of food texture on movements, some implications are worth considering.

Jaw movements can be viewed profitably in two planes. From the side, sagittal views of the movements show that they are very stereotyped in form and vary little with the type of food that is chewed (*Figure 26.10*). From the front, however, jaw movements appear very different. They are variable even in one sequence and opening movements follow a different path to closing movements (*Figure 26.11*). Typically, the opening path is closer to the midline and is less variable than the more laterally positioned early part of the closing path (Mongini, Tempia-Valenta and Benvegnu, 1986). Only a very tiny part of the movements involved in a chew, at the point of

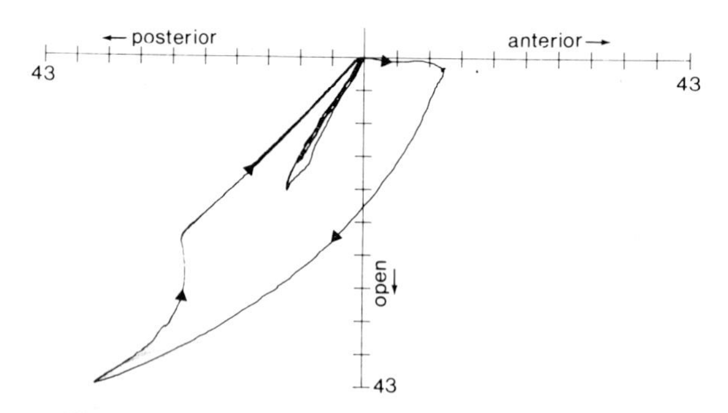


**Figure 26.10** Movements of a lower incisor viewed from the side. The arrows indicate the extreme (border) movements which are limited by joint anatomy and, superiorly, by tooth contact. Movements during chewing in this plane follow a very restricted repeatable path. Units = mm.

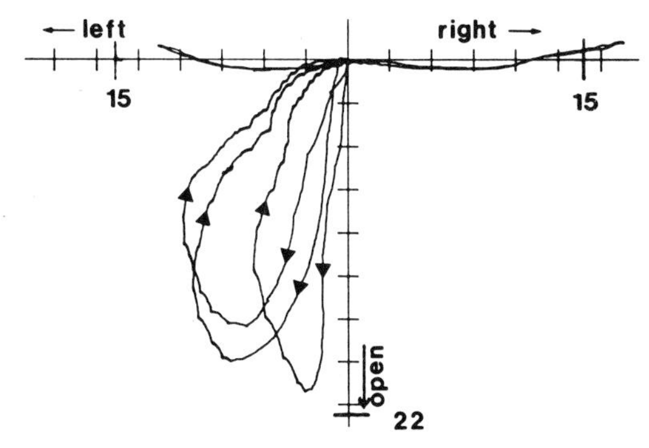


**Figure 26.11** Incisor movements viewed from in front show great variability even within are sequence. Arrows indicate the near vertical movement that is common during opening. The near horizontal line running close to the horizontal axis is the occlusal border of movement. Units = mm.

tooth contact, are limited by anatomical constraints (Wickwire *et al.*, 1981). An obvious prerequisite for the fracture of a particle in a 'jawed' system is that the jaws be opened wide enough. It makes sense to clear particles by varying the vertical amplitude of movement (*Figure 26.12*). A pile of such particles, defined as a volume, is normally collected by the tongue and transferred to the teeth.

The fragmentation of particles cannot be varied by simple changes in the vertical amplitude of movement. The fracture properties of many foods could, however, be changed by varying the lateral amplitude of movement and possibly also the rate of deformation (Shama and Sherman, 1973). Thus, although the vertical amplitude of movement does not influence fracture, the vertical velocity of jaw movement may do so.

The striking feature common to all studies has been the repeatability of the general pattern exhibited by each person (Jemt and Hedegard, 1982). This is particularly true

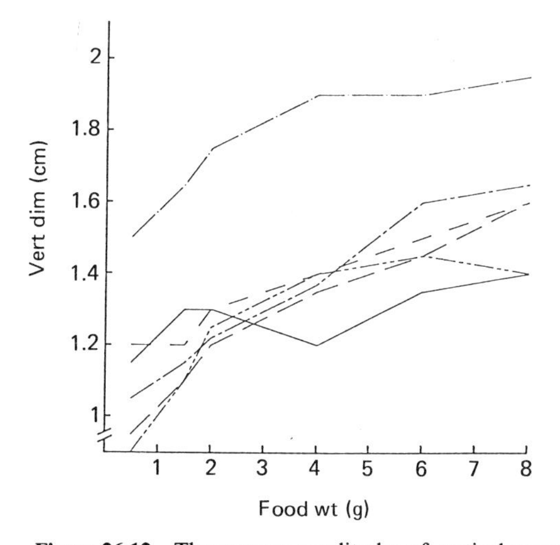


**Figure 26.12** The average amplitudes of vertical movement made by each of six subjects chewing different food weights. Each subject showed a strong tendency to a larger amplitude with larger food weight.

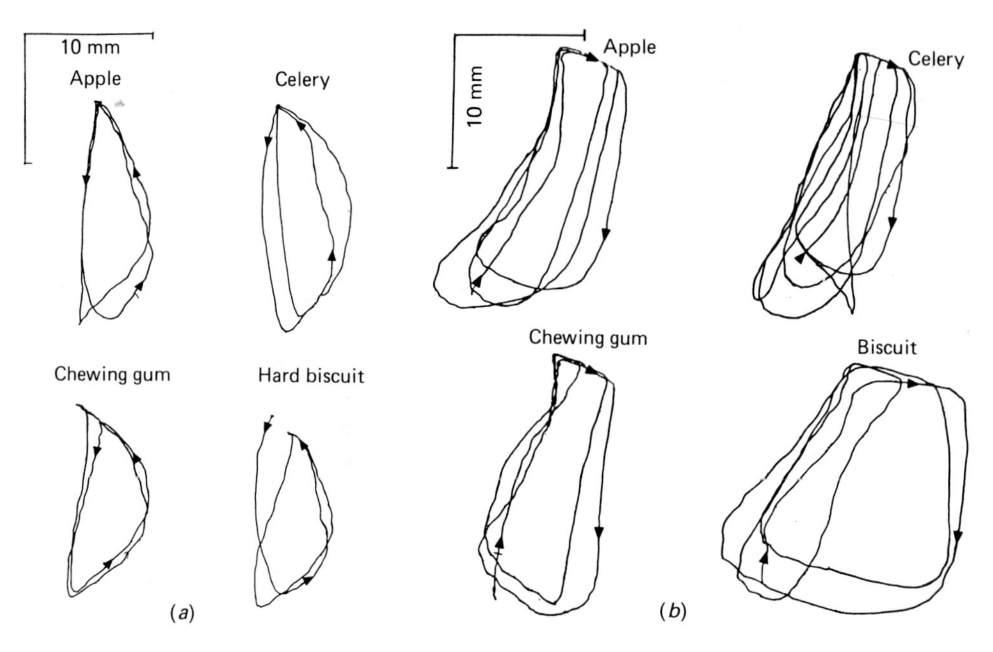


**Figure 26.13** (a) Incisor movements of one subject viewed from in front. The first two cycles for four different foods are shown. The arrows indicate the direction of movement during the first cycle. (b) Incisor movements of a different subject who frequently chewed bilaterally (i.e. on both sides of the mouth at once). Note that although there are differences between foods, contrasts between subjects are more striking.

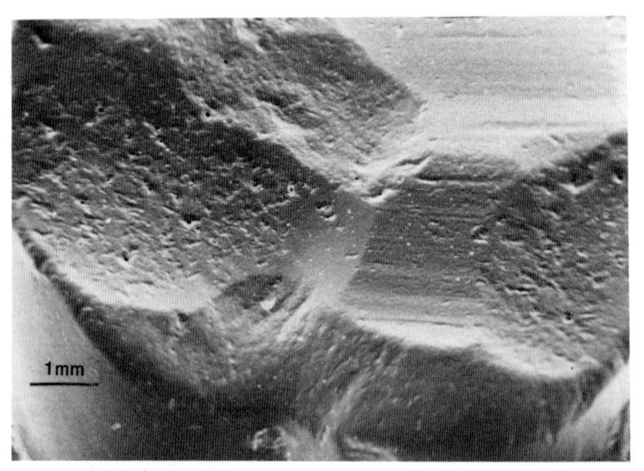


**Figure 26.14** A scanning electron microscopic image of the worn chewing surface of an acrylic denture tooth. Three facets show pitting due to food abrasion. Two facets have relatively smooth surfaces with linear scratching showing the direction of articulation of the upper and lower teeth.

with a homogeneous viscous food such as chewing gum. Comparison between different persons chewing a range of foods suggests that the variation between people is much larger than between foods (*Figure 26.13*). Individual patterns are thought to become established early in life (Ahlgren, 1966).

A recently developed technique for studying wear of acrylic teeth on dentures can indicate which teeth are used for chewing abrasive foods and even, in some cases, the direction of jaw movement (Heath, 1986) (*Figure 26.14*). It is hoped that this will complement other methods of the study of food comminution.

When individual cycles for particulate foods are examined it becomes obvious from the variability that modulation of the basic pattern does occur, presumably related to variation in the assemblage of particles between the teeth (Lucas *et al.*, 1986). It is equally obvious that averaging movements within sequences will lose any relationships that might indicate textural perception. The responsiveness of the pattern of movement to oral stimuli is clearly shown by the straightforward increase in vertical amplitudes of movement with increased volume of peanuts in experimental mouthfuls (Lucas *et al.*, 1986) (*Figure 26.12*). This considerable responsiveness goes a long way to explaining the failure of many studies seeking to demonstrate simple generalizations with averaging techniques.

## Swallowing

This is arbitrarily divided into three stages based on the position of the food bolus (Jenkins, 1978). Stage I is in the mouth and is initiated voluntarily. Stage II is in the pharynx and lasts only about 1–2 s. Stage III is in the oesophagus and ends only at the entrance to the stomach. The initiation of swallowing is often supposed only to depend on conditions surrounding Stage I, but information based on experience concerning Stages II and III may well influence this decision. For example, it is known from fluoroscopic studies of bariumized tablets that certain materials get stuck in the pharynx and oesophagus more readily than others (Hey *et al.*, 1982). The peristaltic muscular waves of the pharynx/oesophagus may drive softer particles down to the stomach more rapidly than harder particles (*cf.* diarrhoea versus constipation). These factors may influence the timing of swallowing to allow of a more appropriate particle size, wetness and/or plasticity as perceived orally. The effectiveness of this monitoring process is indicated by the rarity of choking.

Dental research has mostly been concerned with the association of food particle size and the decision to swallow. The distribution of particle sizes that is normally swallowed has been termed the 'swallowable threshold' by Manly and Braley (1950) and the 'swallowable composition' by Jiffry (1983). It is not clear whether particle size *per se* could be a cue for the initiation of swallowing. Particle size reduction curves are exponential after a small number of chews (about 10) (Lucas and Luke, 1983a; Olthoff *et al.*, 1984; Sheine, 1979). A common feature of such curves, of course, is that they do not possess any turning points or inflexions and thus, in themselves, present no feature that could influence the decision to swallow. The physiological limit of perception of particle sizes is far below those sizes that have been found to be swallowable (Utz, 1983; Utz and Wegmann, 1985). The particle sizes that are swallowable by a subject depend on the volume of food in the mouth, at least for peanuts. Larger particles are swallowed with larger mouthfuls (Jiffry, 1983; Lucas and Luke, 1984; Yurkstas, 1965).

The sizes that are swallowable by different subjects are very variable and may depend on the rate of particle size breakdown that the person can achieve (Dahlberg,

1942; Yurkstas, 1951). The tendency is for people with dentures to chew less well but swallow larger particles rather than fully compensate with more chewing strokes.

Among other factors that might contribute to the decision to swallow are the lubrication of the food (by saliva and expressed moisture) and the intensity (peak or trough) of taste. The rates of hydration, melting and expression of moisture are also likely to follow an exponentially declining form with respect to time.

## Conclusions

Despite the complexity of oral function some inferences can be drawn about the oral perception of food textures. That such perception is used in the modulation of patterns of chewing, salivary flow and swallowing is relatively easy to demonstrate. By combining physiological measurement with textural studies, some fundamental understanding of the masticatory process and commercial applications could be achieved. Fundamental research will undoubtedly need interdisciplinary cooperation, including dental and food texture science.

## Acknowledgements

The authors would like to thank the various authors and publishers for generously allowing the use of *Figures 26.1, 26.2, 26.3, 26.4, 26.8* and *26.9.*

## References

AHLGREN, J. (1966). Mechanism of mastication. *Acta Odontologica Scandinavica*, **24** (Supplement 44), 109

ANDERSON, D.J. (1953). A method of recording masticatory loads. *Journal of Dental Research*, **32**, 785–789

ANDERSON, D.J., HANNAM, A.G. and MATTHEWS, B. (1970). Sensory mechanisms in mammalian teeth and their supporting structures. *Physiological Review*, **50**, 171–195

BATES, J.F., STAFFORD, G.D. and HARRISON, A. (1975a). Masticatory function—a review of the literature. I: The form of the masticatory cycle. *Journal of Oral Rehabilitation*, **2**, 281–301

BATES, J.F., STAFFORD, G.D. and HARRISON, A. (1975b). Masticatory function—a review of the literature. II: Speed of movement of the mandible, rate of chewing and forces developed in chewing. *Journal of Oral Rehabilitation*, **2**, 349–361

BEARN, E.M. (1973). Effect of different occlusal profiles on the masticatory forces transmitted by complete dentures. *British Dental Journal*, **134**, 7–10

BOYER, M.M. and KILCAST, D. (1986). Electromyography as a novel method for examining food texture. *Journal of Food Science*, **51**, 859–860

CASH, R.M. and LINDEN, R.W.A. (1982). The distribution of mechanoreceptors in the periodontal ligament of the mandibular canine tooth of the cat. *Journal of Physiology*, **330**, 439–447

DAHLBERG, B. (1942). The masticatory effect. *Acta Medica Scandinavica*, **112** (Supplement 139)

DE LAAT, A. (1985). Masseteric reflexes and their relationship towards occlusion and temporomandibular joint dysfunction. Doctoral Thesis, University of Leuven

DIXON, A.D. (1963). Nerve plexuses in the oral mucosa. *Archives of Oral Biology*, **8**, 435–447

DRAKE, B. (1965). On the biorheology of human mastication: an amplitude–frequency–time analysis of food-crushing sounds. *Biorheology*, **3**, 21–31

EPSTEIN, B. (1947). The mathematical description of certain breakage mechanisms heading to the logarithmico–normal distribution. *Journal of the Franklin Institute*, **244**, 471–477

GARDNER, R.P. and AUSTIN, L.G. (1962). A chemical engineering treatment of certain breakage mechanisms leading to the logarithmico–normal distribution. In *Zerkleinern Symposion* pp. 217–248. Dusseldorf, Verlag Chemie

GIBBS, C.H., MAHAN, P.E., LUNDEEN, H.C., BREHAN, K., WALSH, E.K. and HOLBROOK, W.B. (1981). Occlusal forces during chewing and swallowing as measured by sound transmission. *Journal of Prosthetic Dentistry*, **46**, 443–449

HANNAM, A.G. (1969). The response of periodontal mechanoreceptors in the dog. *Archives of Oral Biology*, **15**, 971–978

HEATH, M.R. (1986). Functional interpretation of patterns of occlusal wear on acrylic teeth. *Restorative Dentistry*, **2**, 100–107

HEATH, M.R., FERMAN, A.H. and MOLLARD, R. (1980). Electropalatolography: thermoformed pseudopalates and the use of three-dimensional display. *Journal of Biomedical Engineering*, **2**, 145–147

HECTOR, M.P. (1984). Evidence for the involvement of periodontal mechanoreceptors in the control of parotid secretion. *Journal of Dental Research*, **63**, 490

HEY, H., JORGENSEN, F., SORENSEN, K., HASSELBALCH, H. and WAMBERG, T. (1982). Oesophageal transit of six commonly used tablets and capsules. *British Medical Journal*, **85**, 1717–1719

INGERVALL, B., BRATT, C.M., CARLSSON, G.E., HELKIMO, M. and LANTZ, B. (1971). Positions and movements of the mandible and hyoid bone during swallowing. *Acta Odontologica Scandinavica*, **29**, 549–562

JEMT, T. (1984). Masticatory mandibular movements. *Swedish Dental Journal, Supplement*, **24**, 35–41

JEMT, T. and HEDEGARD, B. (1982). Reproducibility of chewing rhythm and of mandibular displacements during chewing. *Journal of Oral Rehabilitation*, **9**, 531–537

JENKINS, G.N. (1978). *Physiology and Biochemistry of the Mouth*, 4th Edn, pp. 532–537. Oxford, Blackwell

JIFFRY, M.T.M. (1983). Analysis of particles produced at the end of mastication in subjects with normal dentition. *Journal of Oral Rehabilitation*, **8**, 113–119

KLINEBERG, I.J. (1980). Influences of temporomandibular articular mechanoreceptors on functional jaw movements. *Journal of Oral Rehabilitation*, **7**, 307–317

KLINEBERG, I.J. and WYKE, B.O. (1973). Articular reflex control of mastication. In *Oral Surgery IV*, (Kay, L.W. Ed.), pp. 253–258. Copenhagen, Munksgaard

LAINE, P. and SIIRILÄ, H.S. (1971). Oral and manual stereognosis and two-point tactile discrimination of the tongue. *Acta Odontologica Scandinavica*, **29**, 197–204

LUCAS, P.W. and LUKE, D.A. (1983a). Methods for analysing the breakdown of food in human mastication. *Archives of Oral Biology*, **28**, 813–819

LUCAS, P.W. and LUKE, D.A. (1983b). Computer simulation of the breakdown of carrot particles during human mastication. *Archives of Oral Biology*, **28**, 821–826

LUCAS, P.W. and LUKE, D.A. (1984). Optimum mouthful for food comminution in human mastication. *Archives of Oral Biology*, **29**, 205–210

LUCAS, P.W., OW, R.K.K., RITCHIE, G.M., CHEW, C.L. and KENG, S.B. (1986). Relationship between jaw movement and food breakdown in human mastication. *Journal of Dental Research*, **65**, 400–404

LUND, J.P. and OLSSON, K.A. (1983). The importance of reflexes and their control during jaw movement. *Trends in Neuroscience*, **6**, 458–463

MANLY, R.S. and BRALEY, L.C. (1950). Masticatory performance and efficiency. *Journal of Dental Research*, **29**, 448–462

MATTHEWS, P.B.C. (1972). *Mammalian Muscle Receptors and Their Central Action.* London, Arnold

MOLLER, E. (1966). The chewing apparatus. *Acta Physiologica Scandinavica*, **69** (Supplement 280)

MONGINI, F., TEMPIA-VALENTA, G. and BENVEGNU, G. (1986). Computer-based assessment of habitual mastication. *Journal of Prosthetic Dentistry*, **55**, 638–649

OLTHOFF, L.W. (1986). Comminution and neuromuscular mechanisms in human mastication. Doctoral Thesis, University of Utrecht

OLTHOFF, L.W., VAN DER BILT, A., BOSMAN, F. and KLEISEN, H.H. (1984). Distribution of particle sizes in food comminuted by human mastication. *Archives of Oral Biology*, **29**, 899–903

OWALL, B. (1978). Interocclusal perception with anaesthetized and unanaesthetized temporomandibular joints. *Swedish Dental Journal*, **2**, 199–208

OWALL, B. and VORWERK, P. (1974). Analysis of a method for testing oral tactility during chewing. *Odontologisk Revy*, **25**, 1–10

PENFIELD, W. and RASMUSSEN, T. (1957). *The Cerebral Cortex of Man. A Clinical Study of Localization of Function.* New York, Macmillan

PORTER, R. (1966). Lingual mechanoreceptors activated by muscle twitch. *Journal of Physiology*, **183**, 101–111

RINGEL, R.L. and EWANOWSKI, S.J. (1965). Oral perception: I. Two point discrimination. *Journal of Speech and Hearing Research*, **8**, 389–397

SHAMA, F. and SHERMAN, P. (1973). Evaluation of some textural properties with the Instron Universal testing machine. *Journal of Texture Studies*, **4**, 344–352

SHEINE, W.S. (1979). The effects of variations in molar morphology on masticatory effectiveness and digestion of cellulose in prosimian primates. PhD Thesis, Duke University, North Carolina

SHERMAN, P. (1969). A texture profile of foodstuffs based upon well-defined rheological properties. *Journal of Food Science*, **34**, 458–462

TAYLOR, A., APPENTENG, K. and MORIMOTO, T. (1981). Proprioceptive input from the jaw muscles and its influence on lapping, chewing and posture. *Canadian Journal of Physiology and Pharmacology*, **59**, 636–644

TORNBERG, E., FJELKNER-MODIG, S., RUDERUS, H., GLANTZ, P.O., RANDOW, K. and STAFFORD, G.D. (1985). Clinically recorded masticatory patterns as related to the sensory evaluation of meat and meat products. *Journal of Food Science*, **50**, 1059–1066

UTZ, K.H. (1983). The interocclusal tactile fine sensibility of natural teeth. *Journal of Oral Rehabilitation*, **10**, 440–441

UTZ, K.H. and WEGMANN, U. (1985). The interocclusal tactile fine sensibility of complete denture wearers. *Journal of Oral Rehabilitation*, **12**, 549

VAN DER GLAS, H., VAN DER BILT, A., OLTHOFF, L.W. and BOSMAN, F. (1985). Selection and breakage processes in the comminution of food: II An experimenal determination. *Journal of Oral Rehabilitation*, **12**, 551

VICKERS, Z.M. (1981). Relationships of chewing sounds to judgements of crispness, crunchiness and hardness. *Journal of Food Science*, **47**, 121–124

VOON, F.C.T., LUCAS, P.W., CHEW, C.L. and LUKE, D.A. (1986). A simulation approach to understanding the masticatory process. *Journal of Theoretical Biology*, **119**, 251–262

WICKWIRE, N.A., GIBBS, C.H., JACOBSON, A.P. and LUNDEEN, H.C. (1981). Chewing patterns in normal children. *Angle Orthodontist*, **51**, 48–60

YURKSTAS, A. (1951). Compensation for inadequate mastication. *British Dental Journal*, **91**, 261–262

YURKSTAS, A. (1965). The masticatory act. *Journal of Prosthetic Dentistry*, **15**, 248–260.

YURKSTAS, A. and CURBY, W.A. (1953). Force analysis of prosthetic appliances during function. *Journal of Prosthetic Dentistry*, **3**, 82–87

# LIST OF PARTICIPANTS

|  |  |
|---|---|
| Angold, Dr. R. E. | Marlow Foods, P.O. Box 127, Lancaster House, Lincoln Road, High Wycombe, Buckinghamshire, HP12 3RL. |
| Arnold, Dr. B. | Mars Confectionery, Dundee Road, Slough, Berkshire, SL1 4JX. |
| Attenburrow, Dr. G. E. | Unilever Research, Colworth House, Sharnbrook, Bedfordshire. |
| Autio, Dr. K. L. | Technical Research Centre of Finland, Food Research Laboratory, Biologinkuja 1, 02150 Espoo, Finland. |
| Bagley, Dr. E. B. | United States Department of Agriculture, Northern Regional Research Center, 1815 N University, Peoria, Illinois 61604, USA |
| Barfod, Dr. N. M. B. | Grindsted Products A/S., Edwin Rahrs Vej 38, DK-8220 Brabrand, Denmark. |
| Bean, Ms. H. | Kelco International Limited, Pitwood Park Industrial Estate, Waterfield, Tadworth, Surrey, KT20 5QF. |
| Bell, Dr. A. E. | Department of Food Science, University of Reading, Whiteknights Campus, Reading, Berkshire. |
| Bellekom, Mrs. L. K. | Courtaulds Chemicals, WSPG Building 43, PO Box 5, Spondon, Derby. |
| Blacklock, Dr. R. | Swel Foods, Mash Lane, Boston, Lincolnshire. |
| Blanshard, Dr. J. M. V. | Department of Applied Biochemistry and Food Science, University of Nottingham, School of Agriculture, Sutton Bonington, Loughborough, Leicestershire, LE12 5RD. |
| Blenford, Mr. D. E. | Hildon Associates, West View, Cotswold Road, Cumnor Hill, Oxford, OX2 9JG |
| Boland, Mr. G. | Pedigree Petfoods, Melton Mowbray, Leicestershire. |
| Borch Soe, Ing. J. | Grindsted Products A/S., Edwin Rahrs Vej 38, D-8220 Brabrand, Denmark. |
| Bourne, Dr. M. C. | Cornell University, N.Y.S. Agricultural Experiment Station, Geneva, New York 14456, USA. |
| Bradley, Mr. T. | Department of Applied Biochemistry and Food Science, University of Nottingham, School of Agriculture, Sutton Bonington, Loughborough, Leicestershire, LE12 5RD |
| Brabbs, Dr. W. J. | The Procter & Gamble Company, 6210 Center Hill Road, Cincinnati, Ohio 45224, USA. |
| Brown, Mr. P. K. | General Foods Limited, Banbury, Oxon. |
| Buhl, Mr. S. B. | A/S Kobenhavns Pektinfabrik, Ved Banen 16, DK-4623 Lille Skensved, Denmark. |
| Bush, Mr. P. B. | P. B. Bush and Associates, Private Bag X4, Mayville, Durban 4058, South Africa. |
| Cant, Dr. P. A. E. | Anchor Foods Ltd., Frankland Road, Blagrove, Swindon, Wiltshire, SN5 8YZ |
| Cauvain, Mr. S. P. | Flour Milling and Baking Research Association, Chorleywood, Rickmansworth, WD3 5SH. |
| Challen, Mr. I. A. | Kelco International Ltd., Westminster Tower, 3 Albert Embankment, London, SE1 7RZ. |
| Cheng, Md. L. M. | Food Science Department, Queen Elizabeth College (King's College), Campden Hill Road, London, W8 7AH. |

| | |
|---|---|
| Cogan, Ms. D. E. | Butterworth Scientific Ltd., P.O. Box 63, Westbury House, Bury Street, Guildford, Surrey, GU2 5BH. |
| Cyril, Mr. H. | Department of Applied Biochemistry and Food Science, University of Nottingham, School of Agriculture, Sutton Bonington, Loughborough, Leicestershire, LE12 5RD |
| Darling, Dr. D. F. | Unilever Research Laboratory plc., Colworth House, Sharnbrook, Bedford, MK44 1LQ. |
| Davies, Miss J. | Department of Applied Biochemistry and Food Science, University of Nottingham, School of Agriculture, Sutton Bonington, Loughborough, Leicestershire, LE12 5RD |
| Debruyne, Dr. I. M. J. | Safinco Coordination Center NV., Prins Albertlaan 12, B-8700, Izegem, Belgium. |
| Desmond, Mr. P. | Elsevier Applied Science Publishers, Crown House, Linton Road, Barking, Essex. |
| Dickens, Mr. A. W. | Baker Perkins BCS Ltd., Westfield Road, Peterborough, PE3 6TA. |
| Dickinson, Dr. E. | Procter Department of Food Science, University of Leeds, Leeds, LS2 9JT. |
| Dixon, Mr. T. | Pedigree Petfoods, National Office, Waltham on the Wolds, Melton Mowbray, Leics. |
| Dobraszczyk, Dr. B. J. | Department of Engineering, University of Reading, Whiteknights, Reading, Berkshire. |
| Edgell, Miss A. | Leatherhead Food Research Association, Randalls Road, Leatherhead, Surrey, KT22 7RY. |
| Egelandsdal, Dr. B. | Norwegian Food Research Institute, P.O. Box 50, 1432-AS-NLH, Norway. |
| Eggen, Dr. K. H. E. | Norwegian Food Research Institute, P.O. Box 50, 1432 AS-NLH, Norway. |
| Evans, Dr. G. G. | United Biscuits (UK) Ltd., St. Peters Road, Furze Platt, Maidenhead, Berkshire. |
| Farr, Mrs. V. R. | International Flavours and Fragrances, Duddery Hill, Haverhill, Suffolk, CB9 8LG. |
| Ferdinand, Mrs. J. M. | Institute of Food Research, Colney Lane, Norwich. |
| Fisk, Mr. R. J. | KP Foods, Windy Ridge, Smisby Road, Ashby-de-la-Zouch, Leics., LE6 5UQ. |
| Flack, Mr. E. A. | Grindsted Products Limited, Northern Way, Bury St. Edmunds, Suffolk, IP32 6NP. |
| Flint, Dr. O. | Procter Department of Food Science, University of Leeds, Leeds, LS2 9JT. |
| Forginiti, Miss R. | Oxford Polytechnic, Gipsy Lane, Headington, Oxford, OX3 0BP. |
| Foster, Dr. B. G. | RHM Research Ltd., Lincoln Road, High Wycombe, Buckinghamshire. |
| Fretheim, Dr. K. | The Innovation Centre, P.O. Box 1044, Blindern, N-0316, Norway. |
| Gaisford, Dr. S. E. | Kelco International Ltd., Pitwood Park Industrial Estate, Waterfield, Tadworth, Surrey, KT20 5HQ. |
| Gathura, Mr. M. | Department of Applied Biochemistry and Food Science, University of Nottingham, School of Agriculture, Sutton Bonington, Loughborough, Leicestershire, LE12 5RD. |
| Gedney, Miss S. | Department of Applied Biochemistry and Food Science, University of Nottingham, School of Agriculture, Sutton Bonington, Loughborough, Leicestershire, LE12 5RD. |
| Given, Mr. P. | R. M. Schaeberle Technology Center, 100 Deforest Avenue, P.O. Box 1941, East Hanover, New Jersey, 07936-1941, USA. |
| Gorton, Dr. H. J. | C/o Mr. F. H. Hillyard, Londreco Ltd., Nestles Avenue, Hayes, Middlesex, UB3 4RG. |
| Gough, Miss A. | 34 Walkwood Rise, Beaconsfield, Buckinghamshire, HP9 1TU. |
| Goward, Mrs. Y. | Department of Applied Biochemistry and Food Science, University of Nottingham, School of Agriculture, Sutton Bonington, Loughborough, Leicestershire, LE12 5RD. |
| Green, Mr. S. R. | 53 Kenswick Drive, Halesowen, West Midlands, B63 4QZ. |
| Guy, Dr. R. C. E. | Flour Milling and Baking Research Association, Chorleywood, Rickmansworth, Hertfordshire. |
| Hardiman, Mr. B. | Pedigree Petfoods, Melton Mowbray, Leicestershire. |

| | |
|---|---|
| Harding, Dr. S. E. | Department of Applied Biochemistry and Food Science, University of Nottingham, School of Agriculture, Sutton Bonington, Loughborough, Leicestershire, LE12 5RD. |
| Harris, Dr. P. | Unilever Research, Colworth House, Sharnbrook, Bedfordshire, MK44 1LQ. |
| Hastilow, Mr. P. A. P. | Department of Biotechnology, South Bank Polytechnic, Borough Road, London, SE1 0AA. |
| Hastings, Mrs. R. J. | Torry Research Station, 135 Abbey Road, Torry, Aberdeen. |
| Head, Dr. D. J. | Convatec Biological Research Laboratory, C/o E. R. Squibb (UK) Ltd., Reeds Lane, Moreton, Wirral, L46 1QW. |
| Heath, Dr. M. R. | London Hospital Medical School, Dental School, Turner Street, London E1 2AD. |
| Heathcock, Dr. J. | Cadbury Schweppes plc., The Lord Zuckerman Research Centre, The University, Whiteknights, P.O. Box 234, Reading, RG6 2LA. |
| Henson, Mr. N. P. | RHM Research, Lincoln Road, Cressex Industrial Estate, High Wycombe, Buckinghamshire. |
| Hermansson, Dr. A-M | SIK—The Swedish Food Research Institute, Gothenburg, Sweden. |
| Hey, Dr. M. | Department of Chemistry, University of Nottingham, University Park, Nottingham. |
| Hicks, Professor R. M. | United Biscuits (UK) Ltd., R & D Centre, St. Peters Road, Furze Platt, Maidenhead, Berkshire, SL6 7QU. |
| Hole, Dr. M. | School of Food Studies, Humberside College, Nuns Corner, Grimsby, DN34 5BQ. |
| Horne, Dr. D. S. | Hannah Research Institute, Ayr, Scotland. |
| Howell, Dr. N. K. | University of Surrey, Department of Biochemistry, Guildford, Surrey, GU2 5XH. |
| Howlett, Miss M. C. | Campden Food Preservation Research Association, Chipping Campden, Gloucestershire. |
| Hoyle, Mr. D. R. | United Biscuits (UK) Ltd., Group Research and Development Centre, St. Peters Road, Furze Platt, Maidenhead, Berkshire, SL6 7QU. |
| Imeson, Dr. A. P. | Kelco International Ltd., Westminster Tower, 3 Albert Embankment, London, SE1 7RZ. |
| Isherwood, Dr. D. P. | Department of Mechanical Engineering, Imperial College of Science and Technology, South Kensington, London SW7 2RX. |
| Jeronimidis, Dr. G. | Department of Engineering, The University of Reading, Whiteknights, Reading, RG6 2AY. |
| Johnson, Mr. M. K. | Ch. Goldrei Foucard & Sons Ltd., (Northern Foods)., Brookfield Drive, Liverpool, L9 7AW. |
| Johnson, Mr. S. | Department of Applied Biochemistry and Food Science, University of Nottingham, School of Agriculture, Sutton Bonington, Loughborough, Leicestershire, LE12 5RD. |
| Jolley, Mr. P. D. | AFRC/Institute of Food Research, Bristol Laboratory, Langford, Bristol, BS18 7DY. |
| Kay, Mr. J. A. | Kellogg Co. (GB) Ltd., Park Road, Stretford, Manchester, M32 8RA. |
| Kay, Dr. M. | Henkel Chermicals Ltd., Road Five, Industrial Estate, Winsford, Cheshire, CW7 3QY. |
| Khatib, Miss L. | Department of Food Technology, Food Studies Building, University of Reading, Reading, RG6 1BR. |
| King, Dr. K. | Agricultural & Food Chemistry Division, Newforge Lane, Belfast, N. Ireland, BT9 5PX. |
| Klopping, Miss. D. | Oxford Polytechnic, Gipsy Lane, Headington, Oxford, OX3 0BP. |
| Knight, Mr. M. K. | Food Technology Section, Leatherhead Food Research Association, Leatherhead, Surrey, KT22 7RY. |
| Krog, Mr. N. K. | Grindsted Products A/S, Edwin Rahrs Vej 38, DK-8220 Brabrand, Denmark. |
| Langley, Mr. K. R. | AFRC Institute of Food Research, Shinfield, Reading, Berkshire, RG2 9AT. |
| Larmond, Ms. E. | Room 7111, Sir John Carling Building, 930 Carling Avenue, Ottawa, Ontario, KIA 0C5. |
| Lavender, Mrs. L. | Department of Applied Biochemistry and Food Science, University of Nottingham, School of Agriculture, Sutton Bonington, Loughborough, Leicestershire, LE12 5RD. |

| | |
|---|---|
| Lawrie, Professor R. A. | Department of Applied Biochemistry and Food Science, University of Nottingham, School of Agriculture, Sutton Bonington, Loughborough, Leicestershire, LE12 5RD. |
| Ledward, Dr. D. A. | Department of Applied Biochemistry and Food Science, University of Nottingham, School of Agriculture, Sutton Bonington, Loughborough, Leicestershire, LE12 5RD. |
| Lee, Dr. J. D. | Griffith Laboratories, Cotes Park Road, Somercotes, Derbyshire, DE5 4NN. |
| Leeds, Dr. D. | M.A.F.F., Great Westminster House, Horseferry Road, London SW1P 2AE. |
| Le Grys, Mr. G. A. | Oxford Polytechnic, Headington, Oxford, OX3 0BP. |
| Leppard, Miss J. S. | Spillers Foods Ltd., Research and Technology Centre, Station Road, Cambridge. |
| Leslie, Professor R. B. | Unilever Research, Colworth House, Sharnbrook, Bedfordshire, MK44 1KU. |
| Levine, Dr. H. | Nabisco Brands Inc., Corporate Technology Group, P.O. 1943, East Hanover, New Jersey, 07936–1943, USA. |
| Lewis, Dr. D. F. | Leatherhead Food Research Association, Randalls Road, Leatherhead, Surrey. |
| Lillford, Dr. P. J. | Unilever Research Laboratory, Colworth House, Sharnbrook, Bedfordshire, MK44 1KU. |
| Lindley, Dr. M. G. | Tate & Lyle Group Research and Development, P.O. Box 68, Whiteknights, Reading. |
| Lips, Dr. A. | Unilever Research Laboratory, Colworth House, Sharnbrook, Bedfordshire, MK44 1KU. |
| Loh, Dr. J. | General Foods Corporation, 555 South Broadway, Tarrytown, New York 10591, USA. |
| McClements, Mr. D. J. | Procter Department of Food Science, University of Leeds, Leeds 2. |
| Mackie, Dr. A. | Department of Chemistry, University of Nottingham, University Park, Nottingham. |
| Marnell, Miss E. J. | Pedigree Petfoods, Melton Mowbray, Leicestershire. |
| Marshall, Mr. R. J. | AFRC Institute of Food Research, Reading Laboratory, Shinfield, Reading, RG2 9AT. |
| Martin, Dr. R. | Weston Research Laboratories, 644 Bath Road, Taplow, Maidenhead, Berkshire. |
| Merrick, Mr. W. | PPF International, Lindtsedijk 8, 3336 Le Zwijndrecht, Holland. |
| Milbourne, Dr. K. | Pork Farms, Farnsworth House, Lenton Lane, Nottingham. |
| Mitchell, Dr. J. R. | Department of Applied Biochemistry and Food Science, University of Nottingham, Sutton Bonington, Loughborough, Leics., LE12 5RD. |
| Mlotkiewicz, Dr. J. A. | Spillers Foods Ltd., Research and Technology Centre, Station Road, Cambridge. |
| Morris, Professor E. R. | Silsoe College, Silsoe, Bedford, MK45 4DT. |
| Morris, Dr. V. J. | Institute of Food Research, Norwich Laboratory, Colney Lane, Norwich NR4 7UA. |
| Neale, Dr. R. J. | Department of Applied Biochemistry and Food Science, University of Nottingham, School of Agriculture, Sutton Bonington, Loughborough, Leicestershire, LE12 5RD. |
| Nicolaou, Mr. G. | Birds Eye Walls Ltd., Station Avenue, Walton on Thames, Surrey, KT12 1NT. |
| Norton, Dr. G. | Department of Applied Biochemistry and Food Science, University of Nottingham, School of Agriculture, Sutton Bonington, Loughborough, Leicestershire, LE12 5RD. |
| Oates, Mr. C. | Department of Applied Biochemistry and Food Science, University of Nottingham, School of Agriculture, Sutton Bonington, Loughborough, Leicestershire, LE12 5RD. |
| Oldroyd, Mr. D. | Pedigree Petfoods, Melton Mowbray, Leicestershire. |
| Parker, Dr. R. | Institute of Food Research, Norwich Laboratory, Colney Lane, Norwich. |
| Paynter, Dr. O. I. | Lyons Bakery Ltd., Carlton, Barnsley, S. Yorkshire, S71 3HQ. |
| Pearson, Mr. S. | Weetabix Ltd., Station Road, Burton Latimer, Northamptonshire. |
| Phillips, Miss S. J. | McDougalls Catering Foods, McDougall House, Imperial Way, Worton Grange Estate, Reading Berkshire. |

| | |
|---|---|
| Purslow, Dr. P. | AFRC Institute of Food Research, Langford, Bristol, BS17 8DY. |
| Ranken, Mr. M. D. | Michael Ranken Services, 9 Alexandra Road, Epsom, Surrey, KT17 4BH. |
| Rao, Dr. M. A. | Department of Food Science and Technology, Cornell University, NYS Agricultural Experiment Station, Geneva, NY 14456, USA. |
| Reeve, Miss M. J. | Kings College, Department of Food and Nutritional Sciences, Kings College, Campden Hill Road, Kensington, London W8 7AH. |
| Roberts, Mr. C. | RHM Research Ltd., Lincoln Road, Cressex Industrial Estate, High Wycombe, Buckinghamshire. |
| Rodger, Dr. G. W. | Biological Products, P.O. Box 1, Billingham, Cleveland. |
| Rosenthal, Dr. A. | College of Technology, Humberside. |
| Ross-Murphy, Dr. S. | Unilever Research, Colworth House, Sharnbrook, Bedfordshire, MK44 1KU. |
| Russell, Dr. P. L. | Flour Milling and Baking Research Association, Chorleywood, Rickmansworth, Hertfordshire WD3 5SH. |
| Salisbury, Ms. J. | Quaker Oats, Ridge Road, Southall, Middlesex. |
| Sanderson, Dr. G. R. | Kelco Division of Merck Inc., 8355 Aero Drive, San Diego, California 92123, USA. |
| Sargent, Dr. J. A. | Hexland Ltd., W & G Estate, Faringdon Road, East Challow, Wantage, Oxford, OX12 9TF. |
| Saunders, Dr. M. J. | Cadbury-Schweppes, The Lord Zuckerman Research Centre, University of Reading, Whiteknights, Reading, Berkshire. |
| Schuler, Mr. P. | C/o Hoffman La Roche & Co. Ltd., Department VM/H Technical Service, Building 72/229, CH 4002, Basel, Switzerland. |
| Sheard, Dr. P. R. | Institute of Food Research, Langford, Bristol, Avon, BS18 7DY. |
| Sherman, Professor P. | Kings College, University of London, Kensington Campus, Campden Hill Road, London, W8 7AH. |
| Slade, Dr. L. | Nabisco Brands Inc., Corporate Technology Group, P.O. 1943, East Hanover, New Jersey, 07936–1943, USA. |
| Smith, Dr. J. | University of Strathclyde, J. P. Todd Building, Food Science Division, 131 Albion Street, Glasgow, G1 1SD. |
| Sopade, Mr. P. A. | Food Studies Building, University of Reading, Whiteknights, P.O. Box 226, Reading, RG6 2AP. |
| Stedman, Mr. J. | Department of Catering Management, Oxford Polytechnic, Gipsy Lane, Headington, Oxford, OX3 0BP. |
| Taylor, Dr. A. | Department of Applied Biochemistry and Food Science, University of Nottingham, School of Agriculture, Sutton Bonington, Loughborough, Leicestershire, LE12 5RD. |
| Tolliday, Mr. I. | Pedigree Petfoods, Melton Mowbray, Leicestershire. |
| Tolstoguzov, Professor V. B. | Institute of Organoelement Compounds, USSR Academy of Sciences, 117813 GSP-1, V-312 Vavilov Str. 28, Moscow, USSR. |
| Tornberg, Dr. E. M. | Swedish Meat Research Institute, P.O. Box 504, S-24400 Kavlinge, Sweden. |
| Totte, IR. T. A. | Departement de Technologie et des Industries Agroalimentaires, Faculté des Sciences, Agronomiques de l'etat, 2. Passage des Deportes, 5800 Gembloux, Belgium. |
| Towers, Mr. P. A. | Cadbury Schweppes Ltd., Lord Zuckerman Research Centre, University of Reading, Whiteknights, Reading, Berkshire, RG6 2LA. |
| Townsend, Mr. G. M. | United Biscuits (UK) Ltd., St. Peters Road, Maidenhead, Berkshire, SL6 7QU. |
| Trobridge, Dr. A. R. | Mars Confectionery, Dundee Road, Slough, Berkshire, SL1 4JX. |
| Trudinger, Mr. G. R. | Carrimed Ltd., Glebelands Centre, Vincent Lane, Dorking, Surrey, RH4 3YX. |
| Van Der Velde, Mr. G. | Polymer Laboratories, Essex Road, Church Stretton, Shropshire, SY6 6AX. |
| Van Kleef, Dr. F. S. M. | Unilever Research Laboratory, P.O. Box 114, 3130 AC Vlaardingen, The Netherlands. |
| Vickers, Dr. Z. M. | University of Minnesota, 1334 Eckles Avenue, St. Paul, Minnesota 55108, USA. |
| Viney, Mrs. L. J. | Bohlin Rheology UK Ltd., Business and Technology Centre, Bessemer Drive, Stevenage, Hertfordshire, SG1 2DX. |
| Visser, Dr. J. | Unilever Research Laboratory, P.O. Box 114, 3130 AC Vlaardingen, The Netherlands. |

| | |
|---|---|
| Wainwright, Mr. A. R. | United Biscuits (UK) Ltd., St. Peters Road, Furze Platt, Maidenhead, Berkshire. |
| Waites, Professor W. M. | Department of Applied Biochemistry and Food Science, University of Nottingham, School of Agriculture, Sutton Bonington, Loughborough, Leicestershire, LE12 5RD. |
| Walters, Mr. J. | Pedigree Petfoods, Hylands, Back Lane, Saltby, Melton Mowbray, Leics. |
| Wan, Dr. G. T. Y. | Flour Milling and Baking Research Association, Chorleywood, Rickmansworth, Hertfordshire, WD3 5SH. |
| Wareing, Dr. M. V. | Kelco International Ltd., Pitwood Park Industrial Estate, Waterfield, Tadworth, Surrey, KT20 5QF. |
| Wheeler, Mr. G. | Department of Applied Biochemistry and Food Science, University of Nottingham, School of Agriculture, Sutton Bonington, Loughborough, Leicestershire, LE12 5RD. |
| Whitehouse, Mr. S. | Rowntree Mackintosh plc., Group Products Research, York, YO1 1XY. |
| Whittam, Miss M. A. | Institute of Food Research, Colney Lane, Norwich, Norfolk, NR4 7UA. |
| Whittington, Professor W. J. | Department of Physiology and Environmental Studies, University of Nottingham, School of Agriculture, Sutton Bonington, Loughborough, Leicestershire, LE12 5RD. |
| Wilkinson, Mr. B. E. | Golden Wonder-HP Foods, Technical Centre, Northampton Road, Market Harborough, Leicestershire. |
| Wilkes, Mr. M. S. | United Biscuits (UK) Ltd., St. Peters Road, Furze Platt, Maidenhead, Berkshire. |
| Wills, Mr. G. | Pedigreee Petfoods, Melton Mowbray, Leicestershire. |
| Wilson, Mr. J. B. S. | United Biscuits (UK) Ltd., Technical Information Office, Windy Ridge, Smisby Road, Ashby-de-la-Zouch, Leics. |
| Winwood, Mr. R. J. | Tunnel Avebe Starches Ltd., Avebe House, Otterham Quay, Rainham, Gillingham, Kent, ME8 7UU. |

# INDEX